THE RELEVANCE
OF PHYSICS

THE RELEVANCE OF
PHYSICS

By STANLEY L. JAKI

 THE UNIVERSITY OF CHICAGO PRESS
CHICAGO AND LONDON

International Standard Book Number: 0-226-39143-4

Library of Congress Catalog Card Number: 66-20583

THE UNIVERSITY OF CHICAGO PRESS, CHICAGO 60637
The University of Chicago Press, Ltd., London

Preface

ALTHOUGH PHYSICS has grown into a major force in history, its own history is little known and seldom reflected upon. Yet, like any other human creation, physics cannot be fully understood unless it is viewed in the perspective of its historical development. Indeed, neglect of the study of the recent and remote past of physics is probably a principal cause of the cultural split that in our age has put scientists and humanists in two widely separated camps — the phenomenon C. P. Snow has labeled "the two cultures."

In part because of its marvelous achievements, physics has come to be accepted by many as a set of unchangeable conclusions and definitive truths. Although the possession of truth constitutes the highest human satisfaction, it is not an easily attainable goal even for a science possessing a highly refined set of tools. Actually, physicists have witnessed many drastic revisions of well-established "truths" in physics, and the end of this process is not yet in sight. It is possible that our times are more aware of this than previous generations were. But awareness that physics molds history and culture not only by its discoveries but also by the state of mind it fosters is minimal indeed. Any appraisal of physics as a cultural force that ignores this fact can promote only the aims of pseudohumanism — in which the "infallibility" of physical science is contrasted to the "incurable errors" of humanistic lore.

A desire to help overcome this trend was one of the principal motives that led to the writing of this book. It aims at restoring that feature of the image of physics which bespeaks the highly revisable character of its statements, the never-ending course of its search and the basic incompetence of many of its conclusions in other important areas of human reflection. For if physics has affected the cultural whole of any age in a harmful way, it has done so because some of its results have been stated as rigid dogmas by its cultivators and (especially) by its often incompetent popularizers. The human side of this process is of paramount importance. This side is characterized by the state of mind of the physicist, the chief means through which physics becomes part of the culture of an age. Therefore, in a study concerned with the contribution of physics to the human quest for understanding, a close look at the states of mind of past and present physicists is indispensable.

The physicist of our day, to whom this book is primarily addressed, must see in detail that his predecessors were not only heralds of truth. They were also, like practitioners of other intellectual professions, men who often promoted half-truths, plausibilities, conjectures, and at times plain errors into "indisputable verities." Only by seeing this will the mind of the present-day physicist be properly guarded against yielding too readily to temptations of this kind. To prepare himself for his cultural role, he must ponder the state of mind of his forebears, and it is to assist him in this task that the extensive documentation of this book is offered. The purpose of the book is not to argue but to illustrate and document. Human reflection thrives best on concrete sayings and events. Abstract reasoning is a poor means of recapturing the state of mind of past and present spokesmen of physics.

Along with these men's wisdom, their fallacies and idiosyncrasies are also brought back from the pages of often forgotten writings. This is not to poke fun at physics or to degrade its significance, but to help prevent its becoming the vehicle for that pseudohumanism that has respect only for what is quantitative in human reflection. True, the nature and aim of physical research make it imperative that the physicist in his work restrict his views to the quantitative aspect of things and processes. It is this limitation that makes the science of physics particularly effective in certain areas of inquiry but at the same time renders it conspicuously impotent to cope with a host of vital issues and problems. Awareness of this is clearly a paramount duty that, if ignored, will only quicken the pace of cultural erosion. Recently no less prominent a figure of present-day American science than Vannevar Bush voiced the desperate cultural need for a systematic illustration of the limitations of physical science. "Much is spoken," he noted, "today about the power of science, and rightly. It is awesome. But little is said about the inherent limitations of science, and both sides of the coin need equal scrutiny." To help redress the balance between those two sides is the aim of this book. Its purpose would be fully achieved if it increased in those who cultivate and love physics that component of the wisdom of science of which Maxwell once wrote, "One of the severest tests of a scientific mind is to discern the limits of the legitimate application of scientific methods."

Contents

PART ONE

The Chief World Models of Physics   1

CHAPTER ONE
The World as an Organism   3

CHAPTER TWO
The World as a Mechanism   52

CHAPTER THREE
The World as a Pattern of Numbers   95

PART TWO

The Central Themes of Physical Research   139

CHAPTER FOUR
The Layers of Matter   141

CHAPTER FIVE
The Frontiers of the Cosmos   188

CHAPTER SIX
The Edge of Precision   236

PART THREE

Physics and Other Disciplines 281

CHAPTER SEVEN
Physics and Biology 283

CHAPTER EIGHT
Physics and Metaphysics 330

CHAPTER NINE
Physics and Ethics 371

CHAPTER TEN
Physics and Theology 412

PART FOUR

Physics: Master or Servant? 459

CHAPTER ELEVEN
The Fate of Physics in Scientism 461

CHAPTER TWELVE
The Place of Physics in Human Culture 501

ABBREVIATIONS 535

NOTES 537

NAME INDEX 591

SUBJECT INDEX 599

Part One

The Chief World Models
of Physics

CHAPTER ONE

The World as an Organism

IN THE SIFTING of true cultural gains from spurious ones there is hardly anything more effective than the relentless flow of time. Men, thoughts, and events that are permitted by history to sink into oblivion far outnumber those that survive, and very rare indeed are the achievements that continue to evoke the unreserved admiration of many critical generations to follow. Among such rare accomplishments belong the intellectual feats of ancient Greece. Nothing shows more forcefully their singular excellence than the admiration of posterity, which keeps steadily growing as one generation yields to another. It has been said in the twentieth century that the European philosophical tradition is but a series of footnotes to Plato, and it was the same century that heard J. Burnet define science as "thinking about the world in the Greek way." [1] Burnet's words fit particularly well the age that saw classical physics emerge. In fact, both the protagonists and the opponents of the new physics found it indispensable to trace their respective positions back to one or the other of the two main ways of looking at nature formulated by the ancient Greeks. For the seventeenth-century founders of mechanistic physics there was the "divus Archimedes" as a supreme symbol, whereas the defenders of the traditional organismic physics waged their battle from behind the protective authority of "the Philosopher." The main figures of the new physics were, however, no less eager to claim that Aristotle, if better informed, would certainly side with them. [2]

The presence of Greek scientific thought, so keenly felt in the seventeenth century, was only natural. Commenting on Aristotle's *Physics* and other ancient texts was still part of the education of scientists. It is far more surprising, however, to find prominent representatives of modern physics referring to the Greek roots of concepts

3

that are almost too novel to suggest connections with antiquity. Thus Heisenberg compared, for instance, the "creation" of an elementary particle from a given amount of energy with the actualization of potentia through a form. Similarly the probability wave of quantum mechanics was interpreted by him as a quantitative version of the old concept of potentia in Aristotelian philosophy.[3] Moreover, the "complete understanding of the unity of matter" toward which physics strives, suggested to him "that the forms of matter in the sense of Aristotelian philosophy would appear as results, as solutions of a closed mathematical scheme representing the natural laws for matter."[4] It was also in a renewed study of the Greek way of looking at the world that Schrödinger sought the means of bridging the gap separating the humanistic and scientific components of our own culture. More specifically, in the field of physics proper he expected from the critical reexamination of the classical origins of modern physical concepts the way out of the "inordinately critical situation in which nearly all the fundamental sciences find themselves."[5] As Schrödinger put it, the crisis into which quantum theory had plunged modern physics can be resolved only by "revising its foundations down to early layers," down to the source where error and truth are easier to recognize, where natural and artificial are easier to separate.[6]

Although the scientific achievements of ancient Babylon and Egypt are being more and more recognized, Greek science still constitutes, as A. Reymond wrote half a century ago, "a veritable miracle."[7] Much as the Greeks might have borrowed from others, they alone found the way to fuse these elements with an almost dramatic speed into closely knit scientific systems, exemplified chiefly by the Euclidean geometry. Western science rests on two foundations: the formal, ordered system of knowledge, and the establishment of causal relationship by systematic experimentation. The former of these is the achievement of the Greeks and is of such magnitude that it prompted Einstein's famous remark: "In my opinion one has not to be astonished that the Chinese sages have not made these steps. The astonishing thing is that these discoveries were made at all."[8]

In the field of physics the major efforts of the Greeks fell into two main classes. First, they formed a long array of seminal concepts; second, they created, chiefly in the works of Aristotle, a physical system that along with classical physics and quantum theory is one of the three principal physical explanations of the universe. This system, which held sway for almost two thousand years, pictured the world as an immense organism and formulated accordingly the methods, the range, and the laws of physics. That an organismic type of physics had to precede the mechanistic or Newtonian physics, let alone the modern, mathematical viewpoint, seems almost natural. After all, the inclination to experience the world as a sort of living entity that can and should be contacted as a fellow being is one of the principal attitudes developed by primitive man. He did this not by the studious application of abstract categories that scrupulously distinguished between animate and inanimate, but rather in a spontaneous acceptance of the phenomena of nature as the spoken words of an all-pervading "thou" that can be asked, cajoled, and at times even commanded. The words that nature uttered were for the primitive man vehicles carrying the irrational as well as the rational elements, the inconceivable no less than the easily grasped patterns. It was an approach to nature that in one and the same act could easily accommodate "the validity of several avenues of approach at one and the same time." [9]

The transition toward the separation of these various elements in human reflection was no doubt a long process, the phases of which cannot be described without some hesitation. At any rate, the mythological description of nature in the Homeric poetry already represents a fairly advanced stage. The clear-cut definition of roles Homer assigns to the gods of various ranks to account for the processes of nature seems to indicate the realization that besides the intuitively grasped unity there is a wide gamut of distinctness in nature. For scientific thinking to be born, however, a higher level of generalization had to be reached, and this could be done only at the price of parting with the mythological description of nature. To be more specific and to stay within the Greek cultural ambiance, a step was to be made from Homer and almost the whole of Hesiod to the recognition of a more or less depersonalized interaction of things.

It was a step not only daring but also intoxicating. The richness of new connections revealed in it could easily make one forgetful of insights obtained by careful reflection on one's inner world. And for a long time to come there were neither intellectual traditions nor a genius available to curb a naïve enthusiasm that overtook those who first hit upon the seductive world of the "mechanical" method. As a result, the traditional or *human* understanding of man and his place in the cosmos began to fade away to an alarming degree. That was at least the way the Greeks came to feel after almost two centuries of bold conjectures that showed the Greek thought swinging from one extreme to the other. The theologians and poets, as Plutarch summed up the matter, paid attention only to higher causes. Their motto was the invocation, "Jove first, Jove last, all things spring out of Jove." Those, however, who styled themselves "natural philoso-

phers" ignored higher causes of every sort and explained the theory
of sensation by "necessary" or "physical" causes, that is, by "elements,
conditions of elements, collisions and intermingling of bodies." Both
positions, Plutarch emphasized, were deficient in an essential part;
the one omitted the agent, the other the means and the material.[10]

Not that the Ionian philosophers, Thales, Anaximander, and Anaximenes, parted with everything they inherited from the Homeric
phase of Greek culture. When they did not raise questions about the
beginning and end of the world, they simply imitated Hesiod. Their
assumption about things developing from the primitive state by successive differentiations and by the interaction of opposing forces,
repulsion and attraction, was also a borrowing from poetical cosmog-
onies. Other thoughts of Hesiod, such as the role of mind, love, and
strife in the physical universe, also keep turning up in the statements
of Anaxagoras, Parmenides, and Empedocles. The occurrence of
analogies taken from the organic world is not absent even in the fragments ascribed to Democritus, that resolute champion of a rigidly
mechanistic interpretation of nature. Did he not try to explain the
preferential grouping of the same kinds of units of matter, as if it
were an attraction, by references to associations among animals always restricted to members of the same species?[11] In all this there
was, however, far more of a semantic accommodation than of a return
to a personalistic representation of the universe. After all, the Ionians
and Democritus too had to speak a language in which the same words
still retained human and superhuman, animalistic as well as mechanistic, nuances. How could they avoid using such basic terms as thy-
mos, pneuma, psyche, and a number of others that resisted any strict
definition? Their stylistic inconsistencies notwithstanding, the basic
orientation of both the Ionians and the atomists is unmistakable. Their
principal aim was not the cultivation of the personalistic texture of
the Homeric outlook, but rather an experimentation in bold aphorisms that preferred to speak of the impersonal world of matter, motion, and space. For them, as Aristotle summed up their views, there
was only one permanent element in the universe, a basic material stuff
that they believed to be conserved in every change, so that nothing
was either generated or destroyed.[12]

Thales reduced everything to water, coming very close to formulating the principle of conservation of matter, whereas Anaximander
derived everything from air and spoke of never-ending motion. Qualities were explained by quantitative changes. The density scale of the
air was to account for fire, water, and earth; and for the first time,
mechanistic models were adopted to illustrate the dynamics of the
universe. The sweeping generalizations reached even the heavens

when Anaxagoras described the sun as a flaming stone larger than Peloponnese. There seemed to be simply nothing that could not be reached by a physics that bore all the marks of an unsuspecting, youthful outburst of scientific speculation: from man's soul to the far reaches of the universe there was believed to be only one principle, which produced, governed, and explained everything. "As our soul," contended Anaximenes, "being air holds us together and controls us, so does wind [or breath] and air enclose the whole world."[13] True, Anaximenes was quick to qualify his grandiose statement by allowing in living creatures the presence of something other than simple and homogenous air and wind. In so doing, however, he was merely warning simpletons and not making concessions about the uncondi-tional validity of his air-principle.

The Pythagoreans believed this ultimate principle to be the unit number. Aristotle was probably reading his own distinctions into the views of the Pythagoreans by saying that for them numbers consti-tuted both the matter and the form of things. Yet, when Aristotle says that for the Pythagoreans "the whole heaven is numbers," he is undoubtedly reporting their genuine conviction.[14] How the heavens could be made of numbers presented no great problem to the Pythag-oreans. In their world view lines were derived from points or unit numbers, from lines surfaces, from surfaces simple bodies, from these the elements and the whole world. As an evidence of the explanation of the world by numbers, the Pythagoreans pointed to the strings of musical instruments and to the motions of stars and planets, thereby uniting music, poetry, matter, and mind into a harmonious whole. Or at least this was their ultimate dream. At any rate, theirs was a staunch belief, as Philolaus informs us, that error and deceit are foreign to the world of numbers. Truth, intelligibility, and certitude they held to be cognate to numbers, which they contrasted with the erroneous world of the undefined, uncounted, senseless, and irra-tional. "Falsehood," they said, "can in no way breathe on number. To the nature of number falsehood is opposed as something hostile and irreconcilable, while truth is something proper and native to the race of numbers."[15]

To reveal the true nature of the cultural change underway the word race was very aptly chosen: the race of impersonal numbers was indeed gaining the upper hand over all other races, human or otherwise. This fascination with numbers, measurement, and calcu-lation suffered only a momentary blow when the Pythagoreans stum-bled upon the existence of irrational numbers. The rebound could hardly be more vigorous, and Archytas a hundred years later confi-dently praised the unexcelled power of mathematics in understanding nature. He boldly asserted that it was the exclusive privilege of the mathematicians to "think correctly about the nature of individual things," for they alone possessed the much coveted knowledge about the whole. What Archytas really meant was that mathematicians succeeded in reaching the very roots of being by giving undivided attention to the realm of numbers. Thus he felt entitled to claim that mathematicians "who are concerned with number and size, these two related primitive forms of being," are alone capable of having "a clear insight into the speed of the stars, and their rising and setting, and of geometry, arithmetic, astronomy and of music as well."[16]

Ionian *physikoi*, Pythagorean mystics, and sober geometers had at least one thing in common. They all claimed for their particular approach to nature the exclusive hallmark of intelligibility. Entirely similar was the case with that most influential concept of early Greek science, the atom. The notion of atoms was offered as the bedrock of understanding, and its properties seemed to symbolize the supposedly ultimate form of reasonable questions that could be raised about the universe. The trust Democritus put in his atoms seems to have been limitless. After all he was their father and as such had to be very partial to their paradoxical though obvious shortcomings. Consequently Democritus could describe without any misgiving how his atoms could, through the wizardry of random collisions, get intertwined "one with another according to the congruity of their shapes, sizes, positions and arrangements, stay together and so effect the coming into being of compound bodies."[17] All this suggested a radically mechanistic conception of the world, and Democritus pushed to the extreme, with thorough consistency, the consequences of his basic postulates. In his universe of atoms there is no place left for "forces" such as weight, attraction, or repulsion. Nor could the atomists accommodate in their system apparently non-material factors, which, like the "mind" of Anaxagoras, are hardly more than a figure of speech when mentioned by them. But not only spiritual entities are deprived of distinct existence in Democritus' world. Much of what constitutes man's view of the outside world becomes but a subjective illusion in their treatment. In antiquity nobody noted this more strikingly than Galen, who, unlike Democritus, wanted above all to understand man himself and his relation to his surroundings. How indeed could a man like Galen not notice the harsh consistency with which Democritus divested the physical world of all its humanly pleasing aspects and let colors, sensations, and feelings exist not in truth but merely by convention?[18] Truly, if there was a lesson to draw from the atomistic approach to nature, Democritus himself had formulated it with shocking directness, and this Galen could not allow to pass unremembered: "One must learn by this rule that Man is severed from reality."[19]

But it was not only man's relation to the external world that became meaningless. Man himself as an individuum was also to lose his footing in the whirl of atoms. With the cosmos — the ordered correlation of things — gone, man's meaning in the universe went too. Of course when Galen noted the pessimism such views had generated, the heyday of atomism had been a thing of the past for almost half a millenium. But in its stage of fresh fascination the atomism of Democritus, which really did not explain any observational phenomena, could not fail to have an overriding impact. Hand in hand with sophism it contributed heavily to a thorough upheaval of the whole set of traditional, humanistic values in pre-Socratic Athens.

It was in reaction to such a state of affairs that a fundamental but until then philosophically little-cultivated aspect of the Greek way of looking at nature came to the fore and asserted itself as the leading principle of the study of nature. What this new trend wanted to achieve above all was to secure an organic place for the personal, intuitive experience of man both in the understanding of himself and of the physical world. The most detailed formulation of this new approach to nature is Aristotle's physics and cosmology, but its most telling document is Plato's account in the *Phaedo* of Socrates' search for a science satisfying the needs and aspirations of man.[20] It was a search for perspectives wide enough to support a stand that from the viewpoint of a consistent mechanistic philosophy had to appear the senseless acceptance of one's own death sentence. Clearly, the mechanistic philosophy of nature and man, as advocated by the Ionian *physikoi*, had no explanation and motivation for an attitude that preferred to abide by value judgments that were but noble illusions to an all-inclusive mechanistic interpretation of things, persons, and events.

Socrates was eager to point this out to his friends who begged him to avoid the inevitable course of a blatantly unjust sentence. The arguments of his friends, as Socrates noted, invariably fell back on the mechanistic explanation of generation and decay, and this appeared to him wholly inadequate even to cope with the most elementary organic processes. That explanations of that type were fascinating, he readily admitted. He could in fact refer to his experience as a youth on reading a book by Anaxagoras[21] that offered explanations about such sundry items as generation and decay on the earth, motion of celestial bodies, and production of mental processes in the brain. Yet, on some reflection Socrates had to realize that these explanations left indeed much to be explained. Actually he found them confusing when applied to the elementary question of what is the cause of a man's growth.

It was no accident that Socrates' grappling with the gravest physical and spiritual issues of human life had ramifications that touched on the physical situation represented by the behavior of an organism. It was precisely in problems of this sort that atomism, Pythagoreanism, and the Ionian natural philosophy proved manifestly inadequate. Furthermore, it was because of the mishandling of such questions by some far-fetched generalizations of a "materialistic" physics that the understanding of man's own inner world came to be drastically debilitated. The apparent brilliance of Ionian physics was not only fascinating but also blinding, as Socrates recalled his own experience, and could leave its admirers in a state of total confusion to what was really known. In ultimate analysis it made a shambles of the basic question of human inquiry: how to connect man's own intuitive, immediate judgments and reflections to the processes unfolding in the outside world. With regard to man's behavior, strivings, and findings there were the instinctively applied categories of purposeful and involuntary, good and bad, fitting and unfitting. Could these qualifications be absent in the phenomena of nature, to paraphrase Socrates' agonizing questions, if nature was to be understood or to be truly connected with man in an organic whole? After all the Ionians themselves, as Socrates recalled, seemed at times to hint that nature worked like a man planning to achieve carefully what was best. Anaxagoras, Socrates noted, even spoke of an all-pervading mind as the arranger and cause of all things.

But Anaxagoras' "mind" was not the source of the type of understanding Socrates looked for. While Socrates wanted to know why it was *best* for the earth to remain at rest in the center of the world, the Ionians, Anaximander to be specific, referred rather to the indifference of an object to move at all when equally distant from the extremes in all directions. Similarly deficient in Socrates' eyes were the other explanations offered by the Ionians for the stability of the earth. To hear that the earth had a flat bottom and was thereby securely supported by the underlying air conveyed only keen disappointment to him. He felt the same as he surveyed the views of the *physikoi* on the nature and motion of the heavenly bodies. As to the sun and the planets, Anaximander identified them as small circular openings in a huge rim filled with fire, whereas Anaxagoras reduced the sun to a flaming stone. Both, however, were equally unconcerned why it was *better* for the planets and stars to move and turn in the way they did.

The ultimate motivation of such a stratagem could not escape

Socrates. He saw that the account of the *physikoi* of the stability and motion of the earth and heaven was but a covert replacement of old values with a new deity. For, as Socrates noted, the all-inclusive mechanism was offered as a new Atlas. Its worshippers, the *physikoi*, charged Socrates, "give no thought to the good which must embrace and hold together all things." [22] In this they were at least consistent, as no goodness, or purpose could be predicated about the Atlas of the Ionian physicalism. Little wonder that Anaxagoras' all-pervading "mind" came to appear to Socrates as a mere travesty of the true mind, and the reasons it represented but a camouflage of the real causes. Socrates with good reason called attention to the fact that voices, hearing, and breath were as little a complete explanation of human conversation as was the actual position of his bones and muscles the ultimate reason of his refusal to escape from prison. The upshot of such considerations was, as can be readily guessed, a search for causes in a sense diametrically opposite to the approach taken by the Ionians or by the atomists. To continue in their footsteps, warned Socrates, was to expose one's mental eyes to total and irreparable blindness. Such at least was the uppermost consideration that made him chart the course of his own intellectual orientation on which, unfortunately, hardly a stopover was left for the study of mechanical causes. Extreme and deplorable as this choice was, in the context of the times it appeared inevitable. More important, many generations after him conformed faithfully to the mental attitude expressed so graphically in the *Phaedo*:

After this, then, since I had given up investigating realities, I decided that I must be careful not to suffer the misfortune which happens to people who look at the sun and watch it during an eclipse. For some of them ruin their eyes unless they look at its image in water or something of the sort. I thought of that danger, and I was afraid my soul would be blinded if I looked at things with my eyes and tried to grasp them with any of my senses. So I thought I must have recourse to conceptions and examine in them the truth of realities.[23]

This lofty program was carried on by a genius no less persuasive than Plato. In his works one comes across time and again the basic charge against the *physikoi* that their approach to nature separated man from nature, and nature from the realm of the good and beautiful. There one finds formulated explicitly the paramount issue, as felt by Socrates' disciples, namely, the role to be assigned to the phenomenon of life in explaining nature. What Plato found particularly repulsive in the *physikoi's* system was that the whole gamut of the manifestation of the living, vegetative, animal, and psychical was

taken by them as the chance product of "absolutely inanimate ex-
istences," [24] such as fire and water, earth and air. To this Plato resolutely opposes the primacy of life, which embraces matter and mind
alike, and defends the method of explaining the whole and parts of
the universe in terms of an organism. His principal work touching on
scientific questions, the *Timaeus*, bluntly states that this world "in
very truth [is] a living creature with soul and reason." [25] To this viewpoint Plato accords an unconditional primacy even in matters of detail. Thus when he discusses the working of the human eye, he
deplores the fact that "the great mass of mankind regard [the geometrical and mechanical aspects of the question] as the sole causes
of all things." Against this he opposes the classification of causes into
two groups: the accessory or mechanical causes that are "incapable
of any plan or intelligence for any purpose," and those that "work
with intelligence to produce what is good and desirable." [26] The reaffirmation of the Socratic or organismic approach in science could
hardly be more unequivocal.

Such an emphasis on the concept of organism as the basic framework in which the cosmos is to be explained derived only in part
from factors like the emergence in the fifth century of the Hippocratic
medical theory and practice. The principal factor was a deeper and
more universal one. It was rooted in the Greek nature as such and was
given unchallenged prominence when cultural developments forced
the Greek mind to reflect on the consequences of a mechanistic explanation of the inanimate and animate world including man both
as an individual and as a member of society. The "Greekness" of the
organismic approach can be seen in the fact that they first applied
the term *cosmos* to a patently living thing — a well-ordered society —
and only afterward to the orderliness of the physical world.[27] Rooted
deeply in their personal, cultural inclinations, this organismic approach to reality, once it became the conscious possession of the
Greeks, had never been seriously questioned or abandoned by them.
Single views of the Ionians and atomists continued, of course, to play
seminal roles in Greek science. What is more, once the cultural crisis
evidenced by the activity of the Sophists was over, even the poets
began to take more kindly to the *physikoi*, who for a while were the
principal targets of plays concerned with the source of various cultural evils. At any rate, the *physikoi* ceased to be called in literary
circles, as Plato remarks, "she dogs uttering vain howlings and talking other nonsense of the same sort." [28] This was, however, merely
a concession that could easily be meted out by those who won the
cultural battle. For as Plato could confidently state in the same context, the ascendency of the mechanical views had been checked, or to

paraphrase his words, the case was reversed in favor of the organismic viewpoint.

This reversal, or decisive choice, was made by the Greeks and was at least as Greek in spirit as the so-called scientific rationalism of the Ionians, which some historians of science so desperately tried to present as the only genuine manifestation of the Greek spirit. These historians were no less wide of the target in ascribing this choice to the influence of factors supposedly foreign to the Greek mind, such as religious ideas penetrating Hellas from the eastern shores of the Mediterranean. On the contrary, long before such influence had become momentous, the greatest figures of Greek thought opted for the primacy of the organismic view mainly because this was found by them to provide the type of intelligibility that best satisfied the aspirations of the Greek mind. That such a choice was in keeping with the best in the Greek mentality is brought out most forcefully by the unanimity with which the subsequent phases of Greek intellectual development adhered to it. Aristotle is very typical. A resolute critic of Plato's main philosophical construct, the theory of ideas, Aristotle remained nonetheless faithful to the basic tenet of the Socratic program: nature was to be understood as something like man himself — moving toward goals, striving toward the best possible arrangement, in short, acting like an organism.

In fact, it is in Aristotle's *Physics* that one finds the first simultaneous occurrence of the terms *microcosmos* and *megacosmos* (macrocosmos), terms which were to serve as the characteristic stamp on every organismic theory about the universe until the very modern times. In the eighth book of the *Physics*, Aristotle reviewed several objections to a fundamental doctrine of his, the eternity of motion. According to these objections, motion need not always be caused by another motion but can at times be preceded by absolute rest. For proof one of the arguments refers to the case of living beings, to human consciousness and animal behavior in particular. "If this [transition from complete rest to motion] is possible in an animal," runs the argument, "why not in the universe? Surely, what can happen in a *microcosmos* can happen in the *megacosmos*." [29] Curiously enough, all Aristotle found to criticize in this objection, so contrary to the foundations of his system, was the conclusion alone and not the organismic analogy on which it rested. With the general organismic principle that allows one to move without any qualification from the small world of an animal to the large world of the inanimate universe, Aristotle was not disposed to find any fault. It was a principle that had to appear to him as basically sound, convinced as he was that the universe was supremely a living being both in its entirety and in its parts. As a consequence, Aristotle, in general so reluctant to praise the acumen of his opponents, qualifies the objection based on this principle as one "raising a most formidable problem" for his doctrine of the eternity of motion. Obviously he realized that the objection had carried the battle to his own grounds.

When faced with objections based manifestly on principles diametrically opposite to his own, Aristotle's phrases betrayed hardly any hesitancy or surprise. Thus in a highly revealing passage of the second book of the *Physics* Aristotle drew a sharp contrast between the early physics and what physics really ought to be. According to the former, nature does not work for the sake of something or because it is better one way than another; it rather works of necessity, that is, in a regularly repeated pattern of sequences. The necessity and regularity of his predecessors' physics had, however, been dependent on chance, and Aristotle eagerly seized on this obvious inconsistency. Chance, Aristotle argued, does not repeat things in a regular fashion. Regularity can be assigned only to purposeful action, which, as he contended, always works for an end and originates from the particular nature that is acting. It was at this point that Aristotle reaffirmed the basic goal of Socrates' arguments: the restoration of unity between man and nature in an organic whole. The declaration of a perfect parallelism between the way man acts and the manner in which nature operates could not be more explicit: "As in human operations, so in natural processes; and as in processes, so in human operations (unless something interferes). Human operations are for an end, hence natural processes are so too."[30]

The "physical method," as Aristotle called Democritus' approach to nature, failed in this most essential aspect because it tried to explain things and processes by decomposing them into their parts, leaving aside the aspect of their wholeness. But, as Aristotle insisted time and again, basic information about objects, living and non-living alike, can be obtained only by a method that concentrates on the wholeness in things and processes. Whether an animal or a couch is to be explained, the investigation should take its start from a definition of the whole, a definition that will shed light on the role of organs or parts played in the whole.[31] By stressing the priority of the whole over the parts instead of treating the whole merely as a sum of the parts, Aristotle meant, in fact, that "the formal nature is of more fundamental importance than the material nature."[32] Or, in other words, the study of nature is to be dominated by the idea expressing the coordination of parts in the whole, which is the principle of organism. Believing that animate and inanimate beings alike have coordinated parts, Aristotle lowered the dividing wall between the two domains to such an extent that comparisons between the living and non-living became the most naturally used device in his writings on natural science. Speaking of the merits of the study of animals, Aristotle was highly elated about the teleology displayed in the animal world and concluded: "Just as in discussing a house, it is the whole figure and form of the house which concerns us, not merely the bricks and mortar and timber; so in natural science, it is the composite thing, the thing as a whole, which primarily concerns us, not the materials of it, which are not found apart from the thing itself whose materials they are." [33]

Although modern natural science is reluctant to speak in the same breath of bricks and animals, Aristotle saw a basic justification for doing so in the wholeness allegedly present in any object. Of this holistic approach to the study of nature Aristotle stated in the most categorical terms that it is precisely there that the "physical" method and the "true" investigation of nature part ways. "We need not argue over details," Aristotle tells us when refuting the opinions of Anaximenes, Anaxagoras, and Democritus on the flatness and motion of the earth. "Our quarrel with the men who talk like that about motion does not concern particular parts but an undivided whole." [34] Once the wholeness of an object is grasped, said Aristotle, its properties and parts will readily reveal themselves. The golden key to the wholeness or nature of a thing consists in detecting its spontaneous motion. Implicit in this spontaneity is the goal of the motion, which in turn lays bare the nature of the thing in motion. "The nature of a thing is in some sense the factor which initiates movement and rest within that thing in which it is itself immediately, not incidentally, present." [35] As opposed to products of crafts, the beings "constituted" or formed by nature, animals, plants, earth, fire, air, and water have within themselves the beginning of movement: "All these natural bodies are capable of moving themselves in space; for nature we have defined as the principle of motion in them." [36] Formed by nature they move toward their respective ends, and the end of such a natural motion is identical with the purpose for which the thing exists. "Whenever there is evidently an End towards which a motion goes forward, unless something stands in its way, then we always assert that the motion has the End for its purpose." [37]

Motion, nature, organism, and teleology were therefore but different aspects of one basic viewpoint, in which locomotion, qualitative change, the growth of the living, and the healing of the sick were treated on the same footing. Having rejected the possibility of a regress to infinity, Aristotle thought he could readily show that if there is anything at all to exist, its motion, the primordial motion, should be a natural one. On the other hand, if the first motion "was; natural, careful consideration will show that there must have been a cosmos."[38] This statement, directed against the notion of atoms whirling in space in every direction before the present configuration of things came to take shape, should forcefully intimate the multitude and sweep of conclusions that Aristotle derived from the concept of natural motion. After all, his universe was a "huge nature" of which all parts were moving, striving, and yearning toward their respective ends.

In fact, Aristotle needed only the distinction between the two types of natural motion, circular and straight, to be ready to tell us what it is like to be a cosmos. First, the nature of circular motion proved for him that the cosmos must be finite. Second, this finite universe is divided into two distinct regions, the upper part, or the region of celestial spheres, where the circular motion reigns supreme, and the sublunary region filled with ordinary matter whose nature is to move up or down. Third, since motion reveals the nature or substance of things, the celestial spheres and bodies, stars and planets, must be composed of a material as different from ordinary matter, as circular motion is from rectilinear. The ether, as this heavenly substance is called, is therefore a material whose nature is to issue in a uniform circular motion. Fourth, it also follows from the Aristotelian analysis of uniform circular motion that the ether is unalterable, suffers neither growth nor diminution, and has no beginning or end, which is to say that it can be neither generated nor corrupted. Fifth, to show that only one such substance can exist in the universe, Aristotle drew on the concept of nature and purpose and declared that a second substance of this type would be as purposeless "as a shoe which is never worn. But God and Nature create nothing that does not fulfill a purpose."[39]

The wizardry Aristotle displayed in squeezing out a long list of conclusions from one basic tenet is astonishing but hardly convincing. The contention that the primary body, or ether, cannot be infinitely extended is served up by him in no less than six variations. The great effort expended was superfluous, however, for as it turned out, the "nature" of any of the four elements could provide for Aristotle a specious argument against the infinity of the universe. Characteristically, according to him, one of the reasons for discarding the concept of an infinite number of atoms moving in infinite space was that in infinite space no natural motion can take place, because in infinite space no place or point can be assigned unambiguously as the termination or fulfillment of any object's motion.

Since the nature of a body determines the pattern or the direction of its motion, it follows that all the earth would move of itself toward the same place, which is the center of the universe, and all fire would always move upward toward the circumference of the world. The weighty problem of the multiplicity of worlds is thereby readily "solved" in a reasoning, the parts of which are as tightly interlocking as the parts of any organism: "Either, therefore, the initial assumptions must be rejected, or there must be only one center and one circumference; and given this latter fact, it follows from the same evidence and by the same compulsion that the world must be unique. There cannot be several worlds." [40]

Aristotle took pains to emphasize that distances between worlds, however large, cannot change the nature of things and their natural orientations. In other words quantitative considerations cannot obviate the conclusions derived from the analysis of the behavior of an organism. Now if the world is an immense but finite and unique organism, it is only natural that all matter available in the universe should be coordinated to that "one, solitary, and complete" system, [41] or else nature would have produced something to no purpose, which is inadmissible. Beyond the confines of the universe, therefore, there are neither places, nor void, nor time, for all these presuppose the existence of a natural body distinct from the unique cosmos.

The concept of "nature" secures for the world not only its uniqueness and finiteness, but also a life of perfection. The highest degree of life in the cosmos resides in the sphere of the ether: this "suffers from none of the ills of a mortal body," "its motion involves no effort," and it is "without beginning and end," [42] as should be with a uniform circular motion. But for life to be perfect, it must contain whatever "is present in the lowest stage of animal life." Therefore, like animals, the heavens must have front and back, left and right, above and below. The three-dimensionality of space is, in fact, just a corollary of the motion of living bodies, that is, bodies that have the principle of motion in themselves. The vertical direction is a consequence of growth that is from above, and the two horizontal directions are the results of locomotion that is from right to left (the right being the more noble side) and of the sensations that are from the front. Now since "the heaven is alive and contains a principle of motion . . . we must suppose it to resemble a being of the class whose right differs from their left in shape as well as in function, but which has been enclosed in a sphere." [44] The direction of the revolution of the skies shows, therefore, that the celestial hemisphere seen in the northern latitudes is really the lower half of the universe, the southern pole of the skies being the top of the cosmos. Hence, those who live in the southern regions are "in the upper hemisphere and to the right, whereas we are in the lower and to the left."[45] An absolutely valid coordinate system is indeed one of the consequences of viewing the world an as organism. What is more, the three perpendicular directions, being corollaries of life and motion, are not even of the same "dignity." The vertical direction is the most important, since the growth that manifests it can be found in every living thing, whereas the horizontal directions are of lesser rank because they cannot be discerned as basic direction of motion or growth in every living thing, such as a plant.

In the Aristotelian system the world taken as an organism revealed its features with alarming ease once its basic striving or nature had been defined. With a tour de force almost unparalleled in the history of science Aristotle showed in a single breath-taking page of *On the Heavens*[46] why the world should consist of different parts and why these parts should be the very same parts we actually observe in the world. His seductive, a priori account of the main features of the universe reveals in its true nature the liberties that an organismic type of physics unavoidably takes in its approach to nature. Some details are especially worth considering if one is to obtain a close-up view of the dubious procedures that physical science is forced to adopt when the universe is viewed as an organism and when motion is believed to unravel the nature of things with dazzling ease. Thus Aristotle declared that the outermost shell of the cosmos is by necessity spherical, for it is made of divine substance and whatever is divine must be circular. It must also be perfectly smooth, for otherwise there would be "places" beyond the limiting circle and that would be tantamount to a contradiction in terms. The direction of the revolution of the heavens is not haphazard either, "since nothing which happens by chance or at random can rank as eternal."[47] In other words the actual direction of this revolution should be accepted as a right to left movement, for right is superior to left. But can the fact be an explanation of the fact itself? Very much so, runs Aristotle's answer, "for if the existing state of things is the best, that will be the reason for the fact mentioned."[48]

The invariable speed of the heavenly revolution too is but an aspect of its eternal nature. Since a decrease of speed is a loss of power, Aristotle argued that such a loss could not take place in the heavens composed of the ether, a substance which by definition was not subject to decay of any sort. On a similar basis is decided the question of the composition of the stars too. If there is a body whose nature is to move in a circle, it is only logical to suppose, the argument runs, that the stars, which have a circular motion, are made of that substance, which again by definition is the ether. On such ground one can also readily discern, so Aristotle believed, whether the stars move round the heavens in the manner of progression observable in living beings. To resolve this question so strongly conditioned by the organismic approach, Aristotle, of course, had to resort to organismic analogies to clinch his proof. If the stars moved in such a way, he averred, nature would have provided them with organs of motion closely resembling those of the animals. But nature, which provides so generously for even the lower types of being, seems to have made the stars as different as possible from creatures that are endowed with various organs of motion. Thus the stars are spherical because a sphere, which is best suited for motion in the same place, is the least suited to progression "because it least resembles bodies which are self-moving. It has no separate or projecting parts as a rectilinear figure has and is of a totally different shape from forward-moving bodies." [49]

Not every motion in the heavens, however, has the apparent simplicity and uniformity of the stars. It was in fact impossible to avoid facing the question of fitting the complicated motion of planets into the divine simplicity of the heavens. In an organismic explanation of the world this gravest of all questions that ancient astronomers grappled with presents no difficulty at all. We have only to remind ourselves, warned Aristotle, that we are inclined to think of the planets "as mere bodies or units, occurring in a certain order but completely lifeless; whereas we ought to think of them as partaking of life and initiative. Once we do this, the events will no longer seem surprising." [50] Thus Aristotle, to explain the irregularities of planetary motions, fell back on one of his stock illustrative examples: the various phases in the progress of a sick organism toward health. The closer one is to health, goes the narrative, the fewer steps are needed to reach it. One individual may be healthy without any exercise, another may need only a little walking, a third might have to exercise strenuously, and a fourth may simply never become healthy despite tremendous exertions. Now, since the case of planets is taken without any second thought to be analogous to that of animals and plants striving for health, it is easy to see, contended Aristotle, that the farther a planet is from the sphere of the stars, the region of perfect life, the more cumbersome would be its motion. Clearly, if difficult problems could be solved with such ease, doubt about the merits of the solution could hardly make itself felt.

One must, however, admit that Aristotle was extremely consistent in claiming that the errors of his predecessors on the earth's immobility, place, and shape were due to their ignorance of the concept of natural motion. In other words, he took them to task for not approaching the problems of geophysics from an organismic viewpoint. This was particularly true of the Pythagoreans, who preferred mathematical notions and geometrical patterns to organismic analogies. For them the earth was revolving around the central fire, the function of which was, in their belief, to protect the most noble part of the cosmos: its geometrical center. But Aristotle quickly retorted that "the same should be true of the whole world as is true of animals, namely that the center (or heart) of the animal and the center of its body are not the same thing." [51] Therefore he concluded on a markedly triumphant note that it is not the mathematical center of the world that should hold the place of honor, but rather its true center. Where this true center might be, however, can be established in the framework of an organismic physics only through an analysis based on the natural motion of bodies. On such grounds, an abstract geometrical point obviously could not compete with the massive body of the earth for the central position in the universe. For could an abstract point display "strivings" and "affections," those supreme signs of intelligibility?

The same attitude, emphasizing invariably the primacy of organismic concepts and their primordial intelligibility, dominated Aristotle's refutation of the geophysical views of the other pre-Socratic philosophers. Aristotle claimed that those who like Thales let the earth rest on water evidently did not understand the *nature* of water, which is not supposed to carry any heavier body such as earth. Those who stated that a huge vortex keeps the earth in the middle failed to see, he remarked, that if there is a motion by constraint, there should be a prior natural motion. Those who invoked the principle of indifference to account for the immobility of the earth did not fare much better in Aristotle's eyes. They reasoned that an object situated at the center and related equally to the extremes in every direction can have no impulse to move in any specific direction. In fact, they compared the situation of such an object with that of a man violently but equally hungry and thirsty, standing at the same distance from food and drink and unable to decide in which direction to move. Aristotle, however, rejected this apparently so "organismic" reasoning on the ground that it is not organismic enough. A consistently organismic explanation, he warned, cannot ignore the basic strivings of all types of matter. True, he admitted, a heavy body, such as stone, has such a connatural indifference to moving toward the periphery. Such is not the case, however, with the light bodies, air and fire, which show an innate tendency to move toward the periphery of the universe, all whose parts are equidistant from the earth. Therefore, concluded Aristotle, the earth's rest at the center of the universe should be taken as a consequence of the earth's *nature* and not of its position there.

Aristotle could not leave this subject without making a remark that brings out vividly how the intuitive, all-embracing "intelligibility" of the organismic approach can deprive physics of the best source of understanding, the benefits of judicious observation. One condition for making such an observation is clearly one's ability to concentrate his attention on the thing or phenomenon to be observed — in other words, to isolate it from anything merely accidental or circumstantial. With such a methodical isolation of inanimate things from their surroundings and from one another, the organismic approach to nature can have no patience. For as long as the concept of organism reigns supreme in physical science, things in nature are viewed as organically interconnected, and the notion of an isolated part becomes basically deficient. His opponents, Aristotle claimed, kept considering only isolated aspects of an organic whole and this he found an inadmissible procedure. To quote his harsh strictures: it is "irrational to ask the question why the earth remains at center, but not why fire remains at the extremity." [52]

This was, however, a hasty recrimination that could quickly boomerang. For in the same context Aristotle by his own procedure unwittingly illustrated the disastrous consequence of the application of the organismic method to physical investigation: the role to be played by observational evidence becomes secondary in establishing scientific conclusions. In arguing the cause of geocentrism, Aristotle referred of course to the absence of any observable change in the relative position of the stars. Yet, one feels that his reference to the data of actual observation does not represent the principal part of the proof in an organismic context: observation at best underlines what has already been established on more general grounds.

No different is the case of the earth's shape as discussed by Aristotle. The keynote of his argument is again the same old organismic precept: the earth has to be spherical because only a spherical shape can result when pieces of matter have an innate tendency to move from every direction toward the same point. To some extent Aristotle along with earlier natural philosophers was willing to view this process as a sort of generation of the earth. But whereas they attributed it to external and mechanical compulsion, Aristotle was quick to stress "the true statement that this takes place because it is the nature of whatever has weight to move towards the center." [53] Astronomical observations — of the eclipses of the moon and of the different constellations visible from various latitudes — serve as further proof of the sphericity of the earth, but not as *the* proof. The latter always rests on an almost introspective grasp of the alleged innate strivings of the earth as a *natural* body.

From the shape and general characteristics of the earth, Aristotle proceeded to a discussion of its composition, and this, like any other part of his physical system, rested on his views of motion and nature. The earth, or more precisely, the sublunary world, must be composed, according to him, of simple bodies, because there are simple motions in that region. "Nature," he said, "is a cause of movement in the thing itself," [54] which means that every natural body must have its own proper motion, striving, if not volition. Now, since motions are either simple or composite and since simple motions belong to simple bodies, "it clearly follows that certain simple bodies must exist. For there are simple motions." [55] The number of simple bodies or elements then is obviously determined by the number of simple motions, a number that cannot be infinite because "the directions of movement are limited to two and the places also are limited." [56] Although this theory strictly speaking calls for only two elements, earth and fire, Aristotle made place for four on the ground that nature, which likes to level out excesses and prefers graduated changes, provides the other two elements, water and air, for intermediate bodies. The decisive role of organismic considerations in such a reasoning hardly needs to be pointed out.

The four elements can change one into another, and the place where such transformations take place — mainly under the influence of the variations in the sun's path — is the sublunary world, the realm of the Aristotelian meteorology. Usually dismissed as the least valuable part of Aristotle's studies in natural science, it illustrates better than anything else, however, the totally self-defeating consequences of the attempt to achieve a sound understanding of the workings of inanimate nature on the basis of an organismic concept of the universe. The part of natural science discussed by Aristotle in his *Meteorologica* roughly corresponds to what at present is called atmospheric physics and geophysics. It deals with the part of the physical universe immediately surrounding man, abounding in easily observable (though at times very complicated) phenomena that hardly betray the characteristics of organismic processes. Yet Aristotle remained faithful as always to his organismic viewpoint and embarked on a fanciful and highly arbitrary account of a host of phenomena that range from the Milky Way to the cavities of the earth.

What makes Aristotle's discussions in the *Meteorologica* so glaringly arbitrary is the resolute determination to explain the features of an inanimate part of the world in the terms and categories of life processes. To the *Meteorologica* pertain, Aristotle stated, "all the

affections we may call common to air and water and the kinds and
parts of the earth and the affections of its parts."⁵⁷ Accordingly
these affections are looked for in every nook and cranny of nature, in
the atmospheric turbulences as well as in those occurring within the
earth. This should hardly be the cause of any surprise. For if a method
is devised to find affections (strivings) characteristic of living organisms, its exclusive use will succeed in causing such affections to be
seen even where they are patently absent. Even when used judiciously, such a method will help one discover only affections and little
else. Judiciousness is hardly the forte of those students of physics
who conceive of the inanimate universe as a living organism. On the
contrary, undue preoccupation with the organismic viewpoint is apt
to create a state of mind that is instinctively driven to sweeping
generalizations. Of this a manifest example is the author of the *Mete-*
orologica, who does not tire of repeating the precept of an ironclad
methodological unity in all branches of the study of nature: there
can be no difference in the method with which comets, winds, rivers,
earthquakes, plants, and animals are to be investigated. It is no surprise, therefore, that in comparing Aristotle's attitude in this matter
with that of the pre-Socratic *physikoi,* such a careful scholar as F.
Solmsen could find hardly any precedent for the single-mindedness
with which Aristotle organized the subject.⁵⁸ The often decried rigid
monism of the Ionians seems flexible indeed when contrasted to the
onrush of inferences dominating the *Meteorologica* from beginning
to end. It is not at all unreasonable to assume, as Solmsen does, that
Aristotle considered it dilettantish, or beneath the calling of a philosopher, to admit more than one explanatory principle. There is, however, hardly a place for several viewpoints aiming at a balanced judgment when one attacks a topic "with a will." Therein lies the chief
characteristic of all efforts, ancient and recent alike, that try to discourse on inanimate nature in terms of purposeful inclinations, in
strict analogy to living organisms.

   Aristotle stated that the *Meteorologica* dealt with the material and
efficient causes and named the sun's rays as the efficient cause of whatever happens between the moon and the depths of the earth. In connection with the production of rain, hail, and snow, Aristotle even
described how the heating caused by the sun's rays sets in motion the
circulation of huge masses of air and moisture in the atmosphere. He
also spoke of the cooling and condensation of vapor at higher eleva-
tions and of the evaporation of water heated by the sun's rays. "So
we get a circular process that follows the course of the sun. . . . We
must think of it as a river, flowing up and down in a circle and made
up partly of air, partly of water."⁵⁹ But this great vision of atmos-

pheric mechanism remained undeveloped. The chief efficient cause turns out to be a rather remote agent that merely releases, so to speak, the "actors" of a complex interplay. The "actors" are natures, dry and moist exhalations, produced by the earth, which is described as a huge animal that grows, digests, and produces gaseous residues. The physical factor driving this process is, as in animals, heat, and Aristotle's train of thought moved most naturally from the hot ashes and lye in the earth to the hot bellies and excreta of animals. Thus just as the nature of ether could yield all the necessary answers in the superlunary world, so do these dry and moist excreta or exhalations of the earth supply the solution to almost anything to be found above and below the earth's surface.

Among the atmospheric phenomena that are said by Aristotle to consist of dry exhalations are the "shooting stars." Speaking in the best matter-of-fact style, he stated that the motion of planets sets afire the dry exhalation present in the upper atmosphere, and the burning material, or shooting star, takes on a shape or course that differs, he added in a scientific-sounding tone, "according to the disposition and quantity of the combustible material." [60] Such vague quantitative cadenzas, however, should not mislead anyone as to the true nature of the procedure. An organismic type of physics that relies so heavily on insights, intuitions, and aperçus will impart almost of necessity to the student of nature a self-assurance that hardly knows limits. The result is more often than not a flight of fancy in which the only element smacking of genuine science is the choice of words and phrases. Thus in the *Meteorologica* firelike appearances in the night sky, "torches," "goats," and what not are accounted for with an ease that betrays no doubt and acknowledges no difficulties. In the same vein we are told that thunderbolts simply consisted of hot and dry exhalations of the earth, compressed by surrounding cold air masses and pressed downward with great velocity. Through their speedy motion across the atmosphere these compressed dry exhalations became fiery (lightning), and when they collided with other clouds, violent sounds (thunderclaps) were produced. The bloodred glow of the evening skies was also reduced to the behavior of dry exhalations as were the comets that Aristotle identified as slowly burning shooting stars. Thus the appearance of comets necessarily indicated for him the rise of strong winds, since the latter were made of the same "all-purpose" substance too. Many and dense comets were therefore signs of a dry and windy year.

To what unsuspecting extremes organismic physics can carry the most penetrating mind can nowhere be better seen than in Aristotle's explanation of the origin of the huge meteorite that fell at Aegospotami in 467 B.C. and that was remembered with awe and puzzlement for a long time thereafter in ancient Greece. As stones could not originate outside the earth's sphere in Aristotle's system, he flatly stated that the huge stone "had been carried up by a wind and fell down in daytime — then too a comet happened to have appeared in the west." [61] The explanation of the Milky Way rushed onward with the same confidence. For if the motion of a planet or of a star could set aglow the dry exhalations and produce shooting stars and comets, why should the great constellations in the belt of the Milky Way not be able to produce the same effect on an even larger scale?

Next to the dry and moist exhalations the cold and the hot as distinct natures played a similarly important part in Aristotle's *Meteorologica* and again involved him in a long series of gratuitous statements. The cold and hot, being contraries, should, according to Aristotle's general explanation of the behavior of contraries, have speeded up reactions depending on their presence. Thus Aristotle was impelled to state that "the fact that the water has previously been warmed contributes to its freezing quickly: for so it cools sooner." Unlikely as such a "fact" is, Aristotle found nothing strange in it. What is more he reported in all seriousness that many people "when they want to cool hot water quickly begin by putting it in the sun." Apparently he found this "law" of physics so important that he even mentioned the custom of the people of Pontus who when fishing in wintertime poured hot water around their reeds "that it may freeze the quicker, for they use the ice like lead to fix the reeds." [62]

Since the main goal of organismic physics is to find out the "volitions" of bodies and objects, it is only natural that what Aristotle wanted to know about winds, windy vapors, rivers, and sea was their nature or "strivings." He thus engaged in a relentless search for distinct natures in the physical world (inanimate "individuals," one might almost say) and ended up regarding air as something essentially different from wind. He did this in the most categorical manner. "It is absurd," he said, "that this air that surrounds us should become wind when in motion, whatever be the source of its motion." [63] It was rather "wind that sets the air in motion." [64] To crown the comedy Aristotle was not loath to heap sarcasm on those who, like Hippocrates, "define wind as a motion of the air." The rest of Aristotle's comment on such a position deserves to be quoted in full:

Hence some, wishing to say a clever thing, assert that all the winds are one wind, because the air that moves is in fact all of it one and the same; they maintain that the winds appear to differ owing to the region from which the air may happen to flow on each occasion, but really do not differ at all. This is just like thinking that all rivers are one and the same river, and the ordinary unscientific view is better than a scientific theory like this.[65]

At any rate, this was not the only time that organismic physics had to side with popular beliefs and reject reasonable views. Aristotle, caught in the vicious circle of his organismic physics, at times had to ignore the obvious and state for instance that the dry exhalation from the earth "is the source and origin of all winds."[66] In the organismic framework a motion like that of the winds must proceed from a specific entity or nature. Therefore Aristotle had no alternative but to conclude: "In its generation and origin wind *plainly* derives from the earth"[67] (italics mine). To solve difficulties arising from such definition of the wind's substance, Aristotle fell back on analogy, a procedure particularly suited to an organismic type of explanation. That small, almost imperceptible volumes of dry exhalations escaping from the ground could produce violent winds was as natural for Aristotle as was the case of many small creeks uniting to form a mighty river. "The case of winds," he stated, "is like that of rivers."[68] This analogy "explained" for him why winds were allegedly far removed from marshy regions that were abundant sources of exhalation: springs too grew into rivers only after flowing a considerable distance. On the strength of such analogies Aristotle concluded that just as the various rivers were not the same but each had its own distinct nature, so regional winds too had distinct individualities.

In a similar fashion Aristotle raised the differences between seawater and freshwater into distinctions between substances. He assigned sweetness to the nature of water originating from springs, and from this he concluded that the sea could neither be the origin nor form the main bulk of water on earth. Were it otherwise, rivers too would be salty. In Aristotle's eyes there was, however, an even more basic reason for stating a sharp qualitative difference between seawater and freshwater. The waters on earth, he observed, either flow or are stationary. The stationary waters are in turn either mere collections of waters left standing or come from artificial sources such as wells. This specious classification, moreover, became firm ground upon which to base the statement that since natural standing water from springs was never found on a scale comparable to the large bulk of seawater, the seawater must therefore be essentially different from ordinary water.[69]

Concerning the origin of the sea, the dry exhalation again came to the rescue. This time it was described as an "undigested residue" of that giant, growing entity, the earth. Mixing with water, this undigested residue or "earthy stuff" produced the salinity of seawater.

Such mixture revealed to Aristotle, however, an organismic interplay or tension between heavy saltwater and fresh drinking water. The latter, said Aristotle, when carried by rivers into the sea spread thin on the surface of the sea and quickly evaporated, whereas the heavy saltwater filled the whole seabed, which in reality would be the natural place of the true (non-salty) water. For Aristotle all this was strictly equivalent to an organismic process, and he identified it with the combustion of food in animal bodies.[70]

The role played in Aristotelian physics by the various dry and moist exhalations came to a climax in the production of earthquakes. That exhalations (winds) could cause great upheavals in the crust of earth was argued from the fact that winds were swift, could move through bodies, and thereby had properties of a physical agent capable of exerting a major motive force. The way these exhalations touched off earthquakes was explained by rather crude organismic analogies. "We must think of an earthquake," warned Aristotle, "as something like the tremor that often runs through the body after passing water as the wind returns inwards from without in one volume."[71] Their violence was illustrated for Aristotle by a simile that could not be more biological. Tetanus and spasm, he said, were caused by "winds" within the body; yet they were at times so violent that even the united efforts of many men were often not enough to subdue some patients. "We must suppose," went the stereotype conclusion of organismic physics, "that (to compare great things with small) what happens in the earth is just like that."[72] That earthquakes did not die out suddenly was also explained by Aristotle in the same vein: the throbbings in the human body do "not cease suddenly or quickly, but by degrees according as the affection passes off."[73]

Whatever observations Aristotle could have gathered of earthquakes, they are rendered completely useless by his arbitrary ideas on how these exhalations should act. In a bewildering set of statements Aristotle connected earthquakes with eclipses, with midnight and noon, with spring and fall, and even with long, drawn-out clouds. He was fully confident that a special calm preceded every earthquake, that earthquakes were always local, that they did not occur in the middle of the ocean but only close to the shores. Such conclusions, he stated, followed from the theory that "has been verified by actual observation in many places."[74] This statement, needless to say, has little to commend itself. It was mainly luck that the organismic approach and biological comparisons on occasion came reasonably near to the actual course of events. One such fortunate remark concerned the changes in the respective positions of dry land, rivers, and oceans. Thus from the slow rate of transformation in animal bodies,

Aristotle concluded that the processes causing the shifts of the Nile valley were of an almost imperceptible slowness. Yet, he was only too eager to recall twice in the same context the allegedly vitalistic roots of such processes. Changes on the surface of the earth were due to the fact, he stated, that "the interior of the earth grows and decays like the bodies of animals and plants," although "the whole vital process of the earth" took place at such a slow rate that a lifetime was not enough to observe the effects of this continuous transformation of the earth.[75]

Discussions that are free of preoccupations with the nature and affections of organic bodies, or which are not amplified by references to animal behavior, are the exception in Aristotle's organismic synthesis of the physical world. One of these few exceptions is his account of atmospheric phenomena connected with the reflection and refraction of light rays, such as rainbows, halos. Consequently, the related observations are less colored and distorted by references to preconceived patterns of organismic behavior. As a rule, observations are allowed to play but a rather subordinate role in a physical study that views the world as an organism. Furthermore, the principle that every spontaneous and regularly recurring motion implies the existence of a specific nature or substance, multiplies the number of essentially distinct natures. This in turn unavoidably hampers efforts to recognize basic features common to them. Phenomena that are only gradations on the same scale, hot and cold, for example, will therefore be taken as distinct entities. For the same reason standing and flowing waters will be put in basically different categories, just as standing and moving air (wind), circle and straight line, flesh and bone are classed in the Aristotelian science as basically different entities. It was also Aristotle's preoccupation with primary principles harking back to organismic considerations that led him to conclude that stars and the sun were not hot and that their white glow had nothing to do with fire. Yet, as Theophrastus remarked, Aristotle could easily have observed what had already been noted long before him, namely, that objects which glow white are hotter than those which glow red.[76] Had Aristotle been less engrossed with general, organismic principles, he could have anticipated the sagacious remark of his successor at the Academy, Theophrastus, who certainly did not refer to a new fact in arguing that fire, because of its invariable need of some combustible substance, could not be a primary element.[77] Nor had the author of *Pneumatics*, Hero of Alexandria, to resort four centuries later to novel discoveries to restate the obvious, the identity of wind and air. All he needed was a sober reverence for facts and some measure of freedom from preconceived tenets.

Since an example of organismic behavior easily lends itself to a quick, intuitive grasp by the human observer, organismic physics such as Aristotle's is likely to be given to hasty generalizations. Time and again one encounters the almost stereotyped phrase when Aristotle takes up a new topic: "Let us recall our fundamental principle and then explain our views."[78] The result is that observations that do not fit in the hastily erected framework are brushed aside or not weighed in their true significance. It was mere capriciousness on Aristotle's part to write that "it is actually observable that winds originate in marshy districts of the earth and they do not seem to blow above the level of the highest mountains."[79] As a matter of fact, Aristotle, in his systematization of the physical world, lives up rather poorly to his definition of the good investigator who, as he said, "must be alive to the objections inherent in the genus of his subject, an awareness which is the result of having studied all its differentiae."[80] Ironically enough his speedy generalizations led him into the very pitfall against which he warned so eloquently: "A small initial deviation from the truth multiplies itself ten-thousandfold as the argument proceeds."[81]

The organismic method, apparently secure in its basic assertions, easily produces in the investigator an attitude that Aristotle so severely deplored in others and that he characterized as "excessive simple-mindedness or excessive zeal."[82] Aristotle was quick to observe "the great difference which exists between those whose researches are based on the phenomenon of nature and those who inquire by a dialectical method." In the same breath he chastized "as men of narrow views" all those who indulged "in long discussion without taking the facts into account."[83] Yet, it escaped him that it was the organismic method that made him commit the same error in physics. For only the uncritical reliance on an intuitive grasp of things and processes can acquiesce in the Aristotelian advice about phenomena that are not observable: "We consider a satisfactory explanation of phenomena inaccessible to observation to have been given when our account of them is free from impossibilities."[84] Thus, believing himself to be in the possession of the only true method for handling any eventuality, observable or non-observable, that might arise in scientific investigation, Aristotle was daring enough to state quite a few things about admittedly unobservable phenomena as the following passage shows: "The winter in the north is windless and calm; that is in the north itself; but the breeze, that blows from there so gently as to escape observation, becomes a great wind as it passes on."[85]

In Aristotle's *On the Heavens* and *Meteorologica* there is hardly a word about phenomena defying solution, and when he voiced puzzlement or caution it was mainly lip service. True, we hear him state in the opening pages of the *Meteorologica* that of the phenomena to be discussed "some puzzle us, while others admit of explanation in some degree." [86] But in the tortuous mental labyrinths of this work there is hardly a conclusion striking a note of reservation. Aristotle did not even flinch when he flatly told his readers that no one could live beyond the tropics, that the shape of the polar regions resembled a tambourine, and that imaginary lines connecting those regions to the center of the earth cut out a drum-shaped figure. Facts obviously contrary to his conclusions were grandly talked away, as, for instance, the case of winds that happen to blow while an earthquake is taking place. "It is true that some [earthquakes] take place when a wind is blowing, but this presents no difficulty. . . . Only these earthquakes are less severe because their source and cause is divided." [87] Clearly the escape was sought in a quick distinction for which neither the theory nor the facts provide any basis. The criticism Aristotle leveled at his opponents applied, therefore, wholly to his organismic physics: "They have a wrong conception of primary principles and try to bring everything into line with hard and fast theories." [88] For all too often in handling facts, or rather, in not handling them, Aristotle became guilty of the weakness he found in earlier physicists: "Out of affection for their fixed ideas these men behave like speakers defending a thesis in debate: they stand on the truth of their premises against all the facts, not admitting that there are premises which ought to be criticized in the light of their consequences, and in particular of the final result of all." [89]

The final result proved indeed to be wide of the mark. This was all the more regrettable because even in the field of physics Aristotle at times played the keen observer that prompted Darwin's famous comment on Aristotle, the biologist: "I had not the most remote notion what a wonderful man he was. Linnaeus and Cuvier have been my two gods, though in very different ways, but they were mere schoolboys to old Aristotle." [90] Thus Aristotle, referring to his experiments, stated correctly that not only saltwater but wine and other fluids too, when condensed after evaporation, yield pure water. With regard to the respective weights of saltwater and freshwater, he recalled the fate of heavily loaded ships that sank when moved from the sea into the river's estuary. He also noticed that the star in the thigh of the Dog shows a tail when one glances at it, a remark that is explained only in the light of the not-so-old knowledge that the peripheral parts of the retina are more sensitive to illumination.

In spite of such bright details Aristotle's organismic physics was to prove "worthless and misleading from beginning to end." Yet, E. T. Whittaker, the author of this devastating phrase,[91] was neither an anti-Aristotelian nor was he surpassed by many in his understanding of present and past physics. For his statement Whittaker did not claim originality. Many before him took a similar stance and voiced at the same time their appreciation for what was lasting in Aristotle's thought. What then did Aristotle's physics lack that could have prevented it from becoming a world-picture that steadily moved away from reality with every step and resulted ultimately in an exercise in names and words? It is only highly ironical that a philosophical system stressing so much the world of perceptions as the source of ordered knowledge was to produce a physics where the facts were to fit into a preconceived pattern which could not permit a detached look at the nuances of facts without crumbling as a whole. Being a projection of man's nature into the external world, the organismic physics as developed by Aristotle's physics is not a depersonalized analysis of the world but rather a subjective penetration of nature. Its parameters are those of the realm of human volition: natural, unnatural, violent, or restful. Its scope is not to find the correlation of things, but to achieve an intuitive insight into the nature of things, into their alleged strivings and affections. This is why in an organismic physics there is no proper place for mathematics, measurement, or experiments; and quantitative results can never negate the qualitative conclusions. Not that the real, movable, and moving bodies could not be, according to Aristotle, the objects of mathematics, but mathematics is supposed to study them only in abstraction from movement. For to quantize and measure a movement — this fundamental expression of the nature of things — would have amounted in Aristotle's view to dissecting both movement and the nature underlying it. Could a "nature," however, be cut to pieces in any meaningful way when it stood by definition for an indivisible whole?

Just as mathematics, according to the Peripatetic dictum, cannot study the nature of movement but only abstractions, conversely, "no abstraction can be studied by Natural Science because whatever nature makes, she makes to serve some purpose."[92] This is why a quantitative study of the elements is rejected.[93] This is why Aristotle could oppose the true center of the universe to its mathematical center.[94] It was for the same reason that his few arguments taken from geometry were exploited in support of his organismic views, as, for instance, when the sphericity of the heavens and stars and their motion was discussed and contrasted to the shape of animals. It was on the same account that Aristotle erected a rigid distinction between mathematicians and natural scientists, or between astronomy and physics. This is why Aristotelian physics, in contrast to geometry, which tries to limit the number of its axioms, results in defining natures without end in order to explain all nuances of motions. It is because of this that in organismic physics not all movements are comparable and that "a circle and a straight line are not comparable, so that the motions which correspond to them will not be comparable either." [95] This is why weight for Aristotle is absolute, not relative, and the speed of the motion of a natural body stands in direct proportion to the "ampleness" of its nature. The more there is of fire Aristotle claimed, "the faster it performs its proper motion upward . . ." and "in fact there are many things which move faster downward the more there is of them." [96] This is why Aristotle said that a chunk of earth, after being raised to a given height, "will begin to travel the quicker the bigger it is." [97] Consequently, the distances traversed by falling bodies in equal times must be stated to be proportional to their weight (ampleness of their natures, that is).[98] And this law holds true of motion in any direction, provided the motion is natural: " Any chance portion of fire moves upwards and of earth downwards if nothing else gets in the way and a larger portion moves more quickly with the same motion." [99]

The legend spun around the Tower of Pisa was certainly not needed to recognize that such was not the case. The scaffoldings that had to be erected whenever a new temple or monument was constructed in fourth-century Athens must have provided enough cases of falling tools and stones of every size to show that the time of fall from the same height was much the same for the great majority of objects, whatever their size, shape, or weight. Almost a thousand years after Aristotle, yet still living in the same antique world, Philoponus could state without the slightest exaggeration that in Aristotle's law about the fall of bodies there was "something absolutely false." [100] Judging by the sources Philoponus used, it is very likely that this criticism merely echoed those already voiced within the Peripatetic school itself. One would seek in vain in Aristotle's works traces of even so slight a misgiving about his own conclusions on the laws of physics. This was particularly evident when he discussed the relation of resistance to the moving force. Taken aback by the logical consequences of his premises, Aristotle admitted that a force must be of a certain minimum magnitude to move a given body. It was manifestly impossible to move an oxcart however slowly with one's little finger. But unfortunately even such contradictions could not move him to make even a conditional reappraisal of the validity of his basic assumptions. The organismic, teleological framework through which he saw nature permitted him no other approach to the phenomena and caused him to take lightly what proved to be all too momentous. As a further consequence, the elements of technical skill, or the more general mechanistic considerations, had to remain severed from the study of nature, and not even a faltering step could be made toward what we call kinematics and dynamics. True, a book called *Mechanics* in which the problem of the lever and related questions were discussed was compiled in Aristotle's school. But that was in the time of Theophrastus, who found enough reason to oppose his master in not a few points: "With regard to the view that all things are for the sake of an end and nothing is in vain, the assignation of ends is in general not easy, as it is usually stated to be." [101] In fact, study of the circumstances surrounding the birth and feeding of animals revealed to Theophrastus so many irregularities that he found it impossible to assume in such cases a purposeful action by nature. He could only do what was contrary to the very soul of organismic physics; he called for a program that would "find a certain limit . . . to the final causation and to the impulse to the better." What he hoped to achieve was a better determination of the "conditions on which real things depend and the relation in which they stand to one another." [102]

In the light of Theophrastus' wholesome advice, it is particularly painful to read certain passages of Aristotle that illustrate in a truly dramatic way how wrong was the direction taken by physical studies when nature was identified as an organism. Aristotle had no choice but to reject the mutual attraction of bodies: "If one removed the earth from the path of one of its particles before it had fallen, it would travel downwards so long as there was nothing to oppose it." [103] He had no patience with "the old saying that like moves to like" and even illustrated his position by a far-reaching hypothetical case: "If the earth were removed to where the moon is now, separate parts of it would not move towards the whole, but towards the place where the whole is now." [104] And he could hardly have realized how right he had been when after rejecting the idea of minimum magnitude he added the words that revealed their deepest meaning only after Planck gave his explanation of the black-body radiation: "Suppose for instance someone maintained that there is a minimum magnitude; that man with his minimum would shake the foundations of mathematics." [105] The basic assumptions of Aristotle permitted no distant galaxies or island universes, for the universe as an organism could not exist of apparently independent parts. Similarly the stars and the sun could not be hot, and the Milky Way, the meteors, and comets *had* to be located below the path of the moon. It was also his infatuation with the concept of organism that led him in a direction diametrically opposed to the concept of inertia. And it is no less ironical that he

could assume only as absurd the case of bodies falling with equal
(though infinite) speed in a vacuum.

That Aristotle's views on what physics should be could dominate
the Hellenistic Age so thoroughly is mainly due to the fact that he was
truly Greek in his attitude toward nature. One may admire the intellectual boldness of the Ionian physicists and their bent for mechanistic rationalism; one may even class them as the first representatives of the much admired Hellenic clarity of mind. It nevertheless
remains true that the strong reaction that set in with Socrates was no
less Greek in character and probably even more so. The post-Aris-
totelian phase of Greek science was simply unwilling to part with
this aspect of the Greek vision of the cosmos. Thus the physics of the
Stoics, though putting considerable emphasis on the mechanistic
aspects of the vibrations of pneuma, regarded it, however, as the
basic, organic stuff of the universe, constituting and pervading everything: ordinary matter, living bodies, and the soul itself. Similarly the
great Greek mathematicians and astronomers continued to accept the
dictum that physics and astronomy were two separate fields of investigation and that the world, being an organism, was not to be dissected
by mathematical analysis. In perfect tune with this did Archimedes
restrict himself to statics, turning away from problems that might perhaps have opened to him the realm of infinitesimal calculus. Ptolemy,
too, though unable to find use for the concentric spheres of Aristotle,
firmly upheld the vitalistic principle when it came to explaining the
motion of celestial bodies.

Actually Ptolemy warned his reader in the opening section of his
Planetary Hypotheses that one could secure, by using mechanical
analogies, only a superficial look, hardly a real insight into the nature
of the planetary motions.[106] For, as he explained it in the second book
of his work, to understand the motion of planets one should think of
the case of birds whose motions were due to a vital principle residing
within them. From this source came the impulse that spread into their
muscles and from there to their wings, without disturbing nearby
birds or the surroundings in general. "Of the situation among celestial
bodies we should think in the same manner, and also uphold the
opinion that each of them possesses a vital force, moves of itself and
imparts motion to bodies united with it." [107] True, the master teacher
of epicycles, deferents, and equants could hardly be expected to deny
a basic interconnectedness between the motion of planets. Yet, even
here his most expressive simile was again taken not from the realm
of dead machinery but from the realm of living beings, and this time
of intelligent beings too. Coordination without mechanical connection
or constraint, as displayed by a group of dancers or by soldiers drilling

with weapons, was for Ptolemy what came closest to a true description of the heavenly motions.[108]

Beyond Ptolemy, well into the Middle Ages, the organismic approach to physics held sway. It also rendered physics essentially barren by cutting it off from quantitative experiments in particular and mathematics in general. For such a course of events one cannot place the whole burden of responsibility on one man, Aristotle. Still this is done by not a few historians brooding over the opportunities lost for science in antiquity. It remains, however, to be proved that the discoveries of Galileo and Newton would indeed have been anticipated by one or two millenia had it not been for Aristotle's influence. The Greeks made their choice because, as pointed out before, the organismic approach was so congenial to them. Since it was equally congenial to many others who fell heir to the ponderings of the Greeks in physical science, one must admit the existence of something deeply and universally human in the organismic approach to nature. For greatly varied indeed are the cultural, racial, and religious backgrounds of those who have found it more natural and more intelligible to speak of nature as an organism rather than a mechanism.

Among them were the Romans, whose practicality had no taste for Greek theorizing and who took pride in restricting their study of mathematics to the useful. Yet the Romans, who as Cicero tells us thanked the gods for not being dreamers like the Greeks, showed an unrestrained liking for the organismic concept of the universe, which they had learned mainly from Stoic sources. What is more, in this organismic ideology their juridical mind was quick to make certain statements more precise and some details more elaborate. Thus Cicero presented a passage attributed to Zeno in the following syllogistic pattern: "Nothing that is without a soul and reason can generate of itself anything endowed with life and reason; the world however generates beings with soul and reason; therefore the world is itself living and possessed of mind." [109] In his *Quaestiones naturales* Seneca goes to great lengths to show how the earth is organized like the human body. He compares watercourses to veins, air passages to arteries, geological substances to body fluids, and earthquakes to injuries.[110] All this was, of course, in keeping with his basic view of nature — that "the whole art of Nature is imitation. . . . The place which God has in the world, the soul has in man, that which in the former is matter, is in us body." [111] The organismic interpretation of the physical universe thus came to comply with the Socratic program that aimed at grafting science into a more general philosophico-religious outlook on life. How dominating such an attitude toward nature was in the late Greco-Roman culture hardly needs documentation. It is enough to think of the *Enneads* of Plotinus where the organismic comparisons are extended to the psychological level as well. The whole range of existence, human, biological, and physical, was clearly accepted as the manifestation of an all-embracing body, an idea to which almost everybody in one way or another assented. Augustine himself tells us in the *Confessions* that for some time he believed himself to be a fragment of a vast and bright body, which he equated with God.[112]

Outstanding as was the appreciation shown by the Arabs toward the mathematical and experimental aspects of the Greek scientific heritage, they too saw a redeeming value in the organismic concept of the world. The famous *Encyclopaedia* prepared by the Brethren of Sincerity at the close of the first millenium was compiled with no aim other than to serve the reconciliation of faith with the sciences. What the Brethren of Sincerity needed to this end was a unified world view, and they believed they had found this in a pan-organic concept of the world. Another great medieval attempt to show by philosophical and scientific arguments the reasonableness of faith, *The Guide of the Perplexed* of Moses ben Maimon, dwells also at great length on a comparison between the cosmos and the purposeful functioning of the human organism and personality. To see the universe as a living being was for the Great Moses, as Aquinas referred to him, "indispensable . . . for demonstrating the unity of God; it also helps to indicate the principle that He who is One has created only *one* being." [113] Still man is a special part of the created whole. It is in him that the purposeful aspect of nature is realized in full measure and this is why, as Moses was eager to point out, only man — no lower animal — had ever been called a *microcosmos*. No less did the parallel between microcosm and macrocosm appeal to the medieval German and Christian, typified in Nicholas of Cusa. In his thought the idea of organism was supposed to bring about a synthesis of the world in which even such Christian mysteries as the Trinity were included.

With the thinking of an age so permeated by the concept of organism, science could hardly be expected to strike out on a new path overnight. Late medieval science was characterized by trends of thought existing side by side — one affirming without reservation the concept of organism, the other trying to find new ways, while continually falling back unawares on the heritage of organismic concepts. The most colorful representative of the former trend was Paracelsus, who always went beyond generalities and carried his organismic tenets into the field of practical applications. A firm believer in the dependence of anatomy on astronomy, Paracelsus never stopped urging his fellow physicians to keep astronomy in mind when diagnosing a sickness.

It was, however, far easier to see the futility of such a procedure than to shake off completely the outlook that generated it. Whatever there was of a revolutionary turn in Copernicus' work, there was still an enormous distance separating his animistic allusions from the air of classical mechanics. For Copernicus still refuted objections against the motion of the earth by purely Aristotelian considerations. Nor did Cardan's mathematical prowess prevent him from devoting a full chapter of his *Arcana aeternitatis* to a discussion of the likeness between man and the world. John Dee's famous preface to the English edition of Euclid's *Elements* (1570) speaks in the same vein of man's body as the great synthesis of the results of sciences and arts. For Tycho Brahe it was both senseless and impious to deny the influence of stars on terrestrial organisms. The skies, as he put it, had a far more universal purpose than just keeping the time, and he pointed to a long list of things on earth that seemed to wax and wane with the phases of the moon. Air, water, and other elementary things were, in his belief, fed by the sky itself, and for the same reason he asserted that man, owing to his basic dependence on the influence of the sky, "acquires an incredible affinity to the related stars."[114] Organismic ideas were just as well in the foreground in the various writings of Bruno where seminal insights were evenly mixed with mysticism and obscurantism. No less deeply steeped in organismic thinking was William Gilbert, the author of *De magnete,* who assigned the attractive power of magnets to a self-moving soul. True, he regretted that such an explanation could not be tested; nevertheless, he obviously found great solace in the intuitive intelligibility he attributed to the biological similes that abound in his book.[115]

Gilbert's emphasis on the intuitive intelligibility of the organismic view is well worth noting. For when a generation later the pendulum swung to the other extreme, another era in physics was ushered in, and men of science swore just as enthusiastically by the perfect and exclusive intelligibility of the mechanical explanation of nature, all too often without any regard for the experimental verifiability of the statements made. The new scientific orientation was asserting itself rapidly. From the early seventeenth century onward, relapses like Gilbert's into the labyrinths of organismic thinking could no longer escape scathing criticism. The writings of the two chief proponents of the new scientific program, Bacon and Galileo, showed this only too well. The former tried on various occasions to excoriate the use of the concept of the microcosm in scientific writings,[116] and the latter chastised Kepler for clinging to occult and organismic ideas.[117] For great as Kepler was as a trailblazer in physics, his works teemed with animalistic references.[118] He enjoyed speaking of the earth as a great animal whose breathing depended upon the sun and whose sleeping and waking caused the rise and fall of the oceans. Thus, whatever his contributions to the clarification of the new concepts needed by the new physics, Kepler echoed the past in his meticulous account of the sundry signs allegedly indicating the presence of a soul operating in the earth.[119]

With Kepler, however, an era essentially came to an end. After him one would hardly find a physicist of stature entertaining seriously organismic views about the realm of inanimate matter or the universe in general. Enormous as it was, this change in the state of mind of the physicists was no less resolute. The writings of Huygens, Newton, and Leibniz on physics contain only on rare occasion an expression harking back to organismic views. And if they indulge in biological similes, they do it mostly in the popular and in the speculative or conjectural sections of their works, a procedure most evident in Newton's *Opticks.* Even with Leibniz, so intent on achieving the great synthesis of human knowledge, the fusion of organismic and mechanistic ideas is clearly characterized by the domination of the latter. True, he spoke with deep conviction of each particle of matter "as a garden full of plants and as a pond full of fishes" and claimed that "there is nothing fallow, nothing sterile, nothing dead in the universe." [120] Still, he stressed even more the need of explaining the living in terms of the mechanical. Living bodies he called "nature's machines," [121] composed of machines even in their smallest parts, and if they formed a class of their own, it was only because they were a kind of divine machine infinitely surpassing all artificial automata.[122]

Leibniz's case also shows that even for a German of the early eighteenth century, it was not altogether impossible to subscribe to the primacy of mechanistic over organismic views. In general, however, it was very difficult to say the least. The line from Paracelsus through Van Helmont and Boehme to the Father of German Romanticism, Herder, runs unbroken and shows vividly the sympathy of the German spirit for a vitalistic or organismic orientation. What this attitude sought to achieve above all was the documentation and implementation of a deep harmony that supposedly existed between natural history and human history. It was this program that Herder tried to prove and elaborate in all details in the first three books of his *Ideas to a Philosophy of the History of Mankind* (1774–75). Herder believed that it was "both anatomically and physiologically true that the analogy of an organism dominates through all the living realm of our Earth," and he presented man as that "finished form in which, as in the finest exemplar, all traits of all begettings are summed up." [123] The key to nature, history, and society lay, therefore, in recognizing the unitary plan in the diversity of forms, and to achieve this one had to rely, according to the romantic precept, on an intuitive, poetical self-identification with the parts and the whole of nature. Such a method, it must be admitted, provided German biologists of the Romantic period with not a few fruitful concepts. When taken as a basis for physical research, however, organismic ideas could only demonstrate once more their basic uselessness.

What happened is extremely instructive, for this time a physics based on vitalistic and organismic views was pitted not against a set of vague conjectures as was the case with Aristotle's opposition to the Ionians and the atomists. It was now pitted against the physics of Newton, which not only represented a superb theoretical construct but also provided amazing tools for man's mastery over nature. Although the campaign waged against Newton was a losing battle from the very start, it threw vivid light on a certain type of scientific strategy. Most important, it revealed in a modern setting the strange features accented in the mind of a "physicist" who had opted for the organismic view. The physicist was none other than Goethe. At least he considered himself a physicist, as he kept stressing that the most valuable of all his achievements were in physics. At the end of a long life rich in superlative literary accomplishments, he was still telling Eckermann: "As for what I have done as a poet I take no pride in it whatever. Excellent poets have lived at the same time with me, poets more excellent lived before me, and others will come after me. But that in my century I am the only person who knows the truth in the difficult science of colors — of that, I say, I am not a little proud, and here I have a consciousness of superiority to many." [124]

Self-taught physicist as he was, Goethe could find nothing strange in his cultivating only one branch of physics, that is, optics. About optics, however, he tried to learn as much as possible, reading almost everything written on the subject from Aristotle to Newton and beyond. The material he collected would have made a nice history of optics had he appreciated the mathematical part of the subject, too. But of mathematics he wanted no part. In science his was a Socratic program that sought to vindicate the wholeness of life against mechanism and its most fearsome tool, mathematics. Reminiscing about his encounter with the world view of mechanistic physics as "represented" by Holbach's *Système de la nature*, he vividly recalled the utter revulsion the book had provoked in him. In Holbach's work Goethe came face to face with a mechanistic philosophy carried to its extremes. His reaction to it betrayed a startling resemblance to that of the young Socrates reading the mechanistic philosopher of his time, Anaxagoras. As Goethe recalled in his *Dichtung und Wahrheit*, he opened Holbach's work with the greatest expectations. He hoped to learn from it something really profound about his idol, nature. What he found there however, was, a "melancholy, atheistical half-night in which earth vanished with all its images, heaven with all its stars." The idea of matter in motion from all eternity and in every direction, which dominated every page in Holbach's work, appeared to him "so dark, so Cimmerian, so deathlike" that he felt simply unable to endure its presence. Holbach's account of the world gave only what there was of constraint and of cold mechanical necessity in the universe. That all this was a part of existence, Goethe readily admitted, but he also insisted that this could hardly be the whole story. Beyond the "necessities of day and night, the seasons, the influence of climate, physical and animal condition" from which no living being could withdraw, there was the factor of overriding importance that no one could ignore: one's own immediate awareness of what appeared to be "a perfect freedom of will." No remark could indeed be more Socratic in tone. The action that followed, though distinctly romantic, was also not alien to the basic aims of the Socratic precepts. To overcome the "mutilation" of nature perpetrated by the mechanistic philosophy Goethe and his friends threw themselves, as he put it, "into living knowledge, experience, action and poetising, with all the more liveliness and passion." [125]

Anyone asserting so resolutely the universality and immediacy of life and man's immersion in it can hardly admit significant exceptions to his tenet, or adopt a method in science that would not depend essentially on the pattern and behavior of living beings. Just as with Aristotle, the study of plants and animals had already set Goethe's attitude in science before he stumbled on the major challenge of his life, the manifestly non-living realm of colors. What he found in the botany of his time left him deeply dissatisfied. In the nomenclature of Linnaeus he could see only the frustration of his own poetic resolutions. With pointed reference to the inability of the quantitative method to grasp nature in its fullness, he asked: "Why are we moderns so distraught? Why are we challenged to demands we can neither attain, nor fulfill?" [126]

The source of trouble was, in Goethe's eyes, the breakdown of organic unity between man and nature, and to restore it man had to recover the ability to see both the entirety and the parts of nature in the light of an organic whole. Given the organic unity of nature, its manifestations had to be, so Goethe thought, an endless variation on basic visual themes or patterns that Goethe called archetypes. Consequently the task of forming a scientific law consisted for him in recognizing and describing those primordial forms and indicating the gradual steps through which they take on a multitude of details. This position entailed for Goethe two consequences: first, various plants and animals were to be reduced to a common type (archetype); second, the parts of a plant or of an animal were also to be considered mere variations on some basic, primordial part of the organism.

The first of these assumptions led Goethe to postulate the existence of the intermaxillary bone in man, the rudimentary form of which he was able to point out convincingly in 1784. It was only natural that such an early success spurred him to pursue the osteological studies that were to hold his intense attention over a number of years. The second consequence of Goethe's principal idea in science was also of some scientific merit. It enabled him to make rather plausible observations on how the shapes of the various parts of plants, such as petals, stamens, and nectaries, could be traced back to the successive differentiation of stem leaves. He made the same reflections on animal structures by indicating, for instance, how the shape of a vertebra or the central segment of an insect can be recognized in other parts of the organism, such as skull, jaws, wings, or even feelers.

In all this it was not the method of measurement but the art of intuitive glance that was to lead the investigator. According to Goethe's account such a flash crossed his mind when looking at a fan palm in Padua, and it was again a chance glance at a broken sheep's skull lying in the sand of the Lido that made him see in it a highly transformed vertebra.[127] Goethe could hardly care less that a "scientific law" left in its embryonic form of artistic intuition (aperçu)[128] will always remain vague and indefinite, as also will be the inferences drawn from it. On this point he wanted no improvement, no more precision. His view was that in science all depends on an aperçu, that is, an esoteric and indefinable insight that cannot be put in a cut-and-dried form. Truly, he sensed that vagueness was of necessity the lifeblood of the organismic explanation of nature. The logic such vagueness could satisfy had to rest, of course, on something that Goethe called "inner truth and necessity." Paradoxically, he identified this "inner truth," which had to be "applicable to all that lives," with strict logic. Thus he insisted that his archetypal plants were "strictly logical plants" and not merely "picturesque and imaginative projections."[129]

A mind like Goethe's, so deeply committed to the organismic, intuitive approach to nature, could hardly suspect the fallacy of such a contention. It was clearly impossible for him to see that "inner truth" could scarcely suffice when one tried to explore the complexities of "external" nature. The failure of Goethe's archetypes is no less evidence of this fallacy than the errors inspired in science by Aristotle's "natures." For these concepts have much in common in spite of the fact that for Aristotle the nature of a being was revealed by observing its motion, whereas Goethe took his start from the shape or pattern of things. Still in both cases there was the firm conviction that through the immediate visual pattern one would obtain a true and wholly satisfactory explanation of the physical world. The real point of satisfaction lay for Goethe and Aristotle alike in their belief that the world, being an organism, could in no way keep its secrets from an attentive observer who felt himself to be part of that organism and who had an intuitive grasp of what it was to be an organism.

Such was Goethe's mind when during his journey of 1786–87 in Italy his interest turned strongly toward the intricacies of color effects. Accordingly, the method to be used in his subsequent studies of optics posed no problem for him. He was convinced, long before he investigated the physics of light, that the phenomena of colors should be studied "from the standpoint of Nature, if one were to gain control over them for the purposes of Art." [130] He obtained no control over colors, although control over nature is the very aim of physics. Nor did he achieve in physics even that small measure of success that was his share in botany and comparative anatomy. As with Aristotle, the organismic approach to nature, which yielded valuable results within the realm of the living, had to fail utterly when applied to inanimate phenomena. It had trapped the greatest of German poets in a labyrinth of errors that stands as perhaps the most pathetic case of stubborn blindness and self-deceit in scientific history. Its beginning could hardly be more innocuous. Pressed to return some optical equipment he had borrowed with a view of conducting some experiments, Goethe decided to look through one of the prisms he had left untouched until then. Naïvely enough, he expected to see the whole wall of his room covered by a bright band of rainbow colors. Instead, the colors appeared only at the edges of the area of the wall illuminated by sunlight. Puzzled as he was, Goethe did not hestitate for a moment to charge Newton with a ghastly error.

The blunder was Goethe's, however, and only the first one in a long series of his illusions as a physicist. He was led by chimeras when he proceeded, true to his concept of science, to the acceptance of light as an *Urphenomenon* that could not be analyzed any further but that must be viewed as the manifestation of nature in its naked simplicity. Light was for him an elemental entity, an inscrutable attribute of creation, a unique phenomenon that had to be taken for granted. [131] In the eagerness so characteristic of the adepts of organismic physics he had almost of necessity to tip his hand. Along with light he classed even granite as a basic entity in nature not subject to any further analysis, which was obvious nonsense even by late eighteenth-century scientific standards. In the best tradition of organismic science, along with light darkness entered too. The similarity between Goethe's definition of the respective roles of light and darkness and the Aristotelian concept of the *diaphanous* is unmistakable. The diaphanous was a property intimately present in all bodies and substances according to Aristotle, whereas Goethe's light was a streaming of reality that prevaded and united everything to no less a degree. The diaphanous produced light when touching upon a firelike substance, the absence of which caused darkness. With Goethe darkness played the same role in giving contrast to light and without it there would be no light. Such closeness between his and Aristotle's views should not be a cause of surprise. A diligent translator of a work on color by Theophrastus, Goethe saw Aristotle as the greatest of all observers of nature, and as late as 1824 he was still trying to get Eckermann to believe that Aristotle's optics represented the current view in physics.

A stubborn contention like this indicates better than anything else the degree of blindness brought upon Goethe by his addiction to an organismic type of physics. It is indeed tragico-comical to see him pursue the study of optics with a self-imposed filter on his eyes that concealed from him the real nature of the path he was following. Even worse, he was totally unable to suspect that every step he made was in the wrong direction. For, as was the case with Aristotle's approach to physics, once the die had been cast, there was no chance to return, no room in the system for a cautious reappraisal of the course already covered, no opportunity for redressing the fatally upset balance. Having rejected the composite character of white light, Goethe had to look elsewhere for the source of colors. The choice he was left with was not wide, and once more he had to rest his case on a sadly amateurish handling of commonplace phenomena. What he "discovered" was that liquids and opal glasses showed bluish shades in reflected light and yellowish ones in transmitted light. From this he concluded that colors were the product of the medium. The purportedly corroborating evidence offered by Goethe to this conclusion was also based on an equally dilettantish interpretation of another well-known fact, namely, that the mixing of paints of various color produces only gray shades but never a pure white. In fact, the more colors are added the stronger and darker the gray appears. Seeing this, Goethe felt entitled to warn with a simile reminiscent of Aristotle's comparisons drawn from the world of animals: "A hundred greys do not make a single white horse." [132]

Wholly blind to his own errors, Goethe could only find faults with every dissenting voice and contrary evidence. The spectacle could hardly be more pathetic. On the one side there was a long array of physicists and painstaking experiments bringing optics to a level almost rivaling astronomy in accuracy. On the other side was a single figure preaching relentlessly that all physicists were wrong. For this awkward situation Goethe put the blame on the alleged narrowness and malice of the physicists. Their leader, Newton, particularly incensed him. He called Newton a "mere twaddle," judged his views "ludicrous" and admirable only for "school-children in a go-cart." He could little realize that his doggerel verses (*Katzenpasteten*) of 1810, which poured scorn on Newton,[133] were better creations of his mind than his polemics in physics. Nor did he suspect that in his sarcastic sketches of Newton's scientific personality he had drawn his own self-portrait as a physicist:

First he finds his theory plausible, then he convinces himself with excessive haste, even before he realizes how contrived a sleight-of-hand he will need to apply his hypothetical insight to the real order of things. But once he has openly expressed it, he does not fail to make use of all his mind's devices, to see his thesis through. In this he asserts with unbelievable sangfroid before the whole world the whole absurdity as the obvious truth.[134]

In Goethe's view, Newton's basic error lay in his concentration on artificial, narrow cases as opposed to the contemplation of the wholeness of optical phenomena as occur in nature. True to himself, Goethe disdained the use of the darkroom as a narrow place where the bright colorful world of light was mutilated. Out of his antagonism toward darkrooms, Goethe rejected one of the major discoveries of his day, the Fraunhofer lines, for they appeared only when light fell through a narrow slit. For him instruments, prisms, mirrors, and calculations in a sense tortured nature into abstractions, which he viewed as brazenly unnatural. Therefore he dismissed the word refractivity as an abstract term, and of refraction and polarization he said that "both are empty words which tell nothing to those who think and still are repeated all the time by scientific men."[135] Why these and many similar terms were found useful by men of science Goethe knew perfectly well. Behind these words stood exact mathematical formulas for which Goethe, the organismic physicist, had no use whatever. He wrote:

Physics must be sharply distinguished from mathematics. The former must stand in clear independence, penetrating into the sacred life of nature in common with all the forces of love, veneration and devotion. The latter, on the other hand, must declare its independence of all externality, go its own grand spiritual way, and develop itself more purely than is possible so long as it tries to deal with actuality and seeks to adapt itself to things as they really are.[136]

Goethe remained adamant against the advice of friends who wanted to see his theory of colors supported by mathematics. For him the great task was to secure the priority of perception over calculation, "to banish mathematical and philosophical theory from those areas of physics, where instead of promoting knowledge they merely inhibit it, and where mathematical treatment, because of the one-sided development of the newer scientific education, has found such ill-advised application."[137] No wonder that he compared mathematicians to Frenchmen: "Talk to them and they will translate it into their own tongue where it at once becomes something altogether different."[138]

Whatever blame can be laid on mathematical physics for making nature "different," nature became more unnatural (divested of itself) in the hands of those who claimed to save nature and its unity with man. A far more inadequate picture of nature was presented by those who argued for the organic identity of man and nature than by those who restricted science to the impersonal, quantitative study of phenomena. On the face of it Goethe's treatment of colors has an immediately human appeal. Still one cannot fail to sense some tragedy when reading Goethe's own evaluation of his *Farbenlehre*. A year before his death he was still speaking with scorn of "certain German professors" who kept warning their pupils against the grave errors of his theory of color. His *Farbenlehre*, he remarked to Eckermann, "is as old as the world and cannot always be repudiated and set aside."[139] With this contention one could hardly agree more, but only in the sense that blind alleys once found ought always to be kept in mind, to make sure that journeys leading nowhere are not undertaken again.

This is one of the reasons why the fantastic features of the organismic physics produced by Schelling and Hegel should also be recalled once in a while. Details of their "physics" are in a sense indispensable for a better understanding of what might seem a rigid reaction produced in the minds of nineteenth-century physicists, especially in Germany. For instead of "giving physics wings," as Schelling pretended, the organismic approach once more tried to put physics on the skids of fancy. Actually Schelling succeeded only in proving that the best anyone can do who holds with him that man is "the most perfect cube" is to avoid discoursing on physics. Otherwise not only will he draw strange parallels between the human body and the solar system, but he will also believe that every law of physics can be squeezed out of the human mind by an easy process of self-reflection.

This is not the place to tell even sketchily how in Schelling's conception the world develops through a threefold cycle: from homogeneous matter through qualitatively differentiated matter into organic processes. It should suffice here to recall some revealing points of Schelling's version of organismic physics. There gravitation reappeared in the male sex, whereas light continued to be acting in the female sex. Animal instinct, operating as it did on the "ground" of existence, was traced back to gravitation, while the life of plants constituted a replica of magnetism, with their lines of force oriented toward the sun. The structure of plants was also said to parallel the arrangement of planets around the sun, and the process of reproduction was a manifestation of magnetism, while human irritability was presented as an outburst of electricity. In the same vein Schelling defined each of the five senses as an embodiment or specification of a distinct physical force. The solution of difficult questions, such as the shape of the planetary orbits, was no problem in Schelling's physics. One need only be aware of the basic law of the polarity of forces to conclude with Schelling that the orbit ought to be elliptic.[140] For Schelling the proof simply rested on the alleged similarity between the two foci of an ellipse and the two poles of a force field, these two poles always being in his view the prerequisites for any motion not produced by impact. One can hardly help thinking, however, that in his unhesitating choice of an ellipse the long-established results of mathematical physics were of some help, to say the least. But cultivators of organismic physics as a rule have little urge to make such admissions. The explanation of the condensation and evaporation of water followed the same easy "natural" pattern.[141] The process was assigned to a tension anchored around two "poles," the earth and the sun. All along the Schellingian physics, if it can be called physics at all, nature yielded with alarming ease its secrets to man who was represented as the perfect synthesis of all forces acting in the universe. For it took only a short reflection to realize that nothing on earth and in heaven could remain hidden from man who, as Schelling wanted to have it, is the center of the universe and is "therefore in immediate inner communion and identity with all things that he is to know. All motions of the great or little nature are concentrated in him, all forms of actuality, all qualities of earth and heavens. He is in a word the system of the universe, the fullness of infinite substance on a small scale — that is the integrated being, man become God."[142] On such a ground, clearly, there is no further arguing.

Many of the details of Schelling's organismic physics had already been worked out when Hegel joined him in Jena in 1801. The meeting was that of truly kindred minds as regards at least their concepts of what physics ought to be. The rapidly produced Habilitationschrift of Hegel, *Dissertatio philosophica de orbitis planetarum*,[143] gave a fair sample of what was to follow. There Hegel defended the profound Kepler (the mystic of course) against the superficial Newton and inveighed against Bode's postulating a planet between Mars and Jupiter. Nature, as Hegel contended, could not follow an accidental sequence of numbers such as given by Bode's Law. The spacing of planets could mirror only the pure creation of human mind, that is, a rational sequence of numbers, an example of which was for Hegel the so-called Pythagorean Series: 1, 2, 3, 4 (2^2), 9 (3^2), 8 (2^3), 27 (3^3). This series not only limits the number of planets to seven, but also retains between the fourth and fifth planet the large distance which by its incongruity prompted astronomers all over Europe to look for a planet between Mars and Jupiter. Their search was ridiculed by Hegel, however, as something contrary to reason, although in a signal lack of rational consistency Hegel did not disdain to substitute 16 for 8 as the relative distance of the sixth planet, Saturn, from the Sun. More consistent with the facts was the acid remark of Duke Ernest of Gotha, who sent Hegel's work to the astronomer Zach with the superscription: *Monumentum insaniae saeculi decimi noni.*[144]

Neither the discovery of Ceres nor the sighting of Pallas, Juno, and Vesta, which came within seven years, could give Hegel second thoughts on the matter. While acknowledging the existence of all four of the then known asteroids, and even admitting some merit in Bode's Law, the author of the *Naturphilosophie* (1816) refused to capitulate. His stubbornness trapped him once more. This time the still unknown satellites of Mars made fun of him. Abandoning his Pythagorean Series, Hegel claimed that all that could be known about the laws underlying the actual outlay of planets was their falling into three groups. The four inner planets with no satellites or with only one, Earth, formed the first group, the asteroids the second, while the third group was formed by the three outer planets, all of which have many satellites or rings like Saturn. This was a specious classification that could hardly conceal its artificiality. Such were the systematizations Hegel wanted to dominate physics and astronomy, for his was a firm belief that "the time will come when these sciences will be governed by the constructs of mind."[145]

The future course of events could hardly be less propitious for Hegel. In 1877, only forty-six years after Hegel's death, Asaph Hall sighted two satellites of Mars. Spared this shock, Hegel could cling all his life to his tenet that astronomers should consult the philosophers if they wanted to know the law of planetary distances. This was indeed the only course for him to take if he were to remain faithful to his premises, for he viewed the wholeness of existence as the unfolding of the Conscious Spirit, and for details he consulted his introspective reflection. What he found there was the immediateness of a personality and the organic coordination of all mental processes in the center of self-awareness. Being a projection of thought, the world, its structure, and its main features had to manifest, in Hegel's system, the very same pattern. Thus the world had to have a universal center, with a main central body (sun) staying absolutely immobile. As to other details, Hegel stated that just as in a living organism the parts carry out their functions in a vital union with the principal organ or center, so the planets move not in a mechanical fashion, but rather go on their ways "in freedom like blessed gods."[146] And since Hegel contended that an organism was highly individual, the path of planets could not be a monotonous circle, the shape of which was always the same, but rather an ellipse "individualized" by its two major axes.

Even if Hegel did not refer to Aristotle time and again, one can easily see that his "physics" was essentially that of *On the Heavens* and of the *Meteorologica* in a different key. At times, however, even the key was the same, as, for instance, when Hegel argued for the elementary status of the four "canonical" elements, air, fire, earth, and water, and rejected chemical substances, such as carbon and oxygen, as elements. In the same vein, the main divisions of Hegel's physics carried a distinctly organismic connotation. The four elements were discussed under the heading of "Physics of Universal Individuality." The four properties, specific gravity, cohesion, sound, and warmth, were explained in the "Physics of Specific Individuality," and the "Physics of Total Individuality" contained topics such as magnetism, electricity, crystallization, and chemistry. The atomic explanation of chemical processes Hegel found, of course, utterly deficient. Atomism was for him mere externality. What made a compound a whole was a vital process, whereby a concrete unity could organize itself from surrounding material and differentiate its wholeness into distinct parts and functions.

Thus the bridge was ready to secure a smooth transition from physics to organics, or the study of living beings. Obviously Hegel made the step with unconcealed joy. Like Aristotle, he too felt more at home among plants and animals and likened the happy move to leaving the prose for the poetry of nature. Not that Hegel had not been poetic enough when discoursing on physics. To the explanation by centrifugal force of the retardation of a pendulum at the equator, he opposed the "true" reason: pieces of cold matter "feel," so he contended, that the direction toward repose is stronger there and this feeling makes them move slower.[147] A reasoning, Aristotle would no doubt have enjoyed immensely. With the same poetic resolution, Hegel called the distinctness of two magnetic poles "the naïveté of nature" and spoke of electricity as "the angry self of a body." Electricity "is a body's own anger, its proper ebullition. No one else is present beside itself, least of all a foreign material. Its youthful spirit strikes out, it raises itself on its hind legs: its physical nature gathers itself against the relationship to something else, and does so as the abstract ideality of Light." [148]

To compare electrical tension to the snarling of hostile dogs was for Hegel's admirers, like J. N. Findlay, merely a "vivid piece of picture-thinking." [149] Again, while acknowledging that even his prosaic summary of what Hegel said on the properties of matter might sound like a fairytale,[150] he continued to stick by his hero. So did the promoters of sundry ideologies and intellectual fancies. Among them was Engels, but his dilettantish eagerness to pontificate on matters of physics earned him no respectable place in scientific history. Nor did organismic physics treat any better those few men of science who, like G. T. Fechner, were lured by the prospects of a new physics based on organismic concepts. A well-deserving pioneer in the use of physical measurement in psychology, Fechner became the hapless victim of a long series of sheer fantasies, of which his fifty-odd analogies between the earth and the human body are perhaps the most revealing.[151]

Yet, for all these blunders, the courting of organismic ideas in the periphery of physics stubbornly persists even to our day. It is highly characteristic, however, that such an attitude is still based on factors like emotional prejudice, philosophical preference, and not least, a lack of proper information in physics. The first of these factors was clearly evident in those sad aberrations of mind that under the instigation of Nazi ideology proposed an Aryan or German physics based on organismic concepts as opposed to "Semitic relativism," "French rationalism," and "British empiricism." The role of the second factor is most obvious in those interpretations of modern physics that take some of its basic notions as a covert reassertion of the organismic viewpoint. It is, however, questionable whether one should speak of an organismic approach when, as is the case with such interpretations, the concept of organism is stretched to the extent as to lose its basic biological meaning. The third factor is at work in those amateurish accounts of modern cosmology in which undigested data of physics are mixed with mythology, astrology, and with copious references to such organismic fantasies as the role of male and female principles in the cosmos. Characteristically enough, the authors of such works like to claim credit themselves for such feats as for instance the alleged prediction of the Van Allen radiation belts several years prior to their discovery. Yet, those obscure passages are as worthless for the physicist of our day, as was in its day the famous page in *Gulliver's Travels* (1726). There Jonathan Swift stated with startling accuracy the existence of two satellites of Mars, including their sizes and the radii of their orbits, a century and a half before their actual sighting. It is more instructive, however, to remember that these present-day fanciers of science do exactly as Swift did when faced with the details of precise explanation. For Swift remained silent about the particulars of that "superior theory" of the astronomers of Laputa, which he contrasted to the "very defective and lame" views of the astronomers of his day whose successes he apparently resented.

Recalling such vagaries of the human mind is in no way to suggest that the concept of organism is an unqualified error. Its indispensability in biology and psychology can be ignored only at great cost. It may even serve philosophy, and its place is no doubt forever assured in poetry and the arts. In physics, however, the organismic approach can safely be said to have no place. True, it helped, especially in the earlier stages of science to strengthen man's belief in the unity of the world, and in the unity and causal correlation of apparently distinct phenomena and processes. Indeed without such belief science can hardly exist and much less could it have come into being. But organismic views were not the only or main source of such conviction. This is true even in the case of Oersted whose discovery is often credited to his leanings toward *Naturphilosophie*. His search for a connection between magnetism and electricity, however, had also motivations distinct from the prompting of organismic speculations.[152] Long before him and *Naturphilosophie*, physicists had clearly recognized that in order to make any meaningful advance in physics, they had to part with the organismic approach. It had not only failed physics in questions of details, but it had fallen conspicuously short of its paramount claim to provide the definitive, exhaustive, and only intelligible understanding of the physical world.

Interestingly enough such was not only the case in the Western intellectual tradition. The history of scientific thought in China shows just as clearly that an intellectual milieu steeped in organismic considerations will hardly give rise to a scientific understanding of inanimate nature.[153] Therefore only part of the story is told in Einstein's remark quoted in the opening pages of this chapter. The inability of the Chinese sages to formulate the modern concept of science is not wholly explained by referring to what is undoubtedly akin to miraculous in the birth of science. Nor is it without reasons of its own that the most enduring achievements of the Greeks in investigating inanimate nature were restricted to fields where organismic concepts could hardly dominate, such as geometrical theory and measurements, and the discussion of the motion of heavenly bodies on the basis of geometry.

The inability of organismic physics to provide satisfactory understanding of inanimate nature has often been noted, and at times with a note of disappointment. Yet, painful as such disappointment might have been to some, even more frustrating had to be the one caused by the consistent failure of organismic physics to secure domination for man over nature. Inborn as man's urge is to "feel" nature and to become enveloped in it, just as inextricable is his striving to control, dominate, and use nature in an effective and systematic manner. Control implies foresight, however, and to secure this about the processes of nature, man had to give up his attempts to find out about the sundry volitions that supposedly provided the dynamics of nature. To foresee the behavior of things, man had to depersonalize his study of the physical universe. It was as if one were to consider the beautiful display on the stage of nature a poetic disguise and look for the ultimate reality in the ugly, soulless mesh of ropes, pulleys, and levers found backstage. However distasteful to humanists of all sorts, this step toward the backstage was taken with the utmost enthusiasm by the men of science. Their enthusiasm pointed to tangible successes, the like of which were simply unattainable by organismic physics. Yet, in ultimate analysis, what changed was only the object of scientific faith, not its degree. For three centuries machines were to be idolized with as little second thought as had been the concept of organism for over two thousand years.

CHAPTER TWO

The World as a Mechanism

THE REDISCOVERY of Greek science by the young nations of the Middle Ages has many unique aspects and unexpected turns. A concourse of two cultures and mentalities so widely separate in time and different in backgrounds was hardly a routine process of cultural assimilation. Of the most arresting cultural traits that can be singled out, perhaps the most surprising is the eagerness with which the then solidifying society of Western Christendom seized upon the information of technical nature available in Greek scientific writings. Though primarily speculative in trend, Greek science could claim to its credit important contributions to technology. Such was the transformation by the Greeks of the practical methods of measuring and calculating into generalized arithmetical and geometrical systems. Another major technological achievement of the Greeks was their explanation of the working of a basic mechanism, the lever. Hero's account of the properties of the five simple machines, the lever, wheel, pulley, wedge, and the screw, remained the basis of all explanations of machinery until the eighteenth century. Nor was there a lack in the Hellenistic culture of fairly extensive use of mechanical contrivances. Nevertheless the machine had never become an integral part of the classical culture. To explain things by analogy to a mechanical device was an exception, rather than a rule, with the classical bent of mind. Although the Greeks had contrivances working at a steady pace, such as water mills equipped with geared wheels to transmit power, they preferred to see behind the regularly recurring phenomena of nature not a machine but an organism.

The appreciation of mechanical contrivances was markedly different in the Middle Ages. Nothing illustrates this so well as the rapid development of mechanical clocks. From the earliest drawing of the escapement mechanism attributed to Villard de Honecourt it took little more than a hundred years for the construction of clock watches driven by springs to be developed. Not many more years separated Roger Bacon's speculations on mechanically propelled carriage and flying machines, burning glasses, gunpowder, and other sundry items from Leonardo's exquisite mechanical drawings. The fascination with machines went so deep that a hundred years before Leonardo, Oresme found it fitting to compare the universe to a clock. And this he did, of all places, in a commentary on Aristotle's *On the Heavens*. For Oresme the skies, the stars, and the planets were no longer moved by desire or by intelligences in charge of each celestial body. The whole configuration and motion of the heavens reminded him rather of "a man making a clock and letting it go and be moved by itself." [1] This was, of course, hardly a commentary in the usual sense of the word, but more a resolute shift from the organismic to the mechanical world view in physics. It signaled the beginning of a gradual abandonment in physics of the "holy and living principles" of the Peripatetics. The world was being spoken of less and less as a huge animal and more and more as an immense machine.

The transition from organismic to mechanistic thinking did not leave the semantics unaffected. Expressions referring to machinery were used more frequently and with greater emphasis. Especially worth noting in this regard is the numerous occurrence of mechanistic concepts in the writings of late medieval authors still committed to the organismic view of nature. Thus expressions like *machina mundi* — a favorite with the Cardinal of Cusa — were becoming commonplace in scientific writings. Instead of a hierarchical structure with the more or less noble parts of an organism, the world was pictured ever more often as a mechanical contrivance with all its parts being of equal rank. Later, during the emergence of the new physics, the term clockwork became a sort of a motto for the new scientific outlook on the world, conveying forcefully as it did the difference between the old and new ways of defining the nature of the material universe.

Arguing in favor of the Copernican system, Rheticus remarked that God's skill certainly could not be second to that exhibited by the clockmakers whose products contained no superfluous wheels. What else is Copernicus' system, he asked, if not the most economical account of the heavenly clockwork? [2] Minds like Kepler's still imbued with organismic ideas, also found in a clockwork the expression through which they achieved their clearest formulation of the new conceptual framework of science. As Kepler wrote in 1605: "I am much occupied with the investigation of the physical causes. My aim in this is to show that the celestial machine is to be likened not to a divine organism but rather to clockwork . . . insofar as nearly all the manifold movements are carried out by means of single, quite simple magnetic force, as in the case of a clockwork all motions [are caused] by a simple weight."[3]

It did not matter that the world picture of classical physics was to resemble ultimately more a billiard ball universe than a clockwork on a cosmic scale; both conformed perfectly to the emerging scientific creed that accepted only mechanical laws valid in every part of the cosmos. It was to this all-inclusive mechanics of the universe that Kepler referred in a letter of 1605, in which he admitted that for the past five years his spare time had been spent on pondering questions of astronomy along the lines of physics. Strongly attracted as he might have been to organismic analogies of the universe, he felt no less keenly that the future of astronomy lay with mechanics. "I believe," he wrote, "that both sciences are so closely bound with one another that neither can achieve perfection without the other."[4] Kepler spoke in fact of nothing less startling than a "physics of the skies," and the reaction of one of his contemporaries clearly indicated that in so doing Kepler had gone far beyond the Copernican interpretation of the heavens. At least it seemed so to the astronomer of Danzig, P. Crueger. According to him Kepler, in trying to base the Copernican hypothesis on physical causes, put forward strange speculations that belonged not in the domain of astronomy but of physics.[5]

The criticism was to the point in that it diagnosed correctly Kepler's intentions. It should not, however, convey the impression that Kepler had progressed too far in formulating the laws of kinematics, or dynamics. Of Kepler's speculations about mechanical laws, only one — the enunciation of the gravitational attraction among all bodies at all times and places — was on the right track. For the most part his mechanical propositions and explanations were hopelessly intertwined with mystical concepts. His famous laws, so indispensable for later progress, were then only pure numerical relations without any hint to their applicability to terrestrial phenomena. As for the physical mechanism underlying his laws Kepler fell hopelessly back on clearly organismic concepts, such as the "magnetic arm" of the sun sweeping the planets around in their orbits.

Lacking sufficient proof, the mechanical concept of the universe was at that time mostly a creed that echoed the Archimedean motto in a mechanistic phrasing, "give me matter and motion and I will construct the universe." Such a program undoubtedly rested on tremendous convictions if not extraordinary ambitions. The clinching proofs of the mechanical philosophy were still far away. Little if any had yet been realized of the great promise of the mechanical creed, namely, the prediction by mathematics of all future events that were to take place in the physical universe. Ambitious programs had, of course, been given publicity by their authors. Galileo, making the most of the extraordinary impact of his *Starry Messenger,* wrote at length to the secretary of state of the Tuscan court, Belisario Vinta, about his scientific program that was centered on mechanics. As the letter stated, Galileo planned "three books on local motion . . . three books on mechanics, two relating to demonstrations of its principles, and one concerning its problems." Actually, of all his contemporaries, he alone lived up to the boastful self-appraisal so fashionable in his day: "Though other men have written on this subject what has been is not one-quarter of what I write, either in quantity or otherwise."[6] The *Mechanics of Galileo,* printed in 1634 on Mersenne's order, fell far short of this goal. What was to become Galileo's decisive contribution to mechanics had emerged from his studies on local motion. The *Discourses on Two New Sciences* intimated for the first time what a book on physics ought to be: a sober, mathematical analysis of experiments followed by deductions. Partial as was Galileo's achievement in mechanics, what he produced was a foundation on which following generations securely built.

This was not the case with most of Descartes' dicta in physics. In fact, as W. Whewell put it long ago: "Of the mechanical truths which were easily attainable in the beginning of the seventeenth century, Galileo took hold of as many, and Descartes of as few, as was well possible for a man of genius."[7] The reason for this rested no doubt on Descartes' unbounded faith in his own version of mechanics that, while expressing well some basic points, glossed readily over details. In his physical universe there were only bodies and motion, and he restricted, at least in principle, his ambitious program of physics to the consideration of an "infinity of motions that last forever in the world."[8] Of the quantity of the total sum of these motions he clearly stated that it could not be diminished or lost.[9] Furthermore, in a passage that in spite of its brevity strikingly anticipates the image of the all-knowing spirit described by Laplace, he argued that "if somebody were to know perfectly what are the small particles of all bodies and what are their movements and their relative positions, he would perfectly know the whole nature."[10] Also, it must be admitted that his writings on physics were in full conformity with his basic tenet rejecting firmly the major preoccupation of organismic physics — the interpretation of physical processes and laws in terms of teleology. "We shall adopt," he wrote, "no opinions whatever about the goals that either God or nature might have set for themselves in producing the things of nature, because we should not arrogantly claim to be privy to their counsels." [11] Although Descartes did not claim to know the intentions of God, he thought that he had succeeded in fathoming the mind of God on at least a few points. Among these points, or absolute truths, he enumerated his rules governing the method of the mechanical philosophy and was convinced that there was no road by which the human intellect could ever discover better ones. About the finality of the truth of the mechanical world view Descartes condoned no doubts whatsoever.

The mechanistic model of the universe that Descartes submitted as absolutely definitive had nowhere gone beyond a qualitative, non-mathematical description of the processes of nature. True, the seventeenth-century discoveries concerning the pressure of air and liquids had made highly plausible the idea that instead of attractive forces, pressure and impulse produced all phenomena, gravitation included. Yet, almost all the details of the Cartesian physics had to be taken on faith. Such were the three kinds of matter, the subtle, the fine, and the coarse; such were the vortices that functioned as all-purpose agents, keeping all bodies in their proper places, the sun and the planets as well; such were the particles of luminous bodies that alone could move the subtle matter in the pores of transparent objects, producing thereby the optical phenomena. Faith was all the more needed, for whenever Descartes or the Cartesian advocates of the mechanistic creed made statements that were easily verifiable by measurements and observations, things usually went sour. Almost none of Descartes' seven rules on impact was experimentally correct. And although he saw this for himself, he remained haughtily unruffled. His laws, he claimed, were valid for perfectly hard bodies, adding that "we see nothing of this kind in the world." [12] This would have been the correct explanation of the differences between rules and observed facts had the experiments been carried out in such a way as to approximate gradually the ideal case with the elimination of disturbing factors, such as resistance. Descartes' faith in mechanism, however, had too many a priori features to permit him to see the crucial need in physics of approximating on the experimental level the ideal conditions in which certain physical phenomena might take place. Consequently, Descartes' mechanistic explanations did not always have enough respect for the actual way things happened. Explaining the refraction of light by analogy with bouncing balls, Descartes satisfied only the shibboleths of mechanism by using the figure of a hefty tennis player in the illustration of the problem in his *La Dioptrique*;[13] the ball, supposedly having a lesser velocity in the denser medium, was shown as following a path bending away from the normal instead of getting closer to it. Had Descartes experimented with tennis balls he would probably have realized that he should look somewhere else for an illustration of his law. Not being an experimenter, he failed to perceive the irony of the situation into which his "absolute truths" had led him. But he had less trust in the facts than in the tenets of his mechanical philosophy, or rather creed, which prescribed that little if any reliance "should be placed upon observations that are not supported by true reasons."[14] The result was a strange imbalance between evidences and "principles" which comes to light nowhere better than in the manner in which Descartes criticized the experiments of others, and those of Galileo in particular. While praising him for his emphasis on mathematics, Descartes took Galileo to task for his attention to particular cases "without having considered the first causes of nature." According to Descartes' sweeping judgment, Galileo "has built without a foundation. Indeed because his fashion of philosophizing is so near the truth one can the more readily recognize his faults."[15] Needless to say the shoe was on the wrong foot.

The Cartesians were no more fortunate when it came to the mathematical results of non-Cartesian physicists. The famous case is Galileo's law of the free fall, which neither Descartes nor Mersenne would accept. Mersenne's remark on this point is particularly characteristic, showing as it does the highly arbitrary character of some formulations of the mechanical creed in the first part of the seventeenth century. "I doubt that Signor Galileo performed the experiments of the inclined planes, because he makes no mention of them and because the proportion he gives often contradicts experience."[16] True, Mersenne wished that as many as possible would perform the experiment and with all the necessary precautions to see "if enough light could be drawn from them for constructing a theory."[17] This was a skeptical proviso, however, for Mersenne held that experiments were not able to give rise to a science. On the contrary, scientific knowledge was to be derived according to the Cartesians from postulates and "necessary conclusions" that were to be considered, to hear J. Rohault — the most widely read expositor of Cartesian physics — not as arbitrary hypotheses but rather as truths that "follow necessarily from the motion and division of the parts of the matter." This, he added, "experience teaches us to recognize in the universe."[18] Such a halfhearted reference to the role of experience was genuinely Cartesian and Rohault probably did not even care to soften much the glaring arbitrariness of the dicta of the Cartesian version of mechanical philosophy.

To claim too much, however, ultimately creates suspicion even within the fold. A Cartesian himself, Huygens could not help seeing some non-scientific motivation behind Descartes' contentions. That Descartes, as Huygens put it, was jealous of the fame of Galileo, and wanted to be revered in the schools as Aristotle had been was only too human and certainly forgivable. Descartes' absolute statements were, however, a different matter, for, as Huygens observed, science had to suffer the consequences. Descartes, he wrote, "put forward his conjectures as verities, almost as if they could be proved by his affirming them on oath . . . and claimed to have revealed the precise truth, thereby greatly impeding the discovery of genuine knowledge." [19]

But whatever cracks and faults there might have been in Descartes' physics, of its basic assumptions, the mechanical explanation of the world, Huygens entertained no doubts. When speaking of the nature of light did Huygens not voice wholeheartedly the mechanical creed? For him the focusing of light by concave mirrors was "assuredly the mark of motion, at least in the true philosophy, in which one conceives the causes of all natural effects in terms of mechanical motions." There were no alternatives, for, as he exclaimed, "this we must necessarily do, or else renounce all hope of ever comprehending anything in physics." [20] Huygens had no particular inclination to search for the basis of the certainty of the mechanical approach to nature. His was a skeptical mind both uneasy and disdainful when facing questions that were more philosophical than physical. Geometrical propositions apart, he admitted only degrees of probability, but never certainty. "We know nothing," he wrote to Pierre Perrault in 1673, "very certainly, but everything only probably, and the probability has degrees that are widely different." [21]

Far less skeptical than Huygens regarding things invisible, Boyle kept trying to give a metaphysical foundation to the mechanical creed all his life. He viewed the shape of the universe as a consequence of the motion and mechanical properties with which God had endowed the particles of matter in the beginning. On this fundamental tenet he based the mechanical philosophy that in his words, "teaches that the phenomena of the world are physically produced by the mechanical properties of the parts of matter, and that they operate upon one another according to mechanical laws." [22] In Boyle's eyes the validity of these laws knew no exception. Even an angel, he said, could not produce any real change in the world without doing it through mechanical means. Not that this meant to impose a sort of constraint on celestial beings. As pure intelligences, the angels had to operate on extended matter along the only way that befitted correct reasoning: the mechanical way. For, as Boyle stressed, the mechanical philosopher can accept only such physical agents as intelligible which can be reduced "to matter and some or other of those only catholick affections of matter." [23] Such emphasis on the *exclusive intelligibility* of the mechanical principles cannot be stressed enough if one is to have a genuine insight into the classical physicist's state of mind. This holds of Kelvin's age as well as of Boyle's. Not that Boyle had been the initiator of the conviction that equated intelligibility with mechanism: his excellence lay rather in his ability to serve as the literary spokesman of the scientific aspirations of his age. For Boyle, prolific as his writings were, did not say much that was new. Yet, in all that he said his contemporaries instinctively recognized their own minds and tastes. As Leibniz commented on Boyle's works in 1691: "In his books, and for all the consequences that he draws from his observations, he concludes only what we all know, namely, that everything happens mechanically." [24]

Leibniz was not alone in such an appraisal of Boyle's efforts. He also recalled with unreserved concurrence the puzzlement of Spinoza, who, as Leibniz tells us, was unable to comprehend why Boyle had not derived "from an infinity of beautiful experiments" conclusions other than the one "which could have been taken as a principle, namely, that everything in nature is effected mechanically; a principle which can be made certain by reason alone and never by experiments however numerous they may be." [25] Leibniz was, of course, the prince of those who during this early phase of classical physics gave precedence to the a priori approach as opposed to a rigorous inductive, experimental procedure. This choice of theirs, however, did not set them apart from people like Hooke and Boyle, as regards that basic tenet which equated intelligibility with mechanism.

The mechanistic physics had to be content, for a while at least, to emphasize the principal point: all genuine explanations in physics had to be molded on some mechanical pattern. The details of the pattern were of secondary importance provided they were strictly mechanical. Thus, in explaining the behavior of air in a pump, Boyle spoke of the spring of air without committing himself to either of the two mechanical models applicable to the problem.[26] The air, he said, might resemble a heap of little bodies lying upon one another like a fleece of wool, or it might just as well be a mass of flexible particles agitated by heat. If the latter is the case, the particles of air would have no "structure requisite to springs," still the principle of mechanical explanation is safeguarded, for what is a chain of impacts if not a form of mechanism?

A rapid glance at the titles of Boyle's works shows that in most of the topics he investigated the exact mechanism of the processes responsible for the phenomena could not be known in Boyle's time. All the same, Boyle spoke undauntedly of the mechanical causation of

all the phenomena perceived by the senses.[27] In the world of phenomena hardly anything missed his attention, and he recounted with
untiring patience an immense list of observations and experiments
relating to the mechanical production of heat, cold, tastes, odors,
volatility, corrosiveness, chemical precipitation, magnetism, and electricity, to name his main topics alone.[28] True, Boyle was aware of
the fact that the gross mechanical processes he spoke of were only
in distant relation to the basic mechanism of nature operating through
the most minute parts of bodies. What this mechanism was like one
could only infer by imagination, helped in no small measure by those
recently devised "outward instruments" as Hooke called the telescope
and the microscope. Machinery, or miniature clockwork, was seen
neither by Hooke nor by others peering through the microscope, but
they saw enough small geometrical patterns in a wide range of materials investigated. They hardly asked for more on behalf of their
faith in mechanism. As Hooke, the most celebrated early reporter on
the world of the microscope, put it: "Those effects of Bodies, which
have been commonly attributed to Qualities, and those confess'd to
be occult, are perform'd by the small Machines of Nature . . . [and]
are . . . the mere products of Motion, Figure and Magnitude."[29] In
all this, however, the rhetoric of conviction was far ahead of the sober
appraisal of the evidence at hand. Little did this disparity between
claims and proofs worry his contemporaries. Instead, they praised
such efforts unreservedly, as Locke did in speaking of Boyle: "He
was thought to go farthest in an *intelligible* explanation of the qualities of bodies"[30] (italics mine).

The situation with regard to the physics of the macroscopic world
was much better. There the mechanistic philosophy claimed important achievements that still left much to be desired, however, as an
overall account of the mechanism of nature. Such an achievement
was Galileo's discovery that the distance covered by freely falling
bodies was a function of the square of the time and that the path of
projectiles was a parabola. To realize how much was contained in
these laws, one had to wait nevertheless for Newton who, with an
intellectual generosity almost unique in those days, acknowledged
that Galileo's laws implied his first two laws.[31] Similar was the case
with Huygens' principal contribution to the edifice of mechanism,
the *Horologium oscillatorium* (1673), a copy of which was sent to
Newton by Huygens himself. The work ended with thirteen theorems
on the magnitude of force resulting from circular motion. They were
first formulated and explained in detail in an earlier work of Huygens,
the *De vi centrifuga*, written in 1659 but still in manuscript in 1673.
The first two theorems contained a gem of physical thought, the

definition of centrifugal force as proportional either to the square of tangential velocity or to the radius. Newton, who had already obtained the same results without publishing them, saw their value better than anyone else, as shown in his letter of July 3, 1673, to Oldenburg: "I am glad we are to expect another discours of the *Vis centrifuga*, which speculation may prove of good use in natural Philosophy and Astronomy as well as Mechanicks." [32] But by the time Huygens' manuscript was published in 1703, eight years after his death, the mechanical philosophy had changed from a fashionable scientific program buttressed by rather disconnected insights into a system of laws valid both for celestial motion and local phenomena.

The feat was achieved single-handedly in a work called modestly by its author a "hard book"; probably no other book in science has ever been more so, that is, if hardness is a measure of unfading achievement. In the *Principia* the mechanical creed received verification beyond the most ardent hopes. Regardless of the role a falling apple might have played in Newton's discovery, in the *Principia* the same law provided explanation for as widely distant phenomena as the weight of an apple and the path of a comet. In all this Newton did full justice to a basic aspiration of the mechanical creed: the propositions — theorems of the *Principia* — were expressed not in vague qualitative statements but in a precise, quantitative fashion along the lines of classical geometry.

On the relation of geometry (which stood for mathematics) to mechanics Newton could not have been more explicit. In the preface to the first edition of the *Principia*, he traced the origin of geometry to the art of mechanics, in the firm conviction that behind the mathematical description of any motion there must be a mechanical pattern. Far from being a mathematical formalist, Newton saw in mathematics the only sensible way to express mechanical relationships. For in Newton's mind mathematical functions and mechanical principles were two sides of the same coin. He offered his work as the mathematical principles of natural philosophy and put the "whole burden of philosophy" to the task of expressing the mathematical relations involved in motions, such method being for him the only means for prying into the properties of forces running the mechanism of nature. These forces, he believed, were ultimately mechanical, producing their effects, attractions, and repulsions "by some causes hitherto unknown." [33] And if all the phenomena of nature were ever to be derived or explained, this was to be accomplished, he stated, through a reasoning from mechanical principles.

Contrary to charges made by Leibniz and Huygens that the *Principia* "deserts Mechanical causes," [34] Newton never wavered in his adherence to the mechanical creed. Although he claimed to know nothing about the causes of gravity, for him it was "inconceivable that inanimate, brute matter should, without the mediation of something else, which is not material, operate upon and affect other matter without mutual contact."[35] In fact, Newton's flights of fancy about the way gravitation might act never parted with the idea of operation by contact. This was made clear in his letter to Boyle of February 28, 1679, and in the final paragraph of the General Scholium, where he speaks of a "most subtle spirit," of "an electric and elastic spirit."[36] To this spirit, which stood for subtle fluid or matter, Newton assigned a universal mechanical role: it was in his eyes the vehicle of gravitational effects, of the propagation of light, of electrical phenomena, and even of physiological perception. Still this subtle matter, the ultimate key in his view to the mechanism of the world, was not to form a part of natural philosophy as long as it escaped accurate determinations and mathematical formulations. Thus Newton never established a dichotomy between mathematical law and mechanical reality, even though the mathematical expression could say nothing about the nature of the latter. On the contrary, he emphasized that his explanation of the planetary motion by "mathematical demonstration grounded upon experiments without knowing the cause of gravity" represented progress toward the ultimate mechanical explanation. At least as good, he claimed, as was one's understanding of the mutual dependence of the wheels of a clock "without knowing the cause of gravity which moves the clockwork."[37] Besides who in 1712 could challenge the intelligibility of a clock's mechanism?

Newton believed so much in the strict one-to-one correspondence of mathematical formalism and mechanical reality that he refused, at least in the *Principia*, to discuss "hypotheses," by which he meant speculations not formulated and verified mathematically or incapable of being subjected to experimental verification. This was why he had no use for Descartes' vortices, however mechanical in appearance, for mathematical analysis did not bear out the mechanical properties Descartes had ascribed to them. The vortex as a mechanical model was useless, declared the General Scholium, which was added to the *Principia* in 1713, because the freedom of motion shown by the widely differing paths of comets contrasted strongly to the very restricted path of a piece of matter caught in a whirl of air or water. It is in this sense that Query 28 rejected the method of "feigning Hypotheses for explaining all things mechanically," since the hypotheses in question had no connection with either mathematics or observations.

Being firmly convinced of the mechanical roots of all the laws of physics when expressed mathematically, Newton gave only secondary attention to the search for the ultimate mechanical causes. Having formulated physical laws mathematically, he felt closer to the mechanical reality. This, however, did not mean that the mathematical formalism could ignore anything that had been directly evidenced by the phenomena of nature. Accordingly, Newton had rejected Huygens' ingenious geometrical construction of the propagation of light as being based on a concept of a wave that lacked the primary characteristic of physical wave motion, periodicity. In this respect Newton was just as exacting when his own laws were brought face-to-face with new observations, or measurements. For Newton, mathematical physics was always to be remodeled after the data of observation. Mathematics, to be sure, was the only idiom through which mechanics expressed itself, but the inner consistency of a mathematical function was not supposed to prejudge the facts. When it failed to accommodate certain facts it was to be revised or to be simply discarded.

Whatever the ultimate fate of Newton's system, there were no doubts in the minds of either his supporters or critics that if it ever was to be replaced by a better one, that new system had to be even more mechanical in outlook. Nowhere was that point expressed with more force than in Leibniz's attacks on Newton, who refused to make more than vague conjectures about the ultimate mechanical causes of gravity or of the absolute hardness of atoms. Newton's reticence — an obvious caution on his part — however, was viewed by others, among them Leibniz himself, as a most illogical, or simply unintelligible, attitude hardly permissible in a "mechanical philosopher." Distorted as Leibniz' views were on Newton's position, the charges he leveled at Newton bring into sharp relief the strict identity between the mechanical and the intelligible professed by classical physics. In his letter to Hartsoeker of February 10, 1711, Leibniz stated that to assume that God produced gravitational effects without using some mechanism was to say that He "produces that Effect without using any intelligible Means."[38] Gravity without mechanism was, as Leibniz put it, "an unreasonable occult Quality, and so very occult, that 'tis impossible it should ever be clear, tho' an Angel, or God himself, should undertake to explain it." To abandon the consistent application of the mechanistic view in any part of physics, Leibniz argued, would leave open only two courses of explanation: one would either have to assume that God acts through perpetual miracles or have to impute to the creator himself a gross logical contradiction. For as Leibniz stated, God could not create a planet that would move of itself any more than he could prevent, short of a miracle, "the Separation of the Parts of the hard Body, if a *Mechanical or Intelligible* Cause does not do it" (italics mine).

Such an insistence on "intelligibility" might, of course, be taken as indicative of a mind highly cautious about its own conclusions. Whatever the degree of Leibniz' caution, when compared with that of Newton, the physicist, Leibniz comes out rather a poor second. For whereas Leibniz claimed to have found the definitive form of the mechanism of nature, Newton, possessed with a deep reverence for facts, was able to write about his own propositions, however successful they might have been: "In experimental philosophy we are to look upon propositions inferred by general induction from phenomena as accurately or very nearly true, notwithstanding any contrary hypotheses that may be imagined, till such time as other phenomena occur, by which they may either be made more accurate or liable to exceptions." [39]

Such exceptional willingness on Newton's part to recognize the basically revisable character of his system could have easily appeared exaggerated for the next two centuries, which saw nothing but triumphs for Newton's system both in the field of observations and in theoretical analysis. No less an authority and resolute critic of classical mechanics than E. Mach wrote of Newton's achievements: "All that has been accomplished in mechanics since his day has been a deductive, formal and mathematical development of mechanics on the basis of Newton's laws." [40] The first major step in this process was the extension of Newtonian mechanics, which originally treated only particles, to massive and rigid bodies. The transition from one to the other was often a very difficult problem, which consisted in the summation of the motions of all particles of a body, and was formulated only generally in 1760 by Euler in his *Theoria motus corporum solidorum seu rigidorum*. Euler's laws gave the key to such problems as the movement of gyroscopes, spinning tops, the nutation of the earth, and a host of related cases. A central step toward the solution of such problems was the definition of concepts like the center of inertia and the moment of inertia, which Euler first calculated for various homogenous bodies.

To extend the Newtonian laws to any system of rigid bodies fell to Lagrange, who in his *Mécanique analytique*, published almost a hundred years after the *Principia*, brought to completion all that was implicitly contained in the Newtonian mechanics. In Lagrange's work there were no geometrical constructions or mechanical models, but, as he put it in the *Avertissement*, "only algebraic [analytical] operations subjected to a uniform and regular procedure." His aim was to provide "general formulas whose simple development gives all the equations necessary for the solution of each problem" in mechanics. This task he achieved with almost perfect success. In all this, Lagrange, an earnest student of the history of mechanics, remained fully aware of being only a beneficiary of Newton's genius. For him Newton was not only the greatest but also the most fortunate genius, "for we cannot find more than once a system of the world to establish."[41]

The conviction that Newton's laws extended as far as there was lawfulness in the physical universe gained even more assurance when it became clear that the powerful principles of mechanics proposed after Newton, such as the principle of virtual displacement and virtual velocities, were but hidden aspects of the Newtonian axioms. What remained for the successors of Newton to achieve in mechanics, was, as Gauss remarked, "to offer only new viewpoints but no new principles." Gauss' principle of least constraint, proposed in 1829, was highly esthetic, uniting as it did both the dynamical and statical aspects of mechanics. About its extent of originality, however, Gauss entertained no illusions. "In the nature of things," he wrote, "there can exist no new principle in the science of equilibrium and motion."[42]

Thus step by step all the particular insights into the laws of mechanics achieved after Newton were eventually shown to have had their origin in Newtonian concepts. This was the case even with the principle of least action that seemed to its author, Maupertuis, to be perhaps the only law "that the Creator and Director of things has established in matter in order to accomplish all the phenomena of the visible world."[43] The law of least action as proposed by Maupertuis was a bold and almost gratuitous extension of Fermat's minimum principle to the field of dynamics. Such a move required a systematic mind with a flair for fantasy. Fantasy, of course, abounded in Maupertuis, who had been musketeer, surveyor, biologist, moralist, linguist, and metaphysician all in one and was a physicist to boot. True, Maupertuis sought a minimum principle not incompatible with Newton's laws of motion, but he believed that he had reached deeper than Newton had. Fortunately for him, he did not live long enough to see Lagrange prove that the quantity of action would be a minimum only if bodies moved according to Newtonian mechanics. In other words, the principle of least action was but a transformation of Newton's laws. Furthermore, the work of Hamilton, by establishing a close analogy between mechanical and optical phenomena on the basis of the law of least action, gave added weight to the general conviction that even the phenomena of light were essentially mechanical in character.

With the full development of the mathematical techniques of the mechanical theory, outstanding problems of great complexity could be solved. The successes were most impressive in the field of celestial mechanics. Thus from Clairaut to the present the study of lunar inequalities have provided only additional evidence in support of the unconditional validity of the inverse square law. Laplace's analysis of the mutual perturbation of planets led to the conclusion that neither the inclinations of the planetary orbits nor their eccentricities could exceed small, finite values. Since in his calculations Laplace neglected all but the principal terms of the infinite series involved, the stability so derived was not absolute, but valid only for an immense period of time. Yet, such a result could not fail to give a powerful boost to the belief that the perfection of Newton's Law and the world-machine were but two aspects of the same reality.

This belief seemed to achieve its final confirmation when shortly later, in 1803, Herschel provided the first evidence that the Newtonian law of gravitation governed the motion of bodies outside the solar system as well. His twenty-five years of observation of double stars, Herschel proudly noted, "will go to prove that many of them are not merely double in appearance, but must be allowed to be real binary combinations of two stars, intimately held together by the bond of mutual attraction."[44] In achievements of this type the name of Newton was continually referred to as synonymous with the mechanical concept of the universe. To pay homage to Newton meant to demonstrate one's allegiance to the mechanical creed on which so many in and outside science based in those days their ultimate certainty. Well-established as such certainty might have appeared in the first part of the nineteenth century, both scientists and non-scientists were overawed by some of its new evidences. Such was the case when Galle directed his telescope to the point of the sky determined by Leverrier and notified him the following day about the sighting of the new planet. If Leverrier's share of glory was enormous, so was that of Newton and of mechanism. Referring to the discovery of Neptune, the Rev. T. R. Robinson, in his presidential address to the British Association in 1849, emphasized the role played by the cooperation among scientific organizations which, as he said, "began with Newton, and he stands like the sun in Heaven; all is luminous after he has risen, all before darkness or twilight."[45]

Whatever the merits of such rhetoric, the clarity that was to make itself felt in man's interpretation of the universe was indeed impressive. It brought into focus that specific feature of any mechanical system on which the possibility of such an enlightenment ultimately rested: any past or future configuration of a mechanical system can in principle always be exactly calculated. In this sense there could be no dark corners in a mechanical system: it was by definition an open book, theoretically at least, with no mysteries, paradoxes, or uncertainties. Though some problems, particularly many-body problems, presented insurmountable difficulties in precise calculation, for a superior intellect, reasoned Laplace, no problem was insoluble. The past and the future alike would be present to the eyes of such an intellect who, Laplace assured his readers, could "embrace in the same formula the movements of the greatest bodies of the universe and those of the lightest atom." [46] In the remaining part of the passage Laplace acknowledged that the human mind offers even at its best only a feeble idea of such intellectual prowess. Great as man's achievements were in astronomy and other branches of science in predicting phenomena "that given circumstances ought to produce," human mind, Laplace admitted, "will always remain infinitely removed" from that "vast intelligence." How sincere and deep this outburst of intellectual humility really was would hardly be a rewarding area of research. At any rate, overconfidence in the limitless potentialities of the human mind and the mechanical method had rather set the tone in the nineteenth-century scientific parlance. That step by step all the physical phenomena could be reduced to the one basic mechanism of matter was a belief that continued to grow by each decade throughout the century. The triumphs of the nineteenth-century physics pointed almost without exception in this direction.

The first of these triumphs was Rumford's series of experiments that put the caloric theory of heat to flight. The idea that heat was a motion of particles was voiced by the Greek atomists and echoed duly in the days of the scientific renaissance. One of the few correct statements made by Bacon on physical problems was his dictum that "heat itself, its essence and quiddity, is motion and nothing else." [47] So spoke Boyle, Hooke, and Newton. But in the course of what Whittaker called the most amazing vicissitude in the history of science,[48] the true theory had to yield temporarily to a concept that pictured heat as an "igneous fluid," a "heat fluid," or "caloric," the latter word having been coined as late as 1787 by Lavoisier and other French scientists. Rumford, however, found that there were heat producing processes that, if the caloric theory were true, should postulate the presence of the "heat fluid" in practically unlimited quantities in any body subjected to continuous friction. There was only one alternative and Rumford put it forcefully: "It appears to me to be extremely difficult, if not quite impossible, to form any distinct idea of anything capable of being excited and communicated in these experiments, except it be motion." [49]

Rumford made no pretense of knowing "by what mechanical contrivance that particular kind of motion in bodies which is supposed to constitute heat is excited, continued and propagated."[50] This problem, however, was only a matter of detail within the overall framework of mechanism. Rumford himself was quick to recall the case of Newton who, while not claiming to know the ultimate cause of gravity, nevertheless considered his discoveries genuine mechanical laws. In the phenomenon of heat Rumford saw a mechanical feature of the universe no less fundamental than the mechanism underlying the gravitational attraction. "The effects," he wrote, "produced in the world by the agency of heat are probably *just as extensive*, and quite as important as those which are due to the tendency of the particles of matter toward one another; and there is no doubt but that its operations are, in all cases, determined by laws equally immutable"[51] (italics mine). If any conjectures were to be made about the mechanical cause of heat, the supporters of Rumford's theory were not in the least reluctant to make an educated guess. Thus the young Humphry Davy defined the mechanical contrivance whereby heat was produced as the "vibration of the corpuscles of bodies."[52]

The caloric theory, however, was a long time dying. The number of scientists who were unable to see in heat a mode of motion decreased but slowly as the nineteenth century approached its midpoint. The actual situation, however, was not reflected in the "Heat" article of the 1860 edition of the *Encyclopaedia Britannica* in which the mechanical theory of heat was dismissed as "vague and unsatisfactory," while the view that "heat or caloric is a material agent of a peculiar nature" was described as the theory most generally accepted among men of science.[53] Actually by 1860 the majority opinion defined heat as the effect of the motion of molecules. True, only fifteen years earlier the caloric theory claimed physicists of note among its supporters. Among them was Philipp Gustav von Jolly, an influential figure in mid-nineteenth-century German physics, who had remarked during a hurried meeting with a somewhat eccentric physician, Robert Mayer, that water should warm up by shaking if Mayer's theory of the mechanical equivalence of heat were correct. To Jolly's consternation, Mayer rushed into his office a few weeks later shouting, "It is so!" It took some explanation on Mayer's part, however, to make Jolly realize that mechanical processes like stirring and shaking do indeed produce a rise in temperature.

These were also the years that saw Joule give the mechanical equivalent of heat and heard Helmholtz, a youth of twenty-six, read a paper entitled "Die Erhaltung der Kraft." In that paper the various forces of the mechanical universe were tied together for the first time in an even more general concept, energy. Gravity, electrical force, magnetic force, and heat all had one thing in common — they could perform mechanical work, and in a measure that remained always the same under similar circumstances. Heat, a necessary by-product of all actually performed work, occupied a central position in this synthesis. As Helmholtz admitted in the fall of 1862, the recognition of the true nature of heat was of decisive importance in establishing the validity of the Law of the Conservation of Force in all natural processes. This was why physicists chose, as Helmholtz put it, to "designate that view of Nature corresponding to the law of the Conservation of Force with the name of *Mechanical Theory of Heat*"[54] (italics mine).

What Mayer, Joule, and Helmholtz established on the macroscopic level, the kinetic theory of gases did on the molecular level: the behavior and properties of gases came to be reduced to the kinetic energy of their molecules. In the kinetic theory of gases the mechanical conception of nature indeed reached its widest development. It expressed a world view in which, to use Planck's words, "all physical phenomena can be completely reduced to movements of invariable and similar particles or elements of mass."[55] It gave the impression of unveiling at last the most fundamental aspect of the world as nothing more than a "billiard ball universe." According to the definition of A. Krönig, one of the creators of the theory, "gases consist of atoms that behave like solid, perfectly elastic spheres moving with definite velocities in void space."[56] Maxwell's paper of 1859, entitled "Illustration of the Dynamical Theory of Gases," which put kinetic theory on solid foundations, voiced the same view "on the motions and collisions of perfectly elastic spheres."[57] His aim was to formulate the laws, the "strict mechanical principles," for the behavior of an "indefinite number of small, hard, and perfectly elastic spheres acting on one another only during impact."

The conviction that such an imagery closely paralleled the physical reality was rooted not only in the apparently unassailable truth of mechanical philosophy, but also in the truly spectacular successes of the kinetic theory. The theory made it possible to calculate quantities like the free mean path of gas molecules, the viscosity of some gases and their specific heat. All these results followed from the same mechanical postulates that treated, for instance, the diatomic molecules as dumbbells that could move, vibrate, and rotate. Mechanics reached its boldest extremes and for a fleeting decade or two was on the threshold of final triumph. So at least the body of physicists preferred to believe.

To be sure, the mechanical approach to nature was amazingly productive. As a case in point one might recall the system of rotating cylinders and small idle wheels that represented for Maxwell the relation of magnetic fields and "particles" of electricity. Such a mechanical representation was not only in accord with the fact that magnetism acted at right angles to the direction of the flow of the current, it accounted also for the phenomenon of induction. It also led Maxwell to the idea of the displacement current, which he first conceived as mechanical oscillations of electric particles in the dielectric about fixed positions. From this it followed that the mechanical effect of the oscillations had to be propagated as waves whose speed, he found, was "within the limits of experimental error, the same as that of light."

True, Maxwell was cautious not to identify the mechanical model as such with reality. "I do not bring it forward," wrote Maxwell, "as a mode of connection existing in Nature." Such a concession would have meant for him anything but a departure from the mechanical view as the only correct interpretation of nature. The mechanical model, though not necessarily an exact replica of the machinery of nature, was for him the indispensable device that "serves to bring out the *actual* mechanical connection between the known electromagnetic phenomena"[58] (italics mine). Such liberality in the details but firmness in the essentials did, in fact, characterize Maxwell's attitude toward mechanism in all that he produced in theoretical physics. This point is all the more important to note, since no nineteenth-century physicist pushed mathematical formalism in physics further than Maxwell did. His famous equations, once called by Einstein the most important event in physics since Newton, were in fact stripped in their final form of all the scaffolding of mechanical analogies. Indeed, the gap between physical representation and mathematical formulas was so enormous in these equations that all efforts aimed at their interpretation in terms of mechanical concepts ended in failure. None of Maxwell's most competent readers could ever feel sure of having grasped the physical significance of these equations. Hertz clearly expressed their puzzlement when he summed up his own feelings in the now famous phrase: "Maxwell's theory is Maxwell's system of equations."[59]

Well coined as the phrase was, it hardly clarified anything. Quite clear, however, was Maxwell's statement about the ultimate truth of the mechanical explanation in physical matters; it was precisely on this note that his great paper of 1865 ended. "In speaking of the energy of the field," Maxwell wrote, "I wish to be understood literally. All energy is the same as mechanical energy, whether it exists in the form of motion or in that of elasticity, or in any other form.

The energy in electromagnetic phenomena is mechanical energy." [60] Nor did Maxwell ever give up his hopes of finding one day an adequate mechanical model to represent electromagnetic phenomena. As he observed in 1875, any success in dealing with such phenomena was due to "following the strict Newtonian path," and this implied for him the possibility of a completely mechanical, wholly visualizable electrical theory. "To form what Gauss called a 'construirbar Vorstellung' of the invisible process of electric action is the great desideratum in this part of science." [61] Electricity thus reduced to mechanism was then but a part of a science dealing with the ultimate constitution of matter, or in other words with the causes of chemical phenomena. As defined by Maxwell, chemistry was the study of the results "of the configuration and motion of a number of material systems," and its essential task was "to determine, from the observed external actions of an unseen piece of machinery, its internal construction." [62]

The latter part of the foregoing sentence contains, in fact, the core of the faith in mechanism. It also brought out forcefully the essential mental step that such a faith implied, a step that classical physics had never rigorously proved, but always assumed. It was a step from the visible to the invisible, from the gross mechanical effects to the hidden contrivances of an infinitesimally minute machinery, of which Maxwell admitted that "it is so small that it cannot be directly observed." [63] The magnitude of this faith had hardly ever dawned on the classical physicist as a physicist. But faith it was and the only course available if one were to stay with the mechanical philosophy. Parting with this in physical science, however, appeared to be nothing short of renouncing intelligibility itself. For the identification of intelligibility with mechanism was an article of faith for classical physics in 1875 just as it had been two centuries earlier. This is worth reading in Maxwell's phrasing:

When a physical phenomenon can be completely described as a change in the configuration and motion of a material system, the dynamical explanation of that phenomenon is said to be complete. We cannot conceive any further explanation to be either necessary, desirable, or possible, for as soon as we know what is meant by the words *configuration, mass* and *force*, we see that the ideas which they represent are so elementary that they cannot be explained by means of anything else. [64]

Here was mechanical science as the final word in the realm of scientific intelligibility, and to a nineteenth-century physicist it was almost a sacred ritual to make solemn statements about the finality of the mechanical creed. Helmholtz stated that the problem of physical science was "to refer natural phenomena back to unchangeable attractive and repulsive forces, whose intensity depends solely upon distance. The solvability of this problem is the condition of the complete comprehensibility of nature." [65] He saw the vocation of physics ending "as soon as the reduction of natural phenomena to simple forces is complete, and the proof given that this is the only reduction of which the phenomena are capable." [66] Helmholtz was even more explicit in identifying physics with mechanics. In an address to physicians and naturalists, delivered in Innsbruck in 1869, he said: ". . . the ultimate aim of physical science must be to determine the movements which are the real causes of all other phenomena and discover the motive powers on which they depend, in other words, to merge itself into mechanics." [67] Hertz voiced the same view just as explicitly and emphatically: "All physicists agree that the problem of physics consists in tracing the phenomena of nature back to the simple laws of mechanics." [68] In 1888 J. J. Thomson, reviewing the principal advances made in fifty years, concluded that "one of their most conspicuous effects has been to intensify the belief that all physical phenomena can be explained by dynamical principles and to stimulate the search for such explanation." [69] Planck noted in 1897 that the aspiration for a uniform theory of nature on a mechanical basis "can never be permanently repressed." [70] A year earlier A. M. Cornu outlined the program of physics in the following way: "The general tendency should be to show how the facts observed and the phenomena measured, though first brought together by empirical laws, end, by the impulse of successive progressions, in coming under the general laws of rational mechanics." [71] But it was E. V. Jamin, another French physicist, who expressed most succinctly these aspirations: "Physics will one day form a chapter of general mechanics." [72]

Characteristic as these statements are of the classical physicists' state of mind, it is even more revealing to see to what extent they were predisposed to retain in some particular problems the appearance at least of a mechanistic explanation. A most instructive case is Helmholtz's interpretation of the periodic table of elements. To see, as Helmholtz did, in each element a specific entity, irreducible to any other element, was nothing short of a fall-back on qualitative, nonmechanistic principles. Actually Helmholtz spoke of "the qualitative immutability of matter," and it might be noted that he clung to this view to the very end of his distinguished career. No less persistent, however, was his instinctive determination to retain mechanism on its hallowed pedestal even in this case. He felt, somewhat naïvely, to have achieved this, by stating that given the qualitatively different units of matter "all changes in the world are changes in the local distribution of elementary matter, and are eventually brought about through Motion."[73]

No less revealing was the case of H. Hertz, the discoverer of electromagnetic waves. He was, beyond any doubt, thinking about mechanical waves — mysterious as the mechanism might have been — when he stated that by knowing the velocity, length, and transversal character of those waves "we know completely the geometrical relations of the motion."[74] The fact that those "geometrical relations" represented the only clear detail in the matter gave, however, to Hertz no second thoughts about the supposedly unshakeable foundations of mechanism. Nor was there any weakening of the belief in mechanism on the part of those nineteenth-century physicists who at one time or another had put more than customary emphasis on the distinction between mathematical formalism and mechanics. Among them were Ampère, Fourier, Rankine, and Kirchoff, but it is important to note that none of them failed to make their obeisance to the mechanical creed. For instance, while Kirchoff defined in a rather positivistic vein the object of mechanics "to describe as completely and as simply as possible the motions produced in nature,"[75] he spoke in the genuinely mechanistic idiom as well: "The highest object to which the natural sciences are constrained to aim, but which they will never reach, is the determination of the forces, which are present in nature, and of the state of matter at any given moment — that is, the reduction of all the phenomena of nature to mechanics."[76]

As a matter of fact, mechanics came to stand as a monument to the simplicity and sweep of Newton's law of the inverse square, the fundamental result of his analysis of a central field of force. It was believed that no physicist could ask for more than to succeed in organizing his findings along lines reproducing this supreme pattern. To achieve the Newtonian degree of perfection was indeed taken as the last word in making any problem in physics intelligible. For the classical physicist understanding a phenomenon meant simply reducing it to the Newtonian laws. It was therefore a source of deep satisfaction for them to learn that the mathematical interpretation of a physical process in which gravitation played no part might show a striking resemblance to the law of gravitation. Maxwell was particularly eager to point this out in connection with the law of the conduction of heat in uniform media.[77] Newton's laws were also the ideal Ampère had emulated in his work with such success that Maxwell was prompted to say: "The whole theory and experiment seems as if it had leaped, full-grown, full-armed, from the brain of the Newton of electricity."[78]

The goal was still the same when toward the end of the century physics began to be dominated by speculations about the ether — a substance whose mechanical properties existed in belief only. P. Drude, the author of *Physics of the Ether on Electromagnetic Basis,* for instance, defined his own method as the strict imitation of the derivation of the Newtonian laws of gravitation from the visible phenomena of planetary motions.[79] About the same time, Rowland, arguing from the assumption that the ether must have the simplest properties, concluded that the waves carried by the ether should obey the simplest of laws, that of the inverse square, since every action at a distance obeyed the same law.[80] How deeply this way of thinking was imbedded in the minds of the nineteenth-century physicists can best be seen in a letter of J. J. Thomson written in 1937. He was still speaking of the ether that he regarded as the working system of the universe, containing all mass, momentum, and energy, the sum of which was constant "so that the Newtonian Mechanics apply."[81]

In the same letter so clearly echoing the nineteenth-century scientific beliefs, Thomson spoke of the paramount necessity in physical research of mechanical models that were clearly visualizable. This was of course the outstanding working method of the mechanical creed, and nineteenth-century physicists spared no effort in thinking up mechanical models for physical phenomena still under scrutiny. Failure to find a suitable model often meant simply abandoning the search. Unable to imagine a mechanism for the propagation of electromagnetic effects, Gauss laid aside the subject.[82] The absence of a mechanism in connection with a mathematically formulated relationship among "forces" more often than not evoked uneasiness in fellow physicists. Thus Joseph Henry wrote: "The theory of Ampère, though an admirable expression of generalization of the phenomena of electromagnetism, is wanting in that strict analogy with known mechanical actions which is desirable in a theory intended to explain phenomena of this kind."[83]

Henry's usage of the word analogy is important, showing as it does that mechanical models were not to be taken in the sense of naïve realism. A systematical use of the mechanical analogy was, however, a rule not to be abandoned in the framework of physical explanation identified with mechanical concepts. This was the position taken by the two principal spokesmen of nineteenth-century physics, Maxwell and Kelvin. To excoriate the "model-worshipers" and to be busily engaged at the same time in constructing new models represented for Maxwell the two necessary sides of the same coin: a judicious but firm belief in mechanism. Entirely similar was Kelvin's attitude. He explicitly warned that the mechanical illustrations of the molecular constitution of solids, which he had proposed in his lectures given at Johns Hopkins University, were "not to be accepted as true in Nature."[84] In disavowing the naïve realism to which we have just referred, Kelvin even went so far as to state that a great deal more could be discovered by algebra than by analyzing mechanical models and that no observation could improve on the mathematical treatment. At one point he described the models as being more of a "help or corrective to brain sluggishness than a means of observation or discovery."[85]

Yet, in the same lectures he equated understanding in physics to the possibility of making a mechanical model of a particular subject. In every model, however crude, he found a core of truth. He readily admitted that his own model of concentric rigid shells connected with springs was too simple to represent adequately a real molecule. Still, Kelvin added that "there must be something in this molecular hypothesis and that as a mechanical symbol, it is certainly not a mere hypothesis, but a reality."[86] In the same manner it was the "mechanical content" in the theories of others that aroused his interest. Thus he did not conceal the fact that his great admiration for Maxwell's electromagnetic theory concerned mainly the mechanical analogies implicit in the theory. Maxwell, he said, "makes a model that does all the wonderful things that electricity does in inducing currents, etc., and there can be no doubt that a mechanical model of that kind is immensely instructive and is a step towards a definite theory of electromagnetism."[87] Kelvin never stopped imagining new models:

I never satisfy myself until I can make a mechanical model of a thing. If I can make a mechanical model I can understand it. As long as I cannot make a mechanical model all the way through I cannot understand; and that is why I cannot get the electromagnetic theory. . . . But I want to understand light as well as I can, without introducing things that we understand even less of. That is why I take plain dynamics. I can get a model in plain dynamics; I cannot in electromagnetics.[88]

The fact that model-making was beset with great difficulties as to the final form of electromagnetic theory was just another strong argument to Kelvin that electromagnetism at no stage should be cut off from its mechanical moorings. Kelvin's confidence in the resourcefulness of mechanics knew no bounds. He viewed his *Baltimore Lectures* as the verification of the universal and exclusive validity of mechanics. The phenomena of light, he said, "can be explained without going beyond the elastic-solid theory. We have now our answer: *every thing non-magnetic; nothing magnetic*"[89] (italics mine).

In his belief that electromagnetic and light phenomena would be reduced one day to a dynamical theory, Kelvin never wavered. His statements made early in his career were just as categorical as the ones uttered decades later. "We *know*," he said in 1860, "that light is propagated like sound through pressure and motion" (italics mine). In the same context he tried to interpret electricity as residual surface tension, defined heat as the motion of electricity, spoke of magnetism as certain alignment of the axes of revolution in the same motion, and concluded that they were "all by one and the same dynamical action."[90] But the fact was that nobody knew of any conclusive evidence that light, for instance, was a pressure wave. Such a conception of light could at best qualify as perhaps the only conceivable hypothesis in the framework of mechanistic physics, but hardly as the conclusion to be stated in terms reserved to well-proved facts. One may therefore wonder to what extent it was "scientific" for Kelvin to say that "it is *absolutely certain* that there is a definite dynamical theory for waves of light, to be enriched not abolished by electromagnetic theory"[91] (italics mine). But could an absolute scientific belief, an absolute scientific conviction, use less categorical terms even if the proofs were still being sought for?

Electromagnetism, particularly in its Maxwellian form, was a newcomer to physics. To verify the universal claims of mechanics, physicists had much older and even more tantalizing problems at hand that asked for a mechanical explanation. Foremost was the puzzle of gravitation which stubbornly defied mechanical and non-mechanical speculations alike. Not that mechanical illustrations, some of them at times very ingenious, had not been proposed from the early days of classical physics to throw light on the matter. One need only recall Plutarch's analogy of a sling revived by Borelli, or the analysis of the behavior of a conical pendulum by Hooke, or the various accounts of the motion of particles of matter in a rotating medium, as described by Huygens in his *Discours de la cause de la pesanteur* (1690). Though clearly inadequate, these analogies were motivated by a deep-seated desire, as can be seen from the opening words of Huygens' work mentioned above, "to find an *intelligible* explanation of gravity" (italics mine).[92]

There had been, of course, prominent physicists who, in view of the enormous difficulties presented by any mechanical explanation of gravitation, chose to make as few conjectures as possible on the subject. Such were, for instance, Newton and Euler. Their reticence, however, meant anything but an acquiescence in the so-called action at a distance divorced from an underlying mechanism. As a final resort there was the ether, unknown though its manner of acting was to science. Yet, as Euler argued characteristically, it was more reasonable to assign gravitational attraction to the action of the ether than to reduce gravity to some unintelligible property.[93]

Little wonder that owing to the persistent impasse the problem presented, the troublesome expression action at a distance became an accepted idiom in the physicists' parlance without raising immediate and passionate objections as it did in Newton's time. Then the expression was viewed, as Mach remarked, as an "uncommon intelligibility" and created heated discussions. But in two centuries minds had got used to it, and the expression turned, to quote Mach again, into a "common unintelligibility."[94] Mach, however, was wide of the mark in saying that "at the present day gravitation no longer disturbs anybody." In 1872 when Mach wrote this he might have already watched the contemporary scene of physics through positivist glasses. He got a rather distorted view.

Actually some of the most prominent nineteenth-century physicists wrestled long with the question as whether gravitation was a primordial law of nature, or only a general effect of an unknown cause. True, they were led nowhere, but lack of success meant anything but lack of interest. Laplace for one spent not a few years speculating on how to explain gravitation mechanically before he admitted defeat: "Here the ignorance in which we are concerning the ultimate properties of matter stops us and removes all hope that we shall ever be able to answer these questions in a satisfactory manner."[95] Similar were the views of Laplace's younger contemporary, John F. W. Herschel, who, like his father, earned his fame as an astronomer. He thought that efforts, like Le Sage's, to explain gravity on a mechanical basis, were "too grotesque to need serious consideration."[96] Le Sage's theory, once called by Maxwell the only consistent theory of gravitation ever proposed,[97] had indeed incurable defects. It could not account for the dependence of gravity on mass and could not save the massive bodies from burning up under the impact of the "ultramundane particles." This Maxwell had clearly shown,[98] without, however, voicing despair about ever finding a mechanical explanation of gravity. Contrary to Mach's assertion, lack of interest in the problem was *not* the attitude of the physicists of his time. Their state of mind was far more accurately described by Maxwell when he said that men of science would gladly devote the "whole remainder of their lives" to the study of a mechanical theory of gravitation if a promising one were available.[99]

A possible mechanical explanation of gravitation was a program the mechanistic creed could not ignore without renouncing itself. Actually no small efforts were spent to find one. That action at a distance was only a *façon de parler*, but hardly an explanation, was only too well known by the nineteenth-century physicists. All of them would have agreed with Helmholtz, who stated that it was of fundamental importance for theoretical science "to understand how apparent actions at a distance really consist in a propagation of an action from one layer of an intervening medium to the next." [100] Consequently, when the action at a distance was proposed as an explanation, the leading exponents of mechanism reacted violently. Kelvin termed it the "most fantastic of paradoxes," [101] and others pointedly recalled that Newton viewed it "so great an absurdity, that . . . no man, who has in philosophical matters a competent faculty of thinking, can ever fall into it." [102]

If at any time — then precisely in the second half of the nineteenth century — physicists were highly reluctant to fall into such a gross error. In full testimony to the unshaken belief in the mechanical program of science, a large number of attempts were proposed to reduce gravity to some mechanical process. In part they were efforts to present old ideas in a mathematically correct way. But here too it was impossible to pour new wine into old skins. In part they were new ideas such as Kelvin's vortex theory that had grown out from the observation of the remarkable behavior of smoke rings. But this too failed to lead anywhere. No more successful were the extensive researches of C. A. Bjerknes on two spheres immersed in a pulsating incompressible fluid, or Maxwell's attempts to include gravitation in his famous memoir of 1864. There Maxwell found that at any place where the gravitational force vanishes, the intrinsic energy of the medium must have an enormously great value. "As I am unable," concluded Maxwell, "to understand in what way a medium can possess such properties, I cannot go any further in this direction in searching for the cause of gravitation." [103] Unsuccessful as were these efforts, they nevertheless evidenced the deep conviction that a mechanical synthesis of the science of physics must include gravitation.

That such an agitated activity in this field came when the mechanical explanation in physics seemed to converge toward its final completion is truly meaningful even in retrospect. Few if any suspected in the late nineteenth century that the mechanistic theory would have to undergo a thorough revision even in its stronghold, the mathematical explanation of the motion of planets. Still, for all the amazing perfection celestial mechanics had attained by the nineteenth century, there were phenomena that defied explanation on the sole basis of the Newtonian formula. One was the advance of the perihelion of Mercury; another was the acceleration of Encke's comet. And at the end of the nineteenth century Newcomb found that the node of the orbit of Venus undergoes a secular acceleration five times the probable error. The most disturbing of these was, of course, the case of Mercury, and when the first consistent solution came with the General Theory of Relativity, there was no place in it for mechanism, or for models, let alone an ether of any sort. A geometry impervious to any visualization superseded the supreme pride of classical mechanics, the Newtonian analysis of planetary motions, to which, in two centuries' belief, the explanation of every physical phenomenon was bound to be reduced.

If the advent of General Relativity meant the abrupt end of serious efforts to account for gravitation on a mechanical basis, similarly sudden was the demise of the ether, the medium that was supposed to carry through immense distances, among other things, the gravitational effect. It was in the Cartesian physics that the ether began to play a universal role, and it remained just as indispensable to the Newtonian philosophers whenever they had to do full justice to a consistent presentation of the tenets of mechanism. The realistic sense in which Hooke, Newton, or Euler spoke of the ether was not second in any way to the solemn dicta about the ether voiced by Kelvin, Maxwell, Rowland, or Michelson, except perhaps for the Victorian tinge that adorned their phrases. In Hooke's presentation of the mechanical theory of gravitation the reality of radiating vibration in an "exceeding fluid" (ether) was submitted without a direct reference to its hypothetical existence.[104] Newton, in the General Scholium, spoke of a "most subtle spirit which pervades and lies hid in all gross bodies," and for him the uncertainties about the ether concerned solely the "accurate determinations and demonstrations of the laws by which this elastic and electric spirit operates."[105] In the same realistic vein the queries of the *Opticks*, so preoccupied with the properties of the ether, were phrased in the negative, so as to permit only one answer: Why, yes of course. Half a century later Euler voiced the same firm conviction as he declared that "we *know* that the whole space which separates the heavenly bodies is filled with a subtle matter called ether"[106] (italics mine). To such a matter-of-fact statement the Victorian physicist could add hardly more than an orotund adjective.

The ether, the vehicle of physical contact between distant bodies, for classical physicists was above all a logical necessity they could not dispense with. After the revival by Young of the wave theory of light, they were forced, however, to give the most detailed attention to the ether, and a feverish production of its various models got under way. The ether was supposed to behave as an elastic solid that obviously had to be extremely rigid and extremely tenuous both to propagate light waves at their enormous speed and not to offer perceptible resistance to the planets moving through it in space. Only faith, enormous faith, could lend plausibility to the existence of two such opposite properties in one and the same substance. This faith mechanistic physicists possessed in an extraordinary degree, most of the time without being aware of it, as shown by their way of speaking in terms so resolutely categorical. They found nothing strange in stretching their analogies to the realms of gratuitous credulity. The most down-to-earth of these analogies was the shoemaker's wax mentioned in 1845 by the future Sir George Gabriel Stokes,[107] who elaborated on it with the typical self-assurance of youth (he was only twenty-six), to whom no difficulties were insurmountable. Through an ordinary piece of shoemaker's wax, as was well known, pieces of metal, or stones, would "sink" in a matter of a few weeks. Why couldn't the ether, asked Stokes, have the properties of this wax in an extreme degree? So it seemed to anyone armed with an unshakable confidence in the mechanistic point of view, and was there any physicist to be found without such armor in 1845? How else could it happen that in the properties of this pedestrian piece of matter a scientific era could find an answer to one of its deepest aspirations? For even at the end of the century a Kelvin would refer to this wax before an admiring audience of physicists as being the most convincing of all analogies on the subject. The ether, he said, "is no greater mystery at all events than the shoemaker's wax." [108]

Even those who were willing to admit candidly as J. MacCullagh did that the constitution of the ether and its connection with other bodies "are utterly unknown," felt impelled to add: "It is *certain indeed* that light is produced by undulations, propagated with transversal vibrations through a highly elastic ether" (italics mine).[109] Such affirmations of the *certainty* or *indisputable reality* of the ether are perhaps the best illustrations in the history of physics that an unreserved confidence in a physical "creed," organismic, mechanistic, positivist as it may be, will seriously weaken the physicist's ability to distinguish between a relatively well-established fact and a hypothesis. For it is one thing to propose an inference as being very plausible and another to assert its reality and in the least uncertain terms at that. In this respect even a Maxwell could not avoid the pitfalls set by an unquestioning faith in mechanism. In his letter to Bishop Ellicott he stated that the ether "is the largest, most uniform and apparently the most permanent object we know." He even volunteered the information that the ether filled interstellar space "without a gap or flaw of 1/100,000 inch everywhere." [110] In his article, "Ether," prepared for the ninth edition of the *Encyclopaedia Britannica*, he was even more emphatic. "There can be no doubt," he asserted categorically, that the ether "is certainly the largest body of which we have any knowledge." [111] Tyndall too professed to know all the sundry items about the ether in a way that transcended all doubts. He ascribed the transparency of bodies to the ether [112] and confidently predicted that the "natural philosophy of the future will *certainly* for the most part consist in the investigation of the relations subsisting between the ordinary matter of the universe and the wonderful ether in which this matter is immersed" [113] (italics mine). But the ether was not so "wonderful" after all and only helped Tyndall make one of the poorest prognostications ever offered by a physicist.

Kelvin, with a penchant for resounding declamations, gave witness just as forcefully to this unconditional faith in the ether. Asking if there was any matter not subject to the law of gravitation, he answered: "I think I may say with *absolute decision* that there is. We are all convinced . . . that ether is matter" [114] (italics mine). It was to Kelvin's credit that on occasion he did not disdain to call mechanism what it truly was, a belief, when taken as an all-inclusive explanation. "Belief that no other theory of matter is possible is the only ground for anticipating that there is in store for the world another beautiful book to be called *Elasticity, a Mode of Motion*." [115] Few of his colleagues followed suit in making such candid admissions. For the word belief was frowned upon in the golden decades of mechanism, and scientists rather preferred to ignore that there was all too much in mechanism that rested on faith alone. And truly robust this faith was. Never in the history of physics did faith ever move mountains of uncertainties or bridge gaps of ignorance with such ease as it did in the heyday of mechanism. The least scientific aspect of this predicament lay no doubt in the semantics of the age that boldly uttered the word evidence about the existence of the ether, when clearly the word assumption was demanded by the evidence at hand. For notwithstanding the ocean of words flowing from the pen of nineteenth-century physicists, the "convincing" information about the ether, let alone its vaunted properties, was at most very indirect. What was more, the difficulties, problems, and contradictions involved in the notion of the ether were without parallel in the annals of physics. Still the tide of unhesitating assertions about its reality moved ahead unabated; still conclusions were being hastily reached with hardly a second thought to the true strength of the bonds that tied them to their premises.

Of this state of affairs the best illustration is found in the comments made by physicists following Hertz's discovery of the electromagnetic waves. What Hertz had demonstrated was the wavelike character of

the electrical disturbances propagated through space as predicted by
Maxwell's equations. On the seemingly unexceptionable reasoning
that if there was a pattern of undulation something must undulate,
Hertz himself saw in his achievement a supreme confirmation of the
existence of the ether. "It is certain," he told his colleagues, "that all
space known to us is not empty, but is filled with a substance, the
ether, which can be thrown into vibration."[116] Yet, the apparently
clinching proof merely strengthened an illusion already firm beyond
ordinary measure. For regardless of his famous experiment, Hertz
could say: "Take away from the world electricity, and light disap-
pears; remove from the world the luminiferous ether, and electric and
magnetic actions can no longer traverse space. This is our assertion."[117]

The word our could hardly have been better chosen, for there was
no physicist unwilling to concur with this type of reasoning. As for
his predecessors in classical physics, Hertz could proudly say of his
assertion that "it does not date from today or yesterday; already it
has behind it a long history. In this history its foundations lie."[118]
As to his contemporaries, their concurrence was just as unanimous.
Thus G. F. Fitzgerald spoke of the year of 1888, which witnessed
Hertz's historic experiment, as one forever memorable, providing as
it did the affirmative experimental decision in favor of the "intervening medium." As Fitzgerald viewed the situation, the rising generation of physicists could only congratulate themselves for having
obtained in the experimentally demonstrated ether the "firm and true
standpoint" for further advances.[119] Rarely indeed had the words
true and firm proved so ephemeral. But Fitzgerald could hardly suspect what was in store for a reasoning that issued in the assertion:
"We have long known that there is an ether, an all-pervading medium,
occupying all known space. Its existence is a necessary consequence
of the undulatory theory of light."[120] What remained now for a state
of mind trapped in the postulates of the mechanistic explanation was
to describe before the festive audience of the British Association the
detection of the ether by Hertz in soaring Victorian prose. This Fitzgerald did with accomplished mastery: "Fire, water, earth, and air
have long been his [man's] slaves, but it is only within the last few
years that man has won the battle lost by the giants of old, has
snatched the thunderbolt from Jove himself, and enslaved the all
pervading ether."[121] The dubious distinction of announcing the cap-
ture of the ether had indeed to be the privilege of truly captive minds.
If there was a difference on this count among physicists, it concerned
only the manner of expression. Even a sober mind like Helmholtz's
was carried away in discussing the bearing of Hertz's experiments

on the existence of the ether. Penning an introduction to Hertz's *Principles of Mechanics* in 1893 (six years after the first flawless execution of Michelson's famous experiment on a possible ether-drag), he firmly stated: "There can no longer be any doubt that light waves consist of electric vibrations in the all pervading ether and that the latter possesses the properties of an insulator and a magnetic medium."[122]

It was rather a pity that one of the very last pieces of writing of such a great mind should miss the mark so widely. For there is a great difference between a very useful hypothesis, which the ether could have been at best in late nineteenth-century physics, and a well-demonstrated phenomenon, which the ether certainly was not. The essential argument for the existence of the ether was an inference that Fitzgerald formulated with vivid concreteness when referring to people who kept asking: Why do you believe in the ether? What is the good of it? "I ask them," said Fitzgerald, "What becomes of light for the eight minutes after it has left the Sun and before it reaches the Earth? When they consider that they observe how necessary the ether is. If light took no time to come from the Sun, there would be no need for the ether."[123] But the true force of any syllogism comes from the conformity of its premises with reality, and if the support of facts fails, even the best logic is no better than an "organized way of going wrong with confidence," as a twentieth-century pragmatist, Charles F. Kettering, once defined the art of logic. True, the fallacy involved in Fitzgerald's argument was not so evident in the late nineteenth century. Still at any rate, could the evidence then available about the ether justify on the level of plain logic the almost lyrical praises of the ether uttered by prominent physicists before solemn scientific audiences?

To dwell for a while on such unfortunate exaggerations certainly ought not to be motivated by a false feeling of superiority over past generations of physicists. There can hardly be a more effective warning about the possibility of similar shortsightedness on the part of the present generation than to take an attentive look at little remembered aspects of past scientific "beliefs." With these "beliefs" in mind it will also be possible to account for the enormous discrepancies that arise at times between the style of physicists appraising a particular phase of physics and the true evidences they are able to marshal. To encounter these discrepancies in their original phrasing and setting can, of course, easily lead to an eerie feeling that is likely to overtake the present-day reader of Rowland's 1899 address to the American Physical Society. In that address, entitled "The Highest Aim of the Physicist," Rowland declared: "Actions feeble and actions mighty from inter-molecular distances through interplanetary and interstellar distances which bound the universe all have their being in this wondrous ether."[124] Michelson, too, identified all the phenomena of the physical world with the "different manifestations of the various modes of motions of one all-pervading substance, the ether." He viewed the proposed reduction of electricity to mechanical stresses in the ether as "one of the grandest generalizations of modern science," which, as he stated, "ought to be true even if it is not."[125]

Being spectators of a post-etherian phase of physics, such appraisals of the ether leave us today either astonished, or bewildered. But how elated was the audience at the London Institution when informed by Lodge on December 28, 1882, that matter was made of the whirls of the ether, that its vibrations constituted light, and that it could even be sheared into positive and negative electricity. They were also told in a tone barring any doubt that it was the ether that transmitted "every action and reaction of which matter is capable."[126] Neither Lodge nor his audience would of course have thought it possible that half a century later a noted physicist would describe the ether as "a medium invented by man for the purpose of propagating his misconceptions from one place to another."[127] The contrast between these two evaluations of the ether could hardly be more striking and should throw a forceful light on the state of mind in which Lodge and his colleagues viewed its role. For them the ether stood for physics itself and for the ultimate synthesis which, as many believed in the closing decades of the century, was not far away. "The ether is all but in our grasp,"[128] declared Poincaré, and with the ether seemed to follow the whole of physics. The feverish tone that animates Lodge's address of 1889 vividly shows the expectations of physicists:

The present is an epoch of astounding activity in physical science. Progress is a thing of months and weeks, almost of days. The long line of isolated ripples of past discovery seem blending into a mighty wave, on the crest of which one begins to discern some oncoming magnificent generalization. The suspense is becoming feverish, at times almost painful. One feels like a boy who has been long strumming on the silent keyboard of a deserted organ, into the chest of which an unseen power begins to blow a vivifying breath.[129]

What Lodge envisaged in this "magnificent generalization" was a total and final synthesis of physics that was expected to include even the solution of such problems as, for instance, the question of the cause of the inertia of matter. The supreme question for physics, as Hertz formulated it, was to learn how all things were fashioned out of the ether, and in this question physics had according to him "the

icy summits of its loftiest range."[130] Precisely what a living nature
was for organismic physics the ether was for mechanistic physics:
an all-inclusive mold within which everything pertaining to physics
had to be accommodated. Hertz did not promise a necessarily quick
success in bringing to a completion this final phase of the history of
physics, but he certainly felt hopeful. And in this hope almost all
doing work in physics shared. Hardly anyone cared to take seriously
the possibility of a discovery that could play havoc with the status
quo of physics. Kirchoff, for instance, explicitly stated that he did
not anticipate the discovery of new facts leading to a thorough revision of fundamental concepts. When A. Schuster, a younger
colleague of Maxwell at Cambridge, told him of a recent English
discovery that light falling on the surface of a selenium bar changed
its electrical conductivity, Kirchoff merely answered: "I am surprised
that so curious a phenomenon should have remained undiscovered
till now." As Schuster recalled it almost three decades later, such an
indifferent reaction represented "the attitude of mind not only of
Kirchoff but of the great majority of physicists at the time."[131]

Yet, in the curious behavior of the selenium bar, there was enclosed
much of the evidence leading to modern physics. It should have been
the cause of great surprise indeed. But the degree of surprise is also
a function of the openness of mind, and it was not always an open
mind that classical physics, as a tightly built system of thought, in
fact produced in its cultivators. Awareness of this fact is of paramount
importance if we are to understand why it took so long for classical
physicists to realize the meaning of their consistent failure to construct mechanical models for the ether. For long-sustained efforts
notwithstanding, no amount of clever manipulation of spheres, rigid
bars, flywheels, rubber bands, and jellies led any closer to the desired
goal.

The moment of sober awakening did not, of course, come at the
same time to all those concerned. Fitzgerald was one of the first to
suspect the hopelessness of the situation, and he was prompted to
remark: "I am afraid, nothing except a complete overthrow of this
whole notion of how the functions of the ether are produced will cure
Sir W. Thomson."[132] To have recognized the fact that all his research
had shed no light whatsoever on the nature of the ether must have
been dismaying to Kelvin. True to his flair for dramatics, Kelvin came
forward with his admission in a way that shocked the festive audience
gathered on June 16, 1896, to celebrate his golden jubilee as a professor of physics. "One word," he said, "characterizes the most strenu-
ous efforts for the advancement of science that I have made perse-
veringly during fifty-five years; that word is failure." To relieve the

sadness that must come of failure, Kelvin could refer only to the *certaminis gaudia* that, in his view, must accompany the efforts of the naturalist in the scientific pursuit and should save him "from being wholly miserable."[133]

The word failure was a haunting one. It seemed to suggest the failure of physics as such. Kelvin himself felt the need to make some clarification, and a month later he explained in a letter that the word failure referred only to his lack of success in learning anything about the ether. Of the ether, he wrote, "I know no more now than I knew 55 years ago."[134] He was, however, quick to add that this should not be construed as doubt on his part in the mechanical theory. "I am as firmly convinced as ever of the *absolute truth* of the kinetic theory of gases" (italics mine). The word absolute was an adjective that a man of science should have used with only the utmost diffidence, but mechanism showed itself highly successful in obliterating such sensitivities in its devotees. Consequently neither Kelvin nor his fellow physicists were able to sense the real import of the negative results of the Michelson-Morley experiment. Admitting as he did the flawlessness of the idea and execution of this experiment, Kelvin saw in it no more than one of the two clouds obscuring "the beauty and clearness of the dynamical theory, which asserts heat and light to be modes of motion."[135]

True, Kelvin was apprehensive of the puzzle presented by the Michelson-Morley experiment. "I am afraid," he said, "we must still regard the Cloud No. 1. as very dense."[136] The cloud, however, was far denser than Kelvin suspected, and besides, its depth was also much greater than he would have imagined. Maxwell's equations, which did not remain invariant when viewed from another inertial frame of reference, were responsible for this. Consequently the speed of light, an integral part of the equations, should also have changed, depending on the relative velocities of the two frames of reference. It was this change in the velocity of light that the Michelson-Morley experiment was supposed to detect, but no positive result was ever forthcoming. To avoid this impasse one had to reject the notion of an absolute frame of reference and this meant discarding the "wondrous" ether. But without the ether there could be no consistent, all-embracing mechanical theory. This was a classic in self-defeat, mechanism itself producing the seeds that were to bring about its radical abandonment. Einstein, who took the drastic step in this sense, rightly pointed out that it was such firm adherents of mechanism as Maxwell and Hertz who demolished, without meaning to, "the faith in mechanism as the final basis of all physical thinking."[137]

Besides gravitation and the ether, there was a third sector in classical physics that saw just as sudden a collapse of the mechanical approach: the problem of the absorption and emission of energy by matter. The kinetic theory that pictured the atoms as little rotators was very successful in determining the value of the specific heat of monatomic gases as a function of the degree of freedom in this hypothetical model of atoms. So great was this success that Kelvin viewed the kinetic theory, created in part by Maxwell, as the "first instalment" of the long awaited comprehensive theory that was supposed to contain the last word in physics. Maxwell's theory was in Kelvin's words "a well-drawn part of a great chart" that was expected to explain chemical affinity, electricity, magnetism, gravitation, and the inertia of mass, all this of course on a strictly mechanical ground.[138] Spurred by the results of the kinetic theory, Kelvin felt encouraged to speak in 1871 of the possibility of an "early completion" of an all-inclusive physical theory, and he did this in a truly sanguine style. Ten lean years, however, somewhat toned down his exuberance. In 1881 he admitted that no guideposts pointing toward the great goal had been discovered "or imagined as discoverable."[139] This he repeated in 1889 with the admission that, all the failures to the contrary, the feeling of the urgency for a comprehensive theory was "growing in intensity every year."[140] His hopes still soared as high as ever, although his timetable had become far less demanding:

This time next year, — this time ten years, — this time one hundred years, — probably it will be just as easy as we think it is to understand that glass of water, which now seems so plain and simple. I cannot doubt but that these things, which now seem to us so mysterious, will be no mysteries at all; that the scales will fall from our eyes; that we shall learn to look on things in a different way — when that which is now a difficulty will be the only commonsense and intelligible way of looking at the subject.[141]

It was not "common sense," or "mechanical intelligibility," however, that was to be used in physics if some of its outstanding problems were to be conquered at all.

With the final theory still nowhere in sight, it must have been all the more discouraging that the kinetic theory had rough going even within its restricted range. As Maxwell pointed out in 1875, the spectroscopic evidence strongly indicated that most of the molecules had many more than six degrees of freedom. Since his theory failed even in the relatively simple case of the specific heat of diatomic molecules, "every additional degree of complexity which we attribute to the molecule," admitted Maxwell, "can only increase the difficulty of reconciling the observed with the calculated value of the specific heat."[142] As time went on the discrepancy between theory and observations had grown wider, and to most of those wrestling with the problem the situation looked desperate. For the modifications the theory had to undergo in order to yield the experimentally correct results involved the arbitrary exclusion of certain degrees of freedom. Such a step, however, as Lord Rayleigh (Robert John Strutt) noted in his paper "On the Law of Partition of Energy" of January, 1900, would have contradicted fundamental assumptions of mechanics. "What would appear to be wanted," mused Rayleigh, "is some escape from the destructive simplicity of the conclusions."[143] Kelvin himself chose to ignore the difficulty that gave rise to what he called Cloud No. 2. In conformity with his unbounded faith in mechanism, he commented on Rayleigh's remark: "The simplest way of arriving at this desired result is to deny the conclusion; and so, in the beginning of the twentieth century, to lose sight of a cloud which has obscured the brilliance of the molecular theory of heat and light during the last quarter of the nineteenth century."[144]

Little did Kelvin suspect, when he uttered these astounding words at the Royal Institution at the end of April, 1900, that the end of mechanism was just around the corner. The summer of 1900 saw Max Planck engaged in an intensive study of blackbody radiation, a problem resting, like that of the specific heat of gases, on the principle of the equipartition of energy among the various degrees of freedom present in a mechanical rotator or an oscillator. In the course of what had truly been an agonizing reappraisal, Planck was forced to admit to himself that in order to obtain the correct theoretical formula, a step diametrically opposite to the principles of classical mechanics had to be taken. The step was the same one that Rayleigh could not bring himself to take: to ignore in a sense certain degrees of freedom, according to various temperature ranges. That such a procedure appeared simply impossible for Rayleigh can easily be understood. His was a mind that produced a treatise on sound,[145] the only monograph covering a full branch of physics where the mechanical principle was victorious over all difficulties.

Planck's mind, however, though no less steeped in mechanics, was better prepared to make hard intellectual decisions. As is known, he chose theoretical physics as a career in spite of the discouragement he received in his younger days from some of the most important figures in German science. Helmholtz for one ignored his doctoral thesis, and Kirchoff disapproved of it. Later, when he was looking for a research topic with a view to qualifying as a university lecturer (*Privatdozent*), it was another noted figure of German physics, Jolly, who told him that "in physics nothing fundamentally new can still be discovered."[146] Several years later his law on the potential difference of the electrolytes, which stood in perfect agreement with the data available, was dismissed by Du Bois-Reymond on the fictitious grounds that the experiments were probably not reliable. In these years of frustration and opposition, Planck was supported by the conviction that the more daring an intellectual step, the greater faith it demanded. "Science," he once said, "demands also the believing spirit. Anybody who has been seriously engaged in scientific work of any kind realizes that over the entrance to the gates of the temple of science are written the words: Ye must have faith. It is a quality which the scientists cannot dispense with." [147] Brahe and Kepler, Planck continued, both had the same facts at hand, but it fell to Kepler to create the new astronomy, for he had a greater faith "in the existence of the eternal laws of creation." Similar was the case in 1900. The data necessary to make the breakthrough were at the disposal of all physicists. To stumble, however, on the pattern underlying them required more than faith in mechanism alone. What was needed was rather a faith in science that did not refuse to go beyond mechanism. For to break with the continuity principle of classical physics and introduce quantization meant precisely to break the confines of mechanism.

In the late summer of 1900 Planck had only a general idea of the boldness of the step he was going to take by introducing the quantum of energy. During a walk through Grünewald, a wood in the suburbs of Berlin, he confided to his son that he was on the verge of a discovery perhaps comparable with those of Newton. Little could he have guessed that his discovery would lead physics into a conceptual framework decidedly different from classical mechanism. After all, physicists hoped to achieve the conquest of the atom not by going beyond Newtonian physics but rather by its most ingenious exploitation. It was in this sense that Kelvin fixed his life's ambition to become the Newton of the atom [148] and, as was customary to say around the turn of the century, to break the code of the world of atoms demanded the genius of a second Newton. This second Newton, whoever he might be, certainly was not expected to set the foundations of physics deeper than had the first one. Like Copernicus, who had broken the circle of the Aristotelian physics without abandoning it, Planck had to break the narrow confines of mechanism without being able to part basically with the mechanistic world picture. For him, as he stated in 1908, the atoms were as real and sharply defined units of the universe as the heavenly bodies. [149]

Yet, before long these sharp contours were to dissolve into a haze. For when physicists in 1913 got their first real glimpse into the "mechanics" of the atom, they were told by Bohr to renounce all attempts to visualize the behavior of an electron during the production of a spectral line. It would have hardly been possible to lay down a precept so much at variance with the cast of mind of any classical physicist. For them the imagination, as equivalent to the visualization of a concept, was the touchstone of the usefulness of any concept in physics, and also the fulfillment of the quest for scientific intelligibility. As Tyndall put it, writing about the ether and ether-atoms, mere numerical relations about the behavior of the ether, such as the frequency of its vibrations, were inherently inadequate to satisfy the desires of human understanding. According to Tyndall, the human mind, owing to its imaginative power, could not stop short of postulating the very existence of the ether. To the question, "Why do we accept the ether?" Tyndall had this to say: "Ask your imagination if it will accept a vibrating multiple proportion — a numerical ratio in a state of oscillation? I do not think it will. You cannot crown the edifice by this abstraction." [150]

In Tyndall's argument we have another classic example of the seemingly unexceptionable reasoning of classical physics, which for almost three centuries professed a naïve realism about notions referring to primary qualities (extension, quantity, etc.), while rejecting the same realism in reference to the secondary qualities (color, taste, etc.). The one-to-one correspondence that classical physics firmly believed to exist between its basic concepts and reality was not only the principal article of faith for classical physics, but also its principal error. A thorough revision of apparently perfect syllogisms therefore had to come in science, and with this an inevitable parting with long-cherished, nay, sacred convictions. For, to recall Mach's strong strictures, the basic procedure of classical physics rested as much on solid ground as it did on mythology, and the mythology this time was not animistic, or organismic, but of all things, mechanical. [151]

To accept this at the turn of the century was enormously difficult. Of the degree of difficulty involved hardly anyone can form a fair idea today, from the distance of two generations, without recalling almost forgotten words, which, when uttered, struck the most responsive chords in those who heard them. The year 1900 heard especially many of those grandiose statements about the absolute, final perfection of mechanistic physics. Did not Cornu voice the conviction of everyone present at the 1900 meeting of the International Congress of Physics in Paris when he declared that "the more we penetrate into the knowledge of natural phenomena, the more developed and precise is the audacious Cartesian conception of the mechanism of the universe"? [152]

Clearly this was more than a sober evaluation of the state of physics.

It was rather a sort of faith, highly plausible at best and certainly deeply ingrained in the minds of many generations, but it was also a faith that disguised itself in the mantle of "pure" reason, a faith that did not stop short of idolizing the science of mechanics, that supreme epitome of what physics was supposed to be. For there was more than a facetious touch in Boltzmann's remark that for a moment made him seem a high priest rather than a critical student of late nineteenth-century physics: "The god by whose grace the kings rule is the fundamental law of mechanics."[153] Yet, even for those who preferred to cast their lot with more enduring beliefs, the downfall of mechanism, as the last word in physics, seemed, for a while at least, to undermine the very meaning of science. This was especially clear from the reactions that followed the works of such critics of classical physics as Mach and Duhem. For all the exaggerations of their brand of positivism, they succeeded in laying bare the central fallacy of classical mechanics, which equated the results of mechanics with the only possible sound knowledge of the external world. This fallacy, as A. Rey noted long ago, was not a hypothetical proposition about the laws of physics; rather it was a dogma.[154] Since there obviously cannot be a middle ground between upholding or rejecting a dogma, scientific or otherwise, the unmasking of this fallacy meant to most only one thing: science too had failed and was no longer possible. It also meant the end of that wishful thinking for which the imperfections of physics were constantly diminishing; it also meant the end of hopes for an early possession of a scientific vision of the universe in which everything was to be crystal clear.

A rude awakening to the basic shortcomings of a grandiose and powerful scientific theory like mechanism could not, however, be lacking in some wholesome lesson for the future. The general consternation over the sudden collapse of the mechanical theory as the ultimate, self-sufficient scientific theory revealed to what extent a scientific theory could become a creed. Why a theory should ever grow into a creed is hard to answer. For such a development is hardly in keeping with the spirit of scientific inquiry, and to give it at least a thin veneer of reasonability rather stringent conditions must needs obtain. The theory must be able to claim not only impressive successes on the experimental level, but it also must be free of serious difficulties. Mechanism had successes to an astounding degree, but at the same time it carried along through its three-century lifespan serious questions as to the soundness of its basic assumptions. Mechanical theory was at its purest a kinetic theory based on the concepts of mass and motion. Moving masses were supposed to act upon one another by impact, and their ultimate parts were believed to be both absolutely hard and perfectly elastic. They were so hard, said Newton, "as never to wear out or break in pieces." On the other hand, to quote Newton again, colliding bodies of equal mass and with equal but opposite velocities would stop where they met "unless they be elastic and receive new motion from their spring." [155] Such explanations clearly implied conceptual dilemmas that classical physics never succeeded in resolving. The various solutions proposed from the time of Leibniz either pushed the problem a stage further back by an indefinite multiplication of the "ultimate" material medium or sublimated matter, as was the case with Boscovich's atom, into a point of space where one could hardly imagine any mechanism at all. To entertain the idea that nature might express itself through the paradoxical unity of irreducible aspects was for classical physics almost impossible to believe. "Paradoxes," as Kelvin voiced the convictions of classical physics, "have no place in science. Their removal is the substitution of true for false statements and thoughts." [156] It was, however, nothing short of paradoxical when within the span of a few years Kelvin characterized Boscovich's atom as "obsolete" (1884), "infinitely improbable" (1893), and finally "reinstated as guide" (1900). [157]

Apparently handling die-hard paradoxes demanded more than rhetoric. They lurked stubbornly in the background whenever classical physics tried to fathom the microscopic constitution of matter or when it tried to construct theories, such as the various kinetic theories of gravitation, ether, or electricity, in which the ultimate particles of matter were assigned definite, macroscopic properties. It was not, or at least should not have been, a secret to anyone who pondered over this problem that the mechanical theories were gravely contradictory. Still, by and large, classical physicists persisted in speaking of their subject as being suffused only by clear intelligibility and tried to find solutions by emphasizing one aspect of a complex problem at the expense of the other. This single-mindedness was clearly evident in the stiff manner, for instance, in which wave and corpuscular theories of light were set against one another from Descartes to Huygens and Newton and well into the nineteenth century. It is almost amusing to recall how often either one or the other theory was declared to be definitively and finally disproved. For there could be no truce, no compromise between conflicting concepts like waves and corpuscles, because it was of the very essence of mechanism that conceptual explanations must reflect the unitary mode of existence of the real world, which was taken to be mechanical.

At this point mechanism clearly overreached itself by resolutely asserting a specific way for the existence of the universe that was far from being proved in the strict sense of the word. In fact, as soon as physics reached the layer of atoms, the evidence against the rigid, unitary concept of nature formed by mechanism began to grow by leaps and bounds. There had to come the recognition of the equal usefulness of such conflicting concepts as are waves and particles. But this was not the worst shock in store for the mechanistic creed. The wave and particle dualism shed further light on the fact that the absolute determinism and precision of which classical physics professed to know so much was not only unattainable but could not even be demonstrated to exist in nature.

That absolute precision in measurements, on which mechanical determinism rested, was unattainable in practice too, was something the classical physicists should have recognized long before Einstein gave his explanation of the Brownian movement. But classical physics had long succumbed to what Einstein called "dogmatic rigidity." [158] It was beset with the shortsightedness that the mechanical creed, in its inflexible attachment to a particular concept of the physical world, almost inevitably produced. For mechanism was a scientific program nothing short of a creed. Throughout its three-century lifespan it drew on a consciousness which believed that, armed with "the infallible demonstrations of Mechanicks," it "must lay a new foundation of a more magnificent Philosophy never to be overthrown." [159]

Yet, not even three hundred years of magnificent scientific effort could secure this type of conceptual stability. Nature turned out to be anything but a precision machine in the common sense, for below a certain level of magnitude the small parts of the "machine" showed no sharp edges: the fuzziness of indeterminacy enveloped everything. Nothing could remedy this situation, not even the superior intelligence of which Laplace spoke. This spirit could see the world only as a machine, and nature was more than that. The failure of this spirit, and of mechanical physics, whose symbol it was, lay precisely in the circumstance that it pretended to know everything possible about the world on a scientific level. For at the risk of being repetitious, it should be pointed out once more that from Galileo's time to Kelvin's mechanistic physics was not proposed as *a* physical theory but as *the* physical theory. It was as such that classical physics superseded organismic physics, tried to rule philosophy, and influenced even sociology and politics. In physics proper it was again as *the* theory that mechanism produced a peculiar stiffening in the thinking of classical physicists, drew rather narrow boundaries for physics, remained conspicuously unaware of its own limitations, and stayed insensitive to signs that hinted of realms of physics lying beyond.

To say this is in no way to minimize the extraordinary contributions of classical physics to human knowledge and civilization. For all its incompleteness it was a wondrous and highly productive achievement. But like its predecessor, organismic physics, it was a *state of mind* as well. Its beliefs, triumphs, paradoxes, and failures all bear witness to this. So does its sudden demise, which left a whole generation wondering about a long array of "final conclusions," "demonstrated principles," "self-evident truths," and "absolute certainties" that were only imperfect reflections of nature in her true reality.

CHAPTER THREE

The World as a Pattern of Numbers

AT THE 1913 MEETING of the British Association when Lord Rayleigh was pressed for his opinion of Bohr's atom model — the scientific sensation of the day — he replied: "Men over seventy should not be hasty in expressing opinions on new theories." Actually he was reluctant to approve the quantization rules proposed by Bohr, for he could not bring himself to believe "that Nature behaved in that way" and admitted that he had "difficulty in accepting it as a picture of what actually takes place."[1] That a mind like Rayleigh's, so accustomed to the visual clarity of mechanical models, was unable to give its assent is understandable. Bohr's theory appeared too daring even for the man who so effectively steered physics from visual patterns toward the abstract realm of a four-dimensional world. Upon learning from Hevesy about the agreement of the Fowler spectrum and Bohr's theory, "The big eyes of Einstein," so goes Hevesy's letter of October 14, 1913, to Rutherford, "looked still bigger and he told me: 'Then it is one of the greatest discoveries.'"[2] As the letter informs us, Einstein confided to Hevesy that some time before Bohr's paper appeared in the July, 1913, issue of the *Philosophical Magazine* he had had ideas similar to those of Bohr, but had not had the courage to publish them because of their extreme novelty. Another pioneer of modern physics, Rutherford, who so often startled the scientific world with his discoveries and ideas, was no less puzzled when he learned about the theory a few months before its publication. For Bohr's theory left untouched the question of what determined the frequency of the spectral line that was emitted when the electron passed from one stationary state to another. In Rutherford's estimation this was such a serious flaw that he was prompted to write Bohr: "It seems to me that you would have to assume that the electron knows beforehand where it is going to stop."[3]

Daring as Rutherford was in experimental research, he had little taste for theoretical physics, let alone for the paradoxes of which the early forms of quantum theory had a generous share. In all fairness to him, he was no exception in this regard among British physicists of the prewar period. When Bohr's theory appeared, with the quantum of energy as its foundation, quantum theory was still, as Eddington once recalled, "a German invention," which had aroused hardly any interest in Britain. Expert or not in quantum theory, Rutherford, however, easily realized that the Bohr theory provided no mechanism to account for the "jumps" of electrons within the atom from one orbit to another. In 1913 nothing yet had been suspected of the matter and wave dualism or of the principles of complementarity and correspondence that at last tied together in one consistent system all the major puzzles raised by quantum theory.

What then was the irresistible attractiveness of Bohr's theory, which postulated that an orbiting, or accelerated, electron should not radiate energy — an idea so at variance with the very soul of classical electromagnetic theory? First, it was in excellent agreement with the data of the spectrum of hydrogen. Second, it provided a derivation of the empirical formulas and constants worked out previously, such as the Balmer formula and the Rydberg constant. Bohr's theory indeed gave the first glimpse into a problem that had challenged the energies of physicists for over half a century. When the first real crack was finally made in the mystery of atom, the effect among physicists was overpowering. A most impressive account of it was given by N. R. Campbell who, writing from a distance of twenty years, recalled how in the summer of 1913 a copy of the *Philosophical Magazine* had fallen from his bookcase and lay open on the floor.

Some algebraic formulae caught my eye. . . . It was part of a paper by a Mr. N. Bohr of whom I had never heard. . . . I sat down and began to read. In half and [*sic*] hour I was in a state of excitement and ecstasy, such as I have never experienced before or since in my scientific career. I had just finished a year's work revising a book on *Modern Electrical Theory*. These few pages made everything I had written entirely obsolete. That was a little annoying no doubt; but the annoyance was nothing to the thrill of a new revelation, such as must have inspired Keats' most famous sonnet. And I had so nearly missed the joy of discovering this work for myself and rushing up to the laboratory to be the first to tell everyone else about it! Twenty years have not damped my enthusiasm.[4]

For a theory to make such an impact it must have had more than excellent agreement with experimental data. It also had to have some deeper significance, and this the Bohr theory did indeed possess. For shortly after Moseley published the results of his X-ray experiments that for the first time clearly showed the physical meaning of the atomic number, an explanation for it was already at hand in the Bohr theory. What in Mendeleev's periodic table was just a "number" turned out to be the most important characteristic of any element: the principal factor determining the chemical properties of the element in terms of its positive charge. Furthermore, just as Mendeleev predicted the existence of unknown elements, Moseley's diagrams indicated that elements were missing at Z-values of 43, 61, 72, 75, 85, and 87. Of these the element with Z-value of 72 (hafnium) was discovered by Coster and Hevesy in 1923, and the others followed in due time. Moseley also could show that contrary to the sequence given in Mendeleev's table potassium's atomic number was 19, whereas argon's was 18. Moseley's integers also proved very effective in distinguishing and ordering the rare-earth elements that had presented until that time a most tantalizing problem to chemists everywhere.

In contrast to the atomic weights, the atomic numbers were integers and followed a straight sequence at that. This was a point of crucial importance giving as it did a powerful boost to the age-old conviction that the world is after all made up of numbers. Bohr himself saw in such an interpretation of the atomic number "an important step towards the solution of a problem which for a long time has been one of the boldest dreams of natural science, namely, to build up an understanding of the regularities of nature upon the consideration of pure numbers."[5] It is indeed worth noting that those who most improved Bohr's theory readily espoused this belief in the fundamental role of pure numbers in the structure of the physical world. The leader in this respect was Sommerfeld and of his motivations Pauli had this to say in his Nobel Prize acceptance speech in 1946: "Sommerfeld . . . preferred . . . a direct interpretation, as independent of models as possible, of the laws of spectra in terms of integral numbers, following as Kepler once did in his investigation of the planetary system, an inner feeling for harmony."[6] What Pauli said of Sommerfeld's approach to physics was in fact the trademark of modern physics, which conceives of the universe not as a mechanism but as a pattern in numbers, or mathematical construct. One might add, a marvelous construct, bordering on the miraculous. In the twentieth-century physicists it produced a state of mind that reflected unmistakably the awe and admiration due that newly found miracle present in the power of numbers. Einstein for one never stopped wondering how the insecure and contradictory data could enable even a mind of such brilliance as Bohr's to discover the fundamental laws of the spectral lines and electron shells, together with their significance for chemistry. This, he said in 1948 in his Autobiographical Notes, "appeared to me like a miracle — and appears to me as a miracle even today. This is the highest form of musicality in the sphere of thought."[7]

A miraculous feat in physics usually makes an old, complicated puzzle look alarmingly simple, but it may also help one to minimize the true proportions of problems still unsolved. Yet, the early enthusiasm of Bohr and Sommerfeld and their colleagues was no doubt far more justified than the sanguine hopes that animated the pioneer figures of classical physics three centuries ago. Then, too, the belief in the magic correspondence between mathematics and reality ran high. It was the sudden blossoming of an ancient trend of thought that went back as far as Pythagoras or perhaps the most ancient account of Egyptian arithmetic, the *Ahmes Papyrus*, which bears the revealing title *Direction for attaining a knowledge of all secret things.*

Of course not all the "secrets" mathematics was supposed to wrest from nature had scientific merit. This holds not only for the role of numbers in magic but also for the many fanciful speculations like the one recorded by Plato on planetary distances. According to the well-known passage from the *Timaeus*, the demiurge split the space between the earth and the sphere of fixed stars "in six places into seven unequal circles, severally corresponding with the double and triple intervals; of each of which there were three."[8] In viewing with favor such conjectures, Plato only attested his determination to salvage as much as possible from the Pythagorean heritage, although as shown in the first chapter he championed a view of nature that was based more on the concept of organism than on that of pure numbers. It is well to remember, however, that the enthusiasm for a concept of the world built on numbers was not without some merit. With some justification one could see in the numbers the basis for the belief in intelligibility and truthfulness in general, and this belief even weathered such a grave crisis as the unexpected discovery of irrational numbers. This discovery meant that contrary to the original hopes of the Pythagoreans unit numbers could not account for even the most elementary geometrical construct, the right-angled triangle with unit sides. The length of the hypotenuse of such a triangle was not in the realm of rational numbers. Yet Plato, as shown in chapter four, engaged in lengthy speculations aimed at explaining the various

properties of the four elements by identifying them with four of his
perfect geometrical bodies. This was all the more noteworthy, since
by that time the historic option had already been made by the Greeks
in favor of an organismic explanation of the world. Plato's continued
"flirtation" with numbers therefore clearly indicated that the Pythagorean view of the cosmos would survive through the long centuries
dominated by an organismic concept of the world. Thus the Neoplatonists resolutely reaffirmed the Pythagorean conviction that
mathematics had a decisive heuristic value as regards man's search
for the patterns of the physical world. In the early fourth century
Iamblichus credited the Pythagoreans with making "mathematics a
principle for all that can be observed in the cosmos."[9] Defining
mathematics as the "prognostic science of nature," Iamblichus stated
that the search for causes, or the causal approach to nature, consisted
"in positing mathematical things as causes" from which the objects
of the perceptible world arise. His was the Pythagorean belief that
only what was possible in mathematics was possible in the structure
of nature, and nothing could exist that implied a mathematical impossibility. He formulated a program that had a ring strongly reminiscent of some aspirations of twentieth-century physics: "I believe
we can attack mathematically everything in nature and in the world
of change."[10]

I say the twentieth century for much as physicists of earlier times
praised mathematics, they saw it only as the key to the ordering of
the facts of nature, not as the source of the facts of nature. The practice of classical physicists essentially conformed to the Baconian
dictum according to which "inquiries into nature have the best result
when they begin with physics and end in mathematics."[11] Also, their
main idea on the relation of physics and mathematics was happily
anticipated by Lord Verulam, who insisted that mathematics "ought
only to give definiteness to natural philosophy, not to generate or
give it birth."[12] In fact, not even the pre-Baconian representatives of
the crucial role of mathematics in the investigation of nature —
Archimedes, Grosseteste, and Leonardo — sided with the radical Pythagorean view when singing the praises of mathematics. Much less
could a physicist take a thoroughly Pythagorean stance as the experimental method began to assert itself from the seventeenth century on.

Yet, even in those days praises of mathematics apparently indicating the contrary were not lacking. And those praises at times went
so far as to declare that the laws of mathematics represented a direct
glimpse into God's thought. "Our knowledge [of numbers and quan-
tities] is of the same kind as God's," asserted Kepler, "at least insofar
as we can understand something of it in this mortal life."[13] Young

Kepler's faith in the power of mathematics and its significance for man was only strengthened by the difficulties his scientific program faced as the years went by. In his last major work, the *Harmonice mundi,* Kepler in a sense equated God and geometry. He spoke of geometry as having existed before the creation of the world, as being coeternal with the divine Mind. And since everything in God had to be God Himself, Kepler concluded that "geometry is God Himself." Geometry, as he put it, "supplied God with a model for the creation of the world and was implanted into human nature along with God's image and not through man's visual perception and experience."[14] If geometry served God as the mold by which to shape the world, why should not man on his part ascribe an immediate heuristic value to mathematics? The author of *Harmonice mundi* had learned, however, that only observations and the painstaking sifting of data could lead to the actual form of geometry embodied in nature.

A similar reluctance to embrace unreservedly the Pythagorean program can be observed even in Descartes, who was otherwise known for his a priori bent of mind. Not that he failed to utter grandiose statements about mathematics — to which his contribution was indeed crucial. Like Kepler, Descartes too attached a divine dignity to the mathematics he knew and felt that one could use interchangeably the terms "God" and "mathematical order of nature." In writing to Mersenne, Descartes flatly rejected the charges of a Mr. Argues that he had abandoned geometry in physics. "If it pleases him," retorted Descartes, "to consider what I have written of the salt, snow, and rainbow, he will recognize that all my physics is nothing but geometry."[15] Although his discussion of the properties of the salt, snow, and countless other phenomena hardly bore out this claim, there is no question that Descartes liked to claim that he accepted "no principles in physics that are not also accepted in mathematics."[16] He believed that he had furnished the only possible explanation of the world, and with the aid of his mathematics at that. "As for physics," he insisted to Mersenne, "I should think I knew nothing about it if I could only say how things may be without demonstrating that they cannot be otherwise; for having reduced physics to the laws of mathematics, I know it is possible, and I believe I can do it for all the little knowledge I believe I have."[17] In fairness to him it should, however, be noted that in stating this he meant only the basic laws of physics. As to the countless small mathematical patterns in the physical world, he willingly admitted that they could be ascertained only through a long series of experiments. It was his intention to carry out this program and there his complete confidence in possessing the final scientific truth asserted itself anew. He called on men of science everywhere to communicate their results to him, considering himself the only man capable of giving those results a correct interpretation.[18]

Although Galileo had no such pretensions, he voiced his astonishment time and again on seeing how closely natural processes follow the patterns of geometry. That he attributed more geometrical patterns to nature than he could demonstrate worried him little. He blamed the discrepancies between mathematically expressed laws of physics and actual observations upon the shortcomings of the calculator, who was unable to eliminate all the "material hindrances" present in physical phenomena.[19] For him there existed a perfect one-to-one correspondence between the abstract world of geometry and the real world of things. As he put it, the computations and ratios made in abstract numbers had to "correspond to concrete gold and silver coins and merchandise."[20] His famous law defining the distance traveled by falling bodies as a function of the square of the time of fall rested more on geometry than on actual experiments. More often than not he was a geometer rather than an experimentalist and made no secret of his admiration for the Pythagoreans. His was a robust confidence that all truths incorporated in the universe — that great book of true philosophy — were written in the language of mathematics, in characters that were "triangles, circles, and other geometric figures, without which it is humanly impossible to understand a single word of it; without these, one wanders about in a dark labyrinth."[21] Sagredo's statement in the *Dialogue* represented only a variation of such scientific methodology: ". . . trying to deal with physical problems without geometry is attempting the impossible."[22] This was why Galileo considered that "the many new and sound observations" of Gilbert did not yield rigorous conclusions, because, as he put it, Gilbert, not being enough of a mathematician, did not rely on "necessary and eternal scientific conclusions," mathematics, that is.[23]

Yet, the mathematics and especially the geometry that the scientists of Galileo's time held in such high esteem was not considered by them a free creation of mind but rather a pattern to be learned from observation of the actual contours of nature. It was agreed on all hands that only the pattern of nature could give rise in man's mind to the formulation of the Euclidean geometry, which in turn was taken as the ultimate, definitive expression of the true features of the physical universe. Yet, familiarity with the propositions of geometry could not fail to be very effective in strengthening the belief that the world in its entirety and in its parts was something in which the basic forms of geometry appeared most forcefully: a machine. It is important to note, however, that the seventeenth-century physicists did not attempt to derive the actual machine from geometry. In their speculations one invariably recognizes the heuristic priority given to the concrete world machine over the abstract system of geometrical patterns. Newton, as was shown in connection with the mechanistic concept of physics, could hardly have been more explicit on this point. After all, no one in his time surpassed him in his unconditional respect for the data of experiments and observations. Even Leibniz, who once said that experiments "do nothing," recognized in the same context that "nothing exists in nature otherwise than mechanically." [24]

That mathematics had to play a subordinate though vital role in the methodology of the just-emerging classical physics was also intimated in a particular difficulty that must have been obvious then to anyone bent on the mathematicization or geometrization of the physical universe. Thus Kepler and Galileo recognized that the geometrical harmony of the world very often eluded the available mathematical techniques. Furthermore they could hardly ignore that nature's taste concerning various geometrical figures was not patterned after human preferences and fashions. The circle, for instance, did not turn out to represent the harmony in nature in a higher measure than did the ellipse. This was one of Kepler's great lessons. As for Galileo, it was precisely his obsession with the absolute perfection of the circle that prevented him not only from recognizing the value of Kepler's discovery of the elliptical orbits of planetary motions but also from formulating in full Newton's first law.

In its dialogue with classical physics mathematics was a splendid tool but not a divining rod. Well into the nineteenth century mathematics remained by and large an outgrowth of speculations about problems of physics. Physical phenomena of all kinds, especially that of accelerated motion, were chiefly instrumental in stimulating the development of the backbone of classical analysis, the infinitesimal calculus. This is particularly evident in the mathematical work of Newton and Leibniz, but is no less clear in the huge volumes of Euler. What is more, for a long time this higher mathematics lacked inner consistency, and its acceptance rested mainly on the successes it had had in coping with the actual problems of physics. Throughout the eighteenth century the cornerstone of calculus — the concept of limit — under rigorous scrutiny appeared a contradiction in terms, and this gave rise to a long chain of caustic remarks and invective.

In the hands of Bishop Berkeley the difficulty served as a potent weapon to chastise the cocksure confidence of some mathematicians and physicists in the infallible effectiveness of their newly devised tool. In the *Analyst*, or a *Discourse Addressed to an Infidel Mathematician* (1734), Berkeley raised the question: "Do not mathematicians submit to authority, take things upon trust, and believe points inconceivable? Have they not their mysteries, and what is more, their repugnancies and contradictions?"[25] They submitted indeed, and to some this was tantamount to oppression of the rational thought that mathematics was supposed to promote. Berkeley in fact felt called upon to pen his *Defence of Free Thinking in Mathematics* (1735) in which he deplored the signs of authoritarianism in mathematics and blamed the unjudicious admirers of Newton for it. Some of these went so far as to claim that it was a crime even to think that man would be able to "see further or go beyond Sir Isaac Newton."[26] The time was indeed not far away when G. Horne would charge with some justification that "every child imbibes almost with his mother's milk that Sir Isaac Newton has carried philosophy to the highest pitch it is capable of being carried, and established a system of physics upon the solid basis of mathematical demonstration."[27] Yet, for all the marvels of the superstructure, or physics, the ultimate mathematical foundations were far from solid. To stop inquiring into these difficulties was in Berkeley's words "to fix a ne plus ultra," and for him this was tantamount "to converting the republick of letters into an absolute monarchy . . . to introducing a kind of philosophic popery among a free people."[28] Berkeley therefore felt justified in pouring ridicule on the concept of the instantaneous rate of change of functions as being neither a finite quantity, nor a quantity infinitely small, nor yet nothing. He called these rates of change "ghosts of departed quantities" and was eager to remind the "philomathematical infidels" that "he who can digest a second or third fluxion . . . need not, methinks, be squeamish about any point in Divinity."[29]

At times even the most prominent mathematicians joined in the chorus of sarcastic criticism. Rolle, the discoverer of the mean theorem, did not stop pointing out in his lectures that calculus was a collection of ingenious fallacies. Lagrange attributed the success of calculus to the fortuitous effect of errors offsetting one another. D'Alembert used to advise students of calculus to keep on with their studies, assuring them that faith in it would eventually come to them. The successes of calculus were of course too great to permit any serious doubt about its basic correctness in spite of some equally fundamental difficulties involved in it. In final analysis it was, however, a matter of hope that a solution to those difficulties would come in time, although such a time did not arrive until 1821 when Cauchy finally succeeded in eliminating all inconsistencies from the concept of limit.[30]

Regardless of the difficulties involved in the basic concepts of calculus, all through the eighteenth century there was no doubt in

the minds of physicists about the indispensability of mathematics in
scientific work, a view that received its philosophical sanction at the
hands of Kant. In the preface to his *Metaphysical Foundations of
Natural Science* (1786), Kant took the view that while it was possible
to speculate about nature in general without using mathematics, a
genuine natural philosophy about definite things was only possible
through mathematics. And since, he added, "in every study of nature
there can be only so much genuine science as there is *a priori* knowledge, by the same token, natural philosophy will contain genuine
science only to the extent in which mathematics can be applied in
it."[31]

Kant's dictum contained some truth although his apotheosis of
mathematics was no doubt in part due to the enthusiasm of the amateur who failed to wade deep enough in the subject. Kant's expertise
in mathematics did not extend beyond elementary calculus, the basic
problems of which were a closed book to him. Yet, for all his admiration for mathematics, Kant did not attribute a heuristic value to
mathematics in a Pythagorean sense. Furthermore, no physicist of
his time would have supported such a view. That Euler at times believed blindly in the results of his mathematical analysis of certain
problems in physics was a rather exceptional symptom. Nor did the
first generation of "mathematical physicists" in the early nineteenth
century see in their successes a proof of Pythagorean views. This held
even for Fourier who kept reminding his readers in the preliminary
discourse of his *Analytical Theory of Heat* that the equations formulated by Descartes were not restricted to the properties of curves and
surfaces and to the problems of rational mechanics but extended to
all phenomena. To Fourier there was no language "more universal
and more simple, more free from errors and from obscurities, that is
to say, more worthy to express the invariable relations of natural
things." Mathematics, he said, defines all perceptible relations; it
measures times, spaces, forces, and temperatures. In the growth of
mathematics Fourier saw that unique development in which every
principle, once established, is preserved forever. Mathematics, in his
words, "grows and strengthens itself incessantly in the midst of the
many variations and errors of the human mind."[32] Yet, for all that,
for him mathematics was the medium by which to express the correlation of facts and not the source of those correlations. He had, in
fact, so little in common with a Pythagorean attitude toward the
physical world that he criticized Abel and Jacobi for not being primarily interested in the actual problems of physics, such as the convection of heat. This prompted Jacobi's remark to Legendre in 1830:
"It is true that M. Fourier was of the opinion that the chief end of

mathematics was public utility and the explanation of natural phenomena; but a philosopher like him ought to have known that the sole end of science is the honor of the human intellect and that under this head a problem of number is as important as a problem of the system of the world."[33] At this time, however, not every mathematician, let alone the physicists, would have agreed with Jacobi on this point.

Powerful and sweeping as was mathematical analysis after Lagrange's and Fourier's contributions, it was not exactly what Pythagoras or the young Kepler had dreamed of when reflecting on the heuristic value of numbers. Not that classical analysis had failed to alert physicists to a fruitful search for various phenomena unsuspected beforehand. Thus the first hint of the existence of "conical refraction" came when Hamilton submitted the phenomenon of double refraction to a thorough mathematical discussion. His results indicated that if the incident beam of light were to fall under a certain angle on the face of a crystal it should spread out into a cone upon entering the crystal. Just as surprising was Poisson's prediction that on the basis of Fresnel's undulatory theory of light the shadow of a small disk was equivalent to a bright patch surrounded by dark rings. Again, it was mathematics in the hands of Maxwell that predicted that the viscosity of a given gas at a fixed temperature was independent of the pressure.

Yet, for all these and many other examples of the heuristic value of classical analysis, few if any solutions indicated a pattern in integral numbers within the pale of phenomena handled by classical physics. And no sooner were theories involving integral numbers, such as Prout's hypothesis, proposed than they were called into doubt by more precise measurements. Not that attempts were not made to discover in the various fields of physics the same role that integral numbers played, for instance, in determining the properties of vibrating systems. Such an attempt was the Bode-Titius law[34] giving the relative distances of the planets. It can be written as $4+3\times2^{n-2}$, where n takes on the values 1, 2, 3 . . . starting with $n=1$ for Mercury, the innermost planet. Although the formula breaks down for Mercury, it gave with surprising accuracy the distances of all the other planets known in 1772 and even predicted the mean distance of the asteroids and the distance of Uranus. On the other hand the formula failed utterly for Neptune and Pluto.

Just as the skies failed to display a pattern in integers, the periodic table of the elements did not support the hopes of those who expected to stumble on integers in the deeper layers of matter. The atomic weights were not integers and the ordinal number of elements, an integer — known later as the atomic number — was for several decades without any physical significance. When on occasion a formula was devised that involved a sequence of integers and that proved very useful at the same time, it rested first on mathematical wizardry rather than on any insight into the underlying physical process. Such was the case with Balmer's celebrated formula. Its author, a schoolteacher in Basel and an accomplished mathematician and a student of architecture, was possessed of a mind that saw everywhere in the world some manifestation of an exquisitely structured building where numerical relations prevailed throughout.[35] His *Habilitationschrift* presented to the University of Basel in 1865 carried the title that in itself is highly revealing of Balmer's bent of mind: "The Prophet Ezekiel's vision of the Temple broadly described and architectonically explained." In it Balmer tried nothing more than to disclose the secret numerical relations involved in the size and proportions of the temple as given by Ezekiel. Balmer made similar studies of medieval cathedrals and other ancient buildings, using to good advantage his specialty in mathematics: projective geometry. Upon learning from his friend, E. Hagenbach, a physicist, about the puzzling arrays of the spectral lines of hydrogen, the problem had almost of necessity presented itself to him as a problem in architecture, or as a problem in numerical proportions. Slowly the recondite mathematical pattern of the "house" of the hydrogen atom unfolded in his mind's eyes, and early in 1884 he notified Hagenbach that the wavelengths of H_α, H_β, and $H\gamma$ could be expressed in correlated fractions by the formula $\lambda = \lambda_0 m^2 / (m^2 - 2^2)$, $(m = 3, 4, 5 \ldots)$. The formula, first communicated to the Basler Naturforschenden Gesellschaft, was later generalized by Balmer as $\lambda = \lambda_0 m^2 / (m^2 - n^2)$ and greatly stimulated the work of such outstanding workers in spectroscopy as Kayser, Runge, Rydberg, and Ritz. But for almost thirty years the formula was not physics. Its sequences in integrals, however successful in calculating and predicting wavelengths of spectral lines, had no connection with physical reasoning of any sort.

As to the mathematical inspiration of Balmer's work — the admiration for certain interconnectedness of integers — the mathematicians of the day were rather reluctant to base their program of research on it. Kronecker was somewhat a lonely figure in contending that "God made the integers, all else is the work of man,"[36] and hardly anyone accepted as supreme the rule that he laid down, namely, that all results of mathematical analysis should be expressed ultimately in terms or properties of integers. In taking such a position, Kronecker merely hurt the feelings of many mathematicians of the era. Their indignant reaction was voiced by Weierstrass, who claimed that Kronecker declared in effect that "all those who up to now have labored to establish the theory of functions are sinners before the Lord."[37] What is more, the possible impact of Kronecker's program was in the estimation of Weierstrass simply injurious to the cause of mathematics.

Although physicists fascinated by Balmer's wizardry with integers could not count on the support of the then prevailing preferences of most mathematicians, they remained nevertheless undaunted in their efforts to find how the atomic code reflecting a pattern in integers might originate and on what physical principles it rested. Of the various suggestions made, the most seminal were those of A. W. Convay and J. W. Nicholson. In 1907 Convay[38] took the view that in order to solve the problem one must abandon the analogies taken from the production of sound by vibrating bodies. He postulated that only one electron at a time of the atom's many electrons was involved in the production of spectral lines. This was a distinct departure from the general thinking that assumed that just as in the production of sound all parts of an instrument were involved, in the same way all parts of the atom were "vibrating" when a spectral line was produced. A few years later, in 1911, Nicholson[39] first suggested that the production of spectral lines was a quantum phenomenon that should be explained on the basis of the Rutherford atom model published a few months earlier. Nicholson also stated that an atom could exist in different energy states and that its angular momentum could have discrete values.

It took Bohr's genius, however, to recognize that of the many suggestions put forward at that time these were the ones of fundamental value. Furthermore Bohr contributed three insights to the emerging picture. He discovered that the angular momenta of atoms must be integral multiples of $\hbar$; he grasped clearly the principle that two energy levels were involved in the production of a spectral line; and he came forward with a most startling idea, namely, that no attempt should be made to visualize or explain what happens to an electron when jumping from one energy level to another. This daring postulate amounted indeed to stating that in place of some mechanism, whole numbers (the various integral values of m and n) were "generating" so to speak the physical phenomena of spectral lines. At least Bohr's theory appeared thus to some of his colleagues. Sommerfeld was particularly eager to point out that while classical physics and even the relativity theory operated with infinitesimals that allowed for perfectly smooth changes on the scale of numerical values, the quantum theory worked with integers that for him represented a conceptually simpler and more precise tool for physics. Modern physics, as Sommerfeld put it, was grafting itself on arithmetic and in a sense reached back to the Pythagorean number mystique. He likened the spectral lines to the ancient triad of the lyre from which the Pythagoreans inferred the harmony of natural phenomena. But not satisfied with mere analogies, he took the view that "our quanta remind us of the role that the Pythagorean doctrine seems to have ascribed to the integers, not merely as attributes, but as the real essence of physical phenomena." [40]

The classic simplicity of Bohr's theory was, of course, soon superseded by more elaborate theorems to account for the fine structure of the hydrogen spectrum and those of other elements. All these improvements on the Bohr theory, however, had one common characteristic: the consistent use of various selection rules that kept underscoring the role of integers or of sequences of numbers in physical theory. While such sequences were prominent only in the theory of vibrating systems in classical physics, their role was much wider in modern physics. Selection rules gave the first hint of space quantization and led to the now classical experiment of Stern and Gerlach. Their heuristic power at times was nothing short of miraculous, as was clearly shown by perhaps the most important of all the selection rules in physics, the Pauli exclusion principle. This principle revealed the existence of closed electronic shells in the atomic structure of matter. It provided physical science with a definitive system that accommodated not only all the elements found in nature but also the unfinished list of transuranic elements. What is more, it led to the prediction of such varieties of one element as the para- and ortho-types of hydrogen. It also forms the basis of the Fermi-Dirac statistics on which rests our understanding of the behavior of electrons in metals and much of semiconductor physics.

Clearly, it was only natural to expect that the heuristic value of numbers revealed in the exclusion principle would be equally effective when the experimental data gathered about the nucleus had to be systematized. In fact, the most impressive of such attempts, the nuclear shell model, stands in direct debt to the electronic shell model of the atom. For just as certain completed or closed electronic shells pointed to chemically very stable elements, the experimental facts indicated that very stable nuclei resulted when either the number of protons or the number of neutrons was equal to one of the following numbers, 2, 8, 20, 50, 82, 126. A startling result indeed. The editors of a prominent scientific journal were so taken aback by it that they declared the whole matter "not physics but only playing with numbers" and refused to publish the codiscoverer's, J. H. D. Jensen, first communication on it. Yet, the "magic numbers," as E. P. Wigner first called them, wholly vindicated themselves, although they easily created the impression, as J. R. Oppenheimer once facetiously remarked, of "explaining magic by miracles." [41]

Magic as they could appear, these numbers documented both the confidence of modern physicists in the power of integers and the aspirations and procedures of modern theoretical physics. Whatever the limits of our understanding of the nucleus, it was through such numbers that nuclear physics succeeded in establishing selection rules that made most of the advances possible in the interpretation and systematization of nuclear phenomena. Again, numbers and selection rules derived from them form the backbone of the various theories that clearly evidenced their heuristic value by predicting many of the known fundamental particles.

Selection rules are of course the direct consequence of the various quantizations dominating the atomic and subatomic layer of nature. Quantization of energy is one of the main pillars of De Broglie's matter-wave theory, which in turn pointed to a host of unsuspected phenomena in nature. A year after the formulation of De Broglie's theory (1924), Elsasser called attention to the possible existence of a diffraction effect to be exhibited by the interaction of electrons with metallic lattices. Actually, ever since Davisson and Germer had stumbled on such an effect, the wave aspect of matter evidenced itself in a countless number of physical experiments and engineering applications. Electron microscopes work on the principle of matter waves as do the accelerating machines of particle physics.

Undoubtedly the astounding heuristic value of mathematics is best evidenced in modern physics in the various branches of quantum physics. One should remember, however, that relativity theory, which aims at a fundamental mathematical simplicity, bears no less witness to the heuristic value of mathematics in physics. Illustrations are the mathematical roots of the mass-energy equivalence in the Special Theory of Relativity and the geometrical formalism in the General Relativity that ultimately led to intense experimental work on the detection of the gravitational red shift and the bending of light in strong gravitational fields. How physicists looked upon such achievements of mathematical physics was well expressed by H. Weyl, one of the most competent interpreters of the theory of relativity. "Our ears have caught," he stated in the closing sentence of *Space, Time and Matter*, "a few of the fundamental chords from the harmony of the spheres of which Pythagoras and Kepler once dreamed." [42]

Again, Dirac's aim was purely mathematical when he started his now historic search for a satisfactory form of the Schrödinger equation from the viewpoint of relativity. Yet the final result that came to be known as Dirac's equations of the electron, contained something most momentous as regards physical reality. In those equations man had the first hints about the existence of the world of antimatter. This, however, is far from all that can be said about the extraordinary role of mathematics and specifically of numbers in modern physics. The various selection rules could point not only to unsuspected phenomena, but could also provide indirectly some understanding of the universal constants of nature, which turned out to be composed of quantized units of mass, energy, momentum, and electric charge. In classical physics the magnitude of the various constants depended on the arbitrary choice of the units of measurement, whereas in modern physics their numerical values show a close interconnectedness. This interconnectedness is very significant as it leads to numerical relations that stand at the basis of such ambitious theories as, for instance, Eddington's posthumously published *Fundamental Theory*.

Whittaker, who saw Eddington's manuscript through the press, called him the "modern Archimedes"[43] who without assuming any number determined experimentally, deduced quantitative propositions of physics, just as Archimedes derived the value of the number π. Only a few followed Whittaker in his unreserved admiration for Eddington's work. Yet, even those who scoffed at the *Fundamental Theory* or declared it incomprehensible readily admitted that it was a work extremely characteristic of the expectations prompted by the "numerical wonders" found in nature by modern physics. With Eddington these expectations manifested themselves also in his comportment. In cloakrooms he showed a distinct preference for the peg marked with the number 137. For deep in his heart Eddington was a Pythagorean who liked, when opportunity arose, to describe the universe as a symphony played on seven fundamental constants corresponding to the seven notes of the musical scale. Pythagorean was also his confidence in the soundness of his approach to nature. In 1943, a few years before his death, he admitted in his lectures at the Dublin Institute for Advanced Studies that he had never had the smallest doubt about what later became known as the *Fundamental Theory*, since he first outlined it in 1928. Again, in the same vein, he was not reluctant to voice complete confidence in his *Fundamental Theory* when a perceptive reader of its manuscript pointed out several obscure transitions in the mathematical sections. Although admitting that the proofs were not quite clear even to himself, he nevertheless felt that the results were correct.[44]

These details are not of course to suggest that Eddington was steeped in numerology, as were the Pythagoreans of old. There was nothing primitive or naïve in his state of mind that in some respect harked back to Pythagorean views. His method of giving what he hoped was a definitive account of the values of the fundamental constants of nature was only partially mathematical. It did not look so much for a basic mathematical theory as it relied on the analysis of the essential conceptual patterns implicit in the notion of physical measurement. At any rate, Eddington did not profess apriorism as overtly as had young Kepler, who in the estimation of his teacher, Maestlin, freed astronomers "from the necessity of exploring the dimensions of the spheres a posteriori . . . because now the dimensions have been established a priori."[45] Yet, Eddington hoped to show that there was only one possible world composed of 10^{79} protons although he rested his case, unlike Kepler, not on geometry but on mathematical philosophy. Eddington's apriorism had more in common with Descartes' approach to nature. In fact he viewed the Cartesian motto, "Give me matter and motion and I will construct the universe," as the reverse side of his axiom, "Give me a world in which there are relations and I will construct matter and motion."[46] Descartes, for one, held that his three basic philosophical rules permitted "a priori demonstrations of everything that can be produced in the world."[47] Descartes should have stated "of almost everything" so as to remain consistent with his other statements. Strikingly similar was Eddington's position; he held that "there is nothing in the whole system of laws of physics that cannot be deduced unambiguously from epistemological considerations." His contention was that an intelligence unacquainted with our universe but familiar with the ways in which the human mind interpreted the wealth of sensory experience "should be able to attain all the knowledge of physics that we have attained by experiment." Yet, in a characteristic proviso harking back to that of Descartes, he added that all this knowledge of physics did not include the particular events of the physicist's experience but only the generalizations based on them. He believed that the magnitude of the fundamental constants of nature could be calculated without reference to particular measurements made in the laboratory. For him these numerical magnitudes were the deductive consequences of the general principles of metrology as he understood them. Consequently he could claim that it was possible to derive from epistemological considerations "the existence and properties of radium but not the dimensions of the Earth."[48]

It was not only a priori trends that Eddington's *Fundamental Theory* resuscitated. His work also evidenced the persistent inadequacy of such an approach to the study of the physical world. By the time of its publication, the *Fundamental Theory* lagged far behind many new facts known to physics. Moreover, some of those new facts were simply unpredictable and even bewildering for the theoretical physicist. To make the irony of nature complete, these facts came from the realm of the nucleus that in Eddington's view was not supposed to have any surprises in store for physics if his conceptions were to prove themselves. The strange barrenness of a priori systematizations of physics was also evidenced by the fact that unlike other outstanding physical theories, such as Maxwell's electromagnetic theory or Einstein's Theory of General Relativity, Eddington's work did not point to new phenomena to be verified by experimental research. Again, in keeping with the characteristic inconsistencies of a priori physical theories, Eddington did not live up to his contention that his otherwise impressive results were deduced solely from propositions wholly independent of the inductively based propositions of relativity and quantum theory. Yet, for all the failures of the a priori approach, Eddington's effort illustrated that modern physics thoroughly oriented along mathematical lines could hardly avoid the task of accounting for the puzzling numerical connections implied by the values of fundamental constants.

One of Eddington's contentions was to show that the reciprocal of the fine structure constant of Sommerfeld must be exactly 137 and not 137.038, the value derived from measurements. That this constant points to something very fundamental in nature can hardly be doubted. For if the velocity of light and the Planck constant are expressed in terms of the natural units that make them equal to unity, the electrical repulsion between two electrons will assume the form of a/r^2, where a is a pure dimensionless number whose reciprocal equals very nearly 137. A comprehensive theory should be able to account for this fact in much the same way as geometry — Archimedes to be specific — could show that the value of π can be found to any desired degree of accuracy by pure theory without any recourse to measurements.

A similar challenge to modern physics is presented by the groups of pure numbers obtained by the combinations of the various constants of physics. For instance, the ratio of the masses of proton and electron is about 1,900. The ratio of the electrical and gravitational forces between the two is of the order of 10^{39}. If the total mass of the universe calculated on the basis of Einstein's world model is divided by the mass of the proton, one obtains a third pure number, 10^{79}. The peculiar thing is that many pure numbers obtained in a similar way fall into one of these three orders of magnitude. That some fundamental aspect of the physical world might be hidden in such groupings of numbers derived from the values of physical constants was first hinted at in 1919 by H. Weyl, who even said three decades later that "the construction of the world seems to be based on two pure numbers, a and ϵ, whose mystery we have not yet penetrated."[49]

By the late 1940's, when Weyl submitted this view, the pure numbers derived from the basic constants of physics did indeed offer a well-argued possibility for an interpretation of the physical world tying into one not only the atomic and stellar realms but also the static and dynamic aspects of the universe. The six constants that form the basis of such a theory are the velocity of light (c), the gravitational constant (G), the age of the universe (t), the mean density of matter (ρ), the radius of the universe (R), and the rate of recession of the nebulae, or Hubble's constant (k). These constants can be multiplied or divided with one another only in three ways as to yield a pure dimensionless number. These simple relations are R/ct, kt, and ρGc^2t^2. What is nothing short of a numerological miracle, however, is that the value of these three formulae is very nearly equal to unity. Furthermore, if one assumes, in agreement with the basic postulate of relativity, that the value of c is independent of time, and that the number and diameter of fundamental particles remain constant, it follows that the gravitational force should weaken as time goes on. Another startling consequence is that the rate of the electrical repulsion and gravitational attraction between two protons had to be unity at a time when the universe was 10^{-23} seconds old, or about 10^{10} years ago, which is the estimated age of the universe. What is more, one may readily infer from all this that time is not continuous, but consists rather of fundamental units, or quanta, called chronons.

This is not the place to discuss at length this topic so indicative of the Pythagorean fascination that animates modern physics. Still, the few details given above should explain why a most articulate spokesman of the ideas and trends of modern physics, H. Margenau, was prompted not long ago to register the unabated effectiveness in modern physics of the Pythagorean view, in which the world is a construct in numbers. Distrustful as modern science might be of mysticism, he noted, "it stands in unreasoning awe before simple numbers."[50]

Many other examples of this fascination with pure numbers in the various areas of modern physics could be cited. Thus the various numerical data connected with fundamental particles stimulated an animated search for that basic algebraic scheme that might underlie the subatomic structure of matter. It was an "empirical Balmer's law" that Y. Nambu looked for when he observed that the masses of fundamental particles, when expressed in units of 137 electron masses, form a sequence of approximate integers and half integers.[51] Just as characteristic of the modern mathematical concept of physics was the system of fundamental particles proposed by J. Grebe in 1958.[52] It is based on two numbers given by the ratios of the masses of μ and π mesons and of the masses of the proton and Σ hyperon. Each of these ratios equal $\pi/4$. The inverse square root of π divided by four in turn yields a new constant, 1.12888, called g, whose successive exponential values from g^0 on provide a set of symmetrical relations. Adding to these the dimensions of relativistic mass, the results give the values for the masses of twenty-eight elementary particles. As Grebe put it, his scheme might be a completely wrong picture; yet, it must be admitted that some "ingenious juggling" with numbers is at times indispensable in the framework of a scientific outlook that thinks of the world as being neither an organism nor a huge machine, but rather as the embodiment of a mathematical construct. In such a context the goal to find "meaning" in the foregoing sequence or set of numbers is not motivated by a hope of arriving at a neat pictorial representation of the building blocks of the universe. What is looked for is a mathematical law that "generates" this sequence and that at the same time suggests further experiments and predicts observable results. As Whittaker put it, "This prediction is the sole function and capability of science. Mechanicism has been replaced by a pan-mathematical conception of the universe."[53]

Such is not the case on the atomic or nuclear level alone. General Relativity, which deals so successfully with the large scale aspects of the universe, "is essentially a geometrisation of physics," to quote Whittaker again.[54] In a more restricted way so is the Special Theory of Relativity. Discussing before the German Association in 1908 the space-time manifold of relativity, Minkowski could think of no better way to comfort those who found the abandonment of the traditional views on space and time too painful than to remind them "of the idea of a pre-established harmony between pure mathematics and physics"[55] so forcefully displayed by the new theory. It was again precisely this aspect in Einstein's famous paper of 1905 on the electrodynamics of moving bodies that appeared so decisive to Jeans when he remarked that "the study of the inner workings of nature passed from the engineer scientist to the mathematician."[56] For as time went on, it became increasingly evident that both relativity and wave mechanics can only draw a purely mathematical picture about nature that, let it be known, is no picture at all in the obvious visual sense of the word. Such a deep split between the visualizing inclination of man and the belief in the fundamentally mathematical structure of the universe, however, is far from being disastrous as regards the rational understanding of nature. For, as Jeans aptly put it, "nature seems very conversant with the rules of pure mathematics,"[57] with rules that are constructed without any reference to the visual features of the outside world. In fact, this harmony beween mathematics and nature is so deep-going that in a sense the understanding of the universe seems to be open only to the mathematician. This is what prompted Jeans to fancy that "the Great Architect of the Universe now begins to appear as a pure mathematician." [58]

The word pure in this context was particularly well chosen. For much of the mathematics that so well suited the needs of modern physics was worked out long before physics became cognizant of its need of various mathematical theories without which modern physics would be simply unthinkable. The case was distinctly different from classical physics, which developed by and large its own mathematics according to the needs presented by problems of physics waiting for solution. Classical calculus, as is well known, sprang from an age-old desire to handle the problem of infinitesimal changes in physical processes. Logarithms were similarly developed to facilitate trigonometrical calculations of already extant problems in astronomy. The development of the great variety of differential equations in the eighteenth century betrays at every point the concern with actual problems of mechanics. To mention only one of the numerous cases, it was the study of a bar clamped at one end that revealed to Daniel Bernoulli the existence of differential equations of the fourth order. The development of the Bessel functions, so indispensable in almost every department of mathematical physics, tells the same story. Its form with coefficients of order zero was formulated in 1732 by Bernoulli in the course of his investigations of the oscillations of heavy chains. The analysis of the vibrations of stretched membranes by Euler in 1764 led to its type with more general coefficients, and problems of perturbations in celestial motions prompted Bessel to investigate further the properties of the function that bear his name. Again, it was the isoperimetric problems of physics that gave rise to the calculus of variations. In general the development of analysis throughout the eighteenth century betrayed the same motivation, as evidenced by the works of Euler, Clairaut, and Lagrange. Their essential aim in mathematics was to find more refined tools to handle problems ranging from fluid motion to the motion of planets. It was especially these latter problems that spurred the extensive development of perturbation methods. In all this an attitude asserted itself that, as J. Bertrand put it over a hundred years ago, was misled by the successes it had in coping with the phenomena of the physical world into the false belief that "the mathematician, without being longer occupied in the elaboration of pure mathematics, could turn his thoughts exclusively to the study of natural laws." [59]

This relation between mathematics and physics remained essentially unchanged well into the opening decades of the nineteenth century, as witnessed by Gauss's theory of the method of least squares, Laplace's work on the theory of probabilities, Hamilton's work on the quaternions, and Fourier's analysis of heat convection. Mention should also be made of the development of potential theory, the theory of functions of a complex variable, functional analysis and differential geometry, the theory of line, surface, and volume integrals as some of the most notable instances of the creative role of physical problems in mathematical research. Again, it was the speculation on the ether that inspired the extensive work of Poisson, Navier, and Cauchy on the mathematical properties of wave motion. The general theory of coordinates as developed by G. Lamé was an outgrowth of concern with problems of physics. He viewed the introduction of various coordinates — rectangular, spherical, elliptical, general curvilinear — as indicators of the advancing phases of physical science.[60] Yet, in all this it was clearly mathematics that seemed to respond to the promptings of physics. Even Riemann's essay "On the hypotheses which lie at the bases of geometry" (1854), which brought to a culmination Gauss's work on the geometry of curved surfaces, had an eye on physics. As a matter of fact, Riemann insisted that purely theoretical analysis of the notion of space not only freed the study of the geometry of space "from becoming hampered by too narrow views" but also made necessary "the successive changes required by facts which it cannot explain."[61]

The assumed connection between Riemannian geometry and physics, however, could hardly have convinced Riemann's audience with perhaps the exception of Gauss. And if there were some physicists who might have worried about the lack of such a connection, the possible application in physics of a mathematical theorem was no longer the principal aim of mathematicians. Their new attitude was best exemplified by Weierstrass who lent his full authority as the leading mathematician of the age to the view that the basic aim of a science cannot be placed outside that science. Not that he stopped referring in his courses to the problems of mechanics. He even held that it was the glory of mathematics to be indispensable in physics. Yet, as he warned in his famous address of 1857, one should think of "the relation between mathematics and physics in a deeper manner than is the case when a physicist sees in mathematics only an indispensable auxiliary discipline, or when a mathematician is willing to see only a rich source of illustrations for his method in the questions posed to him by the physicist."[62] Just as these words had a prophetic ring, so did the reference of Weierstrass to the fact that the studies of conic sections by the Greeks had to wait until Kepler found their "useful application" in physical science.

Clearly, no one realized in 1857 to what extent this pattern would dominate the future relation of mathematics and physics. At that time one merely noted that mathematics was resolved to strike out on a path of its own and choose its problems regardless of whether they had a physical meaning or not. Little could one surmise that the major achievements of this "independent" mathematics would have to wait for over half a century before physics could find them "useful." Yet, such was the case with Gauss's hypergeometric functions, Hermite's polynomials, and Cayley's theory of invariance. Another major contribution of Cayley to mathematics, the theory of matrices, found its way into physics only in 1925 when Heisenberg realized that the array of spectral terms he had set up was nothing but a matrix. The group concept, which F. Klein once considered as most characteristic of nineteenth-century mathematics, proved itself decades later to be a tool of fundamental importance for quantum physics, since quantum numbers were indices characterizing representations of groups. It was in fact through the group theory that quantum mechanics could reveal, as H. Weyl put it, "its essential features which are not contingent on a special form of the dynamical laws nor on special assumptions concerning the forces involved." [63]

If one recalls that for several decades the only use made of the theory of groups consisted of the description of the symmetry of crystals, one will perhaps have an inkling of how difficult it is to assess the potentialities of a modern mathematical theory in physics. A case in point is the publication in 1924 of the first volume of the *Methoden der Mathematischen Physik* by R. Courant and D. Hilbert, which aimed at showing as completely as possible the applicability in physics of the modern methods of linear transformation, expansion of arbitrary functions, bilinear and quadratic forms, and the like. Yet the richness of the content of the book revealed itself only two years later when Schrödinger derived his now famous equation, of which neither of the authors could have had the slightest premonition in 1924. More recently an algebraic theory developed almost a hundred years ago by Sophus Lie turned up as the mathematical formalism best suited to the "eightfold way," one of the most promising systematizations of the set of fundamental particles. Today there is hardly a branch of the various "esoteric" sections of modern mathematics that has not yet found some use in modern physics. Mathematical theories, like set theory and topology, appeared when first formulated to be highly arbitrary mental constructs based on postulates that had little to do with experience and common sense. Yet, today, they are beginning to play an indispensable role in physics. In view of all this, how could one fail to sense the prophetic truth in the remark of Lobachevsky who held that there was no branch of mathematics, however abstract, that might not some day be applied to the phenomena of the real world.[64]

The way in which the classical physicist viewed the role of mathematics in physics was distinctly different. Searching everywhere for "commonsense machines" in the physical world, his mathematics also had to reflect a common sense, however refined. Exaggerated though it was, Kelvin's dictum of classical mathematical analysis was in keeping with the basic assumptions of the mechanistic concept of physics. "Do not imagine," he declared, "that mathematics is hard and crabbed and repulsive to common sense. It is merely the etherealization of common sense."[65] To most classical physicists mathematics was a very useful and commonsense tool, but hardly a device that could do on its own magic tricks for physics. Compared to the role it was to play in modern physics, the role of mathematics in classical physics was relatively modest. As Rowland summed it up very characteristically: "A mathematical investigation always obeys the law of the conservation of knowledge: we never get out more from it than we put in. The knowledge may be changed in form, it may be clearer and more exactly stated, but the total amount of the knowledge of nature given out by the investigation is the same as we started with."[66]

It must be admitted, however, that even within the framework of late nineteenth-century physics, this was not always the case. Of course the mathematics Rowland spoke of was the ordinary analysis used in nineteenth-century physics and not the "extravagant" creations mathematicians started to produce from the 1840's at an ever-increasing rate. About the "physical merit" of those extravagant theorems such a mathematically minded physicist as Gibbs felt prompted to offer this little aside: "A mathematician may say anything he pleases, but a physicist must be partially sane."[67] At the same time, however, Gibbs in his work on the thermodynamics of chemical equilibrium resorted to steps that might have induced some of his fellow physicists to class him with the "mathematicians." For it was precisely the mathematical boldness of Gibbs that welded thermodynamics and chemistry into one. The partial differential coefficients Gibbs introduced had no physically realizable notion. In the situation that he investigated, the entropy and volume of a system were supposed to remain constant while the mass of the system was changing. Such a procedure, however, is purely mathematical, for there is no experimental way of adding or subtracting mass from a system without changing its entropy. As E. A. Milne noted, Gibbs's step was the first instance of the presence in physics "of an 'unobservable' (namely a partial potential) suggested by mathematics. Once this idea has been worked into the subject, the whole of classical analytical thermodynamics follows." [68]

The work of another giant of classical physics, Maxwell, similarly contained indications about the novel role mathematics was to play in the physics of the future. Maxwell for one was deeply puzzled by the thought that while the molecules in an ordinary body had a mean velocity equal to that of a cannonball and moved through very short distances at high speed in every direction, the body as a whole still remained, as far as observation showed, in its fixed position. Reflecting on this problem, Maxwell was led to an extensive study of stable and unstable phenomena that showed him that unstable configurations in the physical world, such as a rock on a mountain top, or a match starting a forest fire, were actually flaws in the deterministic picture of physics. The conclusion he drew was nothing short of prophetic:

If, therefore, those cultivators of physical science . . . are led in the pursuit of the arcana of science to study the singularities and instabilities, rather than the continuities and stabilities of things, the promotion of natural knowledge may tend to remove that prejudice in favor of determinism which seems to arise from assuming that the physical science of the future is a mere magnified image of the past. [69]

Maxwell's equations of the electromagnetic field were particularly effective to convey the impression to his more perceptive colleagues that certain mathematical equations indeed contained more than what was put into them. No one diagnosed this better than Hertz to whom Maxwell's equations appeared as if they "had an independent life and an intelligence of their own, as if they were wiser than ourselves, indeed wiser than their discoverer, as if they gave forth more than he had put into them." [70] Extraordinary as this might have been, it was not without some perplexing aspects. Poincaré for one could point out in Maxwell's papers on electromagnetism not only algebraic errors but also illogical steps and conceptual inconsistencies. As Poincaré put it, all this made the French reader fond of strict rigor feel rather uneasy. [71] Yet, as Duhem noted, Poincaré too was willing to admit that in view of the incontestable brilliance of Maxwell's work mathematical physics had the right to shake off the yoke of too rigorous logic. [72] The most outstanding mathematical physicists of the day began, in fact, to realize that the mathematical code of the physical world transcended the framework of customary mathematical logic and method. Gibbs himself hinted at this in words worth quoting: "If I have had any success in mathematical physics, it is, I think, because I have been able to dodge mathematical difficulties."[73]

Yet, for all the perplexities, confidence remained on the whole unshaken about the possibility of finding one day that self-consistent, all-embracing mathematical synthesis that would impose itself of necessity on the physicist. The physical universe, however, which modern physics stumbled upon, turned out to be almost despairingly distant from the fulfillment of such hopes. Not that modern physics had not done its utmost to make its world picture thoroughly mathematical. It adopted as its principal tools operators and functions, which all too often were very far removed from the rather common-sense framework of the mathematical physics of yesteryear. Modern physics indeed went all the way in embracing the abstract world of mathematics in renouncing unequivocally the desirability of visualization in the physical method. In fact, the basic incompatibility between visualization and modern physics was recognized as soon as quantum mechanics was developed.

As early as 1925 Bohr noted that quantum mechanics implied far more than the mere modification of mechanical and electrodynamical theories. In quantum theory, he warned, one is faced "with an essential failure of the pictures in space and time on which the description of the natural phenomena has hitherto been based."[74] The unanimity on this point could not be more complete among those who were the chief architects of quantum theory. As Heisenberg put it, "the limits of visualization" have been reached by modern physics.[75] Dirac was no less struck by the novelty of the situation: "There is an entirely new idea involved, to which one must get accustomed and in terms of which one must proceed to build up an exact mathematical theory, without having any detailed classical picture."[76] One had to recognize that pictures like that of the electron as a rotating ball could not be taken literally. Nor could one assign in the framework of quantum mechanics any mechanical meaning to expressions like the "structure of the electron." Modern physics merely wanted to assert when speaking of the spin of the electron that the electron has an inner degree of freedom, but as W. Heitler noted, "no further conclusions should be derived from this picture and questions of what the 'radius' of such ball would be, etc., are void of any physical meaning."[77] It was then only natural that Dirac placed the main objective of physical science not in the provision of pictures, or models, but simply in the formulation of mathematical laws predicting new phenomena. Finding "pictures" is no longer of much concern. "If a picture exists," noted Dirac, "so much the better; but whether a picture exists or not is a matter of only secondary importance. In the case of atomic phenomena no picture can be expected to exist in the usual sense of the word 'picture' by which is meant a model functioning essentially on classical lines."[78]

This unconditional willingness to abandon attempts aimed at visualizing basic physical processes, underlined all the more the seemingly unlimited effectiveness of mathematics in providing an adequate description of physical phenomena. As Heisenberg put it confidently in 1930: "It has been possible to invent a mathematical scheme — the quantum theory — which seems entirely adequate for the treatment of atomic processes."[79] Confidence in mathematics could hardly be more robust, and it was hoped that theoretical physics would soon come into possession of a definitive, all-inclusive mathematical formalism. With the ultimate mathematics in hand, one hoped to trace out the bedrock pattern of the physical world as well.

To attribute such power to mathematics was not a wholly new persuasion. The praises of mathematics in effect kept growing ever more fervid as physical science progressed. As God calculates, so the world is made, was a favorite saying with Leibniz as the progress of physics began to gain momentum. In the late 1870's, when physics seemed to approach rapidly its final stage, W. Spottiswoode found no better way to eulogize mathematics than to stress its one-to-one correspondence with the physical world. As he put it at the Dublin meeting of the British Association in 1878:

Coterminous with space and coeval with time is the kingdom of mathematics; within this range her dominion is supreme; otherwise than according to her order nothing can exist, in contradiction of her laws nothing takes place. On her mysterious scroll is to be found written for those who can read it that which has been, that which is, and that which is to come.[80]

The mathematics of which Spottiswoode spoke was the mathematics defined about the same time by B. Peirce as "the science which draws necessary conclusions."[81] This unsuspecting confidence in and respect for mathematics was reflected of course even more strikingly in the statements of non-mathematicians and humanists in general. "Mathematical truths are immutable and absolute," wrote Claude Bernard, who was firmly convinced that "the science of mathematics grows by simple successive juxtaposition of all acquired truths."[82] Poor anticipation of the future indeed. But for the time being the cracks and perplexing complications in the edifice of mathematics were nowhere in sight. Thus Macaulay, in his famous essay on Ranke's *History of the Popes*, felt perfectly justified in making a reference to mathe-

matics in which, as he put it, "once a proposition has been demonstrated it is never afterwards contested."[83] How little did Macaulay
suspect that a few decades later Weierstrass would initiate what he
thought simply impossible: "a reaction against Taylor's theorem."
E. Everett, the historian, president of Harvard and the first American
to earn a doctor's degree at Göttingen, also described the statements
of pure mathematics as absolute truths that existed "in the Divine
Mind before the morning stars sang together, and which will continue to exist there, when the last of their radiant host shall have
fallen from heaven."[84] The passage had only one lasting merit: it
rendered splendidly the state of mind prevailing at that time among
both scientists and non-scientists about the absolutely superior qualities attributed to mathematics. Although the soaring style of Everett
and Macaulay could not find its way into the twentieth century, the
belief in the absolute perfection of mathematics survived at least for
a while as witnessed, for instance, by Whitehead's statement calling
mathematics "the most secure and authoritative of sciences."[85]

If prior to the 1930's some prominent modern mathematicians
poked fun at mathematics in an unguarded moment, it was never
taken as a serious danger sign indicating hidden cracks in the safe
structure of mathematical thought. No one was alarmed when D.
Hilbert characterized mathematics as a game played according to
certain simple rules with meaningless marks on paper.[86] Hardly a
mathematician saw a sign of crisis in Russell's characterization of
mathematics "as the subject in which we never know what we are
talking about, nor whether what we are saying is true."[87] Mathematics at that time (1901) was still enjoying its pre-Gödel era when
the idea of a self-consistent, universally valid mathematical theory
was firmly adhered to. Such a theory existed, however, only in faith,
and the "stubborn facts" of nature, the data of observations, continued to have their supreme say over the mathematical constructions
of theoretical physics. Speaking of the physicists who found it inconceivable that the principles of mechanics could ever be corrected or
changed, Hertz aptly said "that which is derived from experience can
again be annulled by experience."[88] Long indeed would be the list
of theories, for instance, proposed to explain radioactivity that were
discarded by the facts, and such and similar cases did in fact force on
the modern physicist a considerable detachment about pet theories.
Rutherford once remarked, according to Soddy, that the theory of
nuclear disintegration should be abandoned as soon as a single experimental fact emerged contrary to it.[89] Actually there was no
physical theory that could claim an absolutely definitive confirmation
by experimental proofs, numerous as these proofs were. As J. von

Neumann noted about quantum mechanics: "One can never say that it has been proved by experience but only that it is the best known summarization of experience."[90]

It should also be noted that as regards the strictly mathematical features of a physical theory, neither simplicity nor symmetry are in themselves enough to secure the unconditional validity of a given theory. Not that such features were not immensely fruitful in finding the right path, but it was always the facts and the often completely unforeseen facts that had the last say. After the experimental evidence forced on physics the concept of the electron most physicists naturally adopted the simplest assumption in which the electron appeared as a single point charge surrounded by a structureless medium. Undoubtedly this conception allowed a mathematics much simpler than was the case with other models of the electron. Yet, as J. J. Thomson noted in this connection, the simplicity of mathematics could not be taken as a peremptory proof in favor of a theory, as "there is no evidence that the convenience of the mathematicians has been a dominant factor in the scheme of the universe."[91]

The chief convenience or preference of the mathematical physicist are simplicity and symmetry, but as their modes of formulations are almost unlimited in number, no particular mathematical formalism based on any of them can claim in advance an exclusive and ultimate validity. "The mathematical physicist," as Milne forcefully expressed it, "does not dictate to the world what it must be like. But he is guided by mathematical form to make suggestions to the experimenter. His peculiar role then ends. The experimenter decides."[92] In 1929 when Milne made this remark it must have already been abundantly clear to what a large extent the development of quantum theory depended on unexpected experimental data and on the refinement of experimental technique. In fact, H. Weyl, one of the pioneers who shaped quantum mechanics in terms of the group theory, felt compelled to voice his admiration "for the work of the experimenter and for his fight to wring significant facts from an inflexible Nature, who says so distinctly 'No' and so indistinctly 'Yes' to our theories."[93]

It is in this connection of experimental data and mathematical formalism that an all-important feature of the mathematical concept of physics comes to the fore. For just as the notions of organism and mechanism failed to prove themselves as the final word in the conceptualization of physics, no different is the case with its mathematical formalization. Successful and wide-ranging as this may be, no particular formulation of it can claim to itself on purely intrinsic grounds the glory of being *the* true representation of the structure of physical reality. Einstein's pregnant words refer to this state of affairs: "As far as the laws of mathematics refer to reality, they are not certain; and as far as they are certain, they do not refer to reality."[94] This lack of certainty is not of epistemological nature. It rather indicates that out of the large number of mathematical systems, mathematics itself does not have the criterion to choose the simplest one that at the same time would translate perfectly the assumed basic simplicity of the laws of nature. In other words, the confidence that mathematics might find such a criterion can be supported only by a sort of faith in mathematics not by strict arguments.

It is well to remember, as Einstein put it, that "our experience hitherto justifies us in believing that nature is the realization of the simplest conceivable mathematical ideas."[95] Yet, to use his words again, in the actual carrying out of this program experiments and the observed facts remain "the sole criterion"[96] and "the supreme arbiter."[97] Furthermore as there is no telling what unsuspected experimental data will turn up in the course of physical research, there can be no strict scientific basis to claim that the "greatest of all aims," as Einstein called the physical system of the greatest conceivable unity, will one day be achieved by science. The conviction that nature takes basically the character of a well-formulated puzzle therefore was characterized by Einstein as an "outcome of faith."[98] All the successes achieved by physics up to now could give, he insisted, but "a certain encouragement for this faith."[99] Clearly then, the confidence in the correctness of the mathematical world picture of physics rests ultimately, just as was the case with the organismic and mechanistic conceptions of physics as well, on a state of mind rooted in an intellectual faith.

To all those who at one time or another spoke elatedly of the imminent formulation of an all-embracing mathematical synthesis, the persistent impotence of finding it no doubt must have been frustrating. The final answer was not forthcoming either in the mathematical formalism of quantum mechanics or in the Theory of General Relativity. What is more, it became evident that there are mutually exclusive groups of phenomena at the bottom of these two theories. On the macroscopic level, to which relativity theory mainly applies, the coincidence or collision is a primitive event, in the sense that it defines a point in space-time with the provision that the colliding particles are infinitely small. On the microscopic level, however, where quantum mechanics is valid, the event of coincidence is not defined sharply in space-time. This difference is manifested in the fact that quantum mechanics utilizes the infinite-dimensional Hilbert space whereas the theory of relativity uses the four-dimensional Riemann space. Attempts to find a mathematical theory from which these two theories can be derived as approximations have failed so far. Will such a theory be found in the future? Expressing the prevailing sentiment, E. P. Wigner said: "All physicists believe that a union of the two theories is inherently possible and that we shall find it." Yet he was quick to add with commendable objectivity that "it is possible also to imagine that no union of the two theories can be found."[100]

Which of these two possibilities will materialize no one can tell today. At any rate, for the present, a perplexity of such depth should provide a clear warning against sanguine optimism that assumes that the world taken as an embodiment of a unique mathematical function will contain what the organismic and mechanistic syntheses of physics failed to provide: a definitive, all-embracing explanation of the physical universe. Furthermore, it should not be forgotten that the intimate union of modern physics with mathematics has aspects that are almost embarrassing from the point of view of strict logic. As is well known, quantum electrodynamics has to fall back on the technique of renormalization, which, to use the succinct characterization of Dirac, "has defied all the attempts of the mathematicians to make it sound."[101] Renormalization, to be sure, is a highly successful technique. It helped to make the Dirac theory account not only for the Lamb shift but for all known phenomena falling within its range. Still, it is highly unsatisfactory. It almost amounts to cheating, as it replaces infinite quantities (mass and charge of photon cloud surrounding the electron) arrived at by the theory, with the very small quantity established by observation.

The fact that such an arbitrary procedure gives results that in their agreement with observational data surpass, as Dirac put it, the precision characteristic of astronomy, does not constitute a peremptory argument in its favor. It is of little comfort to recall that classical physics too had to rely, until Cauchy's time, on a calculus, the basic theorem of which, the theory of limits, was lacking an unassailable foundation. The difficulty implicit in the method of renormalization could be something basically different. Indications are that it might defy any attempt to resolve it in a perfectly consistent way. As Dirac commented: "It seems to be quite impossible to put this theory on a mathematically sound basis . . . the remarkable agreement between its results and experiment should be looked on as a fluke."[102]

Harsh as this stricture may appear, the unfinished business of renormalization was recalled quite recently by R. Feynman, one of the physicists who received the Nobel Prize for making the infinities of quantum electrodynamics more manageable. "It may be a funny thing to say after receiving the prize," he noted, "but as far as I am concerned, the math isn't solved completely yet."[103] Consequently, the "understanding" of nature, claimed by a purely mathematical expediency, is to be taken by proper reservations. It still remains true, as I. I. Rabi warned almost twenty years ago, that "the theory does not converge but is made to agree with experiment through systematic mathematical manipulation."[104] Renormalization in quantum electrodynamics is therefore a basically ad hoc procedure, and as such it can offer little in the way of understanding the physical reality.

It is not surprising that such "unorthodox" aspects of mathematical techniques are encountered precisely by that type of physics for which the physical universe is primarily a pattern in mathematics. After all, the baffling nature of a tool comes more readily to light when subjected to the most exacting demands, and modern physics through its problems poses a far greater challenge to mathematics than its classical counterpart. But irrespective of physics mathematics stumbled upon a purely mathematical consideration that doomed the hopes of ever arriving at a fundamental and completely self-consistent mathematical system that could also serve as the bedrock layer both in mathematics and in physics. Interestingly enough this setback came when the infinite-dimensional Hilbert space was being taken more and more as the realization of Hilbert's ambitious program, which "aimed at nothing less than to banish once and for all from the world," as he put it in 1922, "the widespread doubts besetting the certainty of mathematical conclusions."[105]

The mathematicians present at the 1922 meeting of the German Association were well aware of these doubts. The rude shocks by which Weierstrass and his followers had awakened the community of mathematicians "from their dogmatic slumber"[106] were still fresh in their memories. Hilbert felt that it was important to remind his audience of the overthrow of several principles of mathematics that had seemed securely established a few decades earlier. Yet, for Hilbert such reminiscences were by no means a reason to despair of the ultimate success of mathematical research. For, as he insisted in an address given in 1921, mathematics owed its greatest advances to the discussions of its fundamental principles, and he viewed all these advances as steps toward the definitive formulation of the ultimate principles of mathematics. About these his was a conviction that in "mathematical matters there can be basically no room for doubts, half-truths, or truths of essentially different kinds." The foundations of mathematics, as he put it, "are capable of perfect clarity, understanding, and definitive solution," difficult as they may be to obtain.[107]

Whatever the immensity of such a task, how could one doubt its basic soundness in the opening decades of the twentieth century, which was inaugurated for mathematicians with the confident words of Poincaré speaking of the arithmetization of mathematics at the Second International Congress of Mathematics in 1900: "We may say today that absolute rigor has been attained."[108] Such a self-assured appraisal of the status of mathematical rigor created a most favorable atmosphere for a program like Hilbert's. A no less competent observer of such aspirations than H. Weyl has recalled the "optimistic expectations"[109] that prevailed in the late 1920's about the feasibility of such an undertaking. Hilbert in particular tried to fuse Einstein's Theory of General Relativity with G. Mie's work on pure field physics. To many in Hilbert's circle the prospect of formulating a universal law valid both for the structure of the cosmos as a whole and for the structure of the basic units of matter seemed near at hand. Yet the arbitrariness implicit in Hilbert's Hamiltonian function could not be ignored, although no less competent men than Weyl, Eddington, and Einstein tried to eliminate it.

Hilbert's program was no doubt as ambitious as any seen before in mathematics, and the talents engaged in it were no less extraordinary. Hilbert's work was carried on by younger collaborators, two of whom, W. Ackermann and J. von Neumann, succeeded in proving the consistency of that part of arithmetic in which the axiom about the conversion of predicates into sets is not yet introduced. This was a notable achievement, yet the high hopes generated by it were soon dashed. The deluge came without warning. In December, 1930, Hilbert was still confidently telling the Philosophical Society of Hamburg how the ultimate certainty can be achieved in number theory, that fundamental branch of mathematics, by formalizing it in a superior form of mathematics, or metamathematics.[110] Yet, a month earlier Gödel's historic paper had already been submitted for publication.[111] In that paper Gödel proved that the formalism of the *Principia Mathematica*, or any other formal system that is not too narrow, cannot have in itself its proof of consistency. What was so shocking in Gödel's procedure was the fact that his proof concentrated on the domain of whole number arithmetic. To declare, as Gödel did, that even in such a basic and elementary domain of mathematics, a proposition can be undecidable unless extraneous assumptions are made had to be upsetting, to say the least. Attentive readers of his paper could not fail to perceive that such a state of affairs would cast a shadow of inconsistency on any formal logical system that encompassed such basic mental operations as the addition and multiplication of integers and zero. This meant, however, that formal mathematics as a whole had to be regarded from then on as incomplete in substance, a far cry from its exalted position as the most consistent and self-sufficient of all intellectual disciplines.

Gödel for one did his best to soothe the impact of his conclusions. Actually he went on to declare in Proposition XI of his paper that his findings represent "no contradiction of the formalistic standpoint of Hilbert." Little did he suspect how ineffective his disclaimer was. For an essential part of his procedure consisted in showing that the symbols and sequences of formulas in Hilbert's formalism can be enumerated in a way in which the assertion of consistency appears in the form of an arithmetic proposition. But as Gödel found, such propositions can be neither proved nor disproved within the formalism. Invariably, some of the arguments needed to prove the consistency of a given system will have no formal counterpart in that system. One is therefore compelled to conclude that either the formalization of the procedure of mathematical induction is not yet wholly known, or that a strictly definitive proof of consistency is out of man's reach. In short, if the game of mathematics is actually consistent, then the formula of consistency cannot be proved within this game. To prove this consistency one needs another class of games, or a metamathematics. Yet, to prove the consistency of metamathematics one again needs a supertheory or a metamathematics, and there is no end to such steps. One is, in fact, caught in a process of endless regression when trying to formalize a metamathematical theory of proof as a set of symbols manipulated according to specified rules. Each set of rules points beyond itself for its proof of consistency. This is why one has to consider dim the prospect of mathematics ever becoming established as the system of "absolute truths."

Such is, in brief, the gist of Gödel's incompleteness theorem that cast serious doubts on any attempt aiming at an a priori mathematical theory that might claim to be the ultimate mathematical pattern of the physical world on the grounds of its built-in consistency. Such considerations could not fail to dampen some sanguine hopes in the circles of mathematicians and theoretical physicists. H. Weyl once described Gödel's theorem as a "constant drain on the enthusiasm" with which he pursued his work, and expressed the belief that his experience was shared "by other mathematicians who are not indifferent to what their scientific endeavours mean in the context of man's whole caring and knowing, suffering and creative existence in the world." [112]

In that creative existence of man the mathematical and scientific reflections play a prominent part. This is especially true of the mathematics that Einstein viewed as the creative principle of physical science. His was a conviction shared by most of his colleagues: since nature is the realization of the simplest conceivable mathematical ideas, we can therefore discover "by means of purely mathematical constructions the concepts and the laws connecting them with each other, which furnish the key to the understanding of natural phenomena."[113] And in a veiled reference to the Pythagoreans, he stated his belief that in a certain sense "pure thought can grasp reality as the ancients dreamed."[114] In a certain sense to be sure, for Einstein was not blinded by the successes of the antiphenomenological constructive method to its limitations. His belief that we can grasp the physical reality for the purposes of science only indirectly, that is, by speculative means or mathematical formalism, impelled him to state that for this very reason the notions of science about physical reality could never be final. "We must always be ready," he warned, "to change these notions — that is to say, the axiomatic sub-structure of physics — in order to do justice to perceived facts in the most logically perfect way."[115]

It is on the ultimate success of such a quest that Gödel's theorem casts the shadow of judicious doubt. It seems on the strength of Gödel's theorem that the ultimate foundations of the bold symbolic constructions of mathematical physics will remain embedded forever in that deeper level of thinking characterized both by the wisdom and by the haziness of analogies and intuitions. For the speculative physicist this implies that there are limits to the precision of certainty, that even in the pure thinking of theoretical physics there is a boundary present, as in all other fields of speculations. An integral part of this boundary is the scientist himself, as a thinker, with the ever-changing patterns of his various states of mind. For just as the shifting moods of the physicist's state of mind bespeak the imperfect, human character of science, so do the basic modes of his search for certainty demonstrate the same about his achievements. Wizardry, however great, with the techniques and principles of mathematics, is not a magic means "to get away from ourselves," that is, from the not always clear-cut operations of the human intellect, as Bridgman, a Nobel laureate physicist, acknowledged in connection with Gödel's theorem.[116] Yet, only a mistaken rationalism can see in this a cause of despair. For one thing, Gödel's theorem casts light on the immense superiority of the human brain over such of its products as the most advanced forms of computers. Clearly, none of these machines can ever yield an answer comparable in its breadth and depth to Gödel's theorem. For another, despair can grow only in a soil where a rigid rationalism has already killed off the seeds of intellectual humility.

Such a soil cannot nurture the recognition that there is no escape from admitting that in mathematics and a fortiori in physics certainty is not the fruit of a "pure rationalistic" procedure alone.

Gödel's proof clearly indicates in which direction one is to face a blind alley when investigating the ultimate source of what there is of validity and certainty in the concepts of mathematics. The setback suffered by the thoroughgoing formalists in the hands of Gödel's theorem should help prevent our forgetting that the mind thrives on sensory experience and that postulates, however abstract or mathematically esoteric, are rooted somewhere, no matter how remotely, in experience. In a sense, this lesson, underscored by Gödel's investigation, completes what has been clear to all those who, following Einstein, concentrated on the geometrization of physics. For by the time geometry proved itself an indispensable tool for the development of modern physics, it had already become recognized that geometry was far from being that paradigm of unchangeable, a priori principles as claimed by Descartes. The analysis of the foundations of geometry in the hands of Riemann and Helmholtz made short shrift of the Kantian belief that the a priori synthetical cognitions of pure geometry were to be accepted as absolutely certain and fundamental. Riemann warned in effect in his paper of 1854 that "the properties by which space is distinguished from other thinkable three-dimensional continua can only be proved by experience."[117] Before long his conclusion received powerful support in Helmholtz's work on the genesis of geometrical notions, and it became unavoidable to recognize that insofar as the axioms of geometry are the results of observation, they command no greater certainty than the statements of physics.

Mindful of this, Minkowski was careful to emphasize that the new views of relativity that made both space and time as independent entities "to fade away into shadows" have sprung "from the soil of experimental physics and therein lies their strength."[118] In the great strides made since then in the geometrization of physical theories, this awareness of the not-at-all absolute character of a particular form of geometry has been clearly noticeable. Einstein for one carefully distinguished between a mathematics or a geometry, "the laws of which," as he put it in 1921 (still the pre-Gödel era), "are absolutely certain and indisputable," and a mathematics or a geometry that is a branch of natural science. The affirmations of the latter type of geometry rest, in his words, "on induction from experience not on logical inferences only."[119] This is why he insisted so emphatically that a decision about the Euclidean or non-Euclidean geometry of the universe could be made ultimately on an experimental basis alone. Again, it was this "sensory" substratum of the geometry physics has to use, that kept suggesting to him that the scientific explanation of physical reality can never be final.

The awareness of this should be no less vivid today when one can witness the development of geometrodynamics, which by definition is "the study of curved empty space" and is based on the assumption that mass and electricity can in a sense be fashioned out of curved space.[120] Clearly, a physical theory claiming that the starting point for physics at the very small distances is the "vacuum, complex in geometry and rich in dynamics," [121] must be peculiarly aware of its ultimate dependence on the data of experimental evidence. Only on such a basis will it be able to stay within what J. A. Wheeler described as the "traditional modest spirit of theoretical physics with all openness to recognizing and formulating the new concepts which are hidden in it." [122] It is in this spirit of openness that one should also remember that the type of geometry that reaches back to the curvature of abstract vacuum to fashion the description of a concrete world out of it is and must be based ultimately on the common human experience imbedded in the concrete plenum.

The concreteness of nature, however, is rich beyond comprehension in aspects and features. This is why even the most bizarre sets of mathematical postulates and geometrical axioms can prove themselves isomorphic with some portion of the observational evidence and useful in systematizing it. This is why the physicist is apt to find himself time and again, as Wigner noted, "in a position similar to that of a man who was provided with a bunch of keys and who, having to open several doors in succession, always hit on the right key on the first or second trial." [123] This is why the physicist might even be overcome by a mood of skepticism concerning the uniqueness of coordination between his mathematical tools and the actual features of the universe. Again, in view of the extreme richness of the features of the physical world, one should not marvel inordinately how space can be non-Euclidean or how complex numbers can be so successful in dealing with alternating currents and an almost endless array of physical processes. The novelty present in those processes is far from being exhausted. No one would dare assume today that there is nothing new for man to observe in the physical world. Consequently, the formulation of new mathematical theories useful for physics will very likely go on indefinitely. For it is not only himself that the mathematician cannot get away from; he cannot get away from the physical world either. It is there, in an immensely variegated nature and not in his finite intellect where ultimately lies the never-ending challenge for the mathematician.

At any rate, the data being collected through experimental physics today are so variegated and increase in number so rapidly that it is doubtful whether an axiomatic approach in the Hilbertian fashion can ever hope to gain a firm foothold in physics except perhaps in its most consolidated parts. As H. Weyl once noted, "Men like Einstein or Niels Bohr grope their way in the dark toward their conceptions of general relativity or atomic structure by another type of experience and imagination than those of the mathematician, although no doubt mathematics is an essential ingredient." [124] This process of groping in the dark should indicate in itself that the modern, emphatically mathematical conception of physics is, in spite of all its successes, far from intimating that the long road of physical discoveries is near its end. It was mathematics that unveiled such startling aspects of the physical world as the mass-energy equivalence and the uncertainty principle. Yet, when commenting on the grave puzzles raised by these cornerstones of modern physics, Max von Laue felt compelled to remark: "Here we feel with particular intensity that physics is never completed, but that it approaches truth step by step, changing forever." [125] In spite of the immense expansion of the applicability of the laws of physics that took place when the mechanistic concept of physics yielded to the mathematical operationalism of quantum mechanics, "the state of theoretical physics," as Max Born put it in 1943, "is just as problematical as it was at any time." [126] Two decades later the British physicist, Sir Henry Massey, could only dismiss the naïve complacency according to which "all that is necessary [in physics today] is further application of established laws which nature follows exactly." [127]

Only the superficial observer could fall easy prey to spurts of sanguine expectations such as the one that manifested itself in the early 1930's when to some everything in physics suddenly seemed to find an explanation. Those who would not be swayed by specious appearances realized not only that as physics progresses it requires for its theoretical formulation an ever more advanced type of mathematics but also that this mathematics grows more abstract and continually shifts its foundations. As Dirac noted in 1931, this increasing abstraction will make it very unlikely that the future advance of physics will be associated "with a logical development of any one mathematical scheme on a fixed foundation." As for the solution of the then outstanding problems of physics, he forecast a "more drastic revision of our fundamental concepts than any that have gone before." [128] Thirty years later his awareness was just as keen about "the drastic changes," as he put it, to be undertaken in theoretical physics if its new but no less vexing riddles were to be solved. Dirac felt that not only would the principle of indeterminacy fail to survive in its present form under the impact of such changes but that physics would also have to part with the four-dimensional manifold as the fundamental framework of its laws.[129]

The persistently recurring need of drastic changes in the mathematical formalization of modern physics clearly indicates that modern physics, by being shaped along mathematical operationalism, is not spared thereby ever fresh crises. Theories, however successful, can alleviate only temporarily that feeling that Pauli disclosed to a friend five months before Heisenberg's first paper on quantum mechanics appeared. "At the moment," wrote Pauli, "physics is again terribly confused. In any case it is too difficult for me, and I wish I had been a movie comedian or something of the sort and had never heard of physics." Yet, five months later he realized that Heisenberg's theory left many questions unanswered. It was, no doubt, a powerful shot in the arm but not the final panacea. As Pauli put it, "Heisenberg's type of mechanics has again given me hope and joy in life. To be sure it does not supply the solution to the riddle but I believe it is again possible to march forward."[130]

In this forward march of physics the final account of the physical phenomena has not yet been secured by basing physics on mathematical symbolism. So far nothing warrants that a mathematical conception of physics will be more successful than its mechanistic and organismic antecedents were in providing man with a *final* scientific explanation of the material universe, or at least not in the foreseeable future. In view of the crises in which mathematics finds itself, one cannot help feeling with particular force the incongruity present in unsuspecting statements, such as the one uttered by Jeans: "The final truth about a phenomenon resides in the mathematical description of it; so long as there is no imperfection in this, our knowledge of the phenomena is complete."[131] Apart from the philosophical poverty of such a flat declaration about what constitutes the full knowledge of a phenomenon, did mathematics succeed in coming up with its final and definitive form?

Men of science who are only too ready to attribute definitiveness to this or that type of mathematics would do well to ponder a little on how badly Diderot fared in prognosticating an impending completion of the science of geometry: "I almost dare to assert," he stated in 1754, "that in less than a century we shall not have three great geometers left in Europe. This science will very soon come to a great standstill where Bernoullis, Eulers, Maupertuis, Clairauts, Fontaines, d'Alemberts, and La Granges will have left it. They will have erected the columns of Hercules. We shall not go beyond that point."[132] Yet, a century later, geometry had already been steered on an entirely new course by three of the many great geometers of the nineteenth century, Gauss, Bolyai, and Lobachevsky. There are other similar instances worth remembering. Sir William Hamilton, who was a professional mathematician, scored no better than Diderot when he declared his quaternions to be both the last word on the generalization of the concept of number and the master key to geometry and mathematical physics. While he was devoting the rest of his life to the study of this "ultimate" concept in mathematics, a contemporary of his, H. G. Grassmann, was developing the concept of even more generalized numbers of which the quaternions were only a minor subdivision. Again, mathematicians had already been busy with five-, six-, and *n*-dimensional space when many physicists and non-physicists flattered themselves that in the four-dimensional space-time manifold they possessed the definitive, exclusively true structure of the universe. Today the pillars of Hercules, marking the ultimate frontiers of geometry and the universe are located only in wishful thinking, not in facts. Yet, the waves washing the feet of those chimerical pillars are very much in evidence, and they resemble anything but a mighty stream rushing in one direction. Of mathematics as it stood two hundred years after Diderot, H. Weyl said that it is "more like the Nile delta, its waves fanning out in all direction." [133]

What is more, mathematics, for all its success in correlating and predicting recondite phenomena of nature, is a stream that by becoming ever more abstract pulls steadily away from what shall forever remain a basic human goal in the quest of understanding: the description and explanation of nature. It is an unavoidable consequence of the modern mathematical concept of physics that "its only object," to use Dirac's words, is "to calculate results that can be compared with experiment." Undoubtedly this is a statement that squares with the way in which physics is being cultivated at present. Yet, the rest of Dirac's statement deserves further comment. There one is told that "it is quite unnecessary that any satisfying description of the whole course of the phenomena should be given." [134] Clearly in such a statement more is involved than the fact that the basic concepts of modern mathematical physics defy visualization. What is also implied there is that as mathematics grows more effective in coping with the problems of physics, it also becomes more evident how limited is that aspect of the world of phenomena that can be grasped, ordered, and correlated by mathematics.

More than two millennia ago the science of physics was born under the symbol of organism in the hope that man would thereby secure the full intelligibility of the physical world. Not only was this goal not achieved by the organismic approach to the inanimate world, but the approach did not even yield a minimum of control over the external world. Through the thorough mathematization of physics control over nature reached proportions that in at least some respects are not far from the measure of fullness. Yet, to expect the full intelligibility of nature from the mathematical concept of physics would be just as illusory as were the hopes and beliefs entertained about its mechanistic and organismic antecedents. Only a limited range of the full reality of things can ever be accommodated in the molds of mathematics, advanced and esoteric as these might be. Unwittingly, however, even Russell recognized this when he stated that "physics is mathematical not because we know so much about the physical world, but because we know so little: it is only its mathematical properties that we can discover." [135]

That little that mathematics can say about the phenomena of nature was illustrated by Maxwell long ago in a memorable passage. He was struck by the fact that the equations for the uniform motion of heat in homogenous media are identical in form with those of attractions varying inversely as the square of the distance. "We have only to substitute," he noted, "*source of heat* for *centre of attraction*, *flow of heat* for *accelerating effect of attraction* at any point, and *temperature* for *potential*, and the solution of a problem of attractions is transformed into that of a problem of heat." Obviously the two phenomena are not identical although they call for exactly the same mathematical formalism. Yet, as Maxwell noted, "If we know nothing more than is expressed in the mathematical formulae, there would be nothing to distinguish between the one set of phenomena and the other." [136] Of the many examples of this sort only one more will be mentioned: the case of Mathieu's equations handling equally well the vibrations of an elliptical stretched surface and the dynamics of an acrobat balancing himself on the top of a sphere. Is it possible to deny that what these two phenomena have in common is enormously less than in what they differ? Unfortunately, there is no lack of scientists who take a stance that comes very close to such a denial. They are the ones who resolutely keep trying to supplement all physicochemical categories with mathematical equations. Their efforts were duly characterized by F. S. C. Northrop as being as ridiculous as the procedure of a pure mathematician "who would expect to derail the Twentieth Century Limited, by attempting to think the equation for a disembodied switch across its pathway." [137]

How little mathematics can replace reality is perhaps best illustrated by the way in which the concept of the ether has stolen back into modern physics. Discredited two generations ago because of the persistent failure of attempts to detect it, the ether was replaced by the vacuum, and its demise became the much talked about symbol of the advent of an entirely new era in physics. Before long, however, modern physics began to embellish space, which the departure of ether allegedly left totally empty, with properties disguised in mathematical nomenclature. In quantum electrodynamics today, the vacuous space, or the mere emptiness, is the seat of zero-point oscillations of the electromagnetic field, of the zero-point fluctuations of the electric charge, and of several even more recondite "properties." It is in fact this reinstatement of the ether in the modern mathematical framework of physics that best shows how little the physicist can dispense with what are commonly called physico-chemical categories.

Truly, "if we know nothing more than is expressed in the mathematical formulae," to recall once more Maxwell's phrase, one would be lost not only in the real world but in physics as well. Mathematics can translate only one aspect of things. It can count the atoms in an apple, it can tell whether this number is odd or even, but mathematics cannot provide either the apple or its atoms. Much less can it say anything about what are known as the *pleasing* features of an apple. But apart from such "anthropomorphic" although indispensable aspects of reality, it is still true that "There is no valid inference," as Whitehead aptly remarked, "from mere possibility to matter of fact, or in other words, from mere mathematics to concrete nature." [138] For contrary to the dreams and hopes of ancient and latter-day Pythagoreans, numbers depend on the concreteness of things instead of generating those things. And for this elementary reason alone, regardless of the problems and uncertainties besetting modern mathematics, the mathematical concept of physics will remain as incapable of providing a final, exhaustive intelligibility of nature, as were the organismic and mechanistic concepts of physics.

Those who seem to forget that there is immensely more in a thing than its quantitative or numerical aspect will of course speak and write as if numbers could tell the whole story about nature. In such a state of mind it is only natural to project the notion of number into an absolute hallowed image. The Pythagoreans composed sacred incantations to honor and worship the number, as the full intelligibility and source of all things: "Bless us divine number, thou who generatest gods and men," went their misguided admiration for numbers. Today one encounters a similar philosophical poverty in the statements of prominent physicists who define God as a mathematician. [139] Of the mathematical expertise of God no mortal can say much. Much more, however, can be said about a specific lesson provided by the history of physics. Outstanding spokesmen of the three main types of physics time and again claimed that the "perfect intelligibility of nature" represented by the physics of the day had some divine hallmark on it. Aristotle, the supreme architect of organismic physics, referred to that ascending staircase of "affections" that supposedly connected both the strivings of ordinary matter and the affections of the ether to the fullness of life in the Prime Mover. Yet, the perfection of divine life did not guarantee perennial validity for organismic physics. The most persuasive salesman of mechanistic physics, Voltaire, called God "the eternal machinist." [140] Little did he suspect that the infinite mechanical skill of God belonged to an incomparably higher level than all the power of Newtonian physics let alone its irremediable shortcomings. It will not be otherwise with the mathematical concept of physics, which may very well prove itself the ultimate form of physics as a science, but certainly not the ultimate intelligibility of the physical world. Such would still be the case were modern physics to find one day its final mathematical form and synthesis. At present the prospects for this are extremely meager. The persistent failure of a priori syntheses of physics, the evidence of the fundamentally experimental roots of geometry, the basic subordination of the heuristic values of mathematics to the experimental observation, and the relative uncertainty in which mathematics is ultimately enveloped all seem to indicate that the replacement of theories in physics will continue as before. This means, however, that only the kernel of scientific truth will become better defined as time goes on. The great aim of physical science, the overall synthesis of the scientific understanding of the universe will remain for all practical purposes what it has always been, the ever-remote objective of an intellectual faith.

Part Two

The Central Themes
of Physical Research

CHAPTER FOUR

The Layers of Matter

EVER SINCE MAN began to raise scientific questions about the universe, the problem of what is the basic stuff and structure of matter has posed a perennial challenge to scientific pursuit. Still, there seems to be something strikingly new in the way the age-old problem of the ultimate constitution of the material world has been weighed in very recent years. This new way of looking at the puzzle of matter stems from an impression that imposes itself more and more on physicists standing in the forefront of fundamental particle research. It was not long ago that the existence of supposedly the last of these particles was predicted on theoretical ground and was, well in advance, confidently named omega. When the search for this "last word" in the realm of matter came to a successful end, however, L. W. Alvarez, one of the physicists responsible for the discovery, felt impelled to remark: "This means that we scientists will never work ourselves out of a job." [1]

Little over half a century ago the atoms were still pictured as the ultimate, absolutely hard, indivisible building stones of matter. Later the electrons, protons, and neutrons inherited these majestic attributes. Today, however, many a physicist prefers to compare the structure of matter to a succession of layers that, like the layers of an onion, reveal themselves only one at a time. What is more, today's physicist is beset with the premonition that the number of those layers for all practical purposes might turn out to be infinite. The thinking that underlies the imagery of the onion stands in sharp contrast to the expectations science entertained for over two thousand years about one day laying bare the mystery of matter. For the most part, science, or rather the men of science, frowned in the past at the idea that there might be a deep-going chain of puzzles in the structure of

141

matter. Or if admittedly there was to be any, it was either grandly overlooked or relegated to the realm of the unobservable. When Thales designated water as the primordial stuff, he was taking a signal step in the direction of reducing the multiplicity of material phenomena to one basic and universal layer. Among his utterances, however, one would search in vain for the admission that such a position is not without obvious difficulties. Clearly, such a measure of self-criticism would have been too much to expect at the very outset of scientific speculation. It remained for some of Thales' younger contemporaries to point out the shortcomings in his dicta. One of them, Anaximander, submitted the view that what is ultimate and universal cannot have observable properties if it is to be truly ultimate and universal. The task of bringing the problem of primordial matter back to the range of physics fell to another Milesian philosopher, Anaximenes, who replaced Thales' water with air. The apparently infinite number of degrees the density of air could take would then account for all the manifestations of matter. Thus fire is nothing but very rare air, which in denser form gives rise to winds and in a directly evolving sequence to clouds, water, earth, and stones and from these latter to everything else. This attempt to reduce the fluctuations of matter to a quantitative, physical parameter, density, no doubt reveals a stroke of genius. Yet, one cannot help being amazed at the ease with which the Ionians projected limited observations into sweeping, all-inclusive generalizations.

The speculations of the Pythagorean tradition on the constitution of matter closely followed this pattern of carefree discourse on an obviously complicated subject. Thus in sections 53–58 of the *Timaeus* Plato readily identified four of the five perfect, or "Platonic," bodies with the four elements. The tetrahedron stood for fire, the octahedron for air, the icosahedron for water, and the cube for earth. To explain material change Plato therefore had to look for a geometrical feature common to these bodies. This he found in the equilateral triangles bounding the tetrahedron, octahedron, and icosahedron. The transformation of fire, air, and water into one another thus received a common denominator. Since squares bounding the cube could not be resolved into equilateral triangles, Plato excluded any transition from earth to the other three elements. To account for the properties of each element Plato decomposed the equilateral triangles into right-angled ones, which in turn could be divided into smaller and smaller right-angled triangles. This operation permitted him to build "perfect" bodies in any size, reducing thereby the various states in which each of the four elements could exist to a variation of the geometrical size. The shapes of the perfect bodies in turn accounted for the properties of the elements. In the case of fire, for instance, its great mobility was explained by the fact that the tetrahedron is the smallest of all perfect bodies, being bounded by the smallest number of triangles. The penetrability of fire was ascribed to the acuteness of the tetrahedron's solid angles.

With no small ingenuity, Plato applied this system to the problem of decomposition. He believed that when fire decomposed water the result was two atoms of air and one atom of fire, because the twenty equilateral triangles of an icosahedron could form, when pierced by a tetrahedron, two octahedra and one tetrahedron, the total number of triangles being conserved. In spite of such fascinating speculations, Plato's theory on the whole ran into too many difficulties for most of the cases to be explained. Yet, for the first time material properties of bodies were connected with geometrical or numerical relations, and this was truly a seminal insight into the puzzle of matter.

That Aristotle, who was of the view that "it is the business of the empirical observers to know the facts, of the mathematicians to know the reasoned fact,"[2] did not appreciate Plato's procedure is only natural. Furthermore, he could point out that Plato's theory was not unambiguous at all. Aristotle was probably wrong in attributing to Plato's theory a meaning it did not have by taking Plato's triangles as real things instead of as a set of explanatory symbols. On the whole, however, Aristotle could claim that the geometrical theory of matter fell easy prey to the alluring method of little observation and much generalization.

The same stricture could also be applied to the representatives of the mechanical theory of matter, the atomists. Although Democritus saw no reason why a principle of limitation should be assumed with regard to the possible sizes and shapes of atoms, Leucippus realized that the kind of permanency present in the observable world could be explained by atomism only if the absolutely unchangeable units of matter were restricted to the microscopic realm. Atomism, devised in part to overcome Zeno's paradoxes on the impossibility of motion, was by definition inaccessible to experimentation, or observation, optical or mechanical. Also declared forever unknown was the nature of the material of which the atoms are made. Atomism, as was typical of a "final" explanation of matter, had to sink its roots into a layer that could be known not by observational evidence but only by fiducial inference and a rather arbitrary one at that.

The other properties of the atoms, as stated by the atomists, were no less arbitrary though no doubt highly imaginative. The atomists endowed the atoms with small contrivances and described them as having "all sorts of shapes and appearances and different sizes. . . .

Some are rough, some hook-shaped, some concave, some convex and some have other innumerable variations."[3] Although such colorful properties did not explain themselves, they were nevertheless offered as "obvious" explanations of some particular, observable properties of matter. According to Leucippus' contentions, sharpness was the result of angular and many-cornered atoms, while sweetness was the effect of the impact of round atoms on the tongue. To make plausible the connection between two parts of the physical world, that of the unobservable atoms and that of the senses, the atomists referred to a similar dichotomy between the large and the small within the observable world itself. Thus to explain how the motion and the rest of the large bodies could be reduced to atoms rushing in every direction, Lucretius referred to the stationary white patch of a flock of sheep grazing on a distant hillside. It was no problem for Lucretius to find other large-scale effects produced by invisibly small processes. There was the ploughshare of iron getting slightly thinner with every fresh furrow, or the stony pavements being slowly rubbed away by the wayfarer's feet. True, admitted Lucretius, "our niggardly faculty of sight" prevents us from seeing the minute particles that are separated in every moment of this continuous wear and tear; still, he felt entitled to assert as basic truth that "nature works by bodies unseen."[4]

Ingenious as was this type of reasoning, it did not fail to cast doubt on the basic role played by sensory perception in the investigation of the physical world. In this lay, however, the seeds of disastrous consequences that did not escape the great majority of Greek thinkers. At times their reaction was simply violent. They realized that on the atomic theory sweet and bitter tastes, for instance, are not realities but mere conventions and that this holds also for qualities like hot, cold, and for colors as well. Atomism allowed truthful existence only to the atoms and to the void. Opponents of atomism were quick to note that according to atomism nothing was apprehended exactly and that the whole world of perception was but an illusory image of the changes in the condition of man's body and of the things that impinged on it. In all this the cleverness of mind seemed to play an abusive game with the senses, and Galen, for one, let the senses rise in utter defiance: "Wretched mind, do you, who get your evidence from us, yet try to overthrow us? Our overthrow will be your downfall."[5]

Downfall it was and to some extent it was justified. The atomic theory was beset with internal contradictions and these Aristotle was eager to recount time and again.[6] To him Empedocles' doctrine of the four basic elements seemed to be more in keeping with the observational evidence and, last but not least, with the organismic view of the world where purposes and qualities reigned supreme. In Aristotle's presentation, the four elements became the four possible combinations of any two of the four elementary qualities, hot, cold, moist, and dry. Later the Stoics identified the four qualities with the four elements, without departing, however, from the Aristotelian thinking in which the analysis of matter had to remain more of philosophy than physics.

Since in Aristotle's philosophy the nature of the thing and the reason of the fact are identical, the regularly upward motion of fire was grafted to the nature of flame and so was the downward fall of stones to the nature of earth. On such a viewpoint it was of little or no value to analyze and correlate the measurable properties of matter. No wonder the existence of a fifth basic element, the quintessence or incorruptible ether, was postulated with unsuspecting credulity to explain the steady, unchangeable circular motion of heavenly bodies. In Plato's account of the constitution of matter there are traces of tentativeness and caution; for him anyone who offered a better solution was a welcome friend. Plato of course could be generous, since physics was for him a "likely account" at best.[7] For Aristotle, however, physics was the true story that admitted only one version — his own of course — and he offered it as the definitive account. The style of Aristotle leaves no doubt that he was highly reluctant to envision the possibility of future revisions. While he argued against the atomists that their position "cannot help coming into conflict with mathematics and undermining many accepted beliefs and facts of observations,"[8] the doctrine of the four elements was not lacking in grave puzzles either, and this was clear even in his day. Still he spoke of his ideas on matter as *the truth* and in a most emphatic manner, as can be seen, for instance, in the next to last chapter of *On the Heavens*. At any rate he was neither the first nor the last to claim that he had found the ultimate scientific explanation of matter.

It must be noted, however, that for all the neglect of the quantitive aspects in Aristotle's discussion of matter, his teaching was not without some beneficial influence on later scientific developments. In biology, it was used and amplified by Galen in the explanation of the functions and interrelations of various humors in animal bodies. In the field of the study of chemical changes, it stimulated practical experimentation by pointing to the possibility of the transmutation of substances. Although the efforts of alchemists produced no gold, they led to the accumulation of a fair amount of practical knowledge about pure and mixed substances and how to prepare them. The steady development of metallurgy, glass technology, and dyeing techniques in the Middle Ages also stressed the importance of more accurate handling of chemical materials. This need became particularly keen when iatrochemistry, with Paracelsus as its primary and bizarre figure, asserted that the principal goal of the study of matter was not the transformation of metals but the purification of substances for medical purposes. The search for "pure substances" did not fail to deepen ultimately the opposition to the doctrine of the four elements and also enhanced the gradual emphasis on quantitative methods, such as the systematic use of the balance. This in turn led to a tentative recognition of the principle of the conservation of matter (Van Helmont, 1648) and weight (J. Rey, 1630) in chemical changes.

Valuable as these developments were, they formed only a small part of the whole story. Most of the views enunciated on the constitution and properties of matter in the sixteenth and early seventeenth centuries were mere speculations, or simply flights of fancy, presented with no indication of their highly arbitrary character. The boasting manner in which Paracelsus replaced the four elements with his triad — mercury, sulphur, and salt — was clearly not in good taste even by sixteenth-century standards. Still, a high degree of arbitrariness is found in the statements of nearly all those who took up the subject even a century after Paracelsus. Bacon, for one, was loath to admit this, in the manner of all those who keep contrasting the enlightenment of their own days with the errors of past generations. With a pointed reference to Paracelsus, he stated that "the Empirical school of philosophy gives birth to dogmas more deformed and monstrous than the Sophistical or Rational school."[9] At the same time, he was quick to absolve his contemporaries of similar wrongdoings, with the exception of Gilbert of course. Bacon could not have been more wrong. Gilbert was only one of the many whose statements on the constitution of matter were distinguished only by a glaring disparity between evidence and conclusion. Even Bacon belonged to this group, and his shortcoming was all the more deplorable since no one in his day laid down so many wise precepts on how to proceed in the New Philosophy, or the art of inductive reasoning. One of the main goals of this new method was, as Bacon put it in the *New Organon*, the elucidation of the true constitution and texture of bodies, which was supposed to determine all the properties of matter and the laws of chemical transformations. Not only did Bacon fail to come up with anything worthwhile along these lines, but his views on the qualities of matter differed perhaps in style only and not in essence from those of the alchemists.[10]

The decisive break with the Aristotelian, alchemistic, and organismic views on the constitution of matter came in the writings of Descartes, who, though not an atomist, emphasized all along the mechanistic approach to the problem. One thing, however, he had in common with his opponents: he too proposed ultimate explanations of the constitution of matter and he did so with no small conviction. Among the "truths" he laid down about the nature of matter were some very general ones — for example, that there can be only one basic kind of matter in the universe [11] and that matter is identical with the three-dimensional extension.[12] When he came to the details, however, he admitted that only experiment and not speculation can decide which of the countless possibilities was chosen by the Creator in setting up the world machine. Nevertheless he could not resist adding that he did not believe that "one could imagine more simple, more intelligible and more likely principles" than those proposed by him for the explanation of matter. Thus he tells us that when the contiguous pieces of the basic form of matter started their circular motion after the creation of the world, they had to produce infallibly the three particular forms of matter through the elementary process of friction.

According to Descartes, in the outer regions of the huge vortices the great speed rubbed the matter particles into spherical globules, and these constitute the luminous matter of which the sun and the stars are made. What was rubbed off, that is, the corners and edges of the luminous particles, forms the most finely ground material in the universe, the *matière subtile*, as Descartes called it, and of this are made the heavens, or the ether. Where the motion of matter is relatively slow, as in the central regions of the vortices, the luminous particles may aggregate by adhesion, giving birth thereby to the opaque and heavy matter that constitutes the earth and the planets. With no less confidence, Descartes extended his views on matter to particular phenomena, as can be seen from his explanation of why the spirit of wine burns readily. He informed his reader without the slightest trace of reservation in his tone, that the spirit of wine "very easily nourishes flame because it consists wholly of very slender particles." He also professed to know some astonishing details. According to him on these slender particles "there are certain little branches, so short and flexible indeed that they do not adhere to each other . . . but such as may leave very small spaces about them that cannot be occupied by globules of the second element, but can only be occupied by the matter of the first element." [13]

Truly this was a farfetched piece of imagination presented in a bewilderingly matter-of-fact style that obviously did not welcome inquiries as to the justification of patently queer details. With an air of superiority Descartes insisted that there was no need, for instance, to explain the subtle matter more clearly than he had. One need hardly add that three centuries later the problem of explaining matter is just as tantalizing, and perhaps even more so. Premonitions to this effect do not arise, however, in a state of mind that prides itself on possessing all sorts of ultimate scientific truths. This overconfidence in one's view as the final word on a subject was a distinctive feature of almost all Cartesians. Perhaps the most prominent of them, J. Rohault, voiced it in these words apropos the explanation of matter: "Perhaps someone will tell me that in the beginning matter was not divided as I supposed it — this is possible, I agree — but this cannot be brought against me; it matters very little how matter was divided in the beginning; for however it might have been, one cannot doubt, that there are now three sorts of matter, such as I have described."[14]

Such physicists, when choosing their style, might have done well to remember that, as Bacon once put it, facts will ultimately prevail, and we must take care that they be not against us. There is of course no progress without errors, and one gets nearer the truth by saying something than by keeping silent. Scientific work needs daring statements, bold conjectures, and at times a rather firm devotion to a particular hypothesis. Yet, even more, it demands a clear awareness of the true character of the statements made. It cannot dispense with a moderately cautious style that is both an inexpensive and effective means against falling into many an error. The reluctance to recognize this is one of the unfortunate weaknesses of the human mind, and the Cartesian physicists were far from being the last victims of it. The post-Cartesian phase of the search for the definitive understanding of matter was no less plagued by that weakness, which misled so many well-deserving men of science and inspired erroneous beliefs in an immense number of their listeners.

Moderate views on what matter might consist of were rather infrequent. One of the cautious voices was Boyle's, who in his *Sceptical Chymist* (1661) subjected to severe criticism all the "philosophical" theories of matter current in his day. Against the Aristotelian tradition of four elements and the "three principles" of the iatrochemists, he insisted on rigorous respect for the concept of the element. What he meant by elements was, as he put it, the doctrine of "those Chymists that speak plainest,"[15] claiming thus no originality on this score. In fact, Van Helmont, whom he admired, stressed well before him that the resistance of a substance to any known method of decomposition is the chief criterion of a true element.[16] Boyle's definition of an element similarly referred to those "certain Primitive and Simple, or perfectly unmingled bodies which are not being made of any other bodies, or of one another, are the Ingredients of which all those call'd perfectly mixt Bodies are immediately compounded, and into which they are ultimately resolved."[17]

This insistence on the experimental evidence of course did not prevent Boyle from embracing theories, nay, strong scientific beliefs. He was in fact the chief initiator of introducing into the study of matter the main tenets of the emerging mechanistic physics. The word corpuscular was coined by him,[18] and if he agreed "with the generality of philosophers so far as to allow that there is one catholick or universal matter common to all bodies," [19] it was only because such a position squared perfectly with his atomistic and mechanistic convictions. It was on such ground that he supported the view that the shape, motions, and contrivances of the smallest parts of a substance might rearrange themselves in such a way that "one and the same parcel of the universal matter may, by various alterations and contextures, be brought to deserve the name, some times of a sulphureous, and some times of a terrene or aqueous body." [20] Yet, he went to great lengths to keep the "absolutist" tone out of his statements. In fact, he submitted the foregoing sentence as merely "not irrational," and in the opening propositions of the *Sceptical Chymist*, he leaned over backward to avoid making peremptory declarations. Thus he introduced the first proposition, that at the beginning matter might have been actually divided into little parts (atoms): "It seems not absurd to conceive. . . ." "Neither is it impossible," he started his second proposition, that these particles form what we now call molecules. Again of the possibility of separating true elements by fire, he said in the third proposition that he "shall not peremptorily deny" it.[21]

This was no doubt a very modest way of theorizing especially in those times, but it was clearly the only attitude permissible in the face of the endless list of obviously unsolved puzzles that had to be recognized as such even by those who studied the properties of matter with only the limited means of seventeenth-century science. Boyle's was a skeptical mind in the sense that for all his admiration of mechanical philosophy he was on the whole reluctant to talk away the glaring difficulties one encountered when the theory was applied to the particulars. In general he was possessed of a keen awareness of the inadequacy of experimental methods then available, and at times this prevented him from making peremptory conclusions even in cases where evidences to the contrary were lacking. Thus he did not declare gold, mercury, or sulphur to be elements, although they resisted all known chemical analysis. In the same vein he did not reject completely the doctrine of the four elements, which he criticized severely, and much less did he draw up his own list of elements. And finally he made no secret of his apprehensions about the possibility of explaining the immense array of qualitative changes and differences on the sole basis of corpuscular rearrangements, dear as this theory was to him. His aim was to strike rather a middle course between scientific dogmatism and skepticism and he was fully convinced that the latter was as incompatible with true philosophy as the former. When it came to the details of phenomena and experiments, he claimed only modest credit for the corpuscular philosophy, namely, that the things treated by him "may be at least plausibly explicated without having recourse to inexplicable forms, real qualities, the four peripatetick elements, or so much as the three chymical principles." [22]

Plausibility was indeed the word that best suited the generalizations in the study of matter for at least a century. But mere plausibility was not a sufficiently shiny expression to flatter most investigators' egos, and as a result Boyle's modest caution was little emulated. This holds equally well of what has been said after him in both the two main streams of the study of matter: the progressive determination of the true chemical elements and the elucidation of their essentially atomic constitution.

As for the elements, the triad of Paracelsus held its sway in various modifications. Thus John Mayow in his *Five Medico-Physical Treatises* (1674) replaced mercury with the nitro-aerial spirit but retained the other two of the Paracelsian triad, sulphur and salt. True, however, to the fashion of the day, he presented what was essentially the inherited iatrochemistry in the jargon of mechanism. The most marvelous of little engines, Mayow's nitro-aerial particles performed almost everything. When in rapid motion, they produced heat and light, and their rest was said to be the underlying cause of such disparate phenomena as cold temperature and the blue of the sky. Again, their rapid motion in the muscles produced muscular contraction, their knocking on the stomach walls was said to create the pain of hunger, and their mechanical "attrition" caused the rusting of iron. Not only were such declarations a far cry from Boyle's plausibility, but the word "attrition" also showed that to part completely with animism was not easy to achieve. Certainly not in the German iatrochemical school to which chemical investigation gravitated from the late seventeenth century on. There one could see Joachim Becher proposing another triad of fundamental elements, the *terra lapida* (the salts present in all solids), the *terra pinguis* (the oily, combustible, sulphurous matter), and the *terra mercurialis* (fluid matter). It was Becher's *terra pinguis* that was renamed *phlogiston* by Stahl in 1703, or motion of heat and fire, which according to Stahl formed the metals when mixed with calx.

Farfetched as this generalization was, it presented before very

long a clear-cut challenge to quantitative verification, which in turn
led to a revolution in chemistry, a revolution that also meant a parting
with views claiming too much authority in the scientific investigation
of matter. The lion's share of the credit for ushering in this revolution
and a better understanding of the elements should no doubt go to
Lavoisier. He towered above his fellow chemists not only in seeing
more deeply in the jumble of accumulated data but also in having
a keener appreciation of the extent of what still might remain unknown in man's understanding of matter. Lack of caution was indeed
the shortcoming for which he took his predecessors to task. Such
criticism was valid not only of the predecessors and followers of Stahl,
but also of those who, like Peter Shaw, John Friend, Boerhaave, and
others, were prompted to dogmatic statements by their hopes that
the chemistry of their day could be made an exact science along the
lines of Newtonian physics. To both groups applied the remark that
Lavoisier made in the preface to his *Elements of Chemistry* (1789):
"All these chemists were carried along by the influence of the genius
of the age in which they lived, which contented itself with assertion
without proofs; or, at least, often admitted as proofs the slightest
degrees of probability, unsupported by that strictly rigorous analysis
required by modern philosophy."[23] The scope of these harsh words
was, however, not so much a self-righteous indictment of the failure
of his forbears in chemistry, as a warning of the debilitating influence
that the state of mind of individuals or the "genius" of an age might
have on scientific research. To advance science therefore was to break
with inherited ways of thought, a break with blatantly careless reasonings, "scientific" prejudices, and self-flattery, or, in short, to initiate
a revolution. To this he referred as early as 1773 in his laboratory
notebook, where he described his program as one that "seemed destined to bring about a revolution in physics and chemistry."[24]

Still, the prospect of revolutionizing a fundamental branch of
science did not go to his head. He spoke of the safeguards with which
he intended to repeat experiments to establish the real import of
hundreds of experiments performed before him, and he never lost
sight of the most important of his goals, which he stated in 1777 as
follows: "It is time to bring chemistry to a more rigorous way of reasoning."[25] This rigor he achieved in a measure far surpassing any of
the chemists before him. But the price of rigor was a cautious, non-
committal attitude to be taken at junctures where almost anyone else
would have been carried away into making "definitive" statements.
The temptation of doing so must have been high in view of the exciting vista that opened before him once he recognized the role of
the oxygenic principle and turned his back on phlogiston. As he put

it in the same *Mémoire*: "Once this principle is admitted, the chief difficulties of chemistry seem to dissipate themselves and to vanish, and all the phenomena may be explained in astonishing simplicity."[26]

Yet, when chemistry came to be laid on firm ground for the first time in his *Elements of Chemistry*, Lavoisier's tone could not have been more soberingly objective. It might have reminded the reader of Newton, correlating the most disparate phenomena through a single mathematical relation, without committing himself ever so slightly to the nature or cause of gravitation. Similar was the manner in which Lavoisier presented his views on the elements. About their nature and number he wrote that it "can be speculated upon in a thousand different ways, not one of which, in all probability, is consistent with nature."[27] Consequently he contented himself with saying that his definition of an element was a provisional one and depended on the actually available chemical means of decomposing substances. Anything that could not be further reduced was therefore for the time being to be considered an element; or to quote him, an element "is the last point that analysis is capable of reaching."[28] None of the thirty-three elements he listed did he want to endow with an aura of absolute finality, although twenty-three of them took their places in Mendeleev's table. Although he listed the caloric as an element, he added that one is "not obliged to suppose this to be a real substance."[29] In the same vein he explicitly indicated that what he called the "earths" might soon cease to be considered simple bodies. This was a conjecture, however, and Lavoisier felt a duty to advise his reader: "I trust the reader will take care not to confound what I have related as truths, fixed on the firm basis of observation and experiment, with mere hypothetical conjectures."[30]

To emphasize the wide difference between conjectures and experimental evidence was not to be construed as an intent to depreciate theory. Dangerous as the "spirit of systems" proved for science, no less to be feared, according to Lavoisier, was the inordinate accumulation of facts. Long and painstaking efforts deserved, in his view, more than being left in disorder and confusion.[31] Theory, Lavoisier argued, had to have rather a liberating effect on scientific investigation: it had to show the road to clarification without curtailing the freedom of the investigator to follow a new lead, as fresh data came to light. This was a timely reflection, for the process of conquering unknown areas began to accelerate more rapidly than ever in the study of matter. "Chemistry advances towards perfection," wrote Lavoisier, "by dividing and subdividing," and of this process he found it impossible to say "where it is to end."[32] But he too had his moments of weakness as a scientific prognosticator. Contrasting the chasms of the chemistry of yesterday with the vision of a great synthesis suddenly looming ahead, he could not resist the lure of sanguine expectations: "We have ground to hope, even in our own times, to see it [chemistry] approach near to the highest state of perfection of which it is susceptible."[33] His days, however, were not the ones destined to see the completion of chemistry, even if the Revolution had not extinguished in a second the brilliance of a genius, which a hundred years won't suffice, as Lagrange remarked, to reproduce. The extent of the secrets of matter was not to be measured in the small units of complacent hopes.

For one, the number of elements began to increase by leaps and bounds, and this left many investigators uneasy. One of these investigators was Davy, who in 1802 listed forty-two and proved especially successful in separating new elements by the recently developed method of electrolysis. Thus from the potash and soda that were on Lavoisier's list of elements, he obtained in 1807 the elements sodium and potassium, and by the same method he separated the alkaline earth metals — calcium, strontium, and barium. But when in 1810 he discovered chlorine, he could no longer contain his misgivings. In his *Elements of Chemical Philosophy* (1812) he not only complained that there were too many elements, but simply professed complete ignorance about the "true elements belonging to nature."[34] Plainly distrustful of several dozen "elements," he engaged in speculations about a "mother-stuff," or perhaps two or three simple substances, and he even conjectured that the various kinds of matter might simply be the effect of the different arrangements of the same particles.[35] Clearly this was a different tack from that taken by Dalton, who preferred to speak of as many kinds of atoms as there were elements. Davy, however, described his own view as "that sublime idea of the ancient philosophers which has been sanctioned by the approbation of Newton . . . namely that there is only one species of matter, the different chemical as well as mechanical forms of which are owing to the different arrangement of its particles."[36]

Were such the case, the atomic weights of elements ought to show some regular sequence and such "atomic" conviction lurked in the back of the minds of those who tried to discern classes or groups among the elements. The numerical data refused, however, for a long time to do full justice to them or to their opponents. To the discomfort of the latter, some atomic weights were too close to an integer, whereas to the grief of the former some were not close enough. Thus for a while the main reason for siding with one or the other camp depended largely on one's scientific taste, or beliefs. What followed constitutes one of the classic examples of the not-so-straightforward advances in the scientific pursuit. Those who sought to discern classes among the elements chiefly on the basis of atomic weights could hardly foresee that the great codifier of the elements, Mendeleev, would be more embarrassed than helped by the atomic weights then available. This was precisely why the efforts of Mendeleev's forerunners had not sparked much conviction, although they had anticipated him even in making successful predictions to some extent. The first of them, J. Döbereiner, professor of chemistry at Jena, noted in 1817 that the atomic weights of calcium, strontium, and barium were very close to an arithmetic series. In 1826, A. J. Balard at the Sorbonne suggested after discovering bromine that its atomic weight should form with chlorine and iodine an arithmetic series, a prediction that was verified later by J. Berzelius, approximately at least. Further arithmetic connections were discovered by M. Pettenkofer in 1850, and soon after M. Dumas lent his full approval to those who believed in the usefulness of searching for such relationships.

Yet, after the careful measurements of the atomic weights by J. S. Stas, it became evident that the systematization of elements should stress rather the similarities in the chemical properties of the various groups of elements. It was this aspect that A. B. Chancourtois emphasized when he briefly outlined in 1863 a classification of elements that resembled a spiral. In the same year J. A. R. Newlands pointed out recurring octaves among the elements, a suggestion that at first earned him but ridicule. He was, however, fully vindicated by Mendeleev who performed the master stroke by recognizing gaps in the periodic sequence which he identified with missing elements and predicted their properties. Discoveries of such elements came in quick succession and were named scandium, gallium, and germanium. Some enthusiastic contemporaries could only think of the discovery of Neptune to describe such successes. Together with the difficulties met by Prout's hypothesis in the non-integral values of the atomic weights they contributed much to create the belief, in some at least, that the elements, or rather their smallest constituent units, were not reducible to a universal type of matter common to all elements. For one, Helmholtz hailed the periodic table in 1869 as the *conclusive* evidence that "these elements are really indestructible, unalterable in their mass, unalterable also in their properties."[37] He spoke of the "qualitative and quantitative immutability of matter,"[38] and in doing so he clearly expressed Mendeleev's own convictions, which the latter stated in detail twenty years later in his Faraday Lecture of 1889.[39]

The lecture dealt only in part with the antecedents, conclusions, and successes of the Periodic Law. Equally important for Mendeleev was the rebuttal of opinions that tried to seek in the Periodic Law support for the existence of uniform, fundamental building blocks of matter. He had no other choice, for to him the eighty-some elements represented the individuality in nature, the basis for diversity, nay, a new mystery that should inspire the same reverence as, in Kant's words, "the moral law within ourselves and the stellar sky above." For proof he referred to the fact that the Periodic Law did not originate in the idea of a unique matter; that it had evolved independently of any conception of the nature of the elements. A uniform matter, in his view, could exist only in an unlimited number of forms, grades, and transitions. It would permit all sorts of interpolations and make impossible at the same time the emergence of distinct, well-defined characteristics in matter. For him the concept of the unitary matter was the denial of individualization, and he did not hesitate to compare the elements to human individuals. From unitary matter, according to him, could not follow the principal features of the world of elements, where we see "side by side with a connecting general principle — leaps, breaks of continuity, points which escape from the analysis of the infinitely small — a complete absence of intermediate links."[40] Consequently he brushed aside the contention that some recent developments in physics might point toward the existence of a unitary matter. The Periodic Law, he said, "affords no more indication of the unity of matter or of the compound character of our elements, than the law of Avogadro, or the law of specific heats, or even the conclusions of spectrum analysis."[41]

Rarely was the scientific situation so badly diagnosed by a great man of science. Beyond the colorful set of some eighty "individual" elements lay a truly complex but nevertheless unitary world of matter. Of this no other spoke more convincingly in the 1880's than one of Mendeleev's listeners, W. Crookes. His famous presidential address of 1886 to the Chemical Section of the British Association[42] could indeed hardly have been unknown to Mendeleev himself, and the two addresses stand as classical examples in the history of science of how diametrically opposite interpretations can be given by two outstanding men of science to the same set of data. Of the two, Crookes proved to be on the right track and almost every section of his address is prophetic. He spoke, as he put it, in the name of the "most keen-eyed chemists, physicists and philosophers"[43] who refused to accept the reality of elements as the ultimate word in the study of matter. Among those whom Crookes quoted in support of his position were Sir Benjamin Brodie, N. Lockyer, and H. Spencer. But his star witness was the late Michael Faraday, who had told him, as Crookes recalled, after his (Crookes') discovery of thallium: "To discover a new element is a very fine thing but if you could decompose an element and tell us what it is made of — that would be a discovery indeed worth making."[44]

In 1886 Crookes could not conceive of any possible means of breaking up an element, but subsequent developments did full justice to his contention that the barrier of elements "must be overthrown, surmounted, or turned, if chemical science is ever to develop into a definite, an organized unity."[45] Nor was he less right in stating that "the complexibility of our supposed elements is . . . in the air of science."[46] As for the future course of scientific pursuit Crookes judged it imperative to do two things. First, to keep before one's mind the idea of a genesis of the elements, in order to stimulate theories on how the elements might be produced through a common process. Second, to keep in view that there exist in all likelihood "laboratories" in nature where the production and decomposition of elements was actually going on. "We are on the track," he said, "and are not daunted, and fain would we enter the mysterious regions of the Unknown."[47]

The brilliance of Crookes' remarks about what lay ahead for the investigation of matter should impress anyone looking back into scientific history from the vantage point of twentieth-century physics. No less striking are the insights contained in almost all the arguments Crookes submitted in support of his position. The limited number of elements, he said in the first place, ought to be derived from an underlying physical process, or else one should fall back on their being produced by chance. Chance, however, seemed clearly inadequate to account for the regular patterns of the periodic table. Nor did chance seem able to explain, argued Crookes, the peculiar distribution of elements in the earth's crust. The simultaneous occurrence of such elements as nickel and cobalt, or that of the rare earths, reflected anything but uniformity in distribution, which should have been a natural consequence if chance had been the chief factor. Strange patterns in distribution, recalled Crookes, were one of the strongest evidences in favor of the Darwinian concept of the evolution of species and might well have the same bearing on the possible evolution of elements as well.[48] Furthermore, argued Crookes, if atoms bear the stamp of a product manufactured with utmost precision and uniformity, and this was a favorite theme with people like the younger Herschel and Maxwell,[49] the idea of a raw material common to all elements should also present itself of necessity. His main argument centered on the periodicity of the properties of elements as shown in Mendeleev's table, and this he illustrated in an ingenious diagram resembling a pendulous swing. To the superficial observer it was hardly more than a mysterious zigzag curve, but Crookes' words were those of a prophet when he said that "he who grasps the key will be permitted to unlock some of the deepest mysteries of creation."[50] As to what the key might consist in, he expressly indicated the units of electricity, as indeed the case turned out to be. He also spoke of the possible existence of a unit matter common to all elements, which he called *protyle*, and anticipated the idea of what thirty years later came to be called the isotopes of an element. This he did with truly amazing clarity. For, taking calcium as his example, he stated that "while the majority of calcium atoms have an actual atomic weight of 40, there are not a few which are represented by 39 or 41, a less number by 38 or 42 and so on."[51] To make an extraordinary lecture even more so, he ascribed this variation of atomic weights within certain narrow limits to the fact that not all "isotopes" (to use the modern term) of an element might be equally stable.

Since Crookes assigned the production of elements to a cooling process in the originally ultra-hot stars, he saw uranium as the heaviest element possible and thereby made perhaps the only wrong prediction in his lecture. He spoke at the same time of negative elements, in an interesting anticipation of modern speculations on "anti-atoms," and insisted that the production of various elements was not confined to the solar system, but was rather a cosmic process. Even such valuable details apart, the conclusion of his lecture was a gem of truth in itself, a declaration that the problem of the evolution of the elements is not only "the most interesting but also the most important in the entire compass of our science."[52] Clearly, the manner in which Crookes diagnosed the future again illustrated a remarkable pattern: in the study of matter the discoveries to come reveal themselves only to those who refuse to accept the actually existing scientific synthesis, in this case the periodic table, as the stage beyond which one could look in vain for even more exciting vistas.

Crookes' reading of the future implications of the science of his day is all the more remarkable if one recalls the state of scientific knowledge of atoms in the next to last decade of the nineteenth century. In essence, it was a highly plausible inference with only indirect evidence to support it. The speculations on the atomic constitution of matter had yet to achieve the perfection that had become by then the hallmark of the other principal objective of the study of matter during the eighteenth and nineteenth centuries, the search for the true chemical elements. As for the knowledge of atoms, the years in which Crookes spoke still belonged in a sense to the era that started with Gassendi's *Syntagma philosophiae Epicuri* (1649). Whatever the scientific shortcomings of this work, it quickly succeeded in making atomism fashionable. The microscope also helped things in no small measure. Gassendi for one argued that one could obtain some idea of the minuteness of atoms from the smallness of the most minute of animals visible through the microscope.[53] W. Charleton, who championed Gassendi's views in England, connected atoms and microscopes just as readily.[54] In the rather uninhibited reasoning that still characterized much of the scientific thought of his day, he decided, by misreading Archimedes' *Sand Reckoner*, that the smallest body discernible in the microscope contains "ten hundred thousand million" atoms.[55] Less arbitrary in principle were his attempts at trying to calculate the number of atoms in a grain of frankincense from its vaporization, or from the diffusion of a grain of vermillion in water. He carefully distinguished between the "atoms of the vulgar" — such as dust particles visible in sunlight — and the invisible or "true" atoms. Still, on one point he conceded a strict similarity between the two classes: no two atoms in either group could have the same size or shape.[56] It was clearly in keeping with his radical atomism that he declared the "vulgar elements" to be at best the "father-principle" but not the "grand-father-principle" of matter[57] and praised Democritus and Epicurus for basing "their physiology on an hypothesis of a single principle."[58] Evidently all this was not so much physics as mechanical philosophy with its exclusive claim for intelligibility. For as Charleton insisted: "That there are such things as Atoms or Insectible Bodies in Rerum Natura, cannot be long doubted by any judicious man who shall thus reason with himself."[59]

Thus the atoms remained for the time being the subject of interesting but clearly speculative discussion. This held as well for Hooke's *Micrographia*, which first unfolded the marvels of the microscope to the non-scientific world, as of Newton's indulging in various hypotheses in the queries of the *Opticks*. A passage from this work will well illustrate the tone of such conjectures:

All these things being consider'd, it seems probable to me, that God in the Beginning form'd Matter in solid, massy, hard, impenetrable, movable Particles, of such Sizes and Figures, and with such other Properties, and in such Proportion to Space, as most conduced to the End for which he form'd them; and that these primitive Particles being Solids, are incomparably harder than any porous Bodies compounded of them; even so very hard, as never to wear or break in pieces; no ordinary Power being able to divide what God himself made one in the first Creation.[60]

To assume that the world could have come to exist in any way other than through atoms created by God was for Newton a procedure simply "unphilosophical,"[61] an adjective that stood for "unreasonable" or "illogical." The atoms had to be accepted without further ado, and once more the exclusive claim of the mechanical philosophy for intelligibility settled "definitively" an issue over which only experimental evidence could play the role of final arbiter.

Changes of material things were for Newton changes in the relative separations, associations, and motions of these "ultimate" particles, which in contrast to the atomism of the ancients, were endowed by Newton with a variety of forces, or "active principles," by which bodies supposedly acted at a distance. Newton listed these principles as gravity, magnetism, electricity, and fermentation. They showed, according to him, "the Tenor and Course of Nature, and make it not improbable that there may be more attractive Powers than these." [62] Whether atoms were endowed by forces could of course be answered by conjectures only when philosophy, theology, and imagination were the main source of information about the role of atoms in the scientific explanation of the processes of nature. Still, the atoms effectively dominated both imagination and scientific reasoning, because the concept of the atom involved fewer difficulties than the various fluid theories when it came to the explanation of both the obviously mechanical phenomena, like impact and collision, and the distinctly more mysterious "action at a distance." Leibniz therefore represented the minority opinion in insisting on the actual divisibility of any piece of matter into an infinite number of parts: "There are," he wrote, "infinite creatures in any given body whatever. All bodies form a coherent whole. All are separable by force from others, but not without resistance. There are no atoms, or bodies whose parts are never separable by force." [63]

Leibniz was twenty-five when he wrote this and by then he had behind him an early "flirtation" with atoms. Late in his life he described with uncommon force that state of mind which in its eagerness to locate the ultimate overlooks too readily considerations to the contrary:

All those who maintain a *vacuum,* are more influenced by imagination than by reason. When I was a young man, I also gave in to the notion of a *vacuum* and *atoms*; but reason brought me into the right way from what had been pleasing to the imagination. The atomists carry their inquiries no farther than those two things: they fancy, they have found out the first elements of things, a *non plus ultra*. We would have nature to go no farther; and to be finite as our minds are. [64]

To rest one's mind on something *final* is an urge that can be both highly justified and dangerously misleading at the same time. The latter aspect was rather obvious in the ready acceptance of the reality of atoms in both scientific and non-scientific speculations as the eighteenth century went on. Of this Voltaire is one of the unmindful witnesses. To opt for atomism as an alternative with less serious difficulties was one thing; to take it, however, as a demonstrated article of science was clearly an instance of wishful thinking. But this was precisely the fallacy that Voltaire propagated when he summed up the history of atomism in the sweeping sentence: "Epicurus and Lucretius admire the atom and the void, Gassendi supports this doctrine and Newton *demonstrated* it" (italics mine). When or where, one might ask, but as is often the case Voltaire offers only rhetoric in place of proofs.[65] On the other hand, once his word about this "demonstration" is naïvely taken at face value, the rest of his article on "atoms" can be accepted as an accurate description of the state of mind characteristic of an age in which, to hear Voltaire say it: "The plenum is considered today as a chimera; . . . the void is recognized: the hardest bodies are considered sieves and they are such indeed. The atoms are accepted as units that cannot be altered, and to this is due the immutability of elements and species; . . . without these elements of immutable nature one were to believe that the universe would only be a chaos."[66]

To tell the truth, if there was any chaos anywhere, it was in the dicta of eighteenth-century physics on the atoms. It was chaos for both details and proofs. A consensus seemed to exist only on two points. The first was the belief that scientific knowledge had already reached the layer of matter lying next to the ultimate, or atomic, layer of the universe. The second point concerned the general belief in the atomic constitution of matter. The fact that this view was commonly shared was of great encouragement to an investigator of atoms around 1800, as can be seen from the words of the physicist who first succeeded in introducing scientific rationale into man's discourse on atoms. For Dalton was anxious to state that the conclusion he reached "seems universally adopted," namely, "that all bodies of sensible magnitude, whether liquid, or solid, are constituted of a vast number of extremely small particles, or atoms of matter bound together by a force of attraction."[67]

From this secure psychological ground Dalton tried to arrive at quantitative statements concerning the atoms. He was the first to realize that atoms would not truly form a part of physics as long as they were not "weighed" or "counted," and in this realization lies his greatness. For however partial might have been the truth in Dalton's statements, what he said about his "inquiry into the relative weights of the ultimate particles of bodies" cannot be contested in any manner: it was "entirely new,"[68] as he put it in his paper of 1803 "On the Absorption of Gases by Water and Other Liquids." And no less novel was the "Table of the Relative Weights of the Ultimate Particles of the Gaseous and Other Bodies," which he appended to the same paper. By making the concept of weight the keystone of his system, Dalton exposed atomic theory to challenges based on quantitative considerations, and therein lay the key to the ultimate success in the search for atoms. Furthermore, his rule of greatest simplicity for the combination of elements stimulated the formulation of laws like those of Gay-Lussac and of Avogadro, which within less than a decade after the publication of Dalton's theory brought essentially to completion the atomic explanation of chemical compounds.

Such a completion could of course be evident in retrospect. Actually it took half a century to see that. For the time being, however, several major problems blocked the way to a clearer understanding of the situation, and ironically enough, not all of them were real. One of the imaginary objections sprung from the very belief that had helped Dalton so much in his early speculations, namely, that repulsive force acts among atoms of the same type only. Thus he was absolutely reluctant to consider that atoms of the same type might form relatively permanent groups of two. His postulate of a negative force between identical atoms also played a part in his dismissal of Prout's hypothesis, which clearly went straight against Dalton's favorite aphorism: "Thou knowest thou canst not cut an atom." For if Prout's theory had been true, Dalton's atoms would have been composed of the basic units of matter, the hydrogen atoms. Prout, for one, proposed his theory with the greatest diffidence. In fact, fearing excessive criticism, he preferred to remain anonymous for a while at least. But as to the possible import of his conclusions, he clearly stated that "if they should be verified, a new and interesting light will be thrown upon the whole science of chemistry," [69] and as he added in a subsequent correction to his paper, the prime matter of the ancients might be recognized in the hydrogen. [70]

T. Thomson, in whose *Annals of Philosophy* Prout's paper was published, did much to make Prout's views triumph, and for a while "in England the hypothesis of Dr. Prout was almost universally accepted as absolute truth," [71] as was later recalled by J. S. Stas, the Belgian chemist who carried out a crucial series of experiments concerning Prout's hypothesis. Absolute truths, however, are few and far between, and in physics their life span in most cases is rather short. Thus doubts about Prout's hypothesis, owing to the investigations of Berzelius, soon arose on the continent and eventually reached England. In 1833 E. Turner reported to the British Association that Prout's hypothesis did not square with facts precisely enough. As the efforts to measure the atomic weights became more painstaking and as the number of methods for those measurements increased, the discrepancy between theory and experimental results became more glaring. Stas had recognized this as early as 1845, but he remained so undaunted that when he started the last of his many attempts to verify Prout's hypothesis, he still had, as he admitted, "an almost absolute confidence in the exactness of Prout's principle." [72] There was, however, no way of resisting forever the weight of experimental evidence, and Stas had to part with cherished beliefs that, as he put it, proved but illusions. "As long as we hold to experiment," he wrote, "for determining the laws which regulate matter, we must consider Prout's laws a pure illusion." [73] The elements, he added, should be regarded as "having no simple relation by weight to one another," and the cause of the analogy that exists among the properties of certain elements must be sought for in factors not connected with the ratio of their weight. [74]

The failure of Prout's hypothesis to prove itself for the time being could not fail to cast a shadow on atomism itself. Mistrust in the reality of atoms and in their usefulness to scientific speculations made itself manifest on a number of occasions during much of the nineteenth century. Faraday, in a paper on electrochemistry that he read before the Royal Society in January, 1834, spoke of his "jealousy" of the term atom, for although, as he put it, "it is very easy to talk of atoms, it is very difficult to form a clear idea of their nature especially when compound bodies are under consideration." [75] In his *Lectures on Chemical Philosophy* (1836), M. Dumas, the prominent French chemist, went to great lengths to point out all the pitfalls, inconsistencies, and vagaries displayed in the history of atomic theory. [76] Not much different was his appraisal of the evidences offered by the science of his day about the reality of atoms. "Now, you see, gentlemen," goes the conclusion of his seventh lecture on May 28, 1836, "what are we left with of our ambitious excursion into the realm of atoms which we have permitted ourselves. Nothing, at least nothing conclusive." [77] No wonder he wished he could expunge the word atom from the vocabulary of science. In 1857 W. Whewell voiced the sentiments of many, when he took the view as a historian of science, that while Dalton's laws "are truths of the highest importance," the atomic view of matter, so dear to Dalton, "is neither so important nor so certain." [78] In this temporary eclipse of atomism, one could of course easily detect the reverberations of the misfortunes of Prout's theory, which time and again was rejected in peremptory terms. J. T. Merz, the leading historian of the ideas of his century, merely echoed the prevalent view in stating that Prout's hypothesis was without any real scientific benefit and "it rather tends to upset the only firm foundation of modern chemistry, the fixity of the equivalent proportions as we now use and know them." [79]

That much of the dissatisfaction with atomic theory came from chemists can easily be understood. After all theirs was the arduous task of applying theory to facts, and the data of experimental chemistry were despairingly complex. What F. Wöhler wrote of organic chemistry in 1835 was also true in a large measure of inorganic chemistry: it gave "the impression of a primeval tropical forest full of the most remarkable things, a monstrous and boundless thicket with no way of escape and into which one may well dread to enter." [80] In 1861 F. A. Kekulé found it important to call attention to the demarcation line that separates facts and speculations in chemistry. For, as he put it, "as soon as one represents in atomic formulas the composition of bodies, the formulas will include hypotheses, because the relative magnitudes of the atoms are hypothetical." [81] Again, it was mainly the chemists and physical chemists who represented the dissenting voice about the truthfulness of atomic theory in the controversy that enlivened the pages of the first volume of *Nature* in 1870. In 1885 M. Berthelot still preferred expression by equivalents to the atomic notation, distrusting as he did the "hypothetical atoms," [82] and still later the future Nobel laureate of chemistry, W. Ostwald, spoke of atomism as an unnecessary hypothesis. [83]

Yet, it was not chemistry alone that prompted Ostwald to reach such a conclusion. His interpretation of the statements and concepts of physical science had much in common with the position of the late nineteenth-century school of positivism that so eagerly seized upon the difficulties of the atomic theory. Thus J. B. Stallo warned physicists that because of the breakdown of mechanistic theories it was useless to search for units smaller than molecules, that is, atoms. [84] Mach was even more resolute in dismissing atoms as mental artifices, as substances, all of which were "things of thought." If the concept of atoms had any usefulness in physics, it did not go, according to him, beyond the meaning of a mathematical model. To hear Mach say it, the atom had as little to do with reality as circular function with vibration, or squares with the motion of falling bodies. [85] Furthermore, he found an *absolute* contradiction between the properties attributed to atoms and the properties observed in bodies, and on this allegedly incontrovertible evidence he rested his case. At bottom, however, his position was an antiatomist dogma, and this is why he wrote of the opposite view, that he could not "accept the existence of atoms and other such dogmas." [86]

In fairness to Mach, many of his opponents were just as dogmatic in speaking and thinking about atoms. Of this there can be no more telling illustration than the exchange of views (beliefs would be a better word) that took place between Mach and Boltzmann after a lecture given by Boltzmann on atoms and molecules. Mach, who was in the audience, took exception and said in substance: You do not know that molecules exist. Boltzmann's reply was no less categoric: I know that there are molecules. Mach rejoined: You do not. With that the debate ended, but not the search for atoms. At any rate, as Boltzmann pointed out in a well-argued paper,[87] some form of atomic theory was indispensable to make meaningful the phenomenological approach of Mach in physics. For, as Boltzmann contended, equations like those of Fourier on heat conduction or the equations of elasticity could have solutions only "if one first imagined a finite number of elementary particles, which can, according to simple laws, act on one another and then one looked for the limit in increasing their number."[88] "Atomism seems to be inseparable from the notion of continuum,"[89] argued Boltzmann, and he recalled that the founders of nineteenth-century mathematical formalism in physics, Laplace, Poisson, Cauchy, and others, were all inspired by considerations based on the concept of the atom. Boltzmann did not ignore the difficulty that every class of phenomena seemed to call for a kind of atom of its own. Still, according to him, atomic theory alone approached the various problems of physics with well-defined propositions, and he pointed in this connection to the kinetic theory of gases of which few could speak with a competence equaling his.

The success of the kinetic theory of gases was the principal but not the only factor in bringing the problem of the atom to the focus of late nineteenth-century physics. There were other advances as well, which for the first time provided indirect yet reliably informative quantitative data about the atom. Such were the revival of Avogadro's theory at the hand of S. Cannizzaro (1858), the calculation by R. Clausius of the free mean path of molecules in gases (1858), and the determination by J. Loschmidt of the number of molecules in a cubic centimeter of gas at standard temperature and pressure (1865). Even more important was Loschmidt's determination of the order of magnitude of the diameter of a gas molecule, a feat that Maxwell in 1873 called a fact of tremendous importance.[90] In the following years indirect methods of determining atomic dimensions increased so much both in number and accuracy that in 1883 Kelvin referred in a lecture entitled "On the Size of Atoms" to such disparate fields yielding data about atoms as the undulatory theory of light, the phenomenon of contact electricity, capillary attraction, and the kinetic theory of gases. His general conclusion was that "molecules of the ordinary matter must be something like the one ten-millionth or from one ten-millionth to the one hundred-millionth of a centimeter in diameter."[91]

The words ordinary matter are well worth noting. Underlying this ordinary matter was, according to Kelvin and most of the late nineteenth-century physicists, the basic matter or ether of which the atoms and the molecules were believed to be made. While ordinary matter could be dissipated, or fractionated, the basic matter suffered no loss and was not subjected to either friction or transformation of any kind. The ether was the bedrock of the stability of the universe and the impenetrable, indivisible atoms made of it easily lent themselves to the soaring prose of Victorian times. Or as Maxwell put it: "Though in the course of ages catastrophes have occurred and may yet occur in the heavens, though ancient systems may be dissolved and new systems evolved out of their ruins, the molecules, out of which these systems are built — the foundation stones of the material universe — remain unbroken and unworn."[92]

Absolute stability and indivisibility of atoms certainly were not ordinary qualities, but for over two thousand years men of science and philosophers spoke of it with such frequency and assurance that Maxwell's words could hardly have created the impression of novelty. After all, who could be foolhardy enough to challenge in 1873 the definition of an atom given by Maxwell: "An atom is a body which cannot be cut in two."[93] More novel, however, were some properties of the ether of which Maxwell, Kelvin, and others spoke with great emphasis. To some, these properties, or at least the simultaneous presence of them in the very same piece of matter, appeared nothing short of fantastic. Most of these "doubters" were from the ranks of chemists, a rather pedestrian group of men of science, as opposed to the physicists, and burdened with the task of subduing the perplexities of "ordinary" matter. The position of the physicists was distinctly more glamorous. To speculate about the "basic" matter seemed to be far less cumbersome, for those speculations could be confronted with indirect evidences only. "No one has ever seen or handled a single molecule," said Maxwell, and it was therefore rather easy for him to define molecular science as "one of those branches of study which deal with things invisible and imperceptible by our senses, and which cannot be subjected to direct experiment."[94]

From such a position of assurance the physicist found it tempting to dismiss the reservations and doubts of "chemists and many other reasonable naturalists of modern times" who, as Kelvin admitted, lost all patience with the atom and relegated it "to the realm of metaphysics."[95] The chemists pointed, in Kelvin's words, to the "incredible assumptions of infinite strength, absolute rigidity, mystical action at a distance, and indivisibility." But in Kelvin's view it was not these incredible assumptions that were at fault but rather chemistry that, owing "to the hardness of its fundamental assumptions," made itself powerless to deal with the atom and its substance, the ether. What had to be done, according to Kelvin, was to consider the atom, in spite of these "incredible assumptions," as a real "piece of matter of measurable dimensions, with shape, motion, and laws of action, intelligible subjects of scientific investigations."[96] In other words, one had to adopt a state of mind in which all these difficulties would be glossed over and then one could say with Kelvin that fantastic as the properties of the ether might be the "luminiferous ether is no greater mystery than all matter is," and that "we know the luminiferous ether better than we know any other kind of matter in some particulars."[97]

To say this was already presumptuous, but such a state of mind demanded even more. With Kelvin one had to postulate for the ether an extreme simplicity and also a rigidity far surpassing that of steel to account for the speed of light waves. At the same time, the most rigid of all the substances also had to be the most tenuous kind of material ever imagined to let the earth and celestial bodies pass through it without resistance. The ultimate property of ether was defined by Kelvin as incompressibility, and although he admitted that there was still much to be learned about the ether, he claimed that "the natural history of the luminiferous ether is an infinitely simpler subject than the natural history of any other body."[98]

This was hardly a piece of sober reasoning. Still, could a careful weighing of evidence play a role in Kelvin's attempts to establish numerical values for various properties of the ether? Maxwell, who followed Kelvin closely in this bold endeavor, gave the specific density of ether as 9.36×10^{-19}, its coefficient of rigidity as 842.8. In the same context, he also stated that the density of air at an infinite distance from the earth was, under certain assumptions, 1.8×10^{327} times less than the estimated density of the ether.[99] The use of numbers, formulas, and equations always conveys some magical force, particularly for the layman to whom these pages of the famous ninth edition of the *Encyclopaedia Britannica* were written. Such parlance makes him imagine that behind the exactness represented by the prolific use of numbers there is a solidly established experimental evidence and a scientific logic unencumbered by preferences and wishful thinking. But the ether, in spite of the realistic style of the physicists, could not claim a support of this kind. Its basic support rested on a *state of mind* that was even willing to entertain the idea of the existence of a kind of matter not subject to gravitation. To do so had to be almost humiliating to a late nineteenth-century physicist, as it was tantamount to restoring scientific respectability to an imponderable, even if it was the "wondrous ether." After all, was not the discrediting of various imponderables the major pride of nineteenth-century physics? Yet, man's search for the ultimate constitution of matter was willing to take this embarrassing detour. Kelvin in 1901 solemnly retracted the "contempt and self-complacent compassion" with which sixty years earlier he and most of his colleagues had looked upon the ideas of elderly people who spoke of the imponderables. He was ready, as he put it, "to hark back to the dark ages of fifty, sixty or a hundred years ago" and admit "that there is something which we cannot refuse to call matter, but which is not subject to the Newtonian law of gravitation."[100]

If Kelvin and others were so willing to absolve the once despised concept of the imponderable, it was only because the ether with its marvelous properties seemed to give them the prospect of deciphering the "inner mechanism of the atom," which Kelvin termed the "superlatively grand question of physics."[101] This is not the place to review the various models built of the ether that were expected to explain the phenomena of light, electricity, and magnetism. Some of the models were rudely mechanical, others were sophisticatedly "ethereal." Of the latter the most talked about was the vortex atom, which grew out of Helmholtz's paper of 1858, "On Integrals of the Hydrodynamic Equations which express Vortex-Motion." Helmholtz's work was not only outstanding as an achievement in mathematics, but two of its results seemed particularly promising to break new ground in the search for the inner mechanism of atoms. One of these results was the perpetuity of the motion of a vortex ring in frictionless fluid; the other was the indestructibility of its form. It was also shown that if the vortex filaments in a fluid were not circular, that is, ring-shaped, but ellipsoidal, they would vibrate at various frequencies, according to their eccentricity, about the equilibrium position, or circular shape.

That these forward-moving, vibrating fluid rings could be easily illustrated by a simple device producing ammonia smoke rings was certainly not essential to the vortex theory, although no doubt highly satisfactory to any physicist of the 1860's. This is not the place to describe the "acrobatics" performed by the ammonia smoke rings when they impinge upon one another. It may be noted, however, that Kelvin was greatly impressed by seeing them in 1867 in Tait's laboratory. In the same year Helmholtz's paper appeared in English translation in the *Philosophical Magazine*, and for Kelvin the prospect of unlocking the mystery of the atom suddenly seemed an imminent possibility. His imagination was fired as never before, and in his letter of January 22, 1867, to Helmholtz he described the vortex rings of ether as being as permanent and indestructible as Lucretius' solid and hard atoms. In the same breath he also pointed out that the variations and combinations of not only two but a long chain of vortex rings might contain all the possibilities needed for the explanation of all observable properties of matter. Since the soundness of the mathematics of Helmholtz's paper was beyond question, only one thing was needed to make reality of these conjectures and hopes, the existence of a perfect frictionless medium, the ether. And which physicist would not swear in 1867 by the reality of the ether? Thus it was inevitable that Kelvin devoted the better part of the next three decades of his career to fashioning various vortex atom models out of the "wondrous ether." Although none of these proved to be the long expected answer to the "superlatively grand question," they commanded general admiration and inspired physicists like Larmor, Rydberg, J. J. Thomson, and others to speculate along similar lines.

Among the great number of admirers was Maxwell, who said in 1870 that "in the vortex theory we have nothing arbitrary, no central forces or occult properties of any other kind. We have nothing but matter and motion, and when the vortex is once started, its properties are all determined from the original impetus and no further assumptions are possible."[102] But as it turned out physics needed further assumptions, and drastic ones at that. Yet, it took four decades before it dawned on the physicists that the very vehicle of the vortex atom theory, the ether, was a sheer construct of arbitrariness, and what was worse, it simply could not be considered as matter. As late as 1904 J. T. Merz spoke of the vortex atom as "the most exalted glimpse into the mechanical view of matter."[103] In 1907 J. J. Thomson still considered the vortex atom theory of matter as more fundamental than "the corpuscular theory of matter with its assumptions of electrical charges and the forces between them." For as he put it in a manner so characteristic of the state of mind of classical physicists: "On this [vortex atom] theory the difference between matter and non-matter, between one kind of matter and another is a difference between the kinds of motion in the incompressible liquid at various places, matter being those portions of the liquid in which there is vortex motion."[104] If the theory had setbacks, these consisted, according to Thomson, in the handling of the mathematical equations involved in the theory. For in 1907, when he offered this evaluation of the prospects and merits of the vortex atom, it was still possible to expect what P. G. Tait envisioned three decades earlier: the mathematics of the vortex atom

"may employ perhaps the lifetimes, for the next two or three generations, of the best mathematicians in Europe."[105]

Such, however, was not the course in store for mathematics. Twentieth-century mathematicians were spared analyzing further a construct of mathematical physics that after all had no real connection
with the structure of matter at the atomic level. For one, serious
doubts arose on mathematical grounds as to the stability of the ordinary circular rings of Helmholtz. In 1905 Kelvin concluded that "if
any motion be given within a finite portion of an infinite incompressible liquid, originally at rest, its fate is necessarily dissipation to infinite distances with infinitely small velocities everywhere."[106] But it
was not basically the mathematical difficulties that caused the vortex
atom theory to share in the ultimate fate of its symbol, the ammonia
smoke ring, and go up in smoke. The reason was more general and lay
in a fundamental assumption of classical physics.

What classical physics tried to do in its search for the atoms was
to "etherialize" first the properties of ordinary, macroscopic matter
and then to find these properties realized in the atoms. To question
the soundness of such a procedure was unthinkable around 1900, and
it was voiced time and again that all manifestations of matter ought
to be reduced to the ether. Thus Larmor called the electric charge a
"mere passive pole — nucleus of beknottedness in some way — in the
ether."[107] In fact, most scientists of any stature hastened to add their
contribution to the temple of the ether. Mendeleev, for instance,
found place in his periodic table not only for two elements, called x
and y, lighter than hydrogen, but also postulated a third one in that
group that he identified with the ether whose atomic weight he calculated to be between 9.6×10^{-7} and 5.3×10^{-11}. Impressive as such
numerical values seemed, they were as gratuitous as the main argu-
ment submitted by Mendeleev in support of viewing the ether as an
element: the existence of inert gases provided, according to him, a
sufficient basis for speaking of a type of matter, such as the ether,
which does not enter into any sort of reaction with other bodies.[108]
Since then science has had to learn that those "inert" gases were no
more inert than the so-called permanent gases were permanent.

Physicists differed greatly of course about the details of how a particular property of matter could be explained by the "structure" of
the ether, but the agreement was rather general that it was the ether
that was to explain matter and not the other way around. "Matter
may be," wrote Larmor, "and likely is a structure in the ether, but
certainly ether is not a structure of matter" (italics mine).[109] In the
same vein J. J. Thomson identified in 1903 the whole mass of any

body with "the mass of ether surrounding the body which is carried along by the Faraday tubes associated with the atoms of the body." His insistence could not have been more categorical: "In fact, all mass is mass of the ether, all momentum, momentum of the ether, and all kinetic energy, kinetic energy of the ether."[110] In 1912 Poincaré still held that "the ether it is, the unknown, which explains matter, the known; matter is incapable of explaining the ether."[111] In the long history of man's search for the understanding of matter, rarely indeed were the categories of the known and of the unknown so erroneously labeled. No wonder it appeared to many in those years that to give up the search for the ether was to abandon the hope of ever knowing anything fundamental in the physical world and that to part with the ether was to part with physics and the physical world.

As it turned out the ether of the classical physicists did not prove itself to be the basic fiber and structure of matter. The complicatedness of atoms proved to be something entirely different from the complications of a grand piano to which Rowland once likened it. To represent an atom by a grand piano reduced by a factor of ten billion or so would have no doubt been in full conformity with the postulates of classical physics that imagined the same sort of mechanical intelligibility to exist at any range of magnitude of the physical world. The difference between the macroscopic and microscopic structure and properties of matter implied for classical physics but a mere change of the order of magnitude. This was of course an unavoidable consequence as long as the Euclidean space-time relations were accepted as exclusively valid in the universe. It is perhaps useful to note that classical physics was always well aware of this consequence. One need only recall a passage from Laplace's most widely read work, *Exposition du système du monde.* Speaking of the universal and rigorous validity of the inverse square law, Laplace noted that one of the properties of this law was that

if the dimensions of all bodies in the universe, their mutual distances and velocities were to be increased or reduced proportionately, they [those bodies] would describe paths entirely similar to those which in fact they do; thus, the universe, once reduced in such a way successively to the smallest imaginable space, would offer the same appearances to its observers. These appearances are therefore independent of the dimensions of the universe, as they also are, owing to the law of proportionality of force to velocity, independent of the absolute motion they might have in space.[112]

What Laplace said was no doubt true within an immense range of magnitude, both for the size and the velocities of bodies. Physics, however, had to find out that in resolving smaller and smaller dimensions of matter, a range can be reached where entirely new aspects of physical reality begin to unfold and where laws are to be discovered of which the laws of the macroscopic world are but special cases. Contrary to Laplace's views, absolute space and absolute motion were concepts that could not fit the atomic reality. The world of atoms that Maxwell described in 1873 as the "region where everything is certain and immutable"[113] turned out to be a realm where certainty was only statistical and immutability was constantly coupled with transformations that could involve even the whole atom.

When experimental research succeeded in edging down to the minute dimensions of atoms in the cathode-ray experiments of J. J. Thomson, the picture unfolding was in sheer contrast to the preconceived patterns offered by classical physics as to what the atom ought to be. In 1896 Thomson found the e/m ratio of cathode-ray particles to be two thousand times that of hydrogen ions, and he was correct in pointing to the novel character of the results: "We have in cathode rays matter in a new state, . . . a state in which all matter — that is, matter derived from different sources such as hydrogen, oxygen, etc., — is of one and the same kind."[114] Such a conclusion was indeed portentous. It implied, as Rutherford commented a quarter of a century later, that the structure of atoms, if there be any, must be electrical in nature and that optical spectra have their origin not in the mechanical vibrations of bits of ether, but rather in the oscillations of electrons. It also suggested that "the electron was probably the common unit in the structure of atoms which the periodic variation of the chemical properties indicated."[115]

The first atom models built on the principle of a group of positive- and negative-charge carriers were not long in coming. Thomson himself proposed several. In one of them the atom was pictured as a positive-charged body embedded with electrons. In another model the positive-charged main body of the atom was closely surrounded by electrons in successive rings, each ring being made up of eight electrons for reasons of stability. Lenard's experiments on the transmission of electrons through thin foils of various materials showed, however, that for electrons to pass through thousands of atoms so easily, the atoms themselves should be rather empty. This meant that the large dimensions of the positive-charged central body of the atom had to be replaced by a relatively small core. The size of the core was for several years a matter of conjecture, but the idea gave rise immediately to speculations that compared the system of positive and negative charges in the atom to a system like Saturn surrounded by its rings.[116]

In a sense this was a definite departure from the absolute simplicity atoms had to have by definition in classical physics; still, one cannot fail to notice how the concept of the Saturnian atom reflected the view in which the world always looked the same whether investigated on a cosmic scale or on an atomic one. The step from the domain of ordinary matter, even from the range of the best microscopic resolution into the dimensions of the atom, however, was far more than just inching closer to finer details. Actually it represented a move into an altogether different world. The step was made possible mainly by the tools of which physics suddenly came into possession, tools that were characterized by an extraordinary refinement as compared to the microtomes, microscopes, and chemical balances of yesteryear. They were the alpha rays from radioactive sources, Wilson cloud chambers, scintillation screens, Geiger counters, and so forth. No less extraordinary than the new tools were the results themselves. To symbolize perhaps how fast theories about the constitution of matter should in the following decades yield to one another, the first theory to be abandoned was the brand new Saturnian concept of the atom.

The trouble with the Saturnian atom was that it transferred unawares the relative distances of the rings from the planet to the imagined structure of the atom. These rings are rather close to the central body of Saturn, the radius of the outer ring being only a little over the double of the radius of the planet itself. A ring of negative electrons and an equal positive central charge placed so close together, however, could not influence very much the path of a charged particle passing through the atom. This is why Rutherford, in accepting in 1911 the idea of the Saturnian atom, envisaged only small-angle scattering of alpha particles sent through a thin foil of gold. He viewed the probability of a large-angle scattering as practically nil, and at the urging of H. Geiger, he gave permission only reluctantly to a young graduate student, E. Marsden, to look for large-angle scatterings. "I may tell you in confidence," he recalled twenty-three years later in 1936, "that I did not believe that they could be." Marsden set to work, however, and Rutherford experienced two or three days later what he called "the most incredible event" that had ever happened to him. As Geiger informed him with great excitement, not only were there large-angle deflections, but some of the alpha rays, or rather particles, were simply bouncing straight backward. This to Rutherford was, to quote his words, "almost as incredible as if you fired a 15-inch shell at a piece of tissue paper and it came back and hit you."[118]

Through such an "incredible" observation physics had to adopt the idea of an atom with a minute massive center carrying a charge. In the same stroke physics also had to learn how unimaginably minute the atomic nucleus was. Its radius turned out to be ten thousand times smaller than the radius of the atom, the size of which had been inferred by physics with reliability only a few decades earlier. At the same time it also became evident that the "tissue paper" that made the fifteen-inch shells bounce back consisted of such units of matter (nuclei of gold atoms), the density of which had to be at least a hundred million times greater than that of ordinary lead. Such a state of affairs was certainly not merely a replica of ordinary matter on a much smaller scale.

Instead of being an impenetrable, indivisible unit of matter, the atom proved to be a realm of vast empty spaces. By 1913 it had also been recognized that the nucleus was not "unbreakable" either. The alpha particles were clearly bits of it. Thus in the seventeen years following Becquerel's discovery of radioactivity and the subsequent establishment of the four natural radioactive series in heavy elements, science began to sense what F. Soddy, one of the chief pioneers of radioactivity, came to say about the "indivisible" atoms of classical physics: "The atom, for all that, is not Nature's unit, but ours." [119] Such an appraisal of the facts of discovery stood in sharp contrast to the general tone of the statements physicists were making about the atom in the preceding decades. In bringing about this change, the most palpable role was played by the discoveries connected with radioactivity. In sketching the short history of radioactivity, J. J. Thomson felt bound to remind the British Association in 1909 that contrary to the widely shared view "there were never any signs of an approach to finality in science." [120] A critical look at the sanguine hopes of yesteryear was no longer avoidable. The last layer of matter that appeared a few years earlier to be within the reach of science was now nowhere in sight. The real extent of the lack of finality in the scientific pursuit could not be ignored, and the suddenly felt keen awareness of it was hardly less novel than the discovery of radioactivity itself.

The twofold aspect of triumphs and puzzles that runs through man's search for an understanding of matter is strikingly illustrated by the process that started with the division of the "indivisible" atom into electrons and a nucleus. In fact, each major triumph in the understanding of the mysteries of the subatomic world involved the discovery of particles unsuspected before. One such success was the justification of Prout's hypothesis by Aston's determination of atomic weights. It was found that the atoms of all elements display variations (isotopes), the weights of which are integer multiples of the hydrogen atom when the mass defect represented by the packing fraction is taken into account. Yet, in order to preserve the electric balance of the atom, one had to choose between two alternatives: either positing a certain number of electrons in the nucleus to offset the extra positive charge if each mass unit was to be a proton or to postulate the existence of new particles having the same mass as protons but no charge.

The latter course was first conceived by Rutherford who in his Bakerian Lecture of 1920 predicted not only the existence of the neutron but also spoke of the deuteron, triple hydrogen, and triple helium well in advance of their discoveries.[121] In 1932 the neutron was discovered by J. Chadwick, and soon after it became clear both on theoretical and experimental grounds that electrons could not be present in the nucleus. By then the relatively simple triple system of the atomic world represented by electrons, protons, and neutrons, however, was undermined by evidences that nobody had foreseen and that could not be fitted into any previous pattern. Dirac, in trying to set up the Schrödinger equation for electrons in a form that would satisfy the requirements of relativity, ended up with a solution that yielded both a positive and a negative value. Dirac did his best to discover errors in his work, but the negative energy value could not be eliminated. There appeared to be only one way to interpret the solution in a meaningful way: to assign the positive and negative values to a pair of positive and negative electrons. This, however, meant the acceptance of a third fundamental unit with charge, later called positron, and physicists were most reluctant to accept such suggestions. For them a third particle with charge seemed to upset the symmetry present in the fact that the unit of electricity was already represented both in its positive and negative forms by the proton and the electron. The consensus was general that the possibilities offered by the basic units of charge were exhausted by these two particles. Such was the view of Rutherford[122] and Eddington,[123] to name only two of the most outstanding physicists of that era. Thomson in 1930 referred explicitly to the prevailing opinion according to which the repetition of discoveries like radioactivity, shaking again the foundations of physics, appeared very unlikely. Thomson, however, firmly dissented: "I am one of them who think that what has been found is but a small fraction of what there is to find, that the electron and the proton are not the last words in the story of the structure of matter, . . . that still, as Newton said, 'the vast ocean of truth lies undiscovered before us.'"[124]

He was a rather lonely prophet. The consensus was that the electron-proton synthesis was to stay. Describing this unanimity Millikan

stated in 1935 that prior to the discovery of the positron "the fundamental building stones of the physical world had been universally
supposed to be simply protons and negative electrons. Out of these
primordial entities all of the 92 elements had been formed."[125]
Viewed from another angle, this consensus amounted to a powerful
conceptual resistance to the possibility of positive electrons, and thus
it happened that physicists for years had been observing electron
tracks that they described as "curving the wrong way," "falling back
into the source," "moving backwards," or "coming up from the
floor."[126] The real situation was far more simple and far more extraordinary at the same time. But to see this it had to be recognized first
that the possibilities of nature far exceed "reasonable" appraisals.

On the night of August 2, 1932, C. D. Anderson obtained photographs of electron tracks that could be assigned only to an electron
with a positive charge, unless one was to continue with explanations
involving even the laboratory floor. His findings were corroborated
by the experiments of P. M. S. Blackett and G. P. Occhialini (1932),
which showed in a detailed manner that positrons and pair-produc-
tion tracks were the confirmation of Dirac's theory of antimatter. The
suddenness of change in the thinking of scientists about the secrets
of matter could not have been more electrifying. Not only did the
general resistance to the concept of the positive electron melt away
in no time, but intensive speculations about the existence of other
antiparticles also got under way. Before long speculation was joined
by actual work on the 6 Bev synchro-cyclotron in Berkeley, California, which was designed with the explicit purpose of detecting the
antiproton implicit in Dirac's theory. Characteristically enough, by
the time the existence of the antiproton was demonstrated in 1955,
the number of known or actively searched for antiparticles had exceeded a dozen, and the consideration of the existence of entire
galaxies built from antielements had come to be accepted as a most
serious possibility. This state of affairs evidenced itself with striking
concreteness after the discovery in 1965 of the first "anticompound,"
the antideuteron. One of its codiscoverers, L. M. Lederman, actually
voiced the view that owing to this discovery the bearing of the law of
symmetry between world and antiworld now makes it impossible "to
disprove the grand speculation that these anti-worlds could be populated by thinking creatures."[127] However that may be, it seems evident that findings and speculations of this type should suggest that
the complexities of the fundamental structure of matter might be far
greater than is often believed. The words of two eminent physicists
uttered in 1936 summed up concisely what has been repeatedly con-

firmed by subsequent discoveries: "We have never been as far from achieving unity in explanation in the realm of ultimate constituents of matter as we are at present."[128]

The discovery of the positron was only one item in the long list of puzzling complexities that were to be disclosed in the course of probing deeper and deeper into the secrets of matter. In general the world of the atom could be given a meaningful, or simply operational, explanation only at the price of introducing assumptions completely at variance with the principles of classical physics. One of these assumptions, the exclusion principle, implied that electrons in the atom cannot be "tagged" — or, to put this in a form bringing out more sharply the contrast with classical concepts, the electrons have no "individual identity" as parts of an atomic system. In the atom itself quantum mechanics eliminated not only the "orbits" of the electrons, but also the electrons themselves as distinct small spheres of electric charge: the nucleus is no longer thought to be surrounded by little satellites, but rather by electron clouds of fluctuating density and variable spatial distribution. Nor was the motion of material particles of atomic dimensions a small-scale reproduction of colliding billiard balls. It resembled rather the propagation of wave packets in terms of the hardly visualizable De Broglie waves. Thus the transition from the range of the best microscopic resolution to the realm of atomic dimensions saw the "hard bits" of matter turn into elusive wave packets. For it is in the very nature of a wave packet that determination of its location, of its energy content, and so forth, becomes meaningless within a volume of space whose diameter is smaller than the wavelength itself.

Advancing toward the next layer of matter, the world of the nucleus yielded no fewer questions than solutions. The major achievements are well known. Suffice it to mention the harnessing of nuclear energy, the countless scientific, medical, and industrial uses of radioisotopes, solutions of age-old problems of stellar physics and the spectacular feat of producing transuranic elements that several billion years ago were still abundant on the earth and throughout the solar system. Enormous as such triumphs are, the puzzles simultaneously emerging were not less so. The most obvious of them was the nature of forces holding together the protons and neutrons in the nucleus. It had to be a charge-independent force, a force far more powerful than that of electricity, but with an effective range barely exceeding the diameter of the nucleus. This was a radical departure from classical patterns and certainly not in keeping with Rowland's sweeping dictum made in 1899 whereby the ether was said to carry all actions at a distance, and therefore all actions at a distance were said to obey the law of the inverse square.[129] The nuclear force implied the idea of exchange particles peculiar to this force, much as the continuous emission and absorption of photons accounts for the electromagnetic field. From the range and strength of nuclear forces, Yukawa concluded that these exchange particles had a mass equivalent to about three hundred electron mass. When in 1937 C. D. Anderson and S. H. Neddermeyer discovered a particle with about 210 electron mass, it was at first believed that Yukawa's prediction was verified. But the μ mesons, as these particles were called, did not react at any considerable rate with nucleons, which should be the case, however, were they the exchange particles of nuclear forces. It was only in 1947 that C. F. Powell established the existence of particles, called π mesons or pions, having a sufficiently high rate of interaction with nuclei, and thus the problem of nuclear forces achieved a more satisfactory status. Such satisfaction, however, had its unsuspected problem. In the mesons, as will be seen shortly, physics caught in its net a particle that had to do with an interaction as novel as the nuclear forces themselves. Once more the basic pattern repeated itself: parallel with the answering of old problems new ones arose as well.

The nucleus, not long ago the hypothetical ultimate unit of matter, proved to be a "world" in itself. Efforts to build a fully satisfactory model of the nucleus have not been successful, and even today the structure of the nucleus lacks a theory comparable to that of the atoms based on the exclusion principle. It is not at all sure today that a combination of the shell model, liquid drop model, liquid rotator model, and complex potential model will lead to a definitive synthesis. One thing, however, became evident, namely, that in the strict sense of the word there is no such thing as a stable nucleus. The difference between unstable and stable nuclei was found to be not a matter of qualitative difference but only that of a very large numerical difference in the rate of decay. The decay of the nucleus results largely from the fact that one of its principal constituents, the neutron, transforms readily. Yet, what is true of the neutron might also hold for the proton and the electron. Of these two the electron has been considered as representing the absolute ground state in the spectrum of fundamental particles. Yet, there is an active search today for possible evidence of electron decay. At any rate all fundamental particles are unstable in the sense that all of them are subject to a total mass-energy conversion.

These were not the only startling results of probing the world of the nucleus. Research also stumbled on unexpected differences between properties of nucleons such as the values of magnetic moment for protons and neutrons, where such discrepancy was not expected to exist at all. But most important, beside the nuclear force there emerged slowly the presence of another nuclear interaction governed by a force wholly distinct from gravitational, electromagnetic, and nuclear forces. It has been named the fourth force, or the force of weak interactions, and it took about three decades for theoretical nuclear physics to recognize its distinct character and fundamental importance. This is not the place to recount the steps that led to the postulate and discovery of the neutrino, which is the chief clue to this fourth basic force in nature.

Viewed first as a stopgap solution, the neutrino's existence was doubted by many for a number of years and not without grave reasons. Foremost of these was the extremely low rate of interaction of neutrinos with matter. It should be recalled that a lead wall of four meters in thickness is needed for gamma rays of medium energy to interact with matter at the same rate electrons do when passing through a lead plate of one millimeter in thickness. For low-energy neutrinos to have the same degree of probability of interacting with the nuclei of atoms, a lead wall of ten million miles in thickness would be needed. Thus one can understand why R. C. Tolman in 1947 called the neutrino a "horrible little particle" and characterized as "very unsatisfactory" the situation wherein physics failed to detect any direct effect due to the neutrinos. He took in effect the view that "perhaps it would be better to abandon the idea of the conservation of energy rather than to invent a new particle that one can never find." [180]

To declare something forever impossible in experimental physics is not, however, the most rewarding kind of prognostication. The neutrinos proved themselves in the rapidly advancing world of modern physics. Their existence was established experimentally by C. L. Cowan and his co-workers in 1956, and further advances made in their study also made it possible to throw light on the nature of weak interactions and to devise experiments for the detection of the W particles governing the so-called "weak" force. Of the four W particles only the positive has been traced so far with reasonable evidence, but it now seems to be a matter of time only before the W^- and W_0 with its antiparticle will also be found. The fourth, or weak, force has thus taken its place in physics and points to a new realm in the study of matter full of puzzles and unsuspected interconnections, as was the case with the other forces already known by physics.

The realm of weak interactions burst on the field of physics when following the discovery of π mesons the vastly improved nuclear photographic plates began to show V-shaped ionization tracks indicating particles having 960 and 2,100 electron mass, which received the names of κ and λ particles, respectively. Their discovery not only multiplied "unnecessarily" the number of "fundamental" particles, but also made their possible connections even more conjectural. The main reason for this lay in their unreasonably long lifetime. Having their origin in strong interactions, these particles should have decayed in about 10^{-23} sec, but their average life span was of the order of 10^{-9} sec, or in other words, these particles were found to live one hundred thousand billion times longer than they should. Because of this gigantic discrepancy between their expected and actual lifetimes they were dubbed "strange" particles without anybody then suspecting how strange indeed they were.

The first clue in deciphering the peculiarities of these "strange" particles came when A. Pais noticed in 1952 that strange particles are produced only in pairs. In other words some unknown physical property forbids that any of them should be produced alone. What this physical property might be is still a matter of conjecture. The most successful speculation is the "strangeness" theory proposed by M. Gell-Mann. For reasons which cannot be detailed here, every fundamental particle can be given a strangeness number, and as it turns out, the total strangeness number represented by particles entering into interactions is conserved in the strong (nuclear and electromagnetic) interactions but not in the weak interactions. The theory also accounts for the associated production of strange particles that are always the products of collisions between non-strange particles. Since the strangeness of the latter is zero, the total strangeness of the products must also be zero. Therefore at least two should be made of them each time and in such a way that the sum of their individual strangeness would equal zero. What is even more important, the theory predicted the existence of the Ξ, Σ, and K particles, which turned out to have 2,580, 2,230, and 965 electron mass, respectively.

After these particles were identified by experiments performed at Brookhaven Laboratories, the view began to gain ascendency that in the total of thirty-four fundamental particles physics has the basic building blocks of matter. This superficial hope was even strengthened by the fact that for several years no new particles were noticed. Those who were more cautious happened to be closer to reality. One of them was Gell-Mann, the author of the theory, who in 1957 noted that "we are at present very far from having a satisfactory theory of the elementary particles."[131] Another was Oppenheimer who in a lecture given in 1955 voiced his doubt about the assumed completeness of the list of elementary particles, and pointed to the fact that some of them were of very recent arrival. "As in a lecture hall," he remarked, "when people are coming in, if some have come in a minute ago, you think a few may come later." [132] Come in they did, and they made a situation already very strange even more so.

The strangeness theory is strange enough in that it is much more of an algebraic device than an expression of some physical property. It gives an "explanation" of the system of fundamental particles comparable to that provided by Mendeleev's Periodic Law for the elements, or by Balmer's formula for the spectral lines of hydrogen. Such an "explanation" is of course of much lower value than that provided by Pauli's exclusion principle for the systematic build-up of elements of increasing atomic number, or by Bohr's theory for the hydrogen atom. In this sense Oppenheimer's words are very much to the point: "These particles are not understood; they were not anticipated; no one had any idea that they would exist; they are not contained in any known theory. . . . They go way beyond the framework of any speculation we had." [133]

The strange particles proved to be even more a world of their own when it was found that in weak interactions parity is not conserved. Conservation of parity had been for several decades a principle invariably obeyed in nuclear interactions. It says essentially that mirror images are identical just as it makes no difference to a physiologist whether he uses the right hand or the left as a model when explaining its functions. In the same way it was natural to assume that the physical properties of subatomic particles were wholly identical whether the particles had a right-handed or a left-handed spin. It is true that since Pasteur's work on tartaric acid it has been realized that certain molecules exist only in a one-handed state, such as tartaric acid, which occurs only in the right-handed version in fermenting grapes. Nobody dared, however, to assume that such asymmetry might be present at the "ultimate" level of matter. After all, the conservation of parity, a most effective principle in analyzing atomic and nuclear interactions, knew of no exceptions. Yet, at the same time a successful theory can easily produce a state of mind that fails to recognize the presence of proofs to the contrary. This is why the clear evidence of certain nuclear emulsion plates showing the non-conservation of parity in π^+ and μ^- decay went unrecognized for almost ten years. There were simply no "eyes" to see that the number of decay electrons was not the same in the direction of the motion of μ mesons as in the opposite sense. Those numbers should have been identical if parity, that is, reflection symmetry, holds in these decay processes.

It was therefore a very daring step when Lee and Yang questioned the validity of the conservation of parity in the decay of K particles and in any weak interaction in general. Experiments suggested to this end have borne out their contention with the recognition of the fact among other things that only left-handed neutrinos exist in nature. On the other hand, antineutrinos, discovered in 1962, emerge from nuclear interaction like a right-handed screw. Moreover it now seems that mesons, electrons, and neutrinos not only have handedness but that this may also be their most distinctive property. It is still too early to tell what long-range effects may emerge in theoretical physics from the overthrow of the universal validity of the conservation of parity. To illustrate how shocking the changes might be, it is best to recall the reaction of such an outstanding figure of modern physics as Pauli, who simply scoffed at Lee's and Yang's conclusions. In fact he told V. F. Weisskopf in a letter of January 17, 1957, that he did not believe that "the Lord was a weak left-hander," and confidently expected the experiments to yield a symmetric result. Only two days later the outcome of the now classical experiments was announced and copies of three papers discussing the data reached Pauli on January 21. Pauli's next letter to Weisskopf reads:

Now the first shock is over and I begin to collect myself again (as one says in Munich). Yes, it was very dramatic. On Monday 21st at 8:15 p.m. I was supposed to give a talk about "past and recent history of the neutrino." At 5 p.m. the mail brought me three experimental papers: C. S. Wu, Ledermann, and Telegdi; . . . What shocks me is not the fact that God is just left-handed but the fact that in spite of this He exhibits Himself as left-right symmetric when He expresses Himself strongly. In short, the real problem now is why the strong interactions are left-right symmetric.[134]

It is no exaggeration to say that the overthrow of the conservation of parity in weak interactions reveals more about matter than many far more spectacular achievements of modern physics. Some compared its impact to the Michelson-Morley experiment that ultimately showed the futility of looking for the ether and helped the way to entirely new modes of thinking in physics. Its sudden recognition symbolizes that strange irony that runs through the development of physics: almost invariably fateful cracks begin to appear in a major synthesis just before the finishing touch is added. Thus when on August 15, 1963, the *Physical Review Letters* announced the detection of the anti-Xi-zero particle, the last yet unobserved fundamental particle predicted by the strangeness theory, the impressive system of thirty-four fundamental particles was already confounded by facts. In 1961 the existence of two particles, the ρ and ω, each with about 1,540 electron mass and with 10^{-22} sec lifetime, was announced by several research groups. The clue to their existence grew out from R. Hofstadter's work on the radius of the cloud of electrification surrounding the proton and neutron. The radius of proton and neutron arrived at by probing them with high-speed electrons, proved to be smaller than the so-called nuclear size. To explain this discrepancy, the existence of two types of heavy muons was postulated, one of them given in advance the glamorous but very inappropriate name, omega. As for the expression "fundamental particles," the sobering array of facts shows that as the facetious saying goes, the only sure thing about them is that they are not fundamental. The ω particle can aspire even less to being the last word in fundamental particle physics because its characteristics might strongly indicate that it is much more of an "excited state" of another particle. Several other excited states have been found in recent years, such as the ϕ meson with about 2,000 electron mass, but their connection with other particles has not yet been definitively established. Nevertheless such discoveries already show that the once rigid concept of fundamental particle is being broken down and that several of them might prove to be only the excited states of a few units of matter that may not be elementary at all in the strict sense of the word.

At any rate, one should not forget that the distinction between more and less fundamental particles reflects psychological preferences rather than the dictates of mathematical physics. From the viewpoint of mathematical formalism, all particles are on equal footing as all refer to singularities. One singularity can, however, hardly have preeminence over another. What is more, not every singularity corresponds to a particle. This is clearly one of the reasons why a perceptive study of the problem had to emphasize rather recently that "there is no hope at the moment of finishing the elementary particle story in any neat way." [135]

What most physicists feel today in the presence of "fundamental" particles whose number with all the resonances counted exceeded sixty in 1963 and was well over one hundred in 1965,[136] parallels closely the deep dissatisfaction most physicists felt a hundred years ago about the seventy-some apparently irreducible "elements." Thus Harold Ticho of the University of California described the search for fundamental particles as "turning up a mess of disconnected beasts,"[137] while L. Brillouin noted in connection with the uninterrupted flow of new data that "our imagination does not manage to follow this infernal race." [138] So far no theory has proved satisfactory to reduce all the "fundamental" particles to a few, although attempts are not lacking. A very interesting theory is Gell-Mann's "eightfold way," which organized the known resonances into groups determined by eight quantum numbers. It was on the basis of this theory that J. J. Sakurai late in 1962 predicted [139] the existence of ϕ mesons, which were discovered four months later by two groups working at

Brookhaven and Berkeley. An even more impressive confirmation of the theory came with the discovery of a particle that has a negative electric charge and a mass of 1,686 million electron volts. "Dramatic" as this discovery could be, as described by P. T. Mathews writing in the February 19, 1964, issue of *The New Scientist*, high-energy physicists will probably be in for a great shock or two if they continue "walking around with a slightly hysterical look as though they are actually witnessing the apple landing on Newton's head." It is indeed rather doubtful that the future course of high-energy research can substantiate the sanguine comment of Gell-Mann who was quoted upon learning of the discovery, that it seems very unlikely that a new particle in the future might upset the theory of the eightfold way.[140]

Actually, impressive as was the success of the "eightfold way" in predicting the discovery of ω^-, the hopes of a final and definitive scheme of fundamental particles failed to materialize.[141] The year 1965 was already dominated by theories that arranged the fundamental particles in three main groups, clustered around the electron, meson, and proton. One of these theories, known as SU-6, announced at the January, 1965, meeting of the American Physical Society, claimed to its credit the prediction of the ratio between the magnetic moment of the proton and that of the neutron. Also, the three main groups of particles numbering 35, 56, and 70, as postulated by this theory, leave enough gaps for further experimental verification. Yet, this way of classifying fundamental particles is based on a mathematical formalism, which, as a few months later was pointed out by V. F. Weisskopf, seems to suggest the existence of even more fundamental particles of matter that he called "quarks." More recently, in September, 1965, the same physicist took the view that the protons whose indivisibility was a basic tenet in physics for over four decades should be regarded as composed of smaller units.[142]

The inner logic of this most recent development could hardly be more typical of the whole history of man's search for the ultimate building blocks of the universe. No sooner had a final solution come into sight than one could also see a new layer of matter emerging on the scene. Whether the "quarks" will be discovered by the recently proposed 1,000 Bev accelerator is a question that no one can answer at the moment. At any rate, it is interesting to note that the booklet that presents the views of some thirty leading physicists on the desirability of such a gigantic device, clearly attests their awareness of the stubborn elusiveness of a final theory of matter.[143] Thus, R. G. Sachs noted the on-and-off hopes concerning a final solution that characterized the last two decades of fundamental particle research.

Accordingly, he spoke of the 1,000 Bev accelerator as a tool that might provide man with an insight *into but the next layer* of the simplicity of the laws of physics.[144]

It is well to remember that the all too large number of particles and resonances prompted many a physicist in the past few years to ponder whether continued emphasis on discovering more "elementary" particles and grouping them in various manners would really lead science to a deeper understanding of matter. Thus it has been suggested rather frequently in recent years that greater attention should be given to symmetry principles such as conservation of electric charge, of mass, and of energy. "The future theory of matter," said Heisenberg in 1958, "will probably contain, as conceived in Plato's philosophy, only assumptions of symmetry. Already now these assumptions of symmetry can be stated to a large extent; they seem to show that the future theory will be very simple and concise in its fundamentals, despite all complications of its inferences."[145] One should not forget, however, that basic as a given symmetry postulate might appear, one can always look for an even more basic reason underlying it. Such would be the case with an apparently wholly satisfactory theory of matter of which, however, no indications are yet in sight. Heisenberg's own equation of matter, expressed in terms of non-linear spinor theory, as a matter of fact ran into difficulties as soon as it was made public in 1958. A year later, when several new data seemed to fit into place, Heisenberg was prompted to strike a very hopeful note again: "My hope is that in two years I will definitely be able to say yes or no about the theory. When it is finished, then the intelligent layman will be able to understand it just as he understood Copernicus' conclusions about the earth revolving around the sun without being able to follow the mathematics of it all."[146] Yet, the final word seems to be as far away as it has ever been. For it is one thing to hope, as Heisenberg did, that "it is quite conceivable that in the not too distant future we shall be able to write down a single equation from which will follow the properties of matter in general."[147] The implementation of such hopes, however, is quite another thing. And so far the various solutions proposed fell far short of the desired goal.

New elements, new forces, antimatter, asymmetry, and a host of unexpected particles were not the only surprises in store for man reaching into the world of the nucleus. Even more upsetting were the changes that took place in long sacred concepts of classical physics about matter as such. The traditional proportionality between mass and volume had to be modified, since all the subatomic particles regardless of their masses are thought to have a radius of the same order of magnitude. The principle of the constancy of matter also had to be greatly modified, in part owing to the functional dependence of mass on velocity, in part because of processes such as the annihilation of matter and pair production. Not even an approximate "mechanism" of these is surmised if it is at all still meaningful to search for "mechanism" on the nuclear level. Again, the β emission is often spoken of as a potentially creative process, a parlance obviously indicating the great change in scientific thinking. In the same vein the representatives of the steady-state theory dared to introduce into physics the idea of a continuous creation of matter *ex nihilo*. It should be clearly understood, wrote H. Bondi, "that the creation here discussed is the formation of matter not out of radiation but out of nothing."[148] Instead of creating furor and wholesale indignation, as it would have certainly done two or three generations ago,[149] the idea has rather been subjected to various experimental tests, with all the evidences so far obtained going against it. That a proposition like the continuous creation of matter is not being dismissed out of hand is highly indicative of the scientific mood of our times, in which scientists begin to ready themselves even for the "impossible." Clearly gone is the era when the leading chemist of the day could introduce his work on the early history of the study of elements with the complacent statement that "the world is today without a mystery."[150]

Small wonder that present-day physics labors under serious semantic difficulties with regard to the uses of concepts like *matter* and *mass*. The word mass does not connote exactly the same thing in mechanics, in electrodynamics, in gravitation, in relativistic, and in quantum mechanical field theories, and the nuances involved in the various meanings are far from being unified. Even more disturbing are the difficulties that arise from the possible existence of the quantum of length, called *hodon*, which might form, with the velocity of light and the Planck constant, a triad of universal constants in nature. The value of this length quantum seems to be of the same order of magnitude as the radius of the nucleus. If space or length is quantized in such a manner then our usual concepts would apply only in regions of space and time that are large compared with the dimensions of the nucleus. Thus concurrently with the exploration of the region of matter in the range of 10^{-18}cm, a revolution may occur again in physics that might surpass in extent and depth the one following the conquest of the range below 10^{-8}cm. To such eventuality did Heisenberg refer when he voiced the opinion that science should be prepared for "phenomena of a qualitatively new character"[151] upon probing deeper into the structure of matter. Foremost of these new phenomena would be the possible detection of time reversal, which was foreshadowed in the process of renormalization in quantum electrodynamics by such negative quantities that in ordinary quantum mechanics should be interpreted as probabilities.

Even if time reversal should prove in principle unobservable, the challenging problems of the world of the nucleus would still remain many. Why elementary particles have the masses and lifetimes they do is a question that demands an answer. There are a host of similar questions still to solve. Perhaps not all such questions are the right ones to ask, but it is precisely these questions that the periodic table of elements once raised and that were later successfully solved by quantum mechanics. With regard to the world of the nucleus, "this is the kind of order," to quote Oppenheimer, "that serious physicists are bound to hope for."[152] Yet, one can never be sufficiently aware of the fact that the order of nature expressed in the various formulas of physics is but the reduction of given sets of data and formulas to more general ones. There is nothing in such results that would rule out the possibility of new advances along these lines. Also, if past experience is of any value, it appears rather doubtful that in unfolding the layers of matter physics would in the foreseeable future reach solutions that would not point in one way or another beyond themselves. The immense potentiality of particle accelerators, of which more will be said in a later chapter, is in itself a powerful indication that the search for the mysteries of matter will in a sense remain an unfinished business. This is not to say that the study of matter will necessarily go through another series of "particle" discoveries repeating on a "smaller" scale the discoveries that accompanied the splitting of the atom and of the nucleus. It is unlikely that scientific history should rerun a past course in such a literal sense. Man, owing to the power of his instruments, is no longer discovering but simply producing particles. In this circumstance alone one should be able to sight vistas that in their immensity should defy any attempt to map them accurately. Such an appraisal of the present stage of man's search for the understanding of matter will hardly be palatable for those who read science with cocksure confidence and sanguine expectations and speak invariably of an impending answer for all puzzles of matter. But science was never meant to nurture vain hopes and "unscientific" illusions. For "nothing is so preposterously unscientific," as Tait wrote almost one hundred years ago, "than to assert . . . that with the utmost strides attempted by science we should necessarily be sensibly nearer to a conception of the ultimate nature of matter."[153]

Tait's words turned out to be literally true. Whatever has happened since then in the investigation of matter did full justice to his sobering appraisal of the immensity of facts still to be learned:

Only sheer ignorance could assert that there is any limit to the amount of information which human beings may in time acquire of the constitution of matter. However far we may manage to go, there will still appear before us something further to be assailed. The small separate particles of a gas are each, no doubt less complex in structure than the whole visible universe; but the comparison is a comparison of two infinites.[154]

For the illusions he so severely castigated, Tait laid blame on the "quasi-scientific" writers of his day. Quasi-scientific writers, however, were not the exclusive symptom of Victorian smugness. Today their voice is no less alluring. Unfortunately some men of science strike rather often the very same note, no doubt because the illusion of an allegedly "final" solution is a state of mind that will always delight the unwary and the superficial. The alternate choice demands a far greater discipline of mind, for it has to blend in a dynamic synthesis both soberness and enthusiasm. Soberness, which is the result of the realization that the great schema of matter, the ultimate synthesis, the rock-bottom layer of the material world is today probably as far away as it has ever been. Today, it is no less true what Schrödinger admitted more than ten years ago, that our picture of the material reality is "more wavering and uncertain than it has been for a long time." [155] Today, even more than ten years ago, a "mess of formulas," to borrow his words again,[156] surround a subject that grows more puzzling. Soberness, however, should be coupled with enthusiasm deriving from man's abiding faith in the inner harmony of nature. Judging by the tremendous effort that goes in our days into fundamental particle research, this faith is as strong as ever. Happily so, for as one prominent figure on the field warned so eloquently:

We cannot make much progress without a faith that in this bewildering field of human experience, which is so new and so much more complicated than we thought even five years ago, there is a unique and necessary order: not an order that we can tell a priori, not an order that we can see without experience, but an order which means that the parts fit into a whole and that the whole requires the parts.[157]

The Frontiers of the Cosmos

The history of astronomy, said E. Hubble, is a history of receding horizons.[1] Ever since man tried to understand and explain the world, he has also felt an instinctive urge to fathom its size and locate its ultimate boundaries. As these attempts followed one another, there never failed to emerge unsuspected regions beyond the momentary limits of the universe, and new measuring rods were needed to keep pace with the ever-receding frontiers of the cosmos. These measuring rods, replacing one another as the centuries went on, were in a sense divining rods too: each major step in the increase of the known size of the universe unveiled one of its hitherto hidden aspects and provoked in turn revolutionary changes in the way man pictured the structure of the physical world.

It was more than coincidence that as the first quantitative estimates on the size of the universe were formulated among the Ionian philosophers the traditional world view of the Homeric age began to yield to a largely demythologized cosmology. In the world picture of the Ionian philosophers, there was no place for the births, struggles, and dissolutions of gods representing various parts of the universe. In their cosmos, matter, motion, and measure dominated. At the same time the Ionians were also the first to give a quantitative description of the size of the cosmos, which they pictured as a huge sphere, half-filled with the sea upon which floated the earth's disk. The extent of this disk was very likely taken by them as equivalent to the dry land reaching from Gibraltar to the Indian Ocean. The thickness of the disk, as Anaximander guessed, equaled one-third of its diameter.[2] These few thousand miles then were the first unit of length used to measure the universe.

For all its shortcomings, this approach to the understanding of the world was nonetheless scientific, for Anaximander expressed the size of the moon and sun in terms of this first astronomical or cosmological distance unit: the diameter of the earth's disk. According to him, the moon and the sun were huge rims inside of which there was fire. On each rim he pictured one round hole through which the interior fire became visible, and it was these fiery, glowing holes that were, according to him, identified by man as the moon and the sun. For reasons entirely unknown to us, Anaximander also stated that the diameter of the rim of the moon was about nineteen times as large as the diameter of the earth's disk, while the respective factor for the rim of the sun was twenty-seven.[3] Such great dimensions, however, were not to the liking of every early Greek thinker. Heraclitus, for one, fancied a sun with a diameter of one foot.[4] Even in his time, however, this was a peculiar view that failed to show any correlation between terrestrial and astronomical magnitudes. That these two realms could not be left unconnected was again recognized by Empedocles, who put the earth-moon distance at one-third the radius of the crystal sphere enclosing the world.[5]

Puny and sketchy as it may appear today, this world picture was the first milestone on the long road for an ever more inclusive grasp of the structure and shape of the world. The first of the scientific world pictures was of necessity very limited, and this the Greeks themselves recognized before long. The sighting of a much larger universe was made possible by several factors. One of them was the long array of bold speculations produced by the Greek mind during the early fifth century. From the Pythagorean circles spread the idea of a spherical earth, moon and sun, and the various phenomena produced by eclipses obtained a simple explanation. What gave a decisive impact to these speculations, however, was the first major advance made by man in the art of measuring large distances. The science of trigonometry, one of the many brilliant inventions of the Greek mind, was in a sense a precursor of telescopes. It brought faraway objects within the compass of measurement and first made it possible for man to penetrate in a quantitative manner the far reaches of space with the result that the accepted ideas on the structure of the cosmos had to be drastically revised.

Whether Thales was the first to compute by trigonometrical methods the distances of ships still far off the coastline cannot be decided today. At any rate, by the beginning of the fourth century the Greek geometers were already in possession of fairly good estimates of the diameter and circumference of the earth. Aristotle when discussing the size of the earth referred to measurements that had been in progress for considerable time; these in part should be credited to Eudoxus and Archytas.[6] The dimensions of the earth in comparison with Anaximander's flat disk increased in no small measure, and the earlier primitive guesses of the size of the sun also came to be completely discredited.

As to the extent of the dry land, Aristotle submitted, with the proviso of "as far as such matters admit of accurate statements," that its east-west extension exceeded that in the north-south direction by a ratio of more than five to three.[7] On the circumference of the globe, he quoted the calculations of the mathematicians as 400,000 stadia,[8] which, if one stade is taken as 300 cubits, or 516 feet, would be about 39,400 miles, or over one and a half times the present-day figure of 24,902 miles. The size of the sun, according to Aristotle, was larger than that of the earth, "if astronomical demonstrations are correct,"[9] and he took the view that "the bulk of the earth is infinitesimal in comparison to the whole world that surrounds it."[10] He put the distance of the stars at many times that of the sun from the earth for the reason that the earth never cast a shadow on the stars.[11]

This clearly went far beyond the views of the Ionians. Aristotle, however, was merely registering a change that became operative two generations beforehand. To what extent was the need felt for a thorough revision of the traditional views on the size of the universe can best be seen in the efforts of the atomists who confidently argued in favor of the infinity of space, of the infinite number of worlds, and of their continuous formation and dissolution. In science, however, bold conceptions and observational evidence should remain in interdependence, and there was no strict evidence whatsoever to support the cosmological system of the atomists. Greek science, therefore, had to decide for a finite universe, whose structure was first worked out on Plato's suggestion by Eudoxus. It was a closed world, centered on the earth, as Simplicius informs us,[12] and built on a series of concentric crystal spheres that carried in their respective orbital circles the moon, the sun, the planets, and the stars.

Though closed and finite, this universe was anything but small. From the third century on Greek astronomers could calculate the size and distance of the moon and the sun in terms of the new measuring rod of the universe: the diameter of the earth. As regards the moon, the results were surprisingly correct. Aristarchus gave the moon's diameter as one-third that of the earth, whereas modern measurements put it at little over one-fourth. From the analysis of the geometry of the moon's total eclipse, Aristarchus found the earth-moon distance to be equal to 40 earth diameters as opposed to its modern value of almost exactly 30 earth diameters. His calculations of the sun's diameter and distance, though based on a correct method, fell far short of the real values, the reason being that he was able to measure the angle enclosed by the half moon, the earth, and the sun only with an error of almost three degrees. Consequently, while the sun's diameter was nearly equal to 110 and its distance from the earth to about 12,000 earth diameters, Aristarchus obtained 6⅔ and 764 earth diameters, respectively. One hundred and fifty years later, Hipparchus, the greatest observational astronomer of antiquity, improved considerably on Aristarchus' data; he gave the moon's diameter as 0.29, its distance as 30.2 earth diameters, and the corresponding values for the sun as 12⅓ and 1,275 earth diameters. Hipparchus' data concerning the moon differ only by a few per cent from the values accepted today, whereas the earth-sun distance given by him is only about one-tenth of the real value.[13]

The earth's diameter as a measuring unit remained the basis of all attempts to size up the universe until very recent times, which have seen the introduction of radar techniques. Few achievements commend ancient Greek astronomy more than the fact that it was precisely in the determination of the earth's diameter that the distance determinations of the Greeks reached their greatest precision in the hands of Eratosthenes. His result, when interpreted in the most favorable way, was only fifty miles less than the modern value for the polar diameter of the earth, 7,900 miles.[14]

Occasionally it occurred to the Greeks that the sun might be much farther away than ten times the earth-moon distance. Posidonius (*ca.* 100 B.C.), for instance, conjectured, as Pliny informs us,[15] that the earth-sun distance might be equal to 5,000 earth diameters, which is about one-half of the real value. Such a scale, however, would have entailed vast, "unused" spaces among the orbits of planets, and this contrasted sharply with a conception of the universe in which everything had to be arranged in a humanly purposeful manner. Furthermore, the system of celestial spheres, which transmitted the motion to all celestial bodies from the sphere of the stars down to the orbit of the moon, required that all the neighboring spheres, and consequenly their orbits too, should be in contact with one another. As a result, each planet was pictured occupying a spherical shell of space whose thickness was determined by the maximum change of the planet's distance from the earth.

Proclus (*ca.* A.D. 400) referred to this conception as the familiar view among astronomers of previous generations,[16] and it is important to note that this conception of the world implied numerical estimates as well for the size of the universe. From the then known values of the earth-moon and earth-sun distances, from the ratios of the radii of the deferent and epicycle of each planet, and from the assumption that there is no "empty" or "unused" space between their orbital shells, one could in fact derive numerical values for the radii of crystal spheres carrying the planets and the fixed stars. The latter was assumed to be directly adjoining the sphere of Saturn. In all likelihood such computations had been completed well before Proclus' time, but the first known table of planetary distances dates back to the work of the ninth-century Arab astronomer, Al-Farghani of Baghdad, whose data were accepted throughout the Middle Ages.[17] Al-Farghani gave the distance of Saturn and of the sphere of fixed stars as equal to 20,110 earth radii, which is about 11 times smaller than the true mean distance of Saturn from the Sun. The distance of the closest star, Proxima Centauri, (4.16 light years or 2.2×10^{13} miles), is about 300,000 times greater than Al-Farghani's figure.

This was just about the farthest the compass of geocentricism could reach into the universe. It is therefore highly revealing that the first man in antiquity to propose heliocentricism as the true representation of the planetary system sensed unerringly that on such a theory the world — the sphere of the stars to be specific — would have to expand enormously. If Archimedes' reading of Aristarchus in the *Sand Reckoner* is correct, then Aristarchus claimed that the ratio of the radius of the earth to the radius of the earth's orbit was equal to the ratio of this to the radius of the sphere of the fixed stars.[18]

Whatever the merits of such an explanation, it might weaken the principal point in Aristarchus' comparison: the intimation of immense celestial distances implied in the heliocentric system. For the magnitude of any of the ratios mentioned by Archimedes was expressed in the phrasing of Aristarchus as the ratio of the center of a sphere to its surface, which makes sense only if the word "center" is taken in the sense of an exceedingly small magnitude, or radius, as compared with the radius of the larger sphere. Greek astronomy was willing, when speaking of the dimensions of the universe, to accept one such ratio. Ptolemy himself considered the dimensions of the earth as merely a point compared with the distance of the fixed stars.[19] The heliocentricism of Aristarchus asserted, however that one has to apply twice the same immense proportion to reach the boundaries of the world. Shocking as this aspect of heliocentricism might have been, it failed to provoke explicit comments. Other aspects of heliocentricism were found even more obviously shocking, or simply "ridiculous" and "absurd" as Ptolemy called them.[20] For him the respective merits of heliocentricism and geocentricism meant the contrast between irresponsible conjecture and final, absolute truth. As a result the immense distances implied in the heliocentric ordering of the universe had no chance in classical times to stimulate cosmological thought.

What is more, as a younger contemporary of Aristarchus expressed it, the heliocentric system was guilty of impiety in the eyes of many for moving the hearth of the world.[21] Clearly, then as now, man's heart was deeply attached to where the hearth of the universe was supposed to be and to make a sudden break with hallowed sympathies by anticipating centuries, let alone two millennia, was not a step likely to be taken. The closed universe with the not too frightening dimensions assigned to it by a small circle of erudite men in Hellenistic times was to stay until the very advent of the modern age.

It is well to remember that this circle was not only very limited in its influence throughout almost eighteen centuries but was also very forgetful at times about the best available data, incorrect as they were. Even the rather precise measurements of the size of the earth could not result in a sufficiently realistic grasp of its vastness. According to Seneca, anyone with favorable winds could sail westward from Hispania to India in four days.[22] Pliny, the industrious Roman compiler of Greek science, showed time and again great gaps in his information, and the way he filled them was unfortunate oftener than not. Thus, for Pliny, the moon was larger than the earth, because otherwise the sun's eclipse would have no explanation.[23] How little the literary circles were permeated by the picture of the world as presented by Greek science can be clearly seen in the works of Virgil or Horace, with very liberal allowances made for poetical license notwithstanding.

The scientific aspects of the world picture were handled with greater care by the more prominent Christian authors of the patristic age. It was only rarely that the tentlike description of the world given in Genesis was fused in their writings with the ideas of scientific astronomy, and only minor figures, like Lactantius and Theodore of Mopsuestia, believed on the strength of biblical passages the earth to be a flat disk.[24] For the Christian mind, however, there was the even greater temptation of coordinating the system of heavenly spheres with various details of supernatural revelation. How thoroughly one could succumb to this can nowhere be better seen than in Dante's *Divina Commedia*. His presentation of the layers of the earth and of the upper regions, on which the whole poem is framed, strongly illustrates the fact that a poetico-philosophical synthesis of the scientific and theological world views corrupts not only the scientific data, but also greatly strengthens a popular theological world view that is too closely identified with a provisional stage of astronomical thought.

In Dante's day and even much later, at the close of the fifteenth century, nothing had yet indicated that this compact, unified world, so easily traversed by the flight of fancy, was only illusion. Still, would a Columbus, for instance, have had enough courage to set sail into the unknown if his maps, drawn up probably around 1489 by the German cartographer, Henricus Martellus, had not located the island of Cipangu, that is, Japan, only 90° west of the Canary Islands? By October 11, 1492, Columbus had sailed 89° west from the Canaries and landed in the Caribbean, discovering a continent he did not know existed. Such was the unexpected reward of the confidence he and his contemporaries put in the highly revered statements of classical authors who, like Seneca, failed almost totally to grasp the bearing of sound calculations on the estimates concerning the vastness of the earth.

When the classical and medieval world picture came to be challenged by Copernicus' work, the problem of adjusting oneself to suddenly opening up spaces made itself heavily felt. For, if the Earth revolved around the Sun, the apparent immobility of the stars could be reconciled with the estimated experimental error of at least six to ten minutes of arc present in Copernicus' determinations of the position of stars, only if they were at least one thousand times farther than the radius of the Earth's orbit, or at least seventy-five times farther than the distance of Saturn as estimated by Al-Farghani. Of such consequences of the new planetary order Copernicus was fully aware. "The heavens," he wrote, "are immense in comparison with the Earth," and he likened their respective magnitudes to the relation of a point to a body and to that of a finite and an infinite magnitude.[25] That even the radius of the orbit of the earth, had to be viewed "as nothing in comparison with the sphere of the fixed stars," however, was for Copernicus a source of comfort: the earth, in spite of its orbiting around the sun, still remained essentially in the center of the universe.[26] Nor was Copernicus disturbed by the tremendous distance between Saturn and the sphere of the fixed stars. To him this showed the "godlike work of the Best and Greatest Artist" who had carefully set a large chasm "between the moved and the unmoved."[27]

The withdrawn canon of Frauenburg could hardly have suspected that before long, men of science, no less than poets, would lament the coherence that through his work had gone out of the world. He would indeed have been surprised to see the immense gap between Saturn and the stars prevent Tycho from embracing his system. In so reacting, however, Tycho unerringly perceived that the suddenly enlarged world was no longer the home of anthropomorphic purposefulness. From the acceptance of a vast gap between Saturn and the stars, it took but a logical step to assume a world of stars extending into infinity — and an agonizing step it was. Even for Kepler, who was a most ardent champion of the Copernican system, the idea of an infinite empty space was a hideous thought against which he revolted with every fiber of his being. To him it appeared that a universe "to which are denied limits and center and therefore also all determinate places"[28] was a most inhospitable, a directly hostile place. Although he saw some prospect of digesting with "intellectual pills" the "monstrous bite" of the enlargement of the solar system demanded by the Copernican theory,[29] the infinite universe, as he exclaimed time and again on reading Galileo's *Starry Messenger*, was for him unthinkable.[30]

Psychological overtones apart, Kepler pointed out that the universe could not be infinite if both the principle of the homogeneity of space and the known facts of optics were to be accepted. Assuming that the equally bright stars were at equal distances from the earth and that their visible diameter was not an optical illusion (both perfectly reasonable assumptions in those times), Kepler argued that owing to the apparent proximity of certain bright stars the sky would look completely different to an observer located on any of them. In fact the two big stars located in the belt of Orion would appear to an observer located on either one, Kepler argued, as a huge fiery disk whose diameter was five times greater than the sun. Thus the homogeneity of space, inseparable for Kepler from the concept of an infinite universe, seemed to contradict the observational evidence. To resolve the dilemma the concept of the infinite world had to yield to the factual data. Clearly Kepler looked forward to such a conclusion. Willing as he was to transcend the limits of narrow geocentricism, the prospect of man's losing his privileged position in the world pained him,[31] and the flights of fancy with which Bruno embellished the concept of an infinite world filled Kepler with horror. For in Bruno's and his admirers' view there were as many worlds as there were fixed stars, and the new star that appeared in the foot of the Serpentarius signaled for them the sudden birth of a new world. The mere thought of this was for Kepler the harbinger of something dangerously occult, a sign of philosophical insanity. In essence, his was the reaction of Donne, who on one occasion characterized such new speculations as one of the symptoms of that general decay that ever since Adam's sin had been plaguing the world.[32] In Kepler's view only astronomy could provide protection against such a dangerous deviation by bringing man back within the confines of a finite world, prison-like as it may have appeared. "Surely," wrote Kepler, "rambling across that infinity can do good to no one."[33]

On this point neither Tycho nor Kepler spoke the language of progress. Much less was there anything seminal in Kepler's speculation on the width of that spherical shell which, according to him, contained the fixed stars and constituted the absolute boundary of the universe. The traditional world picture shaken at its pivotal points could not be kept closed any longer. Within a short time Galileo's telescope showed that the two minutes of arc diameter attributed to the brightest stars was but an illusion, and with it Kepler's argument against the infinity of the universe was swept away. The sphere of the fixed stars had to be abandoned, a process that started with Digges, who in 1576 had already replaced the sphere of fixed stars with an unlimited region of stars. This infinite realm of stars, however, was identified by Digges with the supernatural heavens. The one who gave the first unequivocal expression of the infinite universe in the modern sense was Bruno. Owing in part to his influence, within a generation or two the world came to be pictured as an infinite Euclidean space where countless stars scattered everywhere represented the immensity of the material universe. In such a world not only did the sun-earth system lose its central position but also there remained no point privileged to play the role of an absolute center. The earth itself became merely one of the countless abodes of life, a circumstance that could not fail to stimulate the imagination. Kepler, who in a finite world could be sympathetic to any distance, vast as it may be, was quick to suggest to Galileo, upon reading the irresistible pages of the *Starry Messenger*, that they should pursue the study of the moon and Jupiter with a view to preparing for trips of future space travelers. For, as Kepler put it, it was not unlikely that the moon and Jupiter had inhabitants and that both would be reached by explorers from the earth as soon as the art of flying was mastered.[35]

Whatever there was of exciting news in the *Starry Messenger*, it had been learned through a new instrument that in a sense gave man the ability to fly deeper into the far reaches of space. Although, as Galileo put it, his first telescope brought the heavens only "thirty or forty times closer to us than they were to Aristotle," seeing even that much farther meant to see the heavens in an amazingly new light. Here was a classic example of the pattern that characterized the whole development of astronomy, and Galileo did not fail to take note of it. With an unconcealed feeling of triumph about leaving Aristotle far behind, Galileo pointed out that with the telescope "we can discern many things in the heavens that he could not see and therefore we can treat of the heavens and the sun more confidently than Aristotle could."[36] The upshot of all this was perhaps the greatest turnabout in man's conception about the nature of celestial bodies and regions. Their incorruptible nature went by the board overnight, and the supposedly perfect, crystal-like surface of the moon had to be described as "uneven, rough, full of cavities and prominences, . . . relieved by chains of mountains and deep valleys."[37] The moon looked so earthlike through the telescope that Galileo boldly said of the great cavity located almost in the center of the moon that it "offers the same appearance as would a region like Bohemia if that were enclosed on all sides by very lofty mountains arranged exactly in a circle." Galileo had solid evidence for stating this. His telescope showed him that no sooner had the sun's rays reached halfway across the central plains of this Bohemia on the moon than the mountain peaks enclosing it on the still dark portion of the moon's face were already bathed in brilliant sunlight.[38] To bring home even more concretely the earthlike character of the moon, Galileo computed by simple trigonometrical methods the height of those "enormous peaks," obtaining a value of about four miles. He could not dream that in 1964, four hundred years after his birth, man would be able, by using Laser beams, to measure directly the height of those peaks and fully confirm his result.

With regard to the stars, Galileo found their number increased by at least tenfold in any given sector of the sky, and the haziness of the Milky Way proved to be but the optical effect of myriads of stars that the naked eye was unable to resolve. What is more, on the momentous evening of January 7, 1610, came the sighting of the Medici stars, the satellites of Jupiter, a Copernican system on a small scale. The world appeared to be shaped everywhere according to the same pattern. The wall of qualitative difference between the earth and the rest of the universe had disappeared and one could freely indulge in speculations about the climate and denizens of distant worlds. (This was an irresistible step to anyone already acquainted with the freshly opened regions of heavens.) Sheer imagination suddenly became a respectable source of information, as evidenced from these times on by the parlance of all armchair space travelers, whether novelists, philosophers, theologians, or, for that matter, astronomers. To compare a great cavity in the center of the moon to Bohemia may seem rather innocuous today, but in 1610 it was nothing less than lifting age-old floodgates in the realm of thinking, and the outpouring that followed was astonishing indeed. Only five years after the publication of the *Starry Messenger*, Ciampoli, a friend of Galileo, had already found it necessary to call Galileo's attention to the dangerous speculations that might be stimulated by phrases referring to a Bohemia on the moon. "Your opinion," wrote Ciampoli, "regarding the phenomena of light and shadow in the bright and dark spots of the moon creates some analogy between the lunar globe and the earth; somebody expands on this and says that you place human inhabitants on the moon; the next fellow starts to dispute how these can be descended from Adam, or how they can have come off Noah's ark, and many other extravaganzas you never dreamed of." [39]

Extravaganzas did come in gigantic volume. Hardly anyone limited himself, as Galileo did in his *Dialogue*, to asserting merely the high plausibility of living beings elsewhere in the universe. First, works of classical authors dealing with extraterrestrial worlds and people were made available in the vernacular. Among them were Lucian's *Vera historia* and Plutarch's *De facie in orbe lunae*. John Wilkins entertained the English with his *Discovery of a New World* (1638), Pierre Borel proved for the French the plurality of the worlds,[40] and Cyrano de Bergerac discoursed at length about the empires and estates of the moon and the sun.[41] Fantasies can hardly move the heavens, but it appeared as if the skies were indeed set in motion to display in response a series of startling phenomena that surely gave ample material for bold flights of fancy. The year 1665 saw a large comet, there was unusual sunspot activity in 1676, four years later Halley's Comet made its debut amid great fear and astonishment, and the year 1684 witnessed an eclipse of the sun. Earlier, in 1659, Huygens discovered a unique and most fascinating feature of the heavens: the ring surrounding Saturn. So novel was this phenomenon that Huygens felt impelled to defend himself against possible charges of irrationality and hallucination for giving "to a celestial body a figure, as none has been found until now, whereas it is held for certain and assumed to be a law of nature that only a spherical shape fits them." [42]

Clearly the climate was prepared for the book that took Europe by storm, *Entretiens sur la pluralité des mondes* (1686), by Bernard de Fontenelle, who was to become for almost half a century the influential secretary of the French Academy. It mattered little that Fontenelle's version of the Copernican universe was distinctly Cartesian. This probably helped only to make his book even more persuasive to the still Cartesian French milieu. For several generations the book stood as the model of that kind of popular scientific explanation that rested more on fashionable "scientific credulity" than on "hard science." In 1737, F. Algorotti, in dedicating his work to Fontenelle praised him precisely for having "softened the savage nature of philosophy," which in its mellowed form could readily be introduced "into the circles and toilets of ladies." [43]

In his ability for "softening" the rigor of science, Fontenelle could hardly be surpassed. Starting from the otherwise respectable principle that living beings conform as a rule to their surroundings, Fontenelle boldly described the characteristics of the denizens of each planet down to the most minute details. Of the inhabitants of Venus, it was stated that they resembled the Moors of Granada, a small black people full of spirit and fire, always ready to make love, busy writing poetry, and who were inventing fiestas, dances, and parades every day. On Mercury, where rivers of liquid gold and silver meandered, people because of the excessive heat were very active, if not hotheads and downright fools. "I believe," stated Fontenelle, "that they have no memory . . . ; that they make no reflection on anything; that they act at random and with sudden motions; and finally that the insane asylum of the universe is on Mercury."[44] This last point is of course debatable unless one is very indulgent in judging the preferences of an age which were shared even by such soberly scientific mind as Huygens, who went to great lengths in describing in his *Kosmotheoros* (1698) the ways of living on the planets of our solar system. On the ground that the same law and the same geometry prevailed everywhere in the cosmos,[45] Huygens felt entitled to conclude what sorts of plants and animals the various planets had and discoursed freely on the various activities of their inhabitants, such as crafts, engineering techniques, and shipbuilding. Furthermore, to justify what might have been branded a sheer flight of fancy, he reminded his reader in a most characteristic remark that no adherent of Copernicus could possibly forego such speculations.[46]

Such was not only the case of the seventeenth-century Copernicans; the scientific state of mind of the eighteenth century was as firmly committed to the belief in a world view that imposed far more than a hypothetical speculation on living beings of faraway planets. Even a critical mind like Berkeley asserted "on the grounds of common sense" that "there are innumerable orders of intelligent beings more happy and more perfect than man."[47] J. H. Lambert, who pioneered in the determination of stellar magnitudes and distances by photometric methods, also could not see how uncritical "common sense" was in entertaining such beliefs. For Lambert, those who left "vacant places" in the universe were "limited in their understanding."[48] J. E. Bode, the leading German astronomer of his time, as late as 1780 took Fontenelle's fantasies rather seriously. To the passage where Fontenelle described the extreme liveliness of the denizens of hot Mercury, Bode, as he translated Fontenelle's *Entretiens* into German, felt it proper to add the following note: "Peculiar! One finds rather with us here [Berlin], that too much heat makes the mind more sleepy and lazy than lively."[49] If professional men of science were captivated by the science fiction of their day, what could one have expected from amateurs like Kant? Not only was Kant convinced that the planets of the solar system were all inhabited, but he also believed that there was a direct relation between the moral and intellectual character of various planetary beings and the distance of their habitat from the sun. According to Kant, it could be stated with "fair probability" that the denizens of Mars stood midway between extremes in their physiology and morals, because Mars occupied a middle position in the series of planets. Of the inhabitants of the more distant planets, Jupiter and Saturn, he stated that they were as superior to Newton as Newton was to a Hottentot or an ape.[50]

At any rate, seeing farther in the universe meant the unveiling of aspects of the universe hitherto undreamed of even if not everybody cared about separating dreams from facts. All in all, nothing in the old world picture appeared safe and sacred any longer, and even the central role of the sun, to which Copernicus and Kepler paid an almost religious homage, began to look more and more arbitrary. Newton himself spoke of his law of gravitation as requiring an infinite amount of matter distributed in infinite space, lest the attraction should pull all matter into one huge bulk. In this connection the question could not fail to arise whether there was absolute direction and motion in an infinite space. An absolute frame of reference was assumed by Newton on the ground of certain aspects of the centrifugal force, but the flaws in his reasoning were soon laid bare by an early forerunner of relativity, Bishop Berkeley.

When the full meaning of the Copernican revolution had been finally absorbed, the way the world was pictured had far outgrown the hundredfold increase of its radius, the minimum requirement in Copernicus' day to meet obvious objections. It was rather the very concept of infinity that began to rule every compartment of scientific pursuit. Having pushed the horizon one stage beyond the sphere of fixed stars, science ended up in dissolving the distinct contour of the world into the haze of infinite distances. This transposition of the scientific vision from the range of finite into the infinite had long been achieved when the astronomers were still struggling to size up with reasonable accuracy the dimensions of the solar system.

On the Copernican theory it was evident that the earth-sun distance should be considerably larger than the data inherited from classical times. Brahe already argued that were the Earth moving around the Sun, a stellar parallax of three minutes of arc should be observable, assuming that the nearby stars were not much more distant than Saturn. Although such a large parallax could have easily been detected by Kepler, who fell heir not only to Brahe's valuable data but also to his great respect for precision in measurements, there was no detectable trace of it. The need of assigning greater dimensions to the solar system therefore had to be satisfied by a reasoning that mixed geometry with organismic and mystical considerations, a procedure that always retained a great appeal for Kepler. He took the view that the earth-sun distance bore a simple relationship to the dimensions of the earth, because the earth was after all the abode of measuring creatures. In particular, he decided that the earth-sun distance exceeded the radius of the earth in the same proportion that the mass of the sun bore to the mass of the earth. The earth-sun distance, which he obtained this way, increased the then accepted value by a factor of three, but it was still only about one-seventh of the true distance.

In the absence of an even approximately good value of the earth-sun distance, the heliocentric ordering of planets could not begin to convey the grandiose dimensions of the solar system. This was all the more regrettable, for in the system of Copernicus the relative planetary distances did not depend on such arbitrary assumptions as the alleged contiguity of the planetary spheres. As Copernicus rightly claimed, the world was shown in the heliocentric ordering to have a "wonderful commensurability and that there is a sure bond of harmony for the movement and magnitude of the orbital circles such as cannot be found in any other way."[51] Thus, while Al-Farghani's table gave the distances of Mercury, Venus, Mars, Jupiter, and Saturn in terms of the earth-sun distance as 0.13, 0.92, 7.25, 11.8, 16.4, (a wholly false picture), Copernicus' values were in extremely close agreement with the modern data given in parentheses: 0.37 (0.38), 0.72 (0.72), 1.52 (1.52), 5.22 (5.20), 9.17 (9.54).

Such relative spacing of the planets was brought out fully in Kepler's third law connecting the periods and mean radii of orbital motions, and Kepler was eager to point out how suitable it was for the business of astronomy to use the earth-sun distance as the natural measuring rod of celestial expanses. In this he again displayed the animistic component of his reasoning by ascribing special dignity to the earth and the sun in the planetary system: "Our measuring rod has two very signal termini."[52] The length of this measuring rod, however, had not become known with reasonable accuracy until about fifty years after Kepler's death when, in 1672, Richer and Picard measured the parallax of Mars. Although the second half of the seventeenth century was already an age accustomed to the concept of an infinite universe, nevertheless the result of Richer's and Picard's work astonished their contemporaries. The value of the parallax of Mars yielded not only its distance, but by Kepler's third law the distances of sun and all the planets followed readily. Almost overnight man's estimates of the dimensions of the solar system increased by a factor of ten. It took more than a day, however, to absorb the incredible fact that the value of the mean radius of Saturn's orbit, which represented at that time the outer limits of the solar system, was not equal to twenty thousand but to more than two hundred thousand Earth radii.

A similar effect also accompanied the attempts directed at the determination of the distance of the nearest stars. As the probable error decreased in the determination of the position of the stars, so did the maximum of possible stellar parallax, and as a consequence the stars were located farther from the earth. Reference has already been made to Copernicus' conjecture about the distance of the sphere of fixed stars, an estimate too short by a factor of about a hundred — or at least a thousand if one keeps in mind that his estimate of the astronomical unit was also only about one-tenth of the true value. To the same order of magnitude belongs Kepler's first estimate of the distance of the stars, which he formulated after realizing that the parallax of the Pole Star must be smaller than 8 minutes of arc.[53] Believing that geometrical speculations could help where observation failed, Kepler redefined in the *Epitome*[54] the distance of the stars as sixty million earth radii. This value he obtained on the wholly gratuitous assumption that the orbit of Saturn was a geometrical mean between the diameter of the Sun and the diameter of sphere of the fixed stars. Brave as this jump was on the distance scale, Kepler was still off by a factor of about a hundred from the true distance of the nearest stars.

Attempts like this brought out forcefully both the inability of the astronomer to cope for the time being with the problem and also the inconceivably large distances to which human imagination slowly had to accustom itself. The inability was clearly frustrating and, as Milton put it, could only prompt "His [God's] laughter at their quaint opinion . . . to model Heav'n and calculate the Stars."[55] The training of the human mind to the new distances was hardly easier. Huygens, who gave the distance of Sirius as 27,664 astronomical units, plainly admitted that such distance was "incredibly large."[56] To bring it closer to human grasp, he called attention to the "fact" that the same bullet that takes twenty-five years to reach the Sun from the Earth would need almost seven hundred thousand years to cover the distance to Sirius. Sirius, however, was for him but the nearest star. Beyond it, warned Huygens, there stretched the limitless realm of stars, their number and distances exceeding any imagination. And why, he asked, could God not create worlds entirely different from ours beyond those immense distances?

Clearly, as the boundaries of the universe expanded, man's conjectures about it had to change as well. That Huygens' method by which he obtained the foregoing figure for the distance of Sirius was not rigorous mattered little. In essence it was not without merits. He assumed that the Sun and Sirius were similar stars, and he looked through various small holes until he found one through which the Sun appeared with a brightness equal to that of Sirius. The value he thus obtained fell short by a factor of about eighteen. Newton, who also attempted to compute the distance of fixed stars on the assumption that the intensity of light is inversely proportional to the square of distance, could not of course be sure how correct were his estimates.[57] Nor could Bradley, whose determinations of the positions of the stars were already good to within ten seconds of arc, and therefore rightly claimed that γ Draconis should be at least four hundred thousand astronomical units, or about six and one-half light years away. At any rate, such distances were so immense as to forbid any meaningful speculation of the material composition of stars. For the mid–eighteenth-century astronomer, this problem had to appear exactly as it had a hundred years earlier: "surpassing the strength of human intellect."[58]

In the context of times such a conclusion could hardly be branded as pessimistic. Yet, the aspirations of human imagination always went far beyond the confines of what may be called a "sober appraisal." It was simply impossible for astronomy to dispense with creative imagination that could fly so easily beyond seemingly absolute barriers. In fact the boldest strain of imagination was needed to push toward much more remote regions of the universe when the realm of stars — a realm far beyond the confines of the solar system — began to move into the focus of interest. Newton's view of the infinite number of stars spreading out evenly in an infinite space was evidently an oversimplification, as the starry sky showed anything but uniformity. Thomas Wright was the first to perceive in the luminous band of the Milky Way a non-uniform pattern of the distribution of stars.[59] To him the Milky Way was but the visual effect of the stars located between two parallel and relatively close planes. Wright further assumed that beside the Milky Way there were similar disk-shaped stellar systems (or creations, as he called them) in the universe and that their number was infinite. The next step was taken by Kant, who happened to read in a 1751 number of the Hamburg newspaper, *Freie Urteile*, a summary of Wright's views. Four years later, in his *Universal Natural History and Theory of Heavens*, Kant identified the nebulous patches long observed in the sky as distant Milky Ways and pointed out that these disklike star systems would appear as a circle or an ellipse, depending on their positions relative to the earth. Arguing from the all-powerfulness of God, Kant put the number of these milky ways as infinite and assumed them to be grouped in systems of higher order.[60] The existence of such supersystems was emphasized even more by J. H. Lambert, who thought independently of Wright and Kant that it would be astonishing if distant milky ways should fail to produce impressions on the eye. He conjectured that the pale light in Orion might be a Milky Way nearer to us than the rest.[61]

By the end of the eighteenth century the assumption that stars are grouped in volumes of space that resemble the shape of a disk became a favorite notion. To substantiate such a view, however, systematic observations and measurements[62] were needed, which in fact increased by a factor of about a thousand the encompassed reaches of the universe. This giant step was initiated by Herschel whose success depended on his idealism as well as on his utter naïveté. Had he surmised at the outset of his ambitious astronomical program the real proportions of the problems involved and his own shortcomings as a physicist, he would have in all probability despaired of the magnitude of the task he had set for himself. For in the days of the young Herschel, the starry sky was for the observational astronomer, in spite of the lengthy speculations about infinite worlds and countless Milky Ways, practically what it had been for the Greeks — a huge concave, spherical surface with an invariable radius. The immense variations in the distances of stars and their possible groupings were before Herschel the object of entranced imagination, not of observation. Herschel wrote in 1784:

Hitherto, the sidereal heavens have, not inadequately for the purpose designed, been represented by the concave surface of a sphere, in the center of which the eye of an observer might be supposed to be placed. In the future we shall look upon those regions into which we may now penetrate by means of such large telescopes, as a naturalist regards the rich extent of ground or chain of mountains containing strata variously inclined and directed as well as consisting of different materials.[63]

But Herschel wanted far more than to learn about the rich details of the realm of stars. "The ultimate construction of heavens," he stated, "has always been the ultimate object of my observations."[64] Aided by bold imagination and extraordinary eyesight, his immense efforts were not to go unrewarded. His catalogues published in 1786, 1789, and 1802 added over 2,500 nebulae to the 103 contained in Messier's list. Similarly fruitful was his search of the binary stars. Yet, for him, the gathering of all these data served only to provide a clue to the overriding question of what statistical laws govern both the distribution of stars and their distances. As regards the former, it became progressively evident for him that the Newtonian assumption of a uniform star distribution was not in keeping with the facts. The stars appeared rather to form clusters, and after more than two decades of data gathering, Herschel concluded: "An equal scattering of the stars may be admitted in certain calculations; but when we examine the Milky Way, or the closely compressed clusters of stars, of which my catalogues have recorded so many instances, this supposed equality of scattering must be given up."[65]

Once more a deeper penetration into the reaches of the universe brought about a pronounced change in the way science pictured the physical pattern of the world. As far as the distances appraised were concerned, the changes were no less surprising. A hundred years earlier Huygens and Newton had to content themselves with educated guesses about the distance of Sirius and first magnitude stars in general. Herschel, on the other hand, was moving out toward the very limits of the Milky Way, which he believed himself to have fathomed with his 20-foot telescope, giving the Milky Way's diameter as 6,000 light years as against the present value of 80,000 light years. But his 40-foot telescope, which in his estimate could reach to 2,300 times the distance of Sirius, soon made him realize that the Milky Way was not to be readily encompassed. It dawned on him that just because some faint patches in the sky could not be resolved, one could not yet assume that the most remote boundaries of the starry kingdom had been reached. He began to perceive the meaning of his invariable experience that

When in one of the observations a faint nebulosity was suspected, the application of a higher magnifying power evinced, that the doubtful appearance was owing to an intermixture of many stars that were too minute to be distinctly perceived with the lower power; hence we may conclude that when our gauges will no longer resolve the Milky Way into stars, it is not because its nature is ambiguous, but because it is fathomless.[66]

Although the ultimate frontiers of the universe eluded man anew, the roughly thousandfold increase in the probings into the distant spaces did not fail to conjure up a new vision of the cosmos, which was not, however, without some ambiguity, and all too often the alternate possibilities of resolving it were proposed with a tone of unquestionable finality. One of the possibilities was a system in which nebulae were taken as stellar systems similar to the Milky Way, dispersed in an orderly and perhaps hierarchical manner throughout the infinite universe. Kant for one could not be more dogmatic on this point. With the naïveté characteristic of all dilettantes in scientific matters, he wrote: "If conjectures, with which analogy and

observation perfectly agree in supporting each other, have the same
value as formal proofs, then the certainty of these systems must be
regarded as established." [67] Clearly Kant's standards were not so demanding in the field of observational evidence as in purely philosophical matters. A similar and no less resolute tack was taken by
Lambert, who rejected the concept of an unhierarchical distribution
of nebulae in the universe on the ground that it would smack of
cosmic democracy, with which he evidently did not sympathize. [68]

As one could expect, men wholly devoted to observation, like
Herschel, were more cautious. For many years a supporter of such
a view, Herschel finally abandoned it, since experimental evidence
seemed for him to prohibit speculating with confidence upon what
might lie beyond the Milky Way. The idea of island universes and
supersystems saw, however, a triumphal comeback in 1849 when
Lord Rosse succeeded with his great 72-inch reflector in resolving
stars in several spiral nebulae. What immediately followed was typical. With the uncautious enthusiasm so characteristic of many historians of science when appraising the discoveries of their day, R.
Grant concluded in 1852 that in principle every nebula consisted of
and could be resolved into stars, provided sufficient advances were
made in observational techniques. [69]

The findings of Lord Rosse, however, could hardly make a dent in
the convictions of those of the opposite camp who saw in the Milky
Way a system including all heavenly bodies and formations that were
observable. Thus Whewell wrote in 1854 that the resolution of some
bright spots in certain nebulae was not of necessity an indication of
stars, nor did the absence of such a resolution constitute proof that
some nebulae were immensely distant. He submitted his conclusion
"with the most complete confidence," and to discredit completely
the idea of "island universes," he added that since nebulae are gaseous formations it is "certain that these celestial objects are not inhabited." [70] How much more scientific would it have been to confess
to the completeness of human ignorance on this point. But it was not
always realized that a degree of certainty such as that exuded by
Whewell was precisely that foreign body that science never assimi-
lates on a permanent basis.

The realization of this certainly would have spared H. Spencer
some gross mistakes that he made when joining the crusade for the
all-inclusiveness of the Milky Way. As he put it in 1858, the evidence
in this regard was "overwhelming," and he asked in a superbly confident tone: "Should it not require an infinity of evidence to show that
nebulae are not parts of our sidereal system?" [71] Spencer's main argument referred to the grouping of nebulae around the galactic poles,

a point that was certainly impressive, but his reasoning glossed over many an aspect of an obviously complicated problem. For, clearly, it took some superficiality to state again in 1898 of the genesis of nebulae, or of Laplace's nebular hypothesis, that "practically demonstrated as this process now is, we may say that the doctrine of nebular genesis passes from the region of hypothesis into the region of established truth."[72]

Subjective and objective truth are not, however, always identical, although all too often one may find them interchanged even in science. Thus it happens that new findings are interpreted through an invincible bias, as was the case with W. Huggins who in 1864 established by spectroscopic methods the presence of diffuse gases as well as stars in many nebulae. Now it was impossible to maintain any longer that the only reason why some nebulae could not be resolved into individual stars was that they were located at such immense distances from our galaxy that they could not form a part of it. Although Huggins' chief interest in his discovery does not touch directly on this point,[73] the way he described those exciting hours of his life from a distance of thirty years forms a classic example of the extent to which an outstanding man of science may distort the real meaning of his discovery:

The reader may now be able to picture to himself to some extent the feeling of excited suspense, mingled with a degree of awe, with which, after a few moments of hesitation, I put my eye to the spectroscope. Was I not about to look into a secret place of creation? I looked into the spectroscope. No spectrum such as I expected. A single bright line only. . . . The riddle of the nebula was solved. The answer which had come to us in the light itself, read: Not an aggregation of stars, but luminous gas.[74]

But the nebulae had and still have many riddles and have besides luminous gas plenty of dark dust as well. In his excitement Huggins took this gas for a new element, christened it *nebulium*, hardly suspecting that it would be identified some sixty years later as highly ionized oxygen and nitrogen.[75] For the time being, however, the observable world seemed to form to all appearances one huge single system centered around the polar axis of the Milky Way. This *definitive* picture of the universe was not lacking in some grandiose features. In it the world mirrored a unity and in a sense was even geocentric again, for the sun appeared to occupy a rather central position among the stars.

The Herschelian universe — the observable world confined to our galaxy — displayed a uniform character not only in its structure but also in its large-scale motions and chemical composition. Studies of the motion of binary stars verified the validity of the laws of gravitation beyond the solar system, and the presence of most of the known elements was detected in the spectral lines of the sun and the stars. For, as Kirchoff and Bunsen remarked in their first paper of October, 1859, chemical processes taking place in the sun could be imitated by a device as simple as a Bunsen burner.[76] They clarified the origin of the Fraunhofer lines and listed the presence of several elements in the solar atmosphere where several years later, in 1868, it even became possible to establish the existence of an element hitherto unknown on the earth, the gas helium, as it was named by its codiscoverer, Lockyer. Before long the new element was liberated from mineral clevite (Lockyer and Ramsay, 1895), and the belief in the uniform composition of matter in the universe obtained its major experimental seal.

There were of course puzzles too among the phenomena observed in the Herschelian universe. Foremost was the problem of the classification of nebulae. Herschel on this point could hardly see much below the surface. For him, the variety of nebulae reflected a richness of forms resembling the variety of plants and animals. To some extent he even suggested the idea of stellar and nebular evolution whose mechanism he was, however, unable to guess.[77] Largely owing to Herschel's influence, nineteenth-century astronomy paid increasing attention to the fact that the stars and nebulae exist not only in space but also in time. Still, these were not the times to recognize in full the crucial importance of the time-parameter for a better understanding of the large-scale features of the universe. Actually, to some, like Kelvin, the problems presented by the evolution of stars, solar systems, and nebulae were not challenging at all. In the "automatic progress of the solar system from cold matter diffused through space, to its present manifest order and beauty," Kelvin saw no more mystery than in "the winding up of a clock and letting it go till it stops." At this point he clearly might have felt some misgivings about the sweep of his statement, for he added an explanatory footnote, which was, however, just as baffling: "Even in this and all the properties of matter which it involves, there is enough and more than enough mystery to our limited understanding. A watch spring is much further beyond our understanding than is a gaseous nebula."[78]

Evidently this was an awkward way of classifying mysteries and difficulties. If one was willing to overlook the puzzle of nebular evolution, however, there seemed to be hardly any positive scientific data at the end of the nineteenth century to challenge the uniqueness of the Herschelian world picture. It was generally admitted that the Milky Way included the whole of the observable world and was surrounded, for all practical purposes, by infinite empty spaces. The doctrine of "island universes" fell into such disrepute that, as A. Clerke, the historian of nineteenth-century astronomy, registered the consensus of the day, "no competent thinker" could consider any single nebula "to be a star-system of co-ordinate rank with the Milky Way." As she appraised the situation, "a practical certainty has been attained that the entire contents, stellar and nebular, of the sphere belong to one mighty aggregation, and stand in ordered mutual relations within the limits of one all-embracing system."[79] If such a conclusion was already highly revealing of a state of mind, no less so was the proviso added to it. There Clerke qualified the expression "all-embracing" with the words: "as far as our capacities of knowledge extend. With the infinite possibilities beyond, science has no concern."

In a sense this is true. Science is not about the unknown, but about what is known. It has to aim at the ordering of available data, not at speculations about the unknown. By 1900, however, too many startling upsets had taken place in the long history of astronomy to justify a greater concern for the unknown and to make one profoundly wary of "all-embracing systems" and definite conclusions about them. For within this all-embracing system, what was then believed to be left for the astronomical research of the future? Today it must be somewhat sobering to see that a state of mind that takes too readily the momentary stage of science for the final one will score even poorer in the always dangerous field of prognostication. According to Clerke, the astronomy of the future was to busy itself for centuries before it might gather some rudimentary information of the laws and revolutions of star clusters such as the Pleiades. Actually far more than rudimentary knowledge had been achieved since then. As to the groupings of nebulae, Clerke recorded the expectation that they would "stimulate and baffle human curiosity to the end of time."[80] On this point too impressive progress has since been accomplished.

The major field of research for twentieth-century astronomy, according to the consensus, lay in the field of stellar spectroscopy. As early as 1869 C. Pritchard expected spectrum analysis, still in its infancy, to shed light ultimately on problems such as the condensation of nebulae, the motion of stars relative to the solar system, the dissipation of energy by stars, and the secular cooling of the sun's envelope.[81] A happy forecast indeed. Actually, by the turn of the century the usefulness of spectroscopy in stellar physics had assumed such proportions that it could be termed a "science the reality of which confounds forecast, yet compels belief" and whose "expansiveness in all directions is positively bewildering." Bewilderment in this context should, however, have referred to the prospective subjugation of still little explored fields under laws and concepts already known. The bewilderment of which A. Clerke spoke was in all evidence the bewilderment of triumph and not the bewilderment that overcomes men of science when a major crisis sets in and when basic "truths" are to be abandoned. Among the various "certain" features that were to be discarded under the impact of spectroscopical research was precisely the concept of the all-inclusiveness of the Milky Way. Further penetration into the universe set at naught the "definitive" conclusion that such a concept rests on "unanswerable arguments," that "it scarcely admits of a difference of opinion," that the contrary view "began to withdraw into the region of discarded and half-forgotten speculations."[83]

As usually happens in physics, problems will appear definitively solved to those who concentrate on what may be termed the convergent component of scientific research. It is far more difficult to grasp the real portent of puzzles that are as innocuously present as the air we breathe, or the darkness of the night sky we often admire. That this latter phenomenon is not at all obvious is realized keenly by twentieth-century astronomy. That the previous two centuries showed themselves distinctly insensitive to the problem constitutes a classic case of the often baffling limitations of scientific insight. The most telling measure of this limitation is probably the almost total absence of reaction with which the scientific world greets at times tentative probings into hidden enigmas. Such was the fate of the discussion of the darkness of the night sky by the German astronomer, Wilhelm Olbers. When his paper, "On the Transparency of Space," saw print in 1826 in German, it created but fleeting interest, as did the French and English translations that shortly followed.[84]

To be sure, Olbers could not wax very dramatic about the enigma of the darkness of the night sky. It was clearly beyond his reach at that time to surmise its most likely source, the expansion of a finite universe. True to the basic convictions of classical physicists, Olbers found it unreasonable to assign boundaries to space; the infinity of space was in his eyes a postulate of sound logic and also an inevitable result of the infinite perfection of the Creator. He believed, as did all his forebears in astronomy from Newton and Halley on, that gravitational balance could be maintained in the universe only if every section of space was, as he put it, "sprinkled over with suns, each accompanied with its train of planets and comets." In such a universe the number of stars was evidently infinite. This meant, however, that every line of sight originating on the surface of the earth had to terminate on the surface of a star. In other words, on such considerations, as Olbers noted, the whole vault of heaven should have appeared as bright as the sun. Clearly, Olbers continued his reasoning, only its dark spots would make the sun distinguishable in such a uniformly blazing sky.

Curiously enough, such a possible state of affairs worried only Olbers the astronomer. He could not help realizing that with a sky bathed in uniform brightness there was but little prospect for astronomical research. The problem of heat generated by a blazing sky distressed him not at all. The Omnipotent, he noted with piety, could have easily found means to put our globe and all life on it, in a condition capable of coping with both blinding brightness and scorching heat. This was, however, a rather non-scientific evasion of the problem even in 1826, that is, a generation or so before the science of thermodynamics came of age. Just as superficial was the scientific solution given by Olbers to the problem. He assumed that the very dilute "homogenous substance," which according to the scientific consensus filled the starry spaces, had the property of absorbing the light radiated by stars. The question of what happens to the absorbing medium was left untouched by Olbers. To pass over such a detail evidenced, however, a rather poor appraisal of the problem even by early nineteenth-century standards. Yet, it was this oversight that gave Olbers confidence to proceed with the calculations that seemed to him to clinch his argument. Assuming that one eight-hundredth of the light coming from Sirius was absorbed in the interstellar medium, he could readily compute that a certain quantity of stars situated at a distance ten thousand times that of Sirius "would require to be accumulated close to one another, before, in a clear and moonless night, our most perfect telescopes could render this group visible as a pale nebulosity." Stars, he added, situated thirty thousand times farther from us, would, for all practical purposes, "not contribute to light the celestial vault."

Such was Olbers' final verdict about the enigma of the darkness of the night sky, and he confidently noted that it could not be "very wide of the truth." This truth was, however, the truth of a somewhat naïve trust in the criteria, accepted by the science of the day, of what was intelligible and what was not. Concerning the respective merits of the intelligibility of a finite and infinite cosmos, classical physics handled those criteria all too often with shocking superficiality. As Kelvin once voiced this attitude in a striking passage:

I say finitude is incomprehensible, the infinite in the universe is comprehensible. Now apply a little logic to this. Is the negation of finitude incomprehensible? What would you think of a universe in which you could travel one, ten, or a thousand miles, or even to California, and then find it come to an end? Even if you were to go millions and millions of miles, the idea of coming to an end is incomprehensible.[85]

Faultless as this type of reasoning appeared, it expressed in fact a routine, if not naïve, way of thinking, which recognized no problems in the Euclidean infinity and was unable to pay notice to what was truly seminal in the enigma of the darkness of the night sky. To sight the very depths of this enigma required boldness that dared to reexamine a basic postulate of the Euclidean geometry on parallel lines. Among Olbers' contemporaries three giants of geometry, Gauss, Bolyai, and Lobachevski, were wrestling with this problem. A corollary of it was the question raised by Gauss: Is the Euclidean geometry really valid over wide distances in space? The experimental method proposed by Gauss to decide the matter consisted in measuring the sum of angles in a very large triangle formed by distant points on the earth, or by correlating the magnitude of angles subtended by stars.[86] The method, as later shown,[87] is inconclusive, but at any rate, its execution, like the possible solution of Olbers' paradox, would have demanded a penetration into the universe much farther than anything previously attempted.

This need for reaching much farther out into space was felt all the more when around the turn of the century it became evident that there were hidden contradictions in the concept of infinite matter spread out in infinite space. On the one hand, as H. Seeliger showed,[88] an infinitely strong gravitational field should arise at any point in an infinite space containing uniformly distributed matter that obeyed the Newtonian inverse square law of gravitation. On the other hand, as Einstein noted,[89] the density of matter cannot be zero beyond a certain distance, for with particles moving at random, a zero density at the boundary would entail, according to a well-known theorem of Boltzmann, a zero density at all points inside. The universe therefore cannot be infinite, nor can it have boundaries. But this is possible only if the geometry of space corresponds to a non-Euclidean type.

Theoretical considerations alone could not lead to a proper choice of the type of geometry of space. Observations reaching far deeper into space could only give the clue, but at the turn of the century a cosmological breakthrough was in general not expected from a substantial increase in the light-gathering power of telescopes. Little appreciated was the lesson of the past: never had a decided advance in the resolving power of telescopes failed to pay startling dividends. Still, the results that were obtained by the huge twentieth-century telescopes more than justified those who had kept stressing the need of increasing man's reach into the depths of space. The 100-inch telescope at Mount Wilson had indeed shed unexpected light on a question that had been considered definitely settled, namely, the all-inclusiveness of the Milky Way.

It is true that the estimated size of our galaxy increased considerably even before the 100-inch telescope began to canvass the skies. The increase resulted from an extensive statistical survey of the distribution and magnitude of the stars by Seeliger, Kapteyn, and Shapley, extending the diameter of the Milky Way first to 23,000, later to 55,000, and finally to 80,000 light years, the later figure now being generally accepted. The "limits" of the Milky Way were understood to be represented by the distance where the star density drops to one-hundredth of its value near the sun. Prior to 1924 it was impossible to assign nebulae with any certainty beyond these "limits." Even the discovery of the Cepheid variables as distance indicators could not clarify matters in this regard, for the light-gathering power of telescopes that existed prior to 1918 could not resolve single stars beyond the arbitrary frontiers set by the statistical survey. The crucial change came in 1924 when Hubble was able to resolve with the 100-inch telescope a Cepheid variable in the Andromeda nebula, and it immediately became evident that its distance must be at least ten times the diameter of the Milky Way. Here astronomy had its first incontrovertible proof of a nebula independent of the Milky Way and similar to it both in size and structure. Before 1924 it could still be maintained that the Milky Way was an all-inclusive system, but after 1924 the cosmos began to look like Lambert's conception: countless independent galaxies grouped perhaps in higher systems. The sudden enlarging of the universe following Hubble's achievement had such an impact on the astronomical world that Jeans, for instance, was inclined at that time to see in the determination of the distance of Andromeda the utmost man could ever achieve in this regard. But as the years went by, the 100-inch telescope reached farther and farther, and today its limits are set at about five hundred times the distance to the Andromeda.

The suddenly increased universe with a radius of about one billion light years was not merely the old universe on a larger scale. First of all, for cosmological thought the stars as basic building blocks of the universe have been replaced by the nebulae. The rich variety of nebulae seen by Herschel, much as meadows viewed by a botanist, became a central problem for cosmology. Of even greater importance was the discovery of a general red shift in the spectrum of nebulae that was found to increase with distance. That the stars are not immobile had long been suspected. Only twenty years after Newton had stated in the *Principia* that the "fixed stars are immovable,"[90] Halley concluded that such conspicuous stars as Aldebaran, Sirius, and Arcturus have a "particular Motion of their own."[91] For, as Halley found in 1718, they stood half a degree farther south than the positions given by Hipparchus and Timocharis, and it was difficult to assume that these careful observers should have been so much in error. Clearly, if either the stars or the solar system as such were moving, the shape of some constellations would be affected, as was pointed out by Bradley in 1748. But neither he nor Christian Mayer, the German astronomer, could establish anything conclusive. In fact Mayer in 1760 concluded against the motion of the sun among the stars, for he did not find the stars opening up around some point in the sky and closing in around an opposite point, as one walking through a forest would see the trees separate in front and draw closer behind. Not only did Mayer expect too great an effect of this type were the sun moving, but he also failed to realize that the proper motion of stars might mask in part at least the expected pattern.

Aware of this difficulty, Herschel was not disturbed that of the motions of twenty-seven stars that he investigated only twenty-two could be interpreted as due to the motion of the sun toward λ Herculis. For him his conclusion rested "on a few though capital testimonies,"[92] a conviction, however, that was not shared by the astronomers of his day.[93] The motion of nearby stars was clearly a problem stretching to the very limits the possibilities of positional astronomy. To learn something about the motion of more distant objects such as stellar clusters and nebulae, a wholly different tool, spectroscopy, was needed. Spectroscopy reached indeed much farther into the universe, but at the same time the picture unfolding moved farther and farther away from the traditional conceptions. First, there was the enormous speed of some nebulae. It came as a major astonishment when V. M. Slipher in 1912 succeeded in determining that the Andromeda nebula was approaching the solar system with a velocity of 190 miles per second. Next, a regular pattern in the motion of nebulae began to emerge when a few years later Slipher again found that of the fifteen observed spiral nebulae, thirteen appeared to be receding from our galaxy at an average velocity of 400 miles per second. The question therefore arose whether or not the recession was a universal phenomenon in the realm of nebulae.

That a red shift indicating a recessional motion was a systematic feature in the spectrum of the nebulae was first spelled out in 1929 by E. Hubble. Interestingly enough, he grew more reluctant, as years went by, to see in a recessional velocity the physical cause of the nebular red shift. Possibly he hoped that further investigations would not raise the distance-velocity relation to the rank of a universal cosmic law. For as he gravely remarked in his historic paper, such a relation, if confirmed, "will lead to a solution having many times the weight."[94] By this he meant that another Copernican turn could be in store for cosmology: the replacement of a static, infinite universe by a finite and dynamic one. Such proved indeed to be the case, although efforts were not spared to save the old world picture of an infinite universe that represented to many scientific minds intelligibility itself.

Of considerable weight in favor of the idea of a cosmic expansion is the fact that relativistic cosmological theory predicts both the fact of the expansion and the distance-velocity relationship. The first to connect the instability of the Einsteinian universe and the recession of the nebulae was Lemaître,[95] whose 1927 paper, however, remained largely unnoticed until 1930, when Eddington and McVittie, working on the same subject, stumbled on it just a few months before completing their investigations.[96] With the appearance of Eddington's paper, Lemaître's idea of an expanding universe of finite radius and finite mass became almost overnight the central point of cosmological thought. The obvious step to make was to trace the process of expansion backward, and as a consequence, the idea of the universe being condensed into a relatively small space several billion years ago presented itself naturally. This stage of the universe, which Lemaître once called the Primeval Atom, stands today as a limit beyond which theoretical analysis of physical processes can hardly be carried with sufficient consistency. To calculate the time needed for the expansion to reach its present stage is rather straightforward in principle, but the actual values are far from uniform, nor are they in sufficient conformity with the data calculated for the age of various parts of the universe, such as a galaxy or globular cluster. Still, the various values are not without some impressive uniformity inasmuch as they fall at present within one order of magnitude, that is, from five to fifteen billion years.

While it would be rather unscientific to see in the distance-velocity law and in its consequences an unequivocal proof of a "beginning" of the universe, or a "creation," to use the term in a rather broad sense (a creation in strict theological sense can have no direct relation to physical research), it would be just as unscientific to be unduly upset by the concept of a cosmic beginning. When some men of science meet "with the most ardent resistance," as Weizsäcker put it,[97] the concept of a universe finite in one or other parameter, they only bear witness to the fact that adherence to one or another type of scientific theory is often determined more by an inherited state of mind than by dispassionate pondering of the available information. Regardless of the choice made by the cosmologist, eternity of cosmic time can no longer claim an exclusive respectability in scientific speculation. The data of cosmological research call for perspectives that broaden in much the same measure as the sweep of our instruments into the far reaches of space has grown by leaps and bounds.

The change concerning the "timelessness" of the world was only one of the major transformations in man's scientific concept of the universe that accompanied the big step best symbolized by the 100-inch telescope of Mount Wilson. The Milky Way, pictured as an all-inclusive system, became almost overnight an average-sized building stone of the universe. The theories of Kant, Laplace, Chamberlin, Moulton, and Jeans about the evolution of the solar system were overshadowed by speculations on the evolution of nebulae involving an immensely larger scale of physical processes. The basically static Herschelian universe had to yield to a cosmic dynamism hitherto unsuspected. The idea of the infinite world that saw its triumph only a few centuries ago over the closed world of antiquity had to be replaced by a world model in which the quantity of matter is finite and whose radius is a function of the density of matter. The most paradoxical aspect of this development is that the step from the concept of the infinite world to the idea of a finite but unbounded universe was the result of man's greatly expanding reach into the vast spaces of the cosmos. Also, the known portion of the supposedly infinite universe proved to be much smaller than the distances that man is able to measure in the new, "finite" universe. As a matter of fact, space could be viewed as flat, or Euclidean, only as long as man's observing eyes were restricted to a rather limited range in it. Until then one could declare as B. Russell did fifty years ago that the homogeneity of space was the most fundamental of all scientific axioms.[98] In other words, it was an article of faith for the classical physicist and astronomer that the universe should look the same on any scale. This is why they found it both sound and fascinating to apply the kinetic theory of gases to the motion of stars in globular clusters. Theirs was a firm conviction that, as Poincaré put it, "to the eyes of a giant for whom our suns would be as for us our atoms, the Milky Way would seem only a bubble of gas."[99] Yet, after man succeeded in taking that giant step from the Milky Way to the realm of countless milky ways, or nebulae, the structure of space suddenly appeared in a very different light. It was a shocking lesson to the complacent state of mind of classical physics already beset with a similar upheaval of its ideas after the ever smaller regions of the atom and nucleus opened up for scientific investigation. The universe turned out to be not the same when one moved either toward the realm of the infinitely small or toward the regions of the infinitely large.

There is no way of telling what is going to be lasting in the conception of an unbounded cosmos whose radius is the function of the finite quantity of matter contained in it. Such conception is, however, a far cry from the passionate and dogmatic statements of Bruno about infinite universe and infinite number of worlds. In part at least, Bruno was right, but the infinity, or rather the unboundedness of the universe turned out to be a puzzle far transcending the flat simplicity that blares forth in either Bruno's or Kelvin's statements about the obviousness of the infinite. Painful as it might be to some to accept the fact that the infinite universe of classical physics was no less a myth than was the closed sphere of fixed stars of ancient cosmology, there is no way of ignoring the changes of evidence. At the same time there is no more assurance for taking the Einsteinian universe as the final word in cosmological investigations. The universe rather seems to keep poking fun at grand sayings, "final" solutions, and "definitive" systems. It is, however, always ready, in the long run at least, to yield some of its secrets to sustained efforts aimed at extending the range of man's observing power. Yet, in the same process man is continually forced to revise or alter drastically many a well-established view.

For in physics theory and observation are in indissoluble interplay. Now the observation demands a new theory, now the theory points to a hitherto unobserved phenomenon. The General Theory of Relativity, for instance, implies the proposition that the quantity of matter in the universe determines the curvature of the geometry of space. The truth of this proposition, however, depends on determining the variation of the distribution of nebulae with distance. Once this is known, the actual curvature of the geometry of space readily follows. The solution of this problem demanded, however, a penetration of stellar spaces far beyond the range of the 100-inch telescope. What the possibilities might be of an instrument with an aperture twice as large was outlined by I. S. Bowen during the dedication ceremonies of the 200-inch Hale telescope on June 3, 1948. As he noted, this telescope was expected to throw light on the nuclear processes taking place inside the stars. Most important, however, it was supposed to help decide which of the various geometries comes closest to the actual geometry of space. Speaking of the work carried out at Mount Wilson, Bowen said: "As often occurs, these investigations raised more questions than they answered. For example, is the universe really expanding or are the observed effects caused by some curvature of space? Is space uniformly populated with nebulae or do we finally reach a distance beyond which their numbers fall off rapidly?"[100]

There were astronomers who expected within a few years a solution to a problem related to the geometry of space: the pending choice between a steady-state or an expanding type of universe. Yet, gigantic as the 200-inch telescope is, its range is only twice that of the 100-inch telescope, and it turned out that even over distances of two to three billion light years the distribution of nebulae still does not show any distinct pattern pointing to a specific curvature of space. It might be noted at this point that in the late 1950's radiotelescopes, instruments even more powerful in several respects than the Hale telescope, were expected to yield the definitive answer to this crucial issue of modern cosmology. As of 1966 the data supplied by radiotelescopes constitute a body of evidence that weighs heavily against the steady-state theory. The final verdict is perhaps not far away. Yet, the perceptive astronomer of our day is well aware that the solution of this problem will have its share of enigmas. Optimistic as he might be about the impending clarification of this momentous issue, his optimism, as A. C. B. Lovell remarked in 1958, speaking of his own feelings, "is tempered with a deep apprehension born of bitter experience, that the decisive experiment nearly always extends one's horizon into regions of new doubts and difficulties."[101]

What the Hale telescope has demonstrated in the 1950's is the verification of the Hubble-Humason law up to distances where the recessional velocity is two-fifths of that of light. Furthermore it unveiled possibly the first traces of supersystems formed by galaxies. Groups of nebulae such as our own Local Group, which includes the Milky Way, the Andromeda Nebula, the Magellan Clouds, and a dozen or so nebulae, have been known for some time. Although such groupings are usually of irregular pattern, supersystems, such as the one outlined by G. de Vaucouleurs, would resemble a disk, much like the average galaxy, but many times larger.[102] According to Vaucouleurs' speculations, this supergalaxy might have a diameter of fifty million light years, with its center located in the direction of Virgo. Much work, however, is still to be done to substantiate its existence and clarify its details.

The various steps of enlarging the radius of the known world were accompanied by the ever-growing realization that the earth, the sun, the solar system, our galaxy, the Local Group, or the local portion of space in general have no privileged central position in the universe. It was at the beginning of this century that the sun's peripheral position in the Milky Way came to be recognized. Two decades or so later, the Milky Way, taken for over a century as the all-inclusive system, turned out to be no more than a member of the Local Group and not a central one at that. Again, if Vaucouleurs' ideas prove correct, our galaxy and our Local Group will be almost at the rim of the supersystem. One may wonder what subordinate position astronomy will assign fifty years hence to this supersystem. In all likelihood not a central and all-inclusive one, if past developments provide any pattern at all.

One therefore cannot fail to realize how incautious is the view widely shared at present, which regards the individual nebula the ultimate background or unit of the universe. Such a position is logical only if it is safe to say that "though no exact definition of effective observability has yet been given, we must nevertheless conclude that our observations cover an appreciable part of the effectively observable region."[103] But the limits of the range of astronomical observations that are often mentioned nowadays are not so fixed as they appear to be. One of these limits would be a consequence of the Hubble-Humason law, which when extrapolated to the neighborhood of seven-to-ten billion light years would indicate a recessional velocity equal to the speed of light. Light from such distant nebulae would never reach us. Still, such an extrapolation might turn out to be very arbitrary, reaching out as it does too far beyond the present-day range of observation. Furthermore the relativistic increase of mass with velocity alone should indicate that the Hubble-Humason law implies a limit that can never be reached by receding nebulae. True, the principle of conservation of energy demands that in an expanding universe the radiation emitted from receding nebulae should decrease in strength with increasing distance. This dimming of the intercepted radiation might be thought equivalent to a dilution of energy in an expanding volume element. If lessons of the past are of any value, however, it might be very unwise to set narrow limits to future improvements in instruments designed to detect extremely weak radiant energy. Recent developments in this field are on the contrary very encouraging.

Absorption of light by cosmic dust is another factor that may set a practical limit to observations. To present-day apparatus, this factor should pose insurmountable difficulties at six to seven billion light years. Such a decrease of the radiant energy by distance, however, is an asymptotic function, leaving practically unlimited opportunity for technical improvements. Until now the atmosphere itself was for astronomy an irremovable barrier and a major source of dimming and damaging distortions. Newton himself, who discussed at some length in the *Opticks* the "perpetual Tremor of the air" as setting a limit to the improvement of telescopes, dared to hope only about erecting telescopes in a "most serene and quiet Air, such as perhaps be found on the tops of the highest Mountains above the grosser Clouds."[104] In 1700, it was still somewhat of a dream to envision observatories on high mountaintops, and clearly it could not occur to him that one day a 36-inch telescope would be lifted by Project Stratoscope to a height of fifteen miles where only 4 per cent of the atmosphere remains between the object glass and the stars.

The possibilities opening up in this respect are simply leaving far behind anything the astronomy of yesteryear could hope for. A 400-inch telescope, if placed in orbit, could detect stars of magnitude thirty and would be powerful enough to sight planets around stars within a radius of sixteen light years from the sun. If the engineering problems of constructing and putting into outer space a 400-inch telescope seem staggering, one might find comfort in the fact that even a 20-inch telescope operating from outer space would match the effectiveness of the ten times larger but earthbound Hale telescope. It is therefore not unlikely that photographs taken of remote galaxies from the moon's surface might open up unsuspected chapters in astronomy.

Lest one should think that all the novelties of the cosmos are confined to its most remote reaches, it is well to recall some recent findings that should dissipate at once the complacency that our own backyard in the universe contains no more surprises. Thus it was found in 1964 that Venus is slowly spinning in a direction opposite to that of the Sun, the planets, and most of the moons. The same year also brought the discovery that Mercury too is spinning at a rate of about fifty-nine days, with the result that all its surface is exposed periodically to the sun's light. Such findings, needless to say, might very well force a major revision of the accepted views concerning the development of the solar system. Again, as indicated by the recently observed red-colored spots on the moon's surface, the moon is far from being a huge, long dead chunk of matter. It seems therefore highly plausible to assume that bringing the moon and the neighboring planets closer to the earth by a factor of 1,000 or so with modern techniques of observation might produce the same revolutionary information that invariably followed when astronomy increased in the past by a similar factor its penetration into the universe. The spectacular photographs taken by Ranger VII, Luna IX, and especially by Surveyor I, of the moon's surface represent such a big step, and when man succeeds in getting a sample of the moon's surface, he might answer the question whether or not the strength of gravitation is weakening with time.

But before man reaches the moon, studies its surface, and erects telescopes on it, extensive use will be made of orbiting astronomical observatories. With them, man's window on the universe will widen as rarely before. While for earthbased telescopes only the visible wave band and a very small portion of the infrared is available, and radio-astronomy can detect only a part of radiowaves reaching the earth, a balloon telescope at fifteen miles above the atmosphere can make full use of the infrared region as well. Yet, a total absence of limitations with regard to the electromagnetic spectrum obtains only for a telescope placed in outer space. Using the shorter wavelengths and unhampered by atmospheric turbulence and background radiation, a 36-inch reflector placed in orbit around the earth may record stars one hundred times fainter than the faintest so far recorded by the 200-inch Hale telescope. Thus the radius of the observable universe will extend much beyond the present range of optical astronomy. Entire galaxies may suddenly appear where none were sighted before, and many star images taken so far as single stars may reveal their binary character. In outer space one might also make use of the so-called X-ray telescopes, which in principle should be able to increase by a thousandfold the resolution that can be achieved by using visible light. Thus stellar distances too might be measured with an accuracy far surpassing that of present-day data. Space telescopes using the ultraviolet band might also detect the presence of interstellar hydrogen in molecular form in an amount equal to the total mass of stars, just as radio astronomy discovered interstellar hydrogen in a similar amount. Again the study of ultraviolet radiation from stars might clarify many of the puzzles still unsolved about the relative abundances of many chemical elements in stars.

Although telescopes focusing X-rays are still in an early planning stage, conventional X-ray detectors sent by rockets high above the atmosphere are already unveiling startling new aspects of the skies. A recent case is the spotting of so-called neutron stars, perhaps the most solid objects physics can picture. They are thought to be the remnants of supernovae, composed entirely of closely packed neutrons. Their mass seems to be comparable to that of the sun, although their diameter does not exceed five to ten miles. This means that a cubic inch of such an ultradense star would weigh ten to one hundred billion tons. The number of neutron stars in our galaxy is estimated at about one hundred million, but according to the theory developed by Hong-Yee Chiu, only a small fraction of them is at such a high temperature as to radiate chiefly in the short X-ray region. It is now believed that the exceedingly strong X-ray source detected in the constellation Scorpio by the Aerobee rocket fired from the White

Sands Missile Range on February 1, 1964, is in fact one of those very rare neutron stars.

Fascinating as such findings are, the most rewarding use of the various detectors and telescopes placed in outer space will probably consist in the study of galactic explosions, a phenomenon that took present-day astronomers completely by surprise. For several decades astronomy was aware of exploding stars, called novae and supernovae, and there had even been conjectures about colliding galaxies. It was, however, in 1963 that nearly a dozen objects were noticed whose energy emission far exceeded that of the brightest nebulae, although the diameter of some of these extraordinary energy sources seems to be a mere fraction of the diameter of our galaxy. A close examination of the spectral red shift of one of them, the 3C 273, showed that it is located at a distance of about two billion light years. Two other of these "quasars," or quasi-stellar radio sources, were located at six and ten billion light years, respectively, far beyond the present-day "limits" of optical observation, although such determinations cannot yet be considered definitive.

Theoretical explanations of these startling objects has hardly passed the preliminary stage, but there cannot be much doubt that they represent one of the most challenging discoveries in the history of astronomy, with far-reaching implications. The urgent need to find an explanation for them prompted the call for an impressive gathering of physicists and astronomers in December, 1963, at the Southwest Center for Advanced Studies in Dallas. There J. R. Oppenheimer referred to these novel phenomena as "incredibly beautiful" and as "spectacular events of unprecedented grandeur," and J. L. Greenstein spoke of them as "perhaps the most bizarre and puzzling objects ever observed through a telescope." Half a year later, at the world conference of astronomers in Hamburg, the perplexity provoked by the quasars was no less noticeable. There such prominent students of the galaxies as Van Oort and Ambartsumian noted that the core of a quasar might be the seat of unknown forms of energy, and even of an unknown state of matter. Quite recently the most authoritative summary of the status of the relativistic theory of gravitational collapse found no escape from the conclusion: "All matter must manifest, however weakly, a new form of radioactivity. This radioactivity is associated with a new and previously unrecognized process in elementary-particle physics, the spontaneous collapse of baryons singly or in groups of a characteristic favored size."[105]

It is now generally assumed that the quasars involve the sudden collapse in the galactic core of masses equaling hundreds of millions or perhaps billions of suns and will certainly stimulate extensive studies of the behavior of ultra-strong gravitational fields, a subject almost wholly unexplored at present. They might also prove to be regularly repeated events in the lives of galaxies, signaling the last phase of the production of heavy elements. It is also possible that the chief source of high-energy X-rays and cosmic rays may be found in them. Finally, they may be the flare signals that will make it possible to penetrate two or three times deeper into space than the present range of three to four billion light years. The plausibility of this is greatly enhanced by the recent discovery of H. Smith, who found that one of these "quasars," the already mentioned 3C 273, has since 1886 been regularly photographed and classified as a thirteenth magnitude star in the constellation Virgo. Photometric analysis of the photographic plates showed clearly that the 3C 273 is pulsating on a thirteen-year cycle. It is therefore possible to assume that one day astronomy might find in the pulsating "quasars" distance indicators as useful as the Cepheid variables proved fifty years ago. At any rate, as in the past, a major extension of man's reach into space will very likely again prove instrumental in unveiling many an aspect, still unsuspected, of the universe and might again force major revisions of accepted theories and concepts.

Actually this is precisely what seems to emerge from the most recent developments in astronomy. Today astrophysics is faced with the problem of finding perhaps entirely new physical processes to account for the immense energy production in some quasars. Astrophysicists are also engaged in a tantalizing search for a mechanism that could explain the pulsation evident in some of them. No less a puzzle is posed for astrophysics by the fact that most of the quasars now known appear to be arranged along a plane that tilts slightly away from the plane of our galaxy. Will astronomy be forced to entertain the possibility that the physical universe has a specific shape? Or will the number and estimated distance of quasars indicate, as suggested in June, 1965, by A. R. Sandage, that the curvature of the universe is zero, and as a consequence the universe is oscillating with a period of eighty-five billion years? Speaking of quasars and of the most recently detected "blue stellar objects" (BSO), he confidently asserted that those two classes of celestial objects "provide us with the long sought keys to determine the size and shape of the universe." [106]

Some of this confidence is based on the fact that the quasar 3C-9, the most distant known as of mid-1965 (with a velocity of recession of about 149,000 miles per second), appears to enable astronomers to look back on about 90 per cent of the time since the present expansion of the universe started. Yet, the very fact that the BSO's seem to outnumber quasars in the ratio of 500 to 1 should urge caution. While enthusiasm and high hopes following spectacular discoveries are understandable, it is especially at such a juncture that the astronomer of the day might profit from a little reflection on the lessons provided by the history of his art. To achieve this he need not even reach back beyond the beginning of the twentieth century. When in 1914 Eddington published his great synthesis, *Stellar Movements and the Structure of the Universe*, it seemed to many, as W. de Sitter later noted, that a resting place had been reached on the road of astronomy and that the direction of future research had been firmly established.[107] Eddington himself felt confident that he was able "to glimpse the outline of some vast combination which unites even the farthest stars into an organized system."[108] Yet, in the same work Eddington hastened to make some somber references to the past lessons of astronomy. Remembering the numerous transformations that marked scientific history, he warned against taking his work as an "unalterable truth." The measure of achievement he claimed for his work was defined by him in a statement which is worth quoting: "But as each revolution of thought has contained some kernel of surviving truth, so we may hope that our present representation of the universe contains something that will last, notwithstanding its faulty expression."[109]

His cautions stood him in good stead. His all-embracing stellar system proved to be equivalent to our puny Milky Way. What is more, only a year later, in 1915, Einstein published his first paper on general relativity and gravitation, which was destined to give cosmology an entirely new conceptual framework. The result was that in 1932, De Sitter felt it important to emphasize the transitory character of scientific concepts bearing on the structure of the universe. He also noted the continuous and at that moment very rapid transformation of cosmological theories. Consequently he found it impossible to predict "how long our present views and interpretations will remain unaltered and how soon they will have to be replaced by perhaps very different ones, based on new observational data and new critical insight in their connection with their data."[110] It seems that words like these would describe the situation perfectly well even in 1966. Of today's speculations some will no doubt survive and enter the permanently acquired core of scientific truths. At the same time, it is hardly plausible to assume that we are within the reach of grasping the final picture of the stellar universe. Today the astronomical horizon is crowded with wholly mysterious objects, the nature of which will perhaps be revealed only by telescopes based in outer space.

Telescopes used outside the earth's atmosphere are only one example of the many new techniques of which astronomy avails itself at a rate unparalleled in its history. Traditional methods are also being steadily perfected to keep that novelty in them which marked their first use in astronomical research. Over sixty years ago the American astronomer, S. Newcomb, could point out that celestial photography made possible what a hundred observers using for a whole generation the appliances of the Lick Observatory could not have achieved: finding the fifth satellite of Jupiter, learning of the cloud forms of the Milky Way, and discovering extraordinary patches of nebulous light all across the sky. But it is enough to mention the wide-angle photography of the heavens by Schmidt telescopes, the use of color-sensitive plates, and so forth, to show that stellar photography saw even greater triumphs after Newcomb's day. What was then so revolutionary in the use of photography in astronomy is paralleled today by the introduction of such devices into astronomical research as automatic guiding systems, servomechanisms, and, of all things, computers, which in a matter of hours or days can perform calculations that otherwise would take thousands of astronomers hundreds of years. Still inexhausted are the possibilities of improving the "speed" of large optical telescopes by devices like Lallemand's electron telescope (1936) and Baum's photon counting photometer (1953). In one way or another, all these techniques will contribute vastly to the extension of man's reach into the universe, and even if larger optical telescopes are not built in the near future, still the possibility of extending the range of those already in use is very promising.

The field where unknown regions in space have been really opening up for the past ten to fifteen years is the new science of radioastronomy. That in 1966 radioastronomy is still a "new" science is one of the oddities of scientific history and illustrates to what extent new possibilities that are capable of taking a branch of research far beyond its presumed range may lie unexploited. Soon after the detection of electromagnetic waves by Hertz, it could reasonably have been inferred that emission from stars was not restricted to the narrow optical spectrum. In fact, Edison in 1890 and Lodge in 1894 tried to detect radiowaves from the sun but without success. Perhaps it was their failure that helped the idea fall into oblivion, although undoubtedly the lack of interest was just as much fostered by the outlandish character of the project, something to which routine scientific thinking is always conspicuously insensitive. It is, however, far more difficult to understand what happened after K. G. Jansky felt impelled to conclude in 1933 that the direction of the arrival of radiowaves that he had detected "is fixed in space: i.e. that the waves come from some source outside the solar system."[111] Although his discovery received wide publicity in America and the hissing sound produced in Jansky's apparatus by radiowaves coming from the galactic plane was replayed in regular broadcasts, only one person, a radioamateur, G. Reber in Wheaton, Illinois, found in Jansky's work inspiration for further research. For over ten years he worked as the only radioastronomer in the world with his homemade steerable parabolic reflector thirty feet in diameter set up in his own backyard. He found that there is a radio map of the sky with points of peak emission located often in areas where according to optical astronomy there was hardly a thing to look for.

A new world for astronomy was emerging but the world of science did not begin to suspect it until Reber's first papers saw print between 1940 and 1942. Expressions like *new world* alone fitted the findings, as can be seen in the remark of the Harvard astronomer, B. Bok, made several years later, on the new vistas opened up by radioastronomy: "The thrill we feel is akin to that of Balboa first sighting the Pacific Ocean. . . . We are just now becoming aware of the vast, yet unexplored body of data which awaits detection, measurement and analysis."[112] When the challenge was finally grasped by science, it was done in a superlative fashion, as witnessed by gigantic, intricate, and almost fantastic-looking radiotelescopes that began to pop up everywhere in the world and kept a new generation of astronomers busy.

With the discovery of an unsuspected major feature of the skies came, as so invariably in the past, a significant extension of man's reach into space. Radiotelescopes helped to cross a barrier in redshift measurements that had begun to look impassable. Unlike optical telescopes, which operate in a narrow wave band and are unable to sight objects beyond two to three billion light years owing to the shift of the whole visible spectrum into the infrared, radiotelescopes operating in a much wider wave band can follow receding objects at far greater distances. Thus once more the unexpected turn of research proved fears of being stopped at a certain range too premature. Nor should the fact inspire diffidence that radio astronomy is still lacking distance determinations of its own. The sudden turn brought about in optical astronomy by the wholly unexpected discovery of the Cepheid variables as distance indicators might inspire confidence that something very similar might very well happen in radioastronomy at any time.

Determination of distances of radio galaxies beyond the range of optical telescopes still rests on statistical methods. This is one of the reasons why the evidence furnished by M. Ryle's work against the theory of a steady-state universe cannot be considered definitive. On the other hand, radioastronomy has already been able to show that by far the major portion of matter in the universe is represented not by stars but by cosmic dust and clouds present everywhere in space. As a consequence, the estimates of the average density of matter in the universe should be revised, with consequences touching on the values of the curvature and radius of the relativistic world models. Such revisions might at times be drastic. Let us recall that Eddington estimated in 1928 the radius of space in a relativistic world model at 100 million light years. "That leaves room," he wrote, "for a few million spirals; but there is nothing beyond. There is no beyond, — in spherical space 'beyond' brings us back towards the Earth from the opposite direction."[113] Then came the discovery of the nebular recession, and in 1942 Eddington added in a footnote to the quotation above that the radius of space is to be increased by a factor of ten, although it is very unlikely to exceed 4,000 million light years. His posthumous work of 1946, *Fundamental Theory*, set the radius of the *uranoid*, as he called his relativistic world model, at about 1,000 million light years. Determinations of the radius of the space were no more convincing on the basis of Einstein's world model, the values being greater than those of Eddington by only two orders of magnitude.

All this clearly indicated a most insecure area in which to speculate and calculate, and it has remained so even today. Cosmologists still give rather discrepant values for the size of the universe, and in this connection a little retrospect into the late 1920's might be very telling. Eddington spoke then of a "feeble gleam of evidence that perhaps this time the summit of the hierarchy has been reached, and that the system of the spirals is actually the whole world."[114] Lemaître, however, who supported the larger size for the universe, noted that even the 100-inch telescope covers only one two-hundredth of the radius of the universe, adding that "the largest part of the universe is forever out of our reach."[115] Perhaps, but this cannot be stated with any more assurance than can the hope be entertained that the whole cosmos is already within the reach of science. Final statements and "definitive" conclusions therefore do not reflect the trend of progress in cosmology. New observations never fail to make short shrift of overconfidence, such as De Sitter's. He believed, for instance, the value of the total mass of the universe, derived from relativistic considerations, to be more precise than calculations of the mass of individual nebulae based on detailed observations. Nor can one rely today on the value of the total mass of the universe proposed by Eddington, who believed it to be exact within one part in one thousand.

In our day, when radioastronomy keeps detecting the presence of so much previously "unweighted" matter in the universe, one should be particularly aware of the precariousness of determinations of this type. As a young science, radioastronomy, in all likelihood still has surprises in store that cannot even be guessed today. Radioastronomy might well decide between an expanding universe and a steady-state universe, but it might equally well unravel aspects of the universe that could render both theories obsolete. This might seem rather far-fetched to state, but radioastronomy had already undertaken projects unparalleled in the history of science, such as the plan for establishing radiocontact with denizens of other planetary systems whose exist-ence is a distinct possibility in view of today's science.

Yet, it is well to remember that on this question the scientific consensus reversed itself twice in the span of only three decades, and this was due in no small measure to the suddenly accelerated rate of the increase of man's penetration into the far reaches of the uni-verse. Prior to the turn of the century, the consensus of astronomers rather willingly endowed almost every star with planets capable of sustaining some form of life. It was only in the twentieth century that more attention began to be centered on the opposite view that con-sidered planetary systems like ours as an extreme rarity in the uni-verse and the possibilities for organic and intelligent life elsewhere as negligibly small. The astronomer chiefly instrumental in making this view prevail was Jeans, who put the finishing touch on a long series of studies going back into the second half of the nineteenth century that dealt with the difficulties of Laplace's nebular theory. As Jeans's Adams Prize-winning work of 1917 showed, the concept of the formation of planets through condensation of material ejected from a rotating star was beset with grave problems, and as a conse-quence the origin of planetary systems had to be ascribed to an ex-tremely rare occurrence in space: a fairly close encounter of two stars. The planetary system, Jeans concluded, is just as exceptional as the earth-moon system and "for ought we know may be unique in the system of the stars." [116]

This conclusion was fully endorsed by Eddington, and for more than two decades they succeeded with their great authority and well-coined phrases in almost codifying it. Thus in 1922, when the Gold Medal of the Royal Society was awarded to Jeans, Eddington con-jured up the prodigality of nature in wasting thousands of acorns to produce one new oak tree, to illustrate the "wastefulness" of nature in scattering "a million stars whereof but two or three might happily achieve the purpose," [117] that is, to provide home for intelligent beings like man. Clearly, hardly anything could be more at variance with the accepted views of the age, the imagination of which had been captured by Flammarion's vivid accounts of the life on Mars, by

Haeckel's *Riddle of the Universe*, which acknowledged in fact no riddles at all, and by Arrhenius' germ cells pushed by the pressure of light from star to star spreading life everywhere in the cosmos. Somewhat to allay the shock, Eddington tried to prevent his audience from drawing the conclusion that in Jeans's view "the solar system is a freak system, unusual and possibly unique in the universe." [118] Jeans only meant to show, Eddington insisted, how shaky was the opposite view. Jeans, however, pressed his conclusion relentlessly, and in his *Astronomy and Cosmogony*, a development of his prize-winning essay, he roundly stated in 1928 that "planetary systems must be of the nature of freak-formations: they do not appear in the normal evolutionary course of a normal star." [119] He even conjectured that we should "probably have to visit 50,000 galaxies to find a civilization as young as our own." [120] Jeans's summary of the problem intended for popular consumption was no less emphatic, and there the layman could read that the "mathematical calculation decides the question definitely." [121] Clearly mathematics was meant as a tribunal beyond which no court of appeal could be imagined. As a result, a rather bitter pill had to be swallowed by those who saw organic and intelligent life as the inevitable product of chemical evolution and who therefore would have felt far more comfortable with the belief in living beings popping up spontaneously in every nook and cranny of the universe. Thus H. G. Wells, J. S. Huxley, and P. G. Wells had no choice but to inform the uncounted readers of their *Science of Life* that according to calculations only one star in a hundred thousand might be surrounded by planets, and therefore "in every respect, — space, time, physical conditions — life is limited to an almost inconceivably small corner of the universe." [122] This was a far cry from views enjoying in popular belief the highest respectability only a decade or so earlier and certainly had nothing in common with the fantasies of Bruno, Huygens, Fontenelle, and many others. Awareness of this contrast forced itself on almost anyone studying the subject, and one of its telling manifestations is preserved in a remark of G. McColley, author of a paper entitled "The Seventeenth-century Doctrine of a Plurality of Worlds" (1936): "We retain today little more than the planetary order upon which these interpretations were in a large measure based. The doctrine of a plurality of worlds is now a pleasant myth and the Principle of Plenitude has no place in modern science." [123] Little did the author suspect that ten years later modern science would again be buzzing with hypotheses about planets and civilizations in the very neighborhood of our solar system.

Actually the change was prepared by O. Struve's work in the 1920's, which carefully pondered the implications of the existence of the thousands of galaxies, each with billions of stars, that the 100-inch telescope was beginning to unveil. Clearly it seemed plausible to assume that given the immense number of stellar populations planetary systems should be numerous even if their production was a relatively rare event. Yet, to put such speculations on solid ground and to bring about the sudden change in the way that the frequency of planetary systems was appraised, a new approach to the question of the formation of planetary systems had to come. This was done in the early 1940's when Alfvén, Whipple, and Weizsäcker put forward ideas that could at least in part overcome the principal difficulties of the older nebular hypothesis and also account for the principal features of the solar system. In his paper of 1944 Weizsäcker showed, for instance, that a star plunging through a dense interstellar gas and dust cloud could attract to itself enormous quantities of material that, because of frictional and gravitational forces, would slowly take the shape of a disk in which eddies would appear inevitably.[124] The larger eddies would absorb the smaller ones, as Kuiper later pointed out, and would slowly condense into planets. The theory is, however, not without serious difficulties, and Weizsäcker admitted in 1951 that his paper of 1944 had idealized the actual situation too much and that, contrary to earlier beliefs, the theory does not yield Bode's Law.[125]

Yet, owing to the promising aspects of Weizsäcker's conceptions, the scientific consensus almost overnight abandoned the view stressing the extreme rarity of planetary systems. Planets are now being thought of as rather regular features of most single stars whose number is estimated at 10^{20} in the hundred billion galaxies present within a one billion light-year radius. True, not all of these planets would possess physical characteristics comparable to those of earth, but even after introducing very restrictive conditions that eliminate as possible candidates almost all planets, the number remaining would still run in the billions. As H. Shapley phrased it:

In a speculative frame of mind let's say that only one in a hundred is a single star, and of them only one in a hundred has a system of planets, and of them only one in a hundred has an earthlike planet, and of them only one in a hundred has its earth in that interval of distance from the star that we call the liquid-water-belt (neither too cold, nor too hot), and of them only one in a hundred has the chemistry of air, water and land something like ours — suppose all these chances were approximately true, then we would find a planet suitable for biological experiment for only one star in ten billion. But there are so many stars! We would still have ten billion planets suitable for organic life something like that on the earth.[126]

And what if Shapley, as many astronomers think, had underestimated the number? Whether one thousand billion or one billion or only one million would be the better approximation matters little. Each number is sufficiently large to make one think of the universe in which life and perhaps civilizations are not as exceptional as astronomers not long ago would have had us believe. Once again a major change in man's conception about the universe followed close on the heels of man's ability to see farther and many more objects in the far and wide spaces.

The magnitude of this change can, however, be fully grasped only by remembering that with it arose, with entirely new urgency, the question whether a contact can ever be established with denizens of other planetary systems. In the last century Gauss proposed the construction of ten-mile-wide geometrical figures by cutting clearings in the Siberian forests to attract the attention of intelligent beings on other planets of the solar system. In the same vein, Littrow, the Austrian astronomer, wrote about digging in the Sahara circular trenches, several hundred yards in width and some twenty miles in diameter, in which kerosene would be burned at night to provide a gigantic flare sign sent into space. Seriously proposed as were these projects, they had little scientific merit and merely serve as topics of amusement today. Indeed they appear almost childish when compared with the methods of radioastronomy that aim at the detection of possible signals sent toward the earth from other planetary systems. The actual search for such signals got under way in 1960 at the National Radio Astronomy Observatory at Green Bank, West Virginia, using the 85-foot parabolic reflector. Known as Project Ozma, the listening program was based on the ideas of P. Morrison and G. Cocconi, who argued in an impressively reasoned paper[127] that radio messages coded probably in number sequences are very likely being sent out on the 21-cm wavelength from other advanced civilizations. Of the several stars in the neighborhood of the sun that might very likely possess such planets, τ Ceti and ϵ Eridani were kept under observation but no positive results have been obtained. It is important to note, however, that although the number of suitable target stars is only a dozen or so within a radius of ten to twenty light years, a 600-foot radio telescope could extend the range of the project to about 70 light years and increase the number of suitable target stars to well over a thousand, thereby greatly enhancing the chances for eventual success. The possibilities appear even more promising with the just-completed radio telescope near Arecibo, Puerto Rico, which has a diameter of 1,000 feet and should be able to establish a radio contact with all possible targets within a radius of over one hundred light years. Such an encouraging outlook for the future of the project seems, therefore, to assure that in spite of the early failures the search will go on, for to quote Morrison and Cocconi, "the presence of interstellar signals is entirely consistent with all we now know, and if signals are present, the means of detecting them is now at hand. . . . We therefore feel that a discriminating search for signals deserves a considerable effort. The probability of success is hard to estimate; but if we never search, the chance of success is zero." [128]

How the eventual success of interstellar communication might affect man's notion of the universe, not even the boldest imagination can safely guess today! When it happens, it will no doubt bring about the most startling phase yet in the process that this chapter has tried to illustrate: as long as man keeps reaching farther and farther into space, his concepts of the universe and his scientific systematization of it — astronomy and cosmogony — will have to undergo continuous revision. At the turn of the century, as Shapley pointed out in 1950, the greatest names in astronomy, Poincaré, Newcomb, Struve, Gill, Kapteyn, to list a few, would have been completely baffled by topics that now figure in the examination of any candidate for a Ph.D. in astronomy. [129] Theories, like relativity and quantum mechanics, were simply non-existent; phenomena, like cosmic rays and radio stars, were not even suspected; and instruments like photomultipliers and radiotelescopes were not even on the drawing boards. Great as the advances were in the first half of this century, they will no doubt be dwarfed by what will be accomplished by A.D. 2000. Will the astronomers at that date, asked Shapley, "bother to use such interesting earth-anchored contraptions as the 200-inch telescope in Palomar Mountain or to be dependent on the photographic plate?" [130] The rate at which entirely new steps are being taken in astronomy is becoming faster with every decade, and it is interesting to note that Shapley, who with some reluctance offered in 1950 a few prognostications as to what astronomy might achieve in the 1950's and 1960's, had no inkling of the coming of devices like orbiting telescopes, Laser beams, and X-ray telescopes, which are already on the point of revolutionizing the astronomy of the 1960's.

If the history of astronomy warns in general against forming at a given time too rigid conclusions about the structure and limits of the universe, the ever-faster development of the science of astronomy in the twentieth century manifests in no less degree the characteristics of a transitional period discouraging definitive statements. In the opening decades of this century, it was the recognition of this that impelled Jeans to characterize as "pure dogmatism" any attempt "to dictate final conclusions on the main problems of cosmogony." [131] Ten years later, he was even more convinced on this point. Far from considering the cosmogony of 1928 a "finished science," he described it rather as "the first confused gropings of the infant mind trying to understand the world outside its cradle." [132] Today, there is every reason to believe that astronomy and cosmology are on the threshold of major breakthroughs that will unveil aspects of the universe hidden hitherto. No doubt the extremely rich outcropping of cosmological theories and discoveries of the past half century has left the speculations of previous centuries far behind. But as one truly in the forefront of present-day astronomical research has warned: "All these cosmogonic and cosmologic hypotheses are most likely to be soon made obsolete by the startling new discoveries which have resulted from the introduction since the Second World War of new and unexpected methods of exploration of the universe." [133]

If lessons of the past have any significance, one could confidently expect that the new regions of space yet to unfold will have their long array of unexpected turns. If there is an ever-returning pattern in the course of astronomy, it is the invariable connection of a radical change in man's views of the universe with every major advance toward the elusive horizons of the cosmos. Nothing in the present phase of science indicates that this process is nearing its end, that it is no longer possible to extend our measuring rods farther and farther into the depths of space. On the contrary, as the pace of this penetration into the far reaches of the universe quickens, so does the perhaps irritating and frustrating urgency of replacing outmoded schemes by more inclusive ones. It is precisely the quickening rate of this process that suggests most convincingly that many further phases of this journey into the unknown may still lie ahead.

Among the few secure conclusions one may draw from the analysis of the past legs of this journey is that the less one knows about the universe the easier it is to explain. In this respect the universe as a whole does not differ from any other scientific subject matter of lesser extent. Still, to the study of the ever-expanding stellar sphere, one may apply with special appropriateness the saying, "The greater the sphere of our knowledge, the larger is the surface of its contact with the infinity of our ignorance." The principal reason why this is so is amply documented in the successive phases of man's penetration into the universe and received its most timely expression at the dedication ceremonies of the Hale telescope in the words of V. Bush: "It is a great truth of science that every ending is a beginning, that each question answered leads to new problems to solve, that each opportunity grasped and utilized engenders fresh and greater opportunities."[134] The sobering fact should always be kept in mind that after more than three hundred years of spectacular advances, the evaluation of the scientific cosmology formulated by Salviati, Galileo's spokesman in the *Dialogue*, is still valid. Astronomy, Salviati admitted, has not yet arrived "at such a state that there are not many things still remaining undecided, and perhaps still more which remain unknown."[135]

Of these undecided problems the astronomy and cosmology of the 1960's have their ample share. The most promising approach to cosmology, the General Theory of Relativity, seems to be at variance with Mach's principle, an equally highly regarded guideline for the modern cosmologist. The Schwarzschild singularity, which fifty years ago seemed but a curious side aspect of General Relativity, looms today as a most grave question mark presented by the problem of gravitational collapse. There are also those discrepancies that R. H. Dicke called "disquieting even to the experts who know the data best."[136] Thus, certain galaxies appear to be older than the first heavy elements. Again, certain globular clusters seem to be twice as old as the age of an evolutionary universe of the closed type. The cosmologist of today has to work on his model of the cosmos with an awareness of the fact, that, as Fred Hoyle recently noted, "extraordinarily little is known about the formation of galaxies,"[137] the generally assumed building blocks of the universe. The cosmologist of today must come to grips with the question of whether he should consider galaxies on equal footing with antigalaxies.[138] He is also faced with the problem of whether in view of the exceedingly slight rate of interaction of neutrinos with other fundamental particles, he should entertain the possibility of interpenetrating though hardly interacting universes. Unlike the astronomer of yesteryear, he must scan horizons immensely wider and perplexingly more complex. In addition he also has to be ready to turn his attention to the crust of the good old earth that might contain some traces of large-scale cosmological processes.[139]

Yet, his principal domain will remain the starry heaven where in all likelihood endless surprises are in store for him. The principal reason for this has been the dominant theme of this chapter. Men of science often stated it, but it bears repeating as each generation is prone to think that it holds only of past decades or centuries. Each advance so far into the farthest regions of space has proved a source of new discoveries and new ideas about the cosmos. As a leading physicist of our day, E. Segrè, noted rather recently with reference to the new science of space exploration: "A change in order of magnitude is a well-known source of surprises."[140] The same truth was stated, however, with no less clarity when Galileo's telescope began to explore outer space. One should only recall Salviati's answer to Sagredo's question whether new observations and discoveries made by telescopes will ever cease. If scientific experience has a lesson to offer, went the memorable reply, "things will be seen which we cannot even imagine at present."[141]

In full support of this, astronomy so far has failed to justify those who identified the limits of the world with the horizons of the day and has invariably overthrown theories that tried to cover the immense complexity of the physical world with narrow schemes ruling out drastic revisions. However numerous and illuminating are the details science learns about the universe, the prospect of achieving an all-inclusive explanation of it remains as remote as ever. While man could establish the sphericity of the earth, its motion around the sun, the shape of the Milky Way, the energy processes at work in the stars, and a great number of other precious truths, man's overall view of the cosmos seems to be as provisional as ever. With regard to the whole of the universe, it appears that the range of physics does not bear the mark of limitedness, nor do its solutions have the seal of absolute definitiveness. The universe today appears to be just as fathomless as the Milky Way proved for Herschel. Such a situation demands a cautious attitude on the part of all those who investigate the fascinating riddles of the cosmos and tell the public about their findings and speculations. As R. C. Tolman, a cosmologist with a truly penetrating mind, warned:

It is appropriate to approach the problems of cosmology with feelings of respect for their importance, of awe for their vastness, and of exultation for the temerity of the human mind in attempting to solve them. They must be treated, however, by the detailed, critical, and dispassionate methods of the scientist.[142]

CHAPTER SIX

The Edge of Precision

In physical science a first essential step in the direction of learning any subject is to find principles of numerical reckoning and methods for practicably measuring some quality connected with it. I often say that when you can measure what you are speaking about, and express it in numbers, you know something about it; but when you cannot measure it, when you cannot express it in numbers, your knowledge is of a meagre and unsatisfactory kind: it may be the beginning of knowledge, but you have scarcely, in your thoughts, advanced to the stage of science, whatever the matter may be.[1]

Among the many precepts advanced by Kelvin the one above is of prime importance. To be sure science may require a variety of mental attitudes that have little or nothing to do with either the abacus or the calculus: faith, imagination, intuition, bold guesswork, and intellectual courage of the sort which may at times lead the scientist into perilous grounds. Nevertheless, to exist at all, science must make equal use of the urge in man to measure things and processes, sometimes with astounding precision. For as the younger Herschel put it so well over a century ago, exactness in measurements has a more significant role than simply keeping science off false trails. Precision, he said, "is the very soul of science and its attainment affords the only criterion, or at least the best, of the truth of theories, and the correctness of experiments."[2] Indeed, hardly in any other respect does physical science show itself so demanding than when it comes to precision. There lies the final touchstone of concepts and hypotheses, some of which, though attractive at first, may not survive exact scrutiny. Preoccupation with precision was in fact indispensable to the mental atmosphere which insured the continuous growth of physics and its range extends only as far as does the preciseness of its instruments.

A preoccupation with exact measurements and a willingness to expend often superhuman efforts to achieve further refinements and resolution may impress the non-scientific mind as an attitude that looks backward rather than ahead to something new. The quest for ever greater precision may even appear at times to be an obsession with accuracy for its own sake. Undoubtedly there is something extravagant in a statement of the nineteenth-century German physicist, A. Kundt. Famous for his exact measurements of the velocity of sound in gases and solids, Kundt once stated that "in the end one might just as well measure the velocity of rainwater in the gutter." What Kundt could find so attractive in this odd proposition, of course, was the prospect of measuring even the flow of rainwater with extraordinary accuracy. This is why such a prominent physicist of the Victorian era as F. Kohlrausch could react to Kundt's words with true congeniality. "I would be delighted to do so," was his enthusiastic comment.[3] It is well to remember that behind these facetious statements lay a solemn awareness of the crucial role which precision has taken in the progress of physical science. In this regard, as we shall see, Kohlrausch's own contribution helped to make history in physics, every page of which is a vivid illustration of what Kelvin once put so concisely: "Nearly all the grandest discoveries of science have been but the rewards of accurate measurement and patient long-continued labor in the minute sifting of numerical results."[4] The contribution of increased precision is present in major breakthroughs in physics; significantly whenever the advancement of physics takes on added momentum the simultaneous availability of more precise techniques is clearly discernible.

Discussion of the causes of the stalling of Greek science — or physics in particular — may go on forever; the lack of consistent appreciation by the Greeks of the role of quantitative precision, however, will always stand out as one of the major factors. There had of course been exceptions in certain areas of Greek science. Characteristically enough, it was precisely in these exceptions that ancient science produced its most spectacular results. The accurate determination by the Greeks of the diameter of the earth and of the earth-moon distance has been discussed in the preceding chapter. The most impressive feat of the Greeks along these lines, however, was the discovery of the precession of the equinoxes, or the motion of the poles of the earth in a circle, the period of which is about twenty-six thousand years. The figure given by Hipparchus for the amount of the precession in a century was 1 degree 23 minutes 20 seconds, only 10 seconds of an arc less than the value accepted at present.

Hipparchus' discovery was a classic case of what entirely unexpected features of the universe can be suddenly unveiled by putting together carefully established data originally gathered for wholly different purposes. What Hipparchus wanted to achieve was a better calendar; to this end he made determinations of the lengths of the tropical and sidereal years. If he had not had at his disposal the observations of the Babylonian astronomers recorded continuously from 747 B.C., his work could hardly have achieved anything extraordinary. After analyzing the long list of Babylonian data, he was forced to conclude that the equinoctial point does not retain the same relation to a fixed star but moves slowly forward due west along the zodiacal belt. His conclusion rested of course on the tacit assumption that the records consulted were trustworthy, made with utmost care by critical, cautious minds.

The presence of such a genuinely scientific attitude in an age so much permeated by myths, superstitions, and dreams was truly remarkable. Not only did it bring to ancient science perhaps its greatest triumph, but it also shows that ancient astronomers possessed the same keen appreciation of precision that animates their modern counterparts. The precision of ancient astronomical observations was such, in fact, that it has helped even modern astronomy. Partly on the basis of ancient observations, eighteenth-century astronomers had grown aware of such problems as the great inequality of Jupiter and Saturn and the secular acceleration of the moon.

The discovery of the precession of the equinoxes should not make one believe that an absolute respect and quest for precision prevailed as much in the physics as in the astronomy of antiquity. Viewing astronomy as a branch of mathematics, or geometry, the Aristotelian philosophy of science, so influential in antiquity, readily recognized the need for exact observations in astronomy. As Aristotle expressed it in the opening pages of the *Nicomachean Ethics*, it would be just as foolish to demand scientific proofs from a rhetorician as it would be to accept probable reasoning from a mathematician. In the Aristotelian system, however, physics was a branch of study not to be classed with mathematics and astronomy. Thus certain basic Aristotelian assumptions about the nature of physics relegated precision to somewhat secondary importance in the cultivation of physical science. Aristotle no doubt set down a wholesome principle in stating that "it is the mark of an educated man to look for precision in each class of things just so far as the nature of the subject admits."[5] Less constructive was his definition of the nature of those phenomena that constituted the subject matter of physics. For how could quantitative precision play a decisive role in a physics that viewed the quantitative analysis of motion as an "unnatural" procedure, a dissection of the wholeness of nature manifested through purposeful motion?

Such a concept of motion was of course not limited in classical antiquity to the Aristotelian school. Although the Neoplatonists, as shown in a previous chapter, called for attacking everything in nature mathematically, they hardly made a step toward implementing a program that by its very nature would have demanded a high degree of precision in measurement. They rather contented themselves with paraphrasing the Aristotelian dictum that one cannot expect the same precision in everything. Iamblichus, for instance, compared the various degrees of precision possible in the various branches of theoretical sciences to the various degrees of precision permitted by the materials used in the different branches of technical arts.[6] Physical science, the method of which he defined so emphatically along mathematical lines, was no doubt intended to occupy an uppermost rank in terms of the degree of precision. Yet, the Neoplatonists, no less than the Aristotelians, recoiled from actually trying to apply such a method to the phenomena of the sublunary world of change.

For the Greeks the changes in the heavens appeared to be more nearly eternal, more permeated with reason, regularity, and numbers. This is why interest in improving precision of measurements remained confined, throughout classical antiquity, almost exclusively to astronomical practice. Unfortunately it is almost impossible to reconstruct today with a fair measure of completeness how precision was being perfected in ancient astronomical observations. Very likely some of the best scientific precision instruments used in antiquity are still unknown and perhaps will remain so. That the attainment of the Greeks in this respect was far greater than had been generally believed is best illustrated by some chance archeological findings. The most revealing of these are the fragments of a mechanism which were recovered from the depths near the island of Antikythera in 1902 and are kept in the National Museum in Athens. As assembled and interpreted recently by D. J. Price, these fragments, dating from the first century B.C., seem to be remnants of a mechanical calendar that was capable of reproducing the motion of planets.[7] If such an interpretation is correct, the Antikythera machine represents a major enigma of antique science, displaying a precision that was not equaled by similar scientific instruments until the seventeenth century.

The same precocity holds true in the precision of ancient positional astronomy. The best observations of ancient astronomers were good to within an error of about two to three minutes of arc, whereas Copernicus had to satisfy himself with a determination of planetary positions in error by at least ten minutes of arc. The appeal of the Copernican system did not lie in the precision it permitted for calculating the position of the planets. In this respect it was at best equal to, but did not surpass that of Ptolemy, in spite of the claims of Rheticus to the contrary. Rheticus of course was duty bound to praise both his master's observations "made with the utmost care"[8] and the results that coincided "to the utmost degree of exactness with the observations of all scholars."[9] The fact, however, was that Copernicus' instruments were almost crude, and that no more than twenty-seven of his observations entered into his epoch-making work. The rest of his data were borrowed from the ancients. Being shut in himself, far removed from the main thoroughfares of the world, Copernicus could hardly realize that ocean navigation and the progress of technology were just beginning to force on science a new and hard look at the exactness and reliability of measurements.

Such needs often remain long submerged until suddenly, owing to some unexpected incident, they break to the surface. The impulse that brought the streams of precision and of science forever together was touched off by a partial eclipse of the sun witnessed by fourteen-year-old Tycho Brahe, who was studying rhetoric and philosophy at the University of Copenhagen. That such phenomena as eclipses occurred as predicted had a simply overpowering effect on Tycho's mind. Tycho threw in his lot irrevocably with astronomy. Three years later, already in possession of all available astronomical tables, he found on the night of August 17, 1563, that Jupiter and Saturn were so close as to be hardly distinguishable. To his great shock both the Alphonsine and the Copernican tables were wide of the mark in fixing the date for this event. The former erred by one month, the latter by several days.

For Tycho this represented an intolerable state of affairs. As he had correctly diagnosed matters, the situation could be remedied only if astronomy developed an absolute dedication to the construction of better instruments. In pursuing this end, Tycho had no equal in his day. Before long his rewards came in ample measure. His huge sextant, equipped with a table of figures indicating the errors involved in his observations, played the decisive role in showing the superlunary position of the nova of 1572 and of the comet of 1577. His long list of carefully taken data[10] delivered a mortal blow to Aristotelian cosmology and established Tycho as the foremost astronomer of his time. Furthermore, because of Tycho's influence, the role of precise prediction of the position of planets began to take on added importance in the disputes about the truth of the Copernican hypothesis.

No one was more conscious of this than Kepler, and no one was more determined or in a better position to wrest Tycho's riches from him. "Tycho possesses," wrote Kepler shortly after his arrival at Tycho's residence in Benatek, "the best observations and consequently, as it were, the material for the erection of a new structure."[11] The new structure was the heliocentric system, and to be its chief architect was Kepler's principal ambition. He was confident that he had what Tycho clearly lacked, the intuitive insight that alone could make one recognize the plan of unity lying "hidden exceedingly deep" under the diversity of phenomena. How "exceedingly deep" the plan lay Kepler found out before long. For shortly after he became familiar with some of Tycho's data, he could not fail to realize the full measure of both the power and challenge of utmost accuracy in measurements. In Kepler this brought about a truly momentous change. The young Kepler could resort in his *Mysterium cosmographicum* even to cheating with observational data, and he was willing to accuse Copernicus or Ptolemy of the same when their data happened to disagree with his rather arbitrary layout of the planets.[12] Tycho's data, however, permitted no such nonsense. So Kepler had to reject his first attempt to define the orbit of Mars, because he found it no longer permissible, as he wrote, to neglect a discrepancy of a mere eight minutes of an arc between theory and observation. To neglect those eight minutes would have been, in his view, nothing short of belittling divine mercy itself, which "has given us Tycho, an observer so faithful that he could not possibly have made this error of eight minutes." To take advantage of the situation was now a sacred duty for Kepler, and he saw those mere eight minutes as suddenly pointing the road to "the reformation of the whole of astronomy."[13] Truly, never before had eight minutes constituted, as Kepler put it, the burden of a major astronomical work, and never before or since had the future of science depended so much on an unconditional reverence for the data of observation. That precision could come into its own in physics and assume its role of supreme arbiter over theories depended also, of course, on the conviction, growing stronger since Copernicus, that the geometry embodied in the physical world must be simple. The concrete form of this geometrical simplicity was henceforth to be extracted not from one's preference for one or another geometrical figure but from the data of precision. In Kepler's situation, the hard fact of eight minutes of arc forced him to discover the true form of planetary geometry, the elliptical orbits.

Had Galileo shown genuine appreciation for Kepler's work, he would have in all likelihood grasped the crucial import of Kepler's achievement and strengthened the cause he pleaded for in his *Dialogue*.[14] Even more important, Galileo, by studying Kepler in depth, could also have developed his colleague's unconditional respect for precision. For the author of the *Dialogue* was somewhat inconsistent in this regard. In discussing, for instance, the position of Tycho's star, a crucial issue in deciding for or against the immutability of the heavens, Galileo spoke glowingly of the "painstaking precision" of astronomers.[15] Yet, the same *Dialogue* gives a value for the acceleration of gravity that is off the mark by a round 100 per cent. According to Salviati's contentions, an iron ball of 100 pounds falls from a height of 100 cubits (166 feet) in five seconds.[16] Not only did Galileo in this connection refer to "repeated experiments" emphasizing the "precision" of this numerical result, but in the same context he boldly claimed that because of the same value of the acceleration of gravity a body would fall from the moon's distance to the earth exactly in 3 hours, 22 minutes and 4 seconds. Although at that time nobody could show that the actual time would be about four days and twenty hours, the value of g implied in the fall of iron balls was a "down to earth" matter that could be checked, and checked it was. The results invariably showed that Galileo's value had little to do with reality — as can be seen from Mersenne's, Riccioli's, and Huygens' comments and experiments, all of which got closer to the value of 32 ft/sec².

To increase the mystery Galileo kept mum about the value of g when describing his experiments with inclined planes in his *Discourses on Two New Sciences*.[17] Correct as Galileo's law for the falling bodies was, its precise measurements in his hands fell far short of the geometrical precision which nature was believed to obey. Much as Galileo spoke of the "inexorable laws" of nature, he did not follow Kepler to the summit of intellectual detachment, where even the most appealing geometrical constructions have to yield to the equally importune demand for precision of measurements. Clearly, in the absence of verification based on the maximum precision available, the conclusions are either to be deferred or to be formulated only with some reservation until a more substantial agreement obtains between theory and observations. To be cautious meant, however, to abdicate some of the glory of having achieved a major discovery, and few scientists in the seventeenth century were willing to make such a sacrifice. But human weakness apart, physics at this time was far more deductive than inductive; this is why in Galileo's *Discourses on Two New Sciences* the pursuit of physical science still emphasized geometrical demonstrations more than the practical recognition of the all-important role which meticulous measurements play.

The book in which science reached the first large-scale fusion of theory, bold generalization, and long arrays of critically weighted experimental data, was the *Principia*, which represented a first in many other respects as well. Newton's accuracy in carrying out his own experiments and his unreserved respect for accuracy was nothing short of extraordinary, even when compared with ideals of precision which twentieth-century science considers important. The very first paragraph of the *Principia* opens with a definition of the constancy of mass which is based on "very accurately made" measurements. References to "more accurate" measurements occur four times in the very short prefaces Newton wrote to the second and third editions of the *Principia*, evidencing the paramount importance Newton ascribed to experimental accuracy in the verification of a theory. That such insistence on precision was far more than a question of style is brought out clearly by Newton's attitude toward particular conclusions he derived. Thus he calculated on theoretical grounds that, owing to its rotation, a uniformly dense earth must be flattened at the poles, giving the earth an ellipticity of $\frac{1}{230}$ — as against the slightly different modern value of $\frac{1}{297}$. Such a figure, however, was highly vulnerable to objections based on experimental grounds since the oblateness of the earth caused a series of phenomena that could be measured in independent ways. These effects included the lengthening of the period of pendulums and the decrease of the weight at lower latitudes, and the precession of the equinoxes. It is only natural that throughout his life Newton pursued with special interest experiments relating to this subject. Problem iv of Proposition XX in Book Three of the *Principia* devotes long pages to the pendulum experiments which Richer and others carried out between 1672 and 1704, subjecting to close scrutiny the reliability of each. While rejecting Couplet's observations as "gross," Newton praises those of Richer highly, remarking: "This diligence and care seems to have been wanting to the other observers. If this gentleman's observations are to be depended upon, the earth is higher under the equator than at the poles, and that by an excess of about 17 miles, as appeared above by the theory."[18] Accurate enough, including as it did the effects of temperature changes in the length of the pendulum rod, the result led to the value of the precession of equinoxes as observed astronomically.

Newton was no less critical when it came to his own experiments. Highly accurate, for instance, were his measurements of the separation of interference rings produced when the curved surface of a plano-convex lens touched a flat glass plate. More than a hundred years later his data could still provide a firm foundation for Young to compute the wavelengths of various colors — values in close agreement with those accepted today. What he wrote about his experiments concerning the equality of gravitational and inertial masses is

The Central Themes of Physical Research

true of almost every part of Newton's scientific work: "By experiments made with the greatest accuracy, I have always found the quantity of matter in bodies proportional to their weight."[19] On this particular point his conclusion was only strengthened as centuries passed by, although the error involved in the experiments of Eötvös and, recently, in those of R. H. Dicke was only one part in one billion and one part in ten billion, respectively.[20]

The position Newton took in another problem involving very slight effects, the motion of the aphelion of the planets, did not stand the test of time as well as his conclusion on the equivalence of the inertial and gravitational mass, but this is understandable. In the absence of perturbations due to other planets, any planet would revolve around the sun with its aphelion fixed. This is a fundamental consequence of Newton's theory on the motion in a central field of force. "The aphelions are immovable," he states,[21] and from this it follows that if the gravitational attraction obeyed a law ever so slightly different from the inverse square law, the orbits of planets would not be reentrant and a progressive motion of the apsis in the plane of the orbit would result. Mars, Earth, Venus, and Mercury do indeed show such an effect, for which the perturbing influence of Jupiter and Saturn appeared to be wholly accountable. Newton considered the advance of 4' 16" of arc made by the perihelion of Mercury in a hundred years to be "so inconsiderable that we have neglected [it] in this Proposition."[22] But once the quest for precision had acquired full rights in science, there was no way of limiting its demands in some arbitrary way simply, say, to protect a theory from a complete overhaul, to say nothing of its being discarded altogether.

Leverrier in 1845 found that the advance per century of the perihelion of Mercury was twice as great as the value accepted in Newton's time and that of the total value of 574 seconds of arc, 42 seconds could not be accounted for by any conceivable pattern of perturbation. Inspired no doubt by his discovery of Neptune, Leverrier in 1859 assigned the unaccountable 42 seconds of arc to the effect of a hypothetical planet having the same mass as Mercury but being at half the distance of Mercury from the Sun. In the same year an amateur astronomer, Lescarbault, claimed to have observed the transit of a small round object across the sun's disc. Some of Lescarbault's observations were promptly accepted by Leverrier, who named the new planet Vulcan and computed its revolution around the sun as twenty days. Despite the fanfare, the most competent astronomers, Leverrier included, were unable to observe Vulcan. Leverrier, however, had already cast his lot with Vulcan. When, on April 4, 1876, Weber announced at Peckeloh that he had seen Vulcan's transit,

Leverrier's confidence was bolstered again, and he fixed the transits of Vulcan for March 22, 1877, and October 15, 1882. On the appointed days nothing was sighted with any certainty. Such was also the case during the total solar eclipse of July 15, 1878. Meanwhile, the true amount of the advance of the perihelion of Mercury was subjected to further observations, and Newcomb in 1884 put the amount unaccountable by any known or conceivable factor at 43 seconds of arc per century. To clear up this puny but stubborn discrepancy between theory and observation, an even more drastic step than finding a new planet was needed to do full justice to a mere 43 seconds of arc that takes a full hundred years to accumulate. Nor was the situation saved by proposals which, although innocuous on paper, were espoused by men of science even more reluctantly than was the assumed existence of an entire planet. Undoubtedly Hall and Newcomb must have had misgiving when suggesting the replacement of the inverse square law with the term $1/r^{2.000,000,16}$. It was hardly easier for Seeliger to suggest in 1895 the addition of an exponential term to Newton's formula. After all, it must have been only too clear to them that the various modifications of the inverse square law submitted with a view to account for the lunar inequalities had turned out to be in the wrong direction ever since Clairaut in 1747 tried to add the term $1/r^4$ to the principal term of $1/r^2$.

When the first consistent solution came, the most universal and most respected of all physical laws, Newton's account of the motion in a central field of force, was viewed in a much different light: it was only the particular case of a law of much wider sweep, the General Theory of Relativity. Nowhere else, perhaps, had the edge of precision influenced so forcefully the course science was to follow. The satisfactory account for 43 seconds of arc per century was actually the first experimental evidence Einstein could claim for his theory. To see his equations so complicated (if judged by the standards of ordinary calculus) yield the exact amount, was for Einstein, as he wrote in his letter of November 28, 1915, to Sommerfeld, one of the most exciting times of his life. "The marvelous thing which I experienced was the fact that not only did Newton's theory result as first approximation but also the perihelion motion of Mercury (43″ per century) as second approximation."[23]

Yet, the more momentous is the success of a theory, the more inexorable is the scrutiny to which it is subjected. This scrutiny consists mainly in measurements aiming at a precision far exceeding previous attainments. Part of the process is a hard new look on the experimental value with which the new theory appears to be in close agreement. Not only are new inquiries made about possible errors vitiating the observations, but a thorough search is also initiated in which one looks outside the theory for other possible causes of the experimental value in question. In the case of Mercury it is now known that 10 per cent of the value of the precession of its perihelion might be due to the flattening of the sun hidden under its envelope of brilliant gas. Clearly until this question is settled, the 43 seconds of an arc arrived at by General Relativity cannot be considered as its definitive proof.

It should be noted that the search for experimental proofs in favor of General Relativity is particularly active nowadays partly because only one of the predictions of the theory, the gravitational red shift has been so far established beyond doubt. Evidently, for the unconditional acceptance of General Relativity, more is needed than its unquestionable logical completeness or the verification of one of its predictions. In fact, only three years after the formulation of the theory, Einstein noted in the November 28, 1919, issue of *The London Times:* "If a single one of the conclusions drawn from it proves wrong, it must be given up; to modify it without destroying the whole structure seems to be impossible."[24] As of today, if the problem of gravitational waves is not counted, at least five predictions of the theory have not been tested or not rigorously enough. All these predictions refer of course to effects produced by strong gravitational fields. They are the bending of light and its retardation, the precession of the perihelion of orbits, the so-called geodetic precession of the axis of a gyroscope, and the Lense-Thirring precession owing to the drag exerted on a gyroscope not coaxial with a neighboring rotating body such as the earth.

Experiments of astounding sensitivity are now under way or in preparation to measure all these effects. The question of the bending of light will probably be settled by an electronic eye that can detect the position of stars in broad daylight. The precession of the orbits and the "geodetic precession" are expected to be measured by gyroscope satellites orbiting the earth. The slowing down of light might be established by radar waves bounced back from Venus that are calculated to show a delay of about two ten-thousandths of a second in a round trip lasting about twenty-five minutes. For the detection of the Lense-Thirring precession an arrangement containing a superconducting cable was suggested. In such a frictionless cable the very minute effects might hopefully accumulate over a long period of time so as to be sufficiently large to be measured.

No one familiar with the role of precision in physical science will be surprised at such extraordinary efforts and plans. One should remember that in the solution given by Einstein to the puzzle of Mercury far more was involved than an ad hoc correction of the inverse square law, which, though it could have solved the puzzle, would have at the same time thrown the rest of the celestial machine into a cocked hat. The curvature of space which the new theory introduced into physics could not be fitted into the easily visualized features of the Euclidean space. It did justice, however, both to the utmost refinement in observation and to the principle of universality in explanation. Moreover, in a genuine sense it did full justice to Newton, too, by leaving his law intact as a first approximation of a more general viewpoint. Actually, Newton himself was fully aware of what further advance in precision might do to any theory. Toward the end of Newton's life, Molyneux, Graham, and Bradley thought that their discovery of a nutation in the earth's motion would mean the end of the Newtonian system. Molyneux took it on himself to break the bad news to Newton, who nonetheless remained unperturbed. In a remark of which Einstein's words quoted above were but a faithful echo, he noted: "It may be so," according to Conduitt's account of the incident, "there is no arguing against facts and experiments." [25]

Aiming at ever greater precision in measurements affects physics in several ways. It may establish and confirm a theory, it may pose the need of further clarification, it may reveal unforeseen cracks, and at times it may force the abandonment of a theory and impose a search for a new one. For most of the course of classical physics, advances in precision brought triumphs to the Copernican-Newtonian system of the world and to mechanical theories in general. True, success at times demanded the sustained efforts of several generations of scientists. To clinch the classic proof of the heliocentric theory, astronomy had to work for almost two hundred years to improve the angular resolution of telescopes. From the time that Hooke, in 1669, made the first systematic attempt to detect the stellar parallax, the effort went on relentlessly in spite of the consistent failures and hasty claims of success that only proved to be errors. Hooke's objective lens fell from the roof and shattered. Roemer appeared to be more successful. He spent more than three years (1701–4) trying to measure the parallax of Sirius and Vega and believed he had detected a small annual change in the angle between the two stars. Flamsteed, the first Astronomer Royal, who reduced considerably the error in measuring angular separations, fancied that he had established a parallax of 40", but in all probability what he observed was the aberration of light. P. Horrebow, who recomputed Roemer's data in 1727, was also convinced that the value of stellar parallax had been found and published a paper entitled "Copernicus triumphans." Around the same time, however, Bradley found out that the yearly variation in the direction of stars, which was of the magnitude of 40″ was due to the annual change of the relative direction of the earth's velocity and the velocity of light. Furthermore, the accuracy of Bradley's observations indicated that if the aberrational ellipse of γ Draconis was complicated by a parallactic ellipse, the latter had to be less than 1″ in amplitude.

Bradley's conclusion was very important providing as it did a solid argument against later claims referring to parallaxes of the order of 4–5″. For even in the early nineteenth century, astronomers like Piazzi, Callandrelli, Brinkley, and others believed they have established stellar parallaxes of between 1″ to 10″. Bessel and Struve kept insisting, however, that such parallaxes were far too large. That their views ultimately prevailed was in great part due to the fact that both were in possession of the most precise astronomical instruments available in the first half of the nineteenth century. Both the Königsberg heliometer and the Dorpat refractor were masterpieces of Fraunhofer and represented an improvement by a factor of at least ten in measuring angular separations. Such an advance in instrumental technique was absolutely needed since the parallax of even the closest star, Proxima Centauri, is 0″.78, while α Lyrae observed by Struve and the 61 Cygni observed by Bessel have parallaxes of 0″.12 and 0″.29, respectively. To establish such small values with sufficient reliability, Fraunhofer's instruments proved indispensable. Henderson who five years before Bessel and Struve by sheer luck picked α Centauri, the second nearest star, and measured its parallax as 0″.92, in rather close agreement with the modern value of 0″.75, could not be sure of the result. His was not a Fraunhofer instrument but a mural circle that failed, as before, to give reliable parallaxes. This circumstance left Henderson in a state of uncertainty, and he could not bring himself to discuss and publish his results until after Struve (in 1837) and Bessel (in 1838) published their findings. Thus came to an end the longest persistent effort in the history of science to measure with precision a crucial magnitude. When presenting the Gold Medal of the Royal Astronomical Society to Bessel in 1841, Herschel called the feat "the greatest and most glorious triumph which practical astronomy has ever witnessed." The choice of the word "practical" was particularly apt. It evoked the power of precision, which alone could overleap the barrier that separated man and the stars, the barrier against which man has "chafed so long and so vainly."[26] It was a triumph that put the long awaited classic seal of approval on the vision first outlined by Copernicus and defined by Newton in the language of mathematical physics.

The role of precision was no less decisive for other aspects of the universal validity of the Newtonian mechanical conception of the physical world. It was the precision of Lavoisier's balance that led to the abandonment of the concept of phlogiston and made possible the reorganization of the study of matter on a basis that was designed to emulate the clarity of the Newtonian system. As the younger Herschel put it, the mistakes and confusion of Stahlian chemistry "dissipated like a morning mist as soon as precision came to be regarded as essential."[27] Phlogiston theory was only one of the various non-mechanical theories that came to be abandoned during the eighteenth and nineteenth centuries, chiefly under the impact of increased precision in measurement. Actually the demise of the phlogiston theory did not signal an immediate break with the more general consideration of heat as an "imponderable" fluid in the class of other notorious "imponderables" such as light, electricity, magnetism, and the ether. "Imponderability," as a term, however, was just as "elastic" and evasive as the substances supposedly possessing this uncommon property. Strict imponderability was, for instance, ascribed to the ether in the late nineteenth century, whereas earlier heat as a substance was considered to have only a negligible weight. To see how negligible this weight might be could not fail to present a challenge in an age that witnessed the continuous triumph of the Newtonian theory of gravitation. Thus well before the full development of the Newtonian theory of gravitation in the hands of Lagrange and Laplace, efforts were made to detect the weight of heat, if indeed it had any. Clearly, strict weightlessness could not fit well enough into the framework of mechanical philosophy the outstanding feature of which was universal gravitation. Boyle for one felt prompted to look into the matter, and in his *New Experiments and Observations Touching Cold*[28] (1665) he reported that the weight of water had shown "not one grain of difference" when frozen and unfrozen. Still, Boyle, on the ground that metals gain weight upon calcination, leaned toward the view that heat might be some special substance. More than a hundred years later, in 1785, G. Fordyce, a British physician, set out to throw light on the problem and found an increase of weight of about one part in one hundred thousand when water was frozen.[29] His experiments, however, were not consistent and benefited science only inasmuch as they provided Rumford the incentive to devote his scientific work to the solution of the problem.

A very careful experimenter, Rumford had a clear realization of the role that instrumental precision was to play in deciding the question. With justification, he boasted of the "excellent balance" he had at his disposal as compared with the imperfect balances used previously.[30] So sensitive and accurate was his balance that if water had really gained in weight upon freezing, he could have detected a change of as little as one millionth part of the weight[31] and could have directly measured a change of 1/700,000 part of that weight. The defenders of the caloric theory could hardly find fault with the care with which Rumford's experiments had been carried out. There was not much exaggeration in Rumford's terming his own experiments "perfectly unexceptionable."[32] His critics could make objections only on grounds other than experimental against Rumford's conclusion, which constituted the first major blow to the caloric theory. The heat substance, Rumford wrote, "must be something so infinitely rare, even in its most condensed state as to baffle all our attempts to discover its weight."[33] Thus he felt entitled to consider as safe the conclusion "that all attempts to discover any effect of heat upon the apparent weights of bodies will be fruitless."[34] There was only one alternative, the mechanical theory, and it was clearly in this direction that the edge of precision steered the progress: "Heat is nothing more than an intestine [internal] vibratory motion of the constituent parts of heated bodies," and "the weights of bodies can in no wise be affected by such motion."[35]

Another important field where progress in precision measurements brought signal confirmation to Newtonian theory consisted of showing the strict validity of the inverse square law, the epitome of the laws of classical physics. Of primary importance in these demonstrations was the torsion balance, invented independently by the Rev. John Michell in England and Coulomb in France. In the hands of Cavendish it yielded the first experimental proof not based on astronomical observations of the gravitational attraction of masses according to the inverse square law. As Cavendish characterized the apparatus of Michell, which he inherited, it was a device that could render "sensible the attraction of small quantities of matter."[36] The instrument was indeed a marvel of sensitivity, even in its primitive form, and the potentialities of increasing its sensitivity are even today far from exhausted. For Cavendish it yielded a value of 5.488 for the average density of the earth as against the modern value of 5.522. In Coulomb's hands it led to the direct establishment of the validity of the inverse square law in electrical repulsion and attraction. For the coarse experiments made in 1760 by Daniel Bernoulli on the electric force between two charged disks could hardly warrant his sweeping conclusion that between the disks "the force varies inversely as the square of the distance." Coulomb, however, could rightly claim that his torsion balance "could be used to make precise measurements of very small forces, for example a ten thousandth of a grain."[37] He could do even better than that. His best torsion balance was sensitive enough to measure 1/100,000 of a grain, providing thereby a firm support for the view that the same inverse square law governs all known forces in nature.

Another experimental method of verifying the inverse square law in electrostatics was first pointed out by J. Priestley[38] and brought to a very high order of accuracy in later times. Characteristically enough, the argument rested on analogy with the absence of gravitational force inside a spherical shell. To Priestley such analogy suggested that a law of attraction obeying the inverse square law should be the reason why pithballs placed inside an electrified metal cup would experience no force on themselves. As could be easily seen, even the slightest departure from the inverse square law would result in observable effects, and this Cavendish tried to ascertain. His result was negative, and from the sensitivity of his apparatus, he concluded that the exponent in the force law was between 2.02 and 1.98. This was one of Cavendish's many important experiments that lay unpublished until edited by Maxwell in 1879. Maxwell himself repeated the experiment with a much greater accuracy and gave 2.00005 and 1.99995 as the limits for the variation of the value of the exponent permissible by the experimental error.[39] This was a thousandfold improvement on Cavendish, but was by no means the last word on the subject. In 1936 S. J. Plimpton and W. E. Lawton improved the accuracy of the experiment by another factor of ten thousand, setting the limits of variation as 2.000,000,002 and 1.999,999,998.[40]

In the 1930's, when Plimpton and Lawton published their results, the quest for precision had a significance much different from that which it had during the complacency of nineteenth-century physics. In twentieth-century physics increased precision meant largely suspense, unexpected turns, at times even complete revolutions in scientific theories and concepts. On the contrary, the quest for precision in nineteenth-century physics led oftener than not to further verification of established, reputedly unchangeable laws. A letter of J. J. Thomson, written in 1937, gives a revealing glimpse of this mode of thought. He speaks of some mathematicians, or mathematical physicists, who resented the introduction into physics of new concepts such as Faraday's lines of force. "The mathematicians of the time were very contemptuous and asked why did one want more than the law of inverse square which had been verified to I forget how many decimals."[41] The remark that physicists should be contented with trying to establish everywhere in physics the validity of an inverse square law was at least not contrary to the fondest dreams of classical physicists. Actually this law stood as a symbol for the unification of all branches of physics that nineteenth-century physicists hoped to establish.

In fact, striving for greater precision provided signal breakthroughs in demonstrating the close connection or identity between phenomena widely disparate at first sight. Thus, the first step in concluding with reasonable certainty that elements found on earth compose the sun's substance, too, was made possible on the basis of measurements accurate to an astounding degree of precision. This achievement, the initial verification of the universality of the same matter in the whole world, goes back to the work of W. A. Miller on the wavelength of the double dark line D of the solar spectrum and the double bright line emitted by sodium. Miller's work was indeed a marvel of precision. As Stokes recalled it years afterward: "Miller had used such an extended spectrum that the two lines of D were seen widely apart, with six intermediate lines, and had made the observations with the greatest care, and found the most perfect coincidence." To Kelvin's inquiry of what he thought of Miller's results, Stokes could only say: "I believed there was vapour of sodium in the sun's atmosphere."[42]

The role precision played in establishing the universal validity of the equivalence of heat and mechanical work was just as crucial. Joule had spent almost forty years in continually improving upon his initial data, repeating his experiments under the most varied circumstances and giving attention to all imaginable side effects that could influence the results. To this program he brought undivided devotion. Even his wedding trip had to make a contribution to it, but the alpine waterfalls where he tried to show a change of temperature produced by motion failed him completely. Such adversities, however, could not weaken his determination: "I shall lose no time in repeating and extending these experiments," he told the somewhat indifferent audience of the British Association in 1843.[43] Determination was what he needed as he realized that only a consistent series of data could break down the resistance or indifference to the revolutionary idea he proposed. Joule's subsequent papers never failed to make explicit reference to the decisive role precision was to play in clinching his argument. On the results he obtained in 1843 through generating heat by electric current, he commented: "I intend to repeat the experiments with a more powerful and more delicate apparatus."[44] Clearly, the result of 838 ft-lb/Btu was something to be improved upon, and the steps in this direction followed in quick succession. In 1845 he obtained 798 ft-lb/Btu by developing heat through condensation. He called attention to the remarkable agreement with his first result, but at the same time he spoke again of "very delicate" future experiments to "ascertain the mechanical equivalent of heat with the accuracy which its importance to physical science demands."[45] In the same year Joule published his results based on developing heat by blades rotating in various fluids. The publication, as he remarked, was delayed owing to a slight alteration that had to be made in the apparatus to assure greater precision. The mechanical equivalent of heat thus obtained was 832 ft-lb/Btu, and referring to the mean value of 817 ft-lb/Btu of the three different experiments, Joule noted: "In such delicate experiments where one hardly ever collects more than half a degree of heat, greater accordance of the results with one another than that above exhibited could hardly be expected."[46] Accordingly, he regarded the value of 817 ft-lb/Btu as provisional "until more accurate experiments shall have been made."[47]

It was again with that same proviso[48] that he submitted the value of 781.5 ft-lb/Btu to the British Association meeting at Oxford in 1847. Joule's was still a lonely battle. He was told by the chairman to be brief and no discussion would have followed, had it not been for a remark of a young professor from Glasgow, the future Lord Kelvin, whose observations, in a matter of minutes, stirred up a great interest in Joule's conclusions.[49] From conversations that took place between Joule and Thomson after the meeting, Kelvin "obtained ideas he had never had before," while Joule heard from Kelvin for the first time about Carnot's theory. It was a different, fully confident Joule, sensing the nearness of a decisive triumph, who pointed out two years later the steady convergence of the results of his new experiments with the old ones. He insisted again that "it appeared of the highest importance to obtain that relation with still greater accuracy."[50] The value he now considered to be the most correct was 772.7 ft-lb/Btu, which differed only slightly from the value of 777.9 as accepted today. On this value obtained by 1850 he could improve very little, although as late as 1878 he was still busy with experiments to enhance the accuracy of his conclusions. His indomitable quest for precision and the undeniable convergence of his results could not fail ultimately to have an overpowering effect. What Robert Mayer, the other pioneer figure of the mechanical theory of heat, stated in 1844 became fully vindicated: "One single number has more real and permanent value than an expensive library full of hypotheses."[51]

Utmost precision in measurement was no less decisive in leading to the recognition of the identity of the propagation of light and electric disturbances. Late eighteenth- and early nineteenth-century attempts to determine the velocity of the propagation of electric effects over long insulated wires showed only that such a velocity is too great to measure. Thus J. Aldini, a nephew of Galvani, reported in 1804 that the current "has an enormous speed surpassing all imagination."[52] But from 1834 on, when Wheatstone obtained a value of one and one-half times the velocity of light, the experimental results began to group closer together within the same order of magnitude. In 1850 Fizeau and Gonnelle, using iron and copper wires, found the velocity of the propagation of electricity to be one-third the speed of light in the former metal and two-thirds in the latter. Such results, of course, suggested the identity of the speed of propagation of electricity and light, but they were not conclusive. The reason for this lay in the fact that a year earlier Fizeau had achieved the first terrestrial determination of the speed of light by the rotating toothwheel method, which yielded a value of 314,858 km/sec, the accuracy of which, although low by modern standards, improved greatly upon the data derived from the motion of the satellites of Jupiter and from the aberration of light. Thus a similar improvement of data pertaining to electricity was imperative to argue conclusively the case of identity between the two phenomena. This came a few years later when Weber and Kohlrausch achieved an essential improvement in the accuracy of the determination of the speed of light based on the electrostatic method. Their result was 310,740 km/sec.

The first to note the remarkable closeness of the two data was Kirchoff, who did not feel it necessary to look deeper into the matter. For Maxwell, however, the close agreement appeared to be of the utmost importance, since his formula for the velocity of disturbance in the electromagnetic media could be reduced to Weber's and Kohlrausch's formula, provided the coefficient of magnetic induction was taken equal to unity. The fact that Maxwell had worked out his formula in the country before seeing Weber's communications could only add to the impact which the close agreement of the two values was to have on Maxwell's mind. The conclusion was almost forced upon him by the "sheer and brute facts" of nature once they were measured with genuine precision: "The velocity of transverse undulations in our hypothetical medium," wrote Maxwell, "calculated from the electro-magnetic experiments of MM. Kohlrausch and Weber, agrees so exactly with the velocity of light calculated from the optical experiments of M. Fizeau, that we can scarcely avoid the inference that light consists in the transverse undulations of the same medium which is the cause of electric and magnetic phenomena."[53]

The instances taken from nineteenth-century physics could be multiplied at some length to illustrate the fundamental importance that increased precision in experiments plays in establishing new laws or theories. Ohm's law, the laws of radiation, the gas laws, to mention only a few, were but triumphs in precision. The establishment of well-equipped physical laboratories, first in German and French and later in British universities, clearly evidenced the general recognition of the extraordinary importance precision has in physics. The rewards were at times spectacular, particularly when unknown entities, such as new elements, were discovered. The case of argon was perhaps the most characteristic, resting as it did on the worries of Ramsay and Rayleigh as to why some samples of nitrogen had a weight of 1.257 grams per liter instead of only 1.256. As it turned out, an unknown element, after its discovery called argon, caused this discrepancy. The identification of other inert gases followed in quick succession.

Great as such successes were, they represented nothing essentially new. Actually they meant rather the discovery of some missing building blocks in an already firmly outlined system. Most of the measurements that were being carried out in the newly organized laboratories related to refinements of older data, and this gave rise to the view that Maxwell in 1871 described in this way: "The opinion seems to have got abroad, that in a few years all the great physical constants will have been approximately estimated, and that the only occupation which will then be left to men of science will be to carry on these measurements to another place of decimals."[54] A younger colleague of Maxwell, A. Schuster, recalled this mentality several decades later in a passage that sketches the "decimal" state of mind with memorable vividness:

I think I interpret correctly the recollection of those who passed through their scientific education at the time, when I say that the general impression they received was that, apart from theoretical work, a reputation could be only secured by improved methods of measurement which would extend the numerical accuracy of the determination of physical constants. In many cases the student was led to believe that the main facts of nature were all known, that the chances of any great discovery being made by experiment were vanishingly small, and that therefore the experimentalist's work consisted in deciding between rival theories, or in finding some small residual effect, which might add a more or less important detail to the theory.[55]

Schuster added, however, that some scientists, like Maxwell, refused to go along with such a pedestrian appraisal of the role played by precision in the progress of physics. In his "Introductory Lecture in Experimental Physics," which he delivered after the opening of the Cavendish Laboratories in 1871, Maxwell pointed out that precision leads to the subjugation of *new* regions, to *new* fields of research, to *new* scientific ideas.[56] In spite of all his conviction about the *new* aspects that precision might unveil in physics, Maxwell could little surmise (if at all) the revolutionary dimensions of the newness that was in store for physics, changes due in no small measure to highly refined experiments carried out in the very same Cavendish Laboratories only a generation later.

Still in this respect too, Maxwell saw farther than most of his colleagues. How the great majority of first-rank physicists in the late nineteenth century looked upon greater accuracy was perhaps best expressed by Michelson. According to him, heightened precision was meant only to clarify apparent discrepancies and point out new applications of already known laws. The chance that these laws would ever be supplanted by new discoveries or by more precise measurements was in Michelson's eyes "exceedingly remote." "The extreme refinement in the science of measurement," in which Michelson had few peers in his day, was for him a tool of putting the final touches on an already firmly established edifice. Although he admitted that "it is never safe to affirm that the future of Physical Science has no marvels in store even more astonishing than those of the past," he voiced the conviction that further advances, future discoveries, indeed the "future truths of Physical Science are to be looked for in the sixth place of decimals." [57]

Characteristic as this statement was of the mentality of the day, it could have hardly sounded more ironical than on Michelson's lips. Such would not have been the case, perhaps, if Michelson's scientific work had not touched upon the problem of the ether but had been confined to his other truly astounding feats in precision. His determination of the standard meter in terms of wavelengths of cadmium had an experimental error of only one part in ten million. His 9.4-inch diffraction grating that contains 117,000 lines still remains the widest grating ever made and represents the highest precision ever achieved in this field. His measurements of the speed of light reduced the probable error by a factor of at least fifty. His stellar interferometer made it possible for the first time to measure the diameters of stars, thereby revealing that some are so huge that a good portion of our solar system could be easily accommodated within their boundaries. Perhaps the best fulfillment of the program Michelson assigned to precision, namely, the verification of old laws in some particular instances was his experimental demonstration of the tidal effects produced in the crust of the earth. It is well to remember that this was a problem that Newton considered to be far beyond the range of observational possibilities. [58]

Michelson, however, devoted much of the prime of his scientific career to the detection of an effect that Maxwell judged to be "quite insensible," "too small to be observed" by terrestrial methods. [59] The effect consisted in the change of the velocity of light due to the ether drag. About the existence of this effect late nineteenth-century physicists admitted no doubt. Yet, as their calculations indicated, the effect had to be exceedingly small. This left them somewhat in a state of despair about its possible detection. Such was the reaction of Maxwell, who showed that the effect, depending on the square of the ratio of the earth's velocity to that of light, would have amounted to about only one hundred millionth part of the whole time of the transmission of the light signal. This was the type of challenge, bordering on the impossible, that strongly appealed to Michelson.

The details of what was aptly called "the greatest negative experiment in the history of science"[60] are too well known to be reviewed here. In itself the acme of precision, the experiment inspired efforts persisting to our day to push its accuracy to almost unbelievable degrees. Michelson's famous experiment as it was carried out five times, from September, 1958, to October, 1959, (when the Maser technique was used), could have detected in the relative motion of the ether a velocity change as small as two hundredths of a mile per second. No shift in frequency, and therefore no change of speed, was observed, although the experiment had an accuracy of one part in a million millionth. This was rightly called the greatest precision ever achieved in an experiment up to that time.[61]

In his work on the ether drag, Michelson used, though unknowingly, that aspect of the edge of precision that alone can cut persuasively through century-old "truths" and unfold hidden layers of reality. Moreover, theories — even the most attractive and most recent ones — remain mere conjectures unless supported by experimental evidence. The amount of faith put into a theory by its author can never carry the day alone. A case in point is the Special Theory of Relativity. As originally proposed by Einstein, it had no reference to Michelson's work, although Einstein was aware of its negative result well before the publication of his historic paper in 1905. Thus, contrary to the often heard version, the Special Theory of Relativity was not formulated to explain what soon afterward came to be known as its principal verification.[62] Not that Einstein had not appreciated subsequently the crucial support given to his theory by Michelson's labors. As he emphasized at a festive gathering on January 15, 1931, in words addressed to Michelson: "Without your work this theory would today be scarcely more than an interesting speculation; it was your verifications which first set the theory on a real basis."[63]

The theory of relativity was only one of the many revolutionary turns physics had to take, mainly under the impact of the edge of precision. Although during the age of classical physics precision typically helped confirm details of well-established laws, in modern physics it proved to be the great source of upsets. Hardly a decade passed in this century when J. J. Thomson referred to a host of new discoveries whose effect on physics he compared to the sudden renewal experienced by arts and letters in the Renaissance. In 1909 he could not keep from noticing a new enthusiasm, a "youthful, exuberant spirit" pervading scientific circles and inspiring men of science to "make with confidence experiments which would have been thought fantastic twenty years ago." In the same breath, however, he could not help spoofing gently the "decimal" slogan. The sudden changes, he pointed out, "dispelled the pessimistic feeling . . . that all the interesting things had been discovered, and all that was left was to alter a decimal or two in some physical constants."[64]

The almost humorous aspect of all this was that the effort to improve on the decimals played a vital role in creating the new atmosphere in physics. Thomson, for one, devised methods that later proved indispensable in measuring the change of mass of the electrons with velocity, a feat that signaled one of the first drastic breaks with hallowed tenets of classical physics. The discovery of cosmic rays, which A. H. Compton viewed as equal in scientific importance to Faraday's law of induction,[65] was on Hess's part as much a feat in precision as it was an example of boldness. In a series of balloon flights carried out with courage, and aided by his ionization chambers of greatly improved sensitivity, Hess found what Gockel a year earlier was unable to detect, that from about 1,800 meters up an increase in ionization is undoubtedly in evidence. When Hess subsequently stated that his observations "are best explained by the assumption that a radiation of very great penetrating power enters our atmosphere from above,"[66] a vista opened in physics that is still limitless and rich in new challenges. The precision of Hess's instruments was the starting point of cosmic-ray measurements that gave the first evidence of the transformation of entire subatomic particles into energy and their reappearance from it. Cosmic rays provided the basic information about the condition of interstellar spaces and prompted the development of entirely new ideas about the nature of physical processes inside stars and galaxies. Even today, more than half a century after Hess's balloon flights, the experiments carried out by high-altitude rockets and artificial satellites are, for the most part, concerned with cosmic ray research.

Furthermore, the greater the energy of cosmic rays, the more startling are the glimpses they reveal both of the far reaches of the cosmos and of the subatomic world. From Hess's small ionization chambers there is a long road to the Volcano Ranch Cosmic Ray Research Center near Albuquerque, New Mexico, where, spread over miles of desert, long arrays of scintillometers record the arrival of cosmic rays. The road is long not only in time but most of all in the precision and sensitivity achieved in instrumentation. The results are truly worthy of the tremendous efforts. One may recall, for instance, the extraordinary ray or particle recorded in 1962 that carried an energy of about 10^{20} electron volts, or three billion times the peak energy of the largest accelerators.[67] The most appropriate reaction to such a discovery was best expressed a few years earlier, in 1957, on a similar occasion, by E. P. Ney: "At this high energy I'd be surprised if something new didn't happen."[68] Indeed, to explain such inconceivably high energies, which man had become aware of through a relentless effort to improve on the detecting instruments, physicists have to think of truly extraordinary events, such as the explosion not of a star but of an entire galaxy. "It is too early to say what will follow from Hess's discovery," said Compton a generation ago,[69] and today it seems that the array of discoveries in cosmic ray research is limited only by man's ingenuity to construct more precise, more powerful instruments.

Instances that show how persistent determination to reduce experimental error demanded major revisions of modern physical theories could be listed to no end. Suffice it to mention only the Lamb-Retherford experiment that grew out of the need of clarifying the experimental evidence that seemed to indicate that the separation of the components of the fine structure "doublet" in Hα is only 96 per cent of that predicted by the relativistic quantum theory of Dirac. Thus S. Pasternack suggested in 1938 that the discrepancy is due to the fact that the $2S_{1/2}$ level of hydrogen might be higher than the $2P_{1/2}$, in contrast to Dirac's theory, which states that the energy levels having the same n and j values must coincide.[70] Although Pasternack's suggestion happened to be on the right track, it managed at first to inspire only tentative attempts to provide theoretical explanation for the discrepancy. This was in a sense understandable for the conclusive evidence concerning the separation of the two energy levels was at that time still lacking and almost a decade had to pass before Lamb could announce the successful completion of an experiment that demanded all the precision modern electronics could provide. As a result, the edge of precision had once more set theoretical physics into an agitated search. Not only did it become imperative for physics to face up to the revision of Dirac's theory — one of the most daring and successful of physical theories ever proposed — but also to come to grips with the full range of tantalizing problems represented by the procedure of field quantization. For quantum field theory runs into

infinities that can be removed only by ad hoc renormalization techniques, a procedure wholly unsatisfactory, although it yields values
in almost perfect agreement with experimental data. Thus the re-
normalized Dirac theory clearly points beyond itself, for its morass of
arbitrarily "tamed" infinities cannot be a satisfactory answer. In fact,
the edge of precision inflicted on some of the best physicists an uneasiness whose depth can be best gauged from Pauli's words: "The
correct theory . . . should not use mathematical tricks to abstract
infinities or singularities, nor should it invent a hypothetical world
which is only a mathematical fiction before it is able to formulate the
correct interpretation of the actual world of physics."[71]

The way around this impasse will in all probability depend as much
on the stroke of genius of a theoretical physicist as on the experimental precision carried to perfection. That theories will come and
go at an even more rapid pace than they did in the past is very likely
in view of the amazing precision techniques developed in the past
few years. Writing in 1957, E. R. Cohen, K. M. Crowe, and J. W. M.
DuMond could point out that because of such techniques as atomic
beams, nuclear magnetic resonance, and the like, a "veritable revolution" has taken place since 1950 in the field concerning the determination of atomic constants.[72] As to the measurements of the e/m
ratio of electrons made before 1945, they came to regard these as
"completely obsolete" as compared with more recent determinations.
And still in 1957 it was not yet their privilege to assess even remotely
the fantastic possibilities afforded by such newcomers to the field of
precision technique, as are the atomic and nuclear clocks, the Mössbauer effect, the Maser and the Laser.

In 1957 Cohen, Crowe, and DuMond could recall with pride that in
seventy-five years physics achieved a thousandfold refinement in
measuring the speed of light. The experimental error of 300 km/sec
present in Cornu's measurement of 1875 was narrowed to 0.3 km/sec
in Bergstrand's experiments of 1951. With the Laser, however, this
precision might be increased by a factor of at least 30, leaving a
margin of error of only 10 m/sec. From the viewpoint of a physicist,
it is a matter of secondary importance that such a degree of accuracy
could give the distance to the moon within 72 feet as against the
previous error of about ½ mile. Nor does the primary significance of the
Laser lie in the fact that when used as a radar beam it is 10,000 times
more accurate than the best comparable system using microwaves.
Thus a so-called Laser-radar could measure with almost "absolute"
precision velocities ranging from 5 miles per second, which is the
orbital injection velocity of a space ship, down to 1/10,000 inch per
second, a virtual stop. Furthermore, once the problem of modulating

a Laser beam has been achieved, an immense new era for telecommunications will open up, since a single Laser-channel can carry many thousand times more independent messages than radio or television channels can. Astounding as such technical possibilities are, they are dwarfed when set beside the implications of the Laser for future developments in theoretical physics. The mere fact that there is available in the Laser for the first time in the history of physics a source of coherent visible light, should be sufficient indication of great surprises to come. Gone are the days when textbooks of physics stated with resignation that "it is not possible to make two light sources coherent or even to make two parts of the same light source coherent."[73] Obviously, the edge of precision has its own uncanny say about what is possible and what is not.

No less revolutionary than the Laser in its possibilities the nuclear clock constitutes a tool by which physics may check those theories of gravitation that indicate a decrease with time of the value of the gravitational constant. Since a nuclear clock is independent of gravitational force, it will, according to such theories, steadily gain over timekeepers (such as pendulums and artificial satellites) that work by gravitation. The rate of this gain will be one part in 10^{10} in a year if the age of the universe is taken as ten billion years. To measure it one should use a nuclear clock that is accurate over a year to one part in 10^{11}. Also needed is a satellite the period of which can be determined with similar accuracy over the span of one year. Both requirements are now technically feasible. Consequently, if the experiment is carried out, and if the rates of such clocks become steadily different, one will have to conclude that the strength of gravitation is weakening. Startling as such possibilities might seem, the fact is that gravitation is far from being a topic settled for good by Newton and Einstein. On the contrary, quite a few ingenious experiments have been proposed in recent years to shed light on questions such as these: Does the locally measured value of G change if we approach the sun? Does the local value of G depend on the velocity of the laboratory relative to distant matter? Does the value of G vary periodically according to the motion of the sun around the galactic center?

Efforts to improve on the precision of various data connected with gravity are indeed at a high pitch today. Of the many instances, one may single out the redetermination of the Potsdam gravity standard and the subsequent plan to redetermine the value of the gravitational acceleration on a worldwide basis. It is worth noting that one of the instruments newly designed for this purpose utilizes a Laser to measure the acceleration of a falling prism. Among the most ambitious and revolutionary experiments in the field of gravity are undoubtedly the ones concerned with the detection of gravity waves. The existence of these waves was predicted by Einstein in 1916 and in greater detail in 1918. Yet, it took all the extraordinary advances in electronics over almost half a century to reach the stage where his predictions could be put to test of experiment with a fair degree of reasonable promise. This circumstance alone should be indicative enough of the degree of precision implied in such endeavors. As an example, one may mention some details of the apparatus of J. Weber and his co-workers at the University of Maryland. Its principal part is a 1.5-ton aluminum cylinder, 10 ft long and 7 ft in diameter, suspended in a large vacuum chamber. It would, however, serve as a successful detector of gravity waves only if it proves possible to detect the exceedingly small periodic relative end-face displacements of about 10^{-14} cm produced by the thermal energy of the Brownian motion of the atoms of the cylinder.[74]

It may very well be that such an attempt will turn out to be infeasible in the long run. Then one may have to try the even more complicated possibilities of using the earth, or the earth-moon system as detectors of gravity waves. At any rate, the verification of the existence of these waves will not be merely a triumph of precision putting the final seal of approval on a long familiar theory. Such a triumph would no doubt open the door to many a new fact. It would bear both on our understanding of the possible original explosion of the universe and on the puzzles of fundamental particles. Clearly, if Einstein's prediction is correct, any accelerated motion of mass should produce gravity waves, and the largest of these ought to have originated when the expansion of the universe suddenly started. On the level of the fundamental particles too the detection of gravity waves might provide entirely new insights into their interactions. Tiny as those particles are, their speed is often in the vicinity of the speed of light, and this means that their relativistic mass far exceeds their rest mass. It is such increases in their masses that might possibly be traced by gravity waves. Clearly the edge of precision appears again to be equally effective in promoting man's understanding of the universe in the realms of both the immensely large and the exceedingly small.

Precision in measurements will no doubt play an essential role in deciding whether the gravitational field is quantized like the electromagnetic field and whether there are geons or gravitons similar to photons and mesons, the quanta of the electromagnetic and nuclear field force. Again very high precision in gravitational measurements might lead to the clarification of such a problem as the possible existence of an inertial anisotropy in the universe. For if Mach's principle holds, the inertial response of terrestrial objects will be a function of the distribution of the total mass of the galaxy. Since this distribution is asymmetrical for an object located at a point far from the galactic center, as the earth is, it is possible that the inertial mass of an object on earth will be at a maximum when accelerated toward the center of the galaxy and a minimum when the acceleration is perpendicular to this direction. Efforts to detect such a gravitational anisotropy are not all recent. Non-isomeric crystals were subjected several times in the past to experiments looking for such effect. In 1899 Poynting and Gray performed experiments to this end which, though not yielding a positive result, achieved a precision of one part in 16,000. Several decades later P. R. Heyl increased the precision of such experiments to one part in a billion, again with no positive result. Very recently, however, the Mössbauer effect seemed to offer the possibility of an experiment in which the precision could be carried to one part in ten thousand billion.[75]

The extraordinary degree of precision achievable by the Mössbauer effect will undoubtedly prove the most influential factor in the progress of modern physics. Through this effect, gamma rays whose wavelength is invariable to within one part in a thousand million million are produced. The way toward accomplishing this feat was shown in 1958 by Mössbauer, who discovered a method to eliminate the recoil of certain nuclei upon emitting a gamma ray.[76] Of the tremendous potentialities of the Mössbauer effect, one indication is the Nobel Prize that honored Mössbauer's paper within a scant three years following its publication. The other is the annual conferences held since 1960 for the sole purpose of comparing ideas on how to make the most of this effect. Thus it was early recognized that the Mössbauer effect provided by far the most accurate experimental verification of the relativistic time dilation and of the gravitational red shift. With regard to the latter, before the availability of the Mössbauer effect, one had to rely on the red shift observable in the spectra of ultradense, or the so-called dwarf stars. Such a red shift was, however, very difficult to distinguish from other effects, and a considerable amount of uncertainty remained in most of the measurements. What was to be measured was the loss of energy of a photon due to its escape from the extremely high gravitational field of a star whose density might be 50,000 times that of water. With the Mössbauer effect it is now sufficient to use a gravitational potential difference smaller by a factor of many million such as that existing between the top floor and the ground floor of a three-story building.[77]

The field where the Mössbauer effect might yield the most startling results is probably in the concept and measurement of time. At this point one again comes face to face with that aspect of the edge of precision that manifested itself in a singular sense in the course of modern physics. It is obvious that in the accuracy of timekeeping enormous progress had been made since the ancient Egyptians, who defined the shortest time as that which a hippopotamus needed to thrust his head out of the water for a quick look around. Except for very recent times, however, that progress had been very slow. Galileo still had to rely on his pulsebeat and waterclocks, for he stumbled on the principle of the pendulum clock only toward the end of his life. But this invention opened the way for true accuracy in timekeeping. Huygens' best pendulum clock was already accurate to within one second in an hour. Navigation in particular kept clamoring for further improvements in timekeeping, an objective hardly promoted by Sir Kenelm Digby's "powder of sympathy," which reputedly caused a dog on shipboard "to yelp the hour on the dot." A more solid contribution was Harrison's fourth marine chronometer which kept time with an accuracy of one-third of a second a day, corresponding to an error of only one part in three hundred thousand.

The following one hundred and fifty years saw the accuracy increase by a factor of about ten thousand. The quartz clock, the best timekeeper of the early twentieth century, had an estimated accuracy of one part in 10^8, corresponding to an error of one second in about three years. So matters stood when atomic, nuclear, and hydrogen clocks entered the picture. In less than twenty years the rate at which precision in timekeeping advanced changed from a gentle slope to an almost breakneck climb. With atomic (cesium) clocks one could achieve an accuracy of one part in 10^{10}, which means losing or gaining only one second in about three hundred years. Even better are the prospects for timekeeping devices that will be based on the Mössbauer effect. The best possibilities at present, however, seem to lie with the so-called hydrogen clock, which would utilize the 21 cm wavelength emitted by hydrogen nuclei when the spin axis "flips over." Such a clock could achieve, theoretically at least, an accuracy of one part in 10^{15}, equivalent to a gain or a loss of one second in about thirty million years.

Of the many practical advantages of such advances in measuring time, the most obvious one is connected with the calibration of quartz clocks. Before the advent of atomic and nuclear clocks, it was necessary to keep quartz clocks operating under carefully controlled conditions for a period of years. Only in such a way could one assure that they were properly synchronized with astronomical time, which is based on the daily rotation of the earth and on its yearly revolution around the sun. Atomic clocks, however, are based on vibrations which do not depend on the earth's motion, and thus the calibration of quartz clocks can now be done in a few minutes. Yet, such practical advantages should not take attention away from far-reaching theoretical implications that might be in store for physics owing to recent breakthroughs in the accurate measurement of time. Truly, if one considers that the advance of precision in timekeeping has covered six orders of magnitude in little more than one decade, then the *chronon*, which on theoretical ground is now accepted as the smallest measurable time interval (about 10^{-24} sec), cannot be viewed as something forever and absolutely removed from the realm of time measurements. How soon actual physical measurements will be affected by the indeterminacy implicit in the concept of *chronon* is too early to tell. But regardless of the outcome of this problem one can safely predict that a sudden increase of the accuracy in the measurement of time will lead to many unexpected turns in physics.

The meaning of the last remark will be even more manifest if one recalls that precision in timekeeping is only one among the various areas of physical measurements to show a truly dramatic rate of advance. The determination of minute changes in mass and weight, for instance, has been marked by notable progress in the last half century. At the end of the nineteenth century such determinations could achieve an accuracy of only one part in a million. Thus H. Landolt clearly could not have succeeded in his attempts to measure the change in weight that he correctly assumed to take place in chemical reactions.[78] The fractional change of mass in chemical reactions is, however, of the order of 10^{-10}, and it was only some forty years later that physics reached this degree of accuracy by various radioactive tracer methods.

It would indeed be interesting to speculate how late nineteenth-century physics or chemistry would have reacted had Landolt succeeded in his task. Not having the necessary theoretical concepts to explain the results, physicists and chemists would probably have ignored or questioned the reliability of Landolt's data. For science can assimilate at a particular stage of its development only a certain degree of experimental precision. Tycho's data would probably have seriously hampered Copernicus in his efforts to formulate the heliocentric theory, and the case with Newton would have been similar had he had at his disposal Newcomb's data on the advance of the perihelion of Mercury. In all likelihood Bucherer and Kaufmann would have been gravely puzzled if they had been the first to attempt to measure with their high voltage apparatus the ratio e/m for the electron. They would have been hard put indeed to give a meaning to values that changed with the velocity of the electrons. Characteristically, in more recent times, Schrödinger had to ignore deliberately certain details of spectroscopical evidence and submit his wave equation as a first approximation. Physics could not achieve in one step the degree of perfection in explanation represented by Dirac's wave equation: the temporary block to progress consisted in the degree of experimental precision, which outstripped for a while at least the contemporary capabilities of physical theory. As for Landolt, he would probably have been stranded with his data had he succeeded, for it was only ten or so years later that the concept of mass-energy equivalence was introduced in physics. Again it was only a decade or so later that Aston's mass spectrograph made possible very accurate determinations of the relative atomic weights. One of the many important results of Aston's method was the finding that the relative chemical atomic weights of hydrogen and oxygen differ by about two parts in ten thousand from the data yielded by other methods. One possible explanation for this discrepancy was to assume the existence of two isotopes of hydrogen, H^1 and H^2, in the ratio of 4500:1. This suggestion, made by Birge and Menzel in 1931, had to wait only a year before the existence of deuterium was established by Urey and his coworkers in 1932. That was the year of the discovery of the positron, an event that ushered in a new branch of mass determinations that led step by step to one of the most puzzling problems of modern physics, the grouping by mass of the fundamental particles.

What happened in the area of mass and time measurements was paralleled by more precise determination of lengths too. The crucial role that the accurate measurement of the length of a degree played in Newton's theory of gravitation only signaled the beginning of an unrelenting quest to measure distances and divisions with ever higher accuracy. Obvious as is the inner dynamism of this quest, at times it can be overlooked even by the keenest minds in science. A case in point is Gauss. In his inaugural lecture of 1808 on astronomy he declared that the accuracy of the astronomical instruments of his day represented the *ne plus ultra* of perfection, beyond which "there is little left to be desired and almost nothing to be hoped for." Change on this point he expected only if entirely new means were discovered in the future, of which, he added dejectedly, "we have no idea, nor foreboding."[79] Yet, how could he know that the twentieth century would see Michelson measure the diameter of nearby giant stars, like Betelgeuse, and that by 1964 radioastronomy would establish the apparent diameter of Vega as 0.0037 seconds of arc, or its absolute diameter as 2,800,000 miles, or about 3.2 times the size of the sun? Nor could he foresee that in the same year various spots of the moon would be probed by Laser beams so highly directional that they spread out only two miles in traveling between the earth and the moon.

All these advances, and general advances in construction of micrometers, microscopes, heliometers, telescopes, interferometers, and other instruments, served essentially one purpose, a better resolution in measuring a given length. Here too the rate of advance in the last fifty years or so was far more rapid than in the preceding three centuries. This rapidity, of which physical science bears witness today, might be explained, in part at least, by the fact that more and more frequently, the progress of research almost overnight creates demands for precision that only a few years ago would have been considered fantastic. Since the late seventeenth century, many painstaking observations and calculations have gone into the determination of the astronomical unit, the distances, and the velocities of the planets. Even until recent times the most refined measurements of the size of the solar system differed from one another by as much as four parts in ten thousand. This meant that a probe aimed at Venus could miss the center of the planet by as much as ten thousand miles, or miss the planet completely by as much as six thousand miles, even if the guidance system of the probe worked flawlessly. The data obtained from the flight of Pioneer V did achieve a substantial improvement in the evaluation of the astronomical unit, although it is known even today only with an uncertainty of about twenty-three thousand miles. To remedy the situation, a satellite system orbiting around the sun and equipped with a combination of Laser and radio equipment has been proposed. This might reduce the margin of error enormously. One can confidently expect that in the next ten or fifteen years the rate of improvement over the older data will be nothing short of spectacular.

As regards the recent advances in the determination of very small distances and separations, one may consider electron-, X-ray, and phase-microscopy, which now make it possible to obtain pictures of the arrangements of molecules in crystal lattices. Or one may recall the recent redefinition of the standard meter as "a length equal to 1,656,763.83 wavelengths in a vacuum of the radiation corresponding to the transitions between the level 2P10 and 5D5 of the atom of krypton 86." Far from being an exercise in pedantry, the introduction of the new standard well documents the long-standing truth that standards and physical science are inseparable and that the road of physics is marked by ever more accurate standards. The introduction of new standards is indeed a fundamental condition for the advancement of physics. The substitution of the wavelength of an orange-colored light for the platinum bar kept at constant temperature in a cellar at Sèvres, near Paris, since 1799 was a step made imperative by the demands of both pure science and technology. When Michelson found in the summer of 1892 that the scratches marking the length of a meter on the standard bar are 10–12 cadmium wavelengths wide, he was performing an experiment the probable error of which did not exceed one part in ten million. Today the same precision is increased to one part in one thousand million. Science is now creeping slowly toward a distant and possibly absolute limit in length measurements, the quantum of length, called the *hodon,* which if theory is correct, is of the order of 10^{-13} cm. Whether linear dimensions or the geometry of space are quantized or not, however will remain mainly a guess, unless here, too, the edge of precision will cut through the Gordian knot presented by theory.

Space of course has many other fascinating puzzles and properties. For instance, the vast spaces of the universe are almost totally empty; by comparison, even the best vacuums achieved in laboratories are rather densely filled regions. Not that there has ever been wanting an effort to produce better vacuums. Indeed, improved vacuum techniques have played an indispensable part in making possible many of the major breakthroughs in modern physics. But the case was not essentially different during the earlier phases of the history of physics. This is easy to understand. Better vacuums mean the elimination of "obstacles," of disturbing factors, and a gradual approximation of the "ideal conditions" which a physical experiment should always emulate. It is there, in the "ideal situation," that unsuspected aspects of physical processes might most readily reveal themselves. At the same time the steps by which better vacuums were achieved and new insights gained show also that major advances in a particular branch of measurement often demand ingeniously designed new instruments.

The air pumps of the seventeenth century were already good enough to refute the Aristotelian view of free fall, by showing that objects of very different weight, like a feather and a stone, fall at the same rate in a glass tube pumped free of air. Two hundred years later the study of electric discharges in rarefied gases took a dramatic turn with the invention of Geissler's mercurial air pump in 1855. Although the vacuum thus achieved (about 0.25 mm) was very poor by modern standards, it nevertheless opened the way to unexpected discoveries. Kelvin himself considered that experiments with electricity in a good vacuum were the most promising means toward understanding the relations between the ether and ponderable matter. Others shared this and similar views. Crookes, for instance, who spent perhaps more time than any other in experimenting with Geissler tubes, believed for a while that the rotation of the vanes of his radiometer inside an evacuated tube was because of the impact of ether waves. When he improved his vacuum, however, the radiometer stopped rotating. At any rate, Crookes was so fascinated by the phenomena he observed that he felt justified in speaking of an "ultragaseous" or "fourth state" of matter. Misleading as his expression of "the fourth state of matter" was, Crookes sensed correctly the immensity of new information that high vacuums could yield about the properties and structure of matter. "The phenomena," he wrote, "in these exhausted tubes reveal to physical science a new world, a world where matter may exist in a fourth state, where the corpuscular theory of light may be true, and where light does not always move in straight lines, but where we can never enter, and with which we must be content to observe and experiment from the outside."[80]

As in many other fields of measurement, the rate of increase of precision as regards high vacuums has taken a rapid rise since the turn of the century.[81] With his rotary oil pumps, W. Gaede first achieved a vacuum of the order of 10^{-2} mm. A few years later, in 1905, by introducing his rotary mercury pump, Gaede attained a pressure as low as 10^{-4} mm. In 1915 his mercury diffusion pump could produce a vacuum of the order of 10^{-7} mm. This was further improved upon by Langmuir's condensation pump in 1916. Perhaps never in the field of physics had similar progress been made in a particular branch of measurement in such rapid time. In hardly twenty years precision had actually been increased by about six orders of magnitude. All this was nothing short of "incredible," and well-known physicists received the news coming from Langmuir's laboratory with startled disbelief. If proof was needed, however, there was the sudden emergence of the electronics industry, which, in order to satisfy the rapidly growing demands for efficient tubes, needed precisely the degree of vacuum achieved by Gaede, Langmuir, and their associates. It is not surprising that such perfection in vacuum technique soon began to help resolve a long list of the mysteries of solid state physics and of matter in general. The first of these was Langmuir's demonstration of the fact that electron emission from an incandescent metal surface at low pressure is not caused by the residual gas, and therefore does not disappear in a very good vacuum. Other major discoveries dependent upon very high vacuum followed in quick succession. One may indeed say that a good part of the progress of atomic and nuclear physics has taken place in space relatively empty as compared with the normal density of the air.

The best of laboratory "emptiness," although equivalent to the rarity of the atmosphere at 300 km above the earth, is, however, still a rather densely populated configuration as compared with what nature can produce in true outer space. For even in the present age of ultrahigh vacuum technique, involving pressures as low as 10^{-12} mm in very small volumes of space, the number of molecules is still about thirty thousand per cubic centimeter. To realize how tremendous such achievement is one should recall that recent estimates put the density of interstellar gas at about one thousand atoms or molecules per cubic centimeter. In other words, by improving the technique by another magnitude or two, one could truly simulate the conditions that obtain in outer space. This feat will involve very elaborate apparatus, however, and will be realized only in a very small volume of space. The advent of the space age, with the possibility of interplanetary travel, here offers a unique opportunity to perform large-scale physical experiments that demand a very high degree of vacuum.

Judging by the past course of science, we may predict that such possibilities will almost certainly be exploited, and the discoveries they yield will be startling indeed. For science usually charts a future course that, viewed from the vantage point of the present, looks incredible more often than not. This is amply confirmed by the invariably sad performance of so many long forgotten prognosticators who tried to outline what was ahead for science at a particular time. In 1916 the vacuums achieved by Langmuir appeared incredible to many observers, but scientific ears would have been even more shocked to hear anyone forecasting that science would be striving within thirty years for vacuums a million times better. It would have appeared even more incredible to them that gauges have already been devised to measure vacuums representing an improvement over the latter figure by a factor of about a thousand. Today, the scientific attitude is no doubt more tuned to the unexpected and the incredible. By and large, men of science are less prone to assume that no new marvels are in store for physics. On the contrary, the marvels are being produced at an ever-increasing rate that closely parallels the rapidly growing effectiveness of the edge of precision.

To know how far precision in physical measurements can be advanced is therefore of paramount importance when one is reflecting on the range and finality of physics. For one thing, subjective error can never be completely eliminated from observations. That observations of the position of the same hairline or pointer by different observers usually yield slightly differing results had been a well-known fact in scientific circles from the late seventeenth century on. Even so, it took some time before it was generally realized that no observer, however skilled or eminent, could claim to have made errorless observations. Until then the discrepancies arising from observations of several observers were at times handled arbitrarily. Thus Kinnebrook, one of Maskelyne's assistants, lost his job in 1796 because his readings of the times of stellar transits differed consistently by almost a second from those made by the Astronomer Royal, who supposedly could make nothing but faultless observations. A more equitable situation developed when Bessel, prompted by the Kinnebrook incident, pointed out the systematic existence of a personal error in every pointer reading and the need for a so-called personal equation to secure the proper evaluation of readings made by any observer.[82] About the same time, it also became customary to compute the average value of all observations. Various methods were developed to procure the optimum result, the most powerful of these being that of the least squares, which had been worked out by Legendre and Gauss. How effective the method was soon became evident. By applying it to Piazzi's data on the new planet Ceres (discovered on the first day of the nineteenth century but lost seven weeks later), Gauss was able to indicate with sufficient accuracy where to look for it again. Still the method did not live up to the expectations of Gauss and Laplace, who tried to develop it, but in vain, into a truly "objective" method of measurement.

During the whole life-span of classical physics nothing conclusive was established about a method of measurement wholly free of subjective error. Not that classical physicists would have therefore entertained even for a moment the idea that uncertainties in measurements might ultimately be due to some fundamental imprecision in nature. This outlook was voiced in 1904 in a memorable phrase by Horace Lamb of hydrodynamics fame in his presidential address to the mathematical and physical section of the British Association. "It is recognized indeed that all our measurements are necessarily to some degree uncertain, but this is usually attributed to our own limitations and those of our instruments rather than to the ultimate vagueness of the entity which it is sought to measure."[83] Consequently, the perfectibility of measuring instruments was taken to be limitless for all practical purposes. It was of course recognized that the actual performance of a measuring instrument usually involved some factors diminishing the accuracy: measuring the thickness of a wire by a caliper, for instance, would cause slight deformation of the wire; when the same measurement is made by light waves, the wire will absorb heat and expand. Yet, such inaccuracies never seemed to present insurmountable obstacles in terms of the degree of accuracy to be achieved in particular experiments. Experimentalists succeeded in overcoming difficulties to such a surprising degree that it became natural to assume that the approximation to the ideal situation can always be improved upon.[84] The continuous progress in achieving precision then came to be interpreted as an ever more perfect realization of the ideal case of strict determinism. Furthermore, as the gradual approach toward the ideal was a reality, classical physicists, with apparently unassailable logic, assumed the factual existence of the ideal case. Such reasoning was at the bottom of classical determinism, and strangely enough, classical physicists never pondered seriously the logical implications of the fact that initial conditions that in theory at least are assumed to be knowable with perfect precision are never so de facto.[85]

The belief in perfect precision on which absolute determinism in physics is largely based was not therefore without a generous element of complacency. Such complacency received its first major shock when the application of statistical mechanics, through the studies of Smoluchowski and Einstein, shed light on the true nature of the Brownian movement. This phenomenon was first observed in 1827 by the botanist, Robert Brown, who found that the pollen grains of *Clarckia pulchella* are in constant irregular motion when suspended in water.[86] During the following decades, it came to be recognized that the phenomenon was independent of temperature irregularities, external disturbances, and even of the chemical composition of the particles suspended. The movement was, as J. Perrin characterized it succinctly, "eternal and spontaneous"[87] and seemed to be a common feature of all liquids and gases.

It is easy to see that such a movement would expose measuring needles to molecular impacts that keep changing irregularly in time and in direction. If, now, the needle is extremely thin, as the case ought to be in an ultrasensitive instrument, and if the magnitude of such molecular impact is sufficiently large, the needle will be kept in an irregular vibration around the position the measurement is to fix. The effect is even more pronounced in the case of a radiometer that is absorbing heat and light. The molecules on the hotter surface of the radiometer vane will have greater energy and will impart to the radiometer an oscillation great enough to make measurements impossible below a certain level. Liquids and gases are not the only media subject to irregularities such as those typified by the Brownian movement. The same holds true of any aggregate of atoms or parts of atoms, such as electrons, and has proved to be of considerable significance in explaining the limitations of electronic measuring devices where one has had to cope with phenomena like the statistical fluctuation of electrons emitted from hot surfaces and the so-called noise and shot effects.

A further basic restriction on the supposedly limitless accuracy in measurements came to light when in 1905 Schweidler showed that the law of radioactive decay formulated by Rutherford and Soddy is strictly statistical in character. In other words, since we cannot determine exactly the number of atoms decaying in a given second in any radioactive substance, the strength of radioactivity in a particular radioactive substance will display statistical fluctuations. This means of course that such data as radioactive half-life and concentration, for instance, can never be measured with absolute precision even though the instruments are theoretically perfect.

As we have seen, the surroundings of a very fine pointer needle, and the material of the needle itself, display an inherent statistical fluctuation that rules out absolutely precise measurements. One of the main signal carrying media, free electrons in a wire, imply a very similar difficulty. To make the picture complete, one should remember that the other main signal-carrying medium, visible light, also is not free of statistical fluctuations. Einstein's theory, formulated in 1917, states precisely that both the emission and the absorption of light are strictly statistical in character. Consequently neither the intensity nor the timing of a light signal can be made with absolute precision; another unavoidable limitation is thereby imposed on the concept of absolutely precise measurement.

Statistical considerations apart, once it was recognized that light consists of photons, the energy of which is equal to the product of their frequency and of Planck's quantum, it became impossible to localize particles and to register their interaction in time with absolute precision. For to obtain information about the position of an electron, a photon must be reflected from it; in this process, the electron must receive an impact that will change either its state of rest or its velocity. The magnitude of this impact is uncertain since the photon can follow any path within the angular opening of the objective lens. This situation obviously pointed to some deeper principle, one that was formulated by Heisenberg in 1927. Known as the "uncertainty principle," it states that magnitudes represented by canonically conjugate variables, or operators, cannot be measured simultaneously with unlimited accuracy. Such paired variables are position and momentum, angular momentum and angular position, moment of inertia and angular velocity. Since the limit of resolution in optics is of the order of the wavelength used, it would be possible to fix the position of the electron with a probable error of about one hundred thousandth of a centimeter by using ultraviolet light. The corresponding minimum uncertainty in the velocity of the electron would then be of the order of one kilometer per second. Conversely, if the velocity of the electron is given within an accuracy of one hundred thousandth of a centimeter per second, the electron could be anywhere within one kilometer. One could of course theoretically use a radiation of much shorter wavelength to achieve greater precision in measuring say the position of the electron, but this would necessarily increase the minimum error involved in measuring its velocity by the same factor as that by which the wavelength of radiation was reduced.

If the uncertainty relation connecting position and momentum, $\Delta x \Delta p \geqq h$ is written as $\Delta x \Delta v \geqq h/m$, then the situation can be grasped even more directly. The ratio h/m for an electron is about unity. Suppose therefore that one desires to measure the classically computed value of the radius of the orbit of an electron in a hydrogen atom with an error of about 10^{-8} cm. This would be of the same magnitude as the radius itself. A larger error would evidently make the measurement meaningless. The corresponding minimum uncertainty in the value of the orbital velocity of the electron would be of the same order of magnitude as the value of the velocity computed classically. A greater precision in measuring the radius could only increase the already large uncertainty regarding the value of the orbital velocity. This essentially means that even with minimum uncertainty achieved, the position of the electron around the nucleus cannot be located. The uncertainty, in fact, is so great that one has to assume that the electron is "spread" around the nucleus. Thus the electron cannot be pictured as moving in a well-defined continuous orbit, or spinning on its axis as a perfect rotating solid. For the same reason such basic constants as the spin and magnetic moment of an electron cannot be directly measured by a magnetometer experiment. Such an experiment has to work with a moving electron, the position and velocity of which cannot be measured with perfect accuracy at the same time. The uncertainty in the velocity gives rise to an uncertainty in the magnetic field induced by the moving charge but the uncertainty thus produced is greater than the magnetic field owing to the spin magnetic moment that was supposed to be measured.

Another area in which precision has reached the ultimate limits set by the indeterminacy principle is the measurement of the width of a spectral line. Three main causes contribute to the broadening of spectral lines. Two of these, the translational motion of the atom and the interaction of atoms due to pressure, can be eliminated to almost any degree, at least theoretically. The third one, however, which produces the so-called natural width, is permanently inherent in any measurement of a spectral line. The energy width produced will be of the order of magnitude given by the ratio of Planck's constant divided by the time needed for the measurement. Since an ordinary excited state of an atom has a lifetime of the order of 10^{-8} seconds, it follows that lines in the visible spectrum will show a widening of about one ten-thousandth of an Angstrom. This is too small to measure, but in the X-ray region the natural width is easily observable. Thus the edge of precision actually opened up for physics what P. W. Bridgman called once in a rather pessimistic vein "the portals of a shadowy domain beyond which no refinement in measurement can carry the physicist."[88]

Still, the limitations set by the uncertainty principle should not be construed as too narrow confines within which the possibilities will soon be exhausted by advances in instrumental precision. What the uncertainty principle essentially means is that the determination of the "state of the world at an instant" is not possible in terms of mechanistic physics. And this limitation holds also for that proverbial "superior spirit" to which Laplace and others liked to refer. It is, however, well to remember that while classical physicists recognized their inferiority to that superior spirit, it was not assumed to be superior to the principles of classical physics. At any rate, modern physics in general, and the indeterminacy principle in particular, made it abundantly clear that the mechanical intelligibility does not exhaust the whole range of intelligibility. Therefore one should not conclude on the basis of the indeterminacy principle that "the world is not a world of reason, understandable by the intellect of man."[89] Those modern physicists who do so fall back unawares on the definition of intelligibility as formulated by classical mechanism. If, however, one is willing to part with this long discredited notion of intelligibility, cherished as it may be even today by some, the future of physics will appear promising not in the least because of the great potentialities of its precision instruments.

There is indeed nothing to indicate that the rate of progress in devising more precise instruments would hit a snag in the near future.[90] To play the prophet in scientific matters is of course rather hazardous, but in the field of precision instruments the future usually vindicates those who keep expecting the almost impossible. The course of events has not justified even the slightest touch of resignation concerning the perfectibility of measurements in physics. How little did a Rowland (a true wizard of precision himself) suspect the degree of progress that was to take place in the field of measurements in the first half of the twentieth century! Not that he was pleased with everything physical science could offer in this respect as the nineteenth century came to its close. In 1899 he noted with some dissatisfaction that the attainment of precision had in some respect left much to be desired.[91] As he surveyed the field, he saw that mechanical rotations in his day were limited to a few hundred revolutions per second, that the greatest pressure achieved did not surpass one hundred tons per square inch, that the sun's temperature was the upper limit in temperature measurements, and that the vacua, to use his own words, were very imperfect indeed.

Subsequent advances in all of these fields have been such that one who demanded them in 1899 would have seemed to be asking for the moon. Actually nothing short of the moon itself was reached in little more than half a century. At the same time, of course, spectacular successes also meant spectacular surprises, some of which were embarrassing. Photographs taken of the far side of the moon threw lunar theories into turmoil. The pictures obtained by Ranger IX, the last of which were taken just half a second before impact, provided ground for widely differing speculations about the structure of the moon's surface. The vigor of those speculations did not diminish after Surveyor I began to transmit thousands of pictures from the moon's surface. Although Mariner II, observing Venus from a vicinity of 21,000 miles in late 1962, did indeed confirm some of the findings of earthbound telescopes, Mariner IV accomplished what usually happens when the resolving power of the instruments of physics is considerably increased. By showing objects as small as two miles in diameter on the surface of Mars, Mariner IV not only surpassed anything achieved along these lines by astronomy, but also left in disorder a firmly established scientific consensus that expected anything but the revelation of a crater-pocked, moonlike Martian land. To be sure, the degree of precision built into the devices of space exploration is very great. Yet, no less astonishing are the results which the edge of precision provides in peeling off layer after layer of the mysteries enveloping the physical universe.

If the present-day search into space brings out sharply the fundamental role of precision in physics, the same kind of enlightenment is being achieved by efforts that aim at penetrating ever further into the structure of matter. In this connection it is well to remember that the chief probing instruments of matter, the accelerators, have not yet reached their ultimate possibilities. The greater the energy a particle acquires through acceleration, the smaller is its De Broglie wavelength and the smaller are the details it can resolve in its target. To "see" the rough details of a molecule, an electron microscope using electrons with an energy of several thousand electron volts is needed. To "see" the nucleus, a sort of "proton microscope," using protons of 2 Mev, is required. To resolve the details of the nucleus, accelerators operating in the Bev energy range have to be used. This is, however, only one aspect of the story behind the huge accelerators that somebody has called the cathedrals of the twentieth century. At sufficiently short wavelengths, particles are not only "seen" but also "created." A ϕ 1 Mev photon has enough energy to create a positron-electron pair. The heavier π mesons can be produced at 150 Mev, and at 1 Bev one reaches the energy needed to create κ and λ mesons. Bevatrons at 6 Bev can produce antiprotons, antineutrons, and at even higher energies Ξ and Σ particles. In fact, more than 30 Bev was needed to verify experimentally the existence of the anti Xi-zero particle and that of the antineutrino, to say nothing of the several hundred thousands of photographs that had to be taken in both cases to provide the few indisputable records of their occurrence. Reflecting on the successive steps, that have improved acceleration machines, one ought to be struck by the fact that a major advance in peak energy led invariably to the unfolding of qualitatively new aspects of the structure of matter. This feature of the development of accelerators was stressed recently with special emphasis by C. N. Yang, the co-discoverer of the non-conservation of parity in weak interactions, who had this to say about the need for more powerful accelerators:

> Sometimes people tend to think that to have greater intensity is only to increase experimental accuracies from $\pm 10\%$ to $\pm 1\%$, or from $\pm 1\%$ to $\pm .1\%$. Let me emphasize that that is not the reason why greater intensity will be badly needed in high energy physics in the near future: greater intensity will be needed because one will have to understand *qualitative* features of elementary particles. . . . At the present state of our knowledge of high energy physics, the danger that all these features will be qualitatively understood to such an extent that a high intensity machine will be left only with a job of adding decimal points to known results is very small indeed.[92]

That many startling discoveries will follow in studies of the basic structure of matter can be confidently expected from the fact that there is hardly any evidence in sight for a fixed ceiling for peak energy attainable in future accelerators. Plans for a 200 Bev accelerator have already reached the stage of selecting its location. Yet, while this enormous machine with a diameter of 4,528 feet is still about ten years away from becoming operational, more and more consideration is already given to accelerators with a peak energy of 600 to 1,000 Bev. Their construction might of course necessitate an international pooling of material and scientific resources. Clearly, to make such progress in little over three decades from Lawrence's first homemade cyclotron to the present-day plans, the execution of which may necessitate a cooperation on an international scale, is a cause for no small satisfaction. At the same time one should not forget that research here runs the full circle of scientific discovery. Bigger machines make new particles that in turn create problems that motivate further increases in the peak energy of accelerators. Still if a survey of past steps, triumphs, hopes, and dreams is any indication, one might say with confidence that the future belongs to those who like to think of possibilities in very bold terms. The late E. Fermi once suggested that the earth be ringed with a vacuum tube; by using the earth's magnetic field, one could obtain in such a way energies three thousand times greater than the peak energy of the largest accelerators now in use. This suggestion, truly fantastic only a few years ago, should appear more realistic since artificial satellites have come of age. A set of satellites, each equipped with injectors and focusing apparatus might make available even one million Bev without a single vacuum pump.

Physics has indeed reached the stage where the demands of precision have made inevitable the construction of machines that are nothing short of gigantic and fantastic. A case in point is the apparatus used for the detection of neutrinos. First devised by F. Reines and C. Cowan in the mid-1950's, these enormous installations provided in 1965 the first evidence of the existence of natural neutrinos. Here, bulkiness represents extraordinary sensitivity. Clearly, only a sensitivity that defies imagination is capable of detecting natural neutrinos, although billions of them bombard each square centimeter of the earth's surface every second. This feat took place under the direction of F. Reines near Johannesburg in a cave 10,492 feet deep.[93] A similar experiment was devised to trap each day three to nine neutrinos out of trillions, and then after a month's time to extract their accumulated by-products, some 270 radioactive argon atoms, from 100,000 gallons of target fluid. All this represents a classic illustration that the more elusive a particle, the more elaborate are the instruments needed for its detection.[94] At the same time instrumental layouts, such as the ones used for the detection of neutrinos, evidence not only an outstanding feat in reasoning but also a fantastic degree of precision in measurement. Thus if neutrinos are indeed carrying as much energy in the universe as do heat, light, and all sorts of cosmic particles, the apparatus described above might prove both one of the sharpest instruments man has yet devised and the most portentous as well in theoretical results. In all likelihood once again the edge of precision will unfold startling new aspects that might give entirely new turns and impetus to science, and might lead once more to a thorough overhaul of scientific concepts.

The quest for ever greater precision keeps science relentlessly on the move. Science is, in a sense, as De Sitter noted, "the struggle for the last decimal place." In his words, "the great triumphs of science are gained, when, by new methods or new instruments, the last decimal place is made into the penultimate."[95] It is, however, precisely there, in the ever-continuing emergence of new digits that the paradox of the quest for precision lies. It is a quest that not only brings triumphs but also necessitates revolutionary reorientations. By turns it confirms and discards theories and points to aspects hitherto unsuspected. The moments when new, more precise instruments become available are usually turning points in science. The future is hidden, but precision will probably affect the future course of science as it has affected science in the past: scientific history should permit one to make that tentative conjecture.

To see something of the full force of scientific precision, one might indulge in pondering a few hypothetical questions. What course would science have taken if the pressure of light could have been demonstrated in the eighteenth century?[96] Or if Faraday had had a sufficiently strong magnetic field to show its effect on atoms emitting light rays? Or if Dalton could have measured with far greater precision the atomic weights? Or if optical technology had permitted the carrying out of the Michelson-Morley experiment half a century earlier? Such questions, which could be multiplied endlessly, would not be altogether a sheer exercise in fantasy. Instead, they should rather bring out forcefully the extent to which the "limits" of physical research depend on the momentary inability to cope with specific experimental problems rather than upon the often illusory definiteness of many a result and conclusion. In the past, the waiting period for the proper tool to cope with a particular puzzle has often taken all too long. Today, however, the verdict of experiments upon the theories and predictions of science, usually come more quickly.[97] Puzzles that loom as unsurmountable barriers turn into routine investigations at an ever faster pace as the edge of precision keeps cutting across what has seemed to be "ultimate" limits to physical research.

Innumerable indeed are those "final" conclusions in physics that the edge of precision has shown to be only imaginary. It is to this relentless edge that the most valuable discoveries of modern physics also owe their existence. These recent findings have no doubt revealed marvelous details about the universe, but they have not discovered everything. Against the fine structure of the tools of physics, nature has so far invariably succeeded in coming up with her own "hyperfine structure." This interplay between the ingenuity of research and the richness of nature will hardly change if past and present patterns should have any meaning for the future. It is in this interplay that the edge of precision reveals most forcefully its true nature as a basic tool of physical research: it is a double-edged knife, as effective in cutting through extant problems as it is in creating new ones. In its twofold role, the edge of precision is truly a symbol and a proof of the unending character of that search which aims at unfolding the immensity of details embodied in the material universe.

That nature can so persistently baffle man in his quest for her ultimate secrets is owing only in part to the unfathomable richness of physical reality itself. Marvelous as are the instruments designed by physics, they can attack at one time only one small aspect of the immense complexity of things. Sharp as the edge of precision might be in one respect, it will remain crudely blunt in others. What Pascal said of justice and truth in general and of our tools for reaching them applies no less to the relation between the tools of physics and the truth of physical reality: "Truth is so subtle a point that our instruments are too blunt to touch it exactly. When they do reach it, they crush the point and bear down around it, more on the false than on the true."[98] Sobering as this reflection might be, it mirrors a perennial aspect of the human condition, which, as it concerns the search for truth, is enveloped in twilight. The edge of precision furthers man's progress toward dimly lit horizons where one can little discern the shape of discoveries to come. To persist in his often superhuman efforts to improve the precision of his tools, the man of science needs more than the reassurance derived from past successes. He needs faith as well. What Millikan once said of Michelson's striving for ever greater precision holds for all his colleagues: "He merely felt in his bones or knew in his soul, or *had faith to believe* that accurate knowledge was important."[99] To be fully aware of all this is perhaps the most valuable lesson one can draw from reflecting on the role of the edge of precision in the great quest of learning.

Part Three

Physics
and Other Disciplines

CHAPTER SEVEN

Physics and Biology

MAJOR ADVANCES IN PHYSICS never failed to stir hope that the laws of physics might ultimately prove as effective in explaining the problems of living organisms as they do in coping with the phenomena of the inorganic world. What is unfolding today in the startling discoveries and ambitious projects of biochemistry and biophysics is in fact but the latest upsurge of the conviction that physics has a fundamental relevance in dealing with the phenomena of life. With no less emphasis the same conviction asserted itself a hundred years ago when physics succeeded in unifying for the first time its various branches under the principle of the conservation of energy. In the same manner, the discoveries that ushered in classical physics were accompanied by many references to what physics may achieve in untangling problems connected with the realm of the living. Actually, reflections of this sort are as old as physical science itself. For the Greeks, the relevance of physics to biology was such a paramount problem that it ultimately led to the codification of the first main type of physics in which organismic concepts became predominant.

Not that the Greeks had been unaware of the marvels of mechanisms and automata. The Hippocratic medical texts refer time and again to mechanical analogies to illustrate the various processes of the human organism. References to automata can be found even in Aristotle's biological writings and the playwrights — Aristophanes and Euripides — compare at times eyelashes to sieves, ears to troughs, the nose to a wall, and passages in the body to channels. Such mechanistic analogies and metaphors, however, were not meant by ancient science and thinking as adequate explanations. The basic science was supposed to be biology, the study of organismic, purposeful, and

283

functional behavior, and for almost two thousand years physics had to search for the strivings, trends, and aspirations lying supposedly at the root of the behavior of inanimate matter. Mechanism was a study to be governed by the concepts of vitalism and not vice versa. Clearly then a new type of physics, differing radically from the organismic one, had to be formulated first to secure proper meaning for the question of how effective the laws of physics may be in explaining the manifold phenomena of life. Different as was the orientation of the new scientific thought compared to the Aristotelian science, it speculated with no less concern about the basic relation between animate and inanimate matter. It was this concern that inevitably produced what has been known for several centuries as the confrontation between physics and biology, or between mechanism and vitalism.

When the art of latitudes — the representation of physical quantities by graphs — started to develop from the early fourteenth century, suggestions to extend the method to the realm of the living were not far off. Oresme applied his theory of impetus to psychology, and moral qualities, such as faith, and other properties of spiritual perfection were discussed in terms of latitudes by Dumbleton and others. Health, too, was reduced to degrees of heat, and it was only natural to do the same with various aspects of Galen's physiology, as can be seen in the writings of Jacopo da Forli. These were, for the most part, vague speculations or sanguine hopes such as the one voiced by Rheticus who, entranced by the sweep of the Copernican system, took the view that medicine might one day reach the same perfection that Copernicus gave to the description of the motion of planets.

The emphasis from vitalism to mechanism of course did not shift overnight. Still, from Leonardo on, the trend toward mechanism was unmistakable, as can be seen in the bold statements of both the precursors and the early figures of classical physics. Thus Leonardo speaks of birds and "flying machines" as things "working according to mathematical law,"[1] and it is therefore not surprising to hear him declare that "mechanical science is very noble and useful beyond all others, for by its means all animated bodies which have movement perform their operations."[2] The year 1543, which saw Copernicus give a new arrangement of the planetary system, also witnessed Vesalius trying to give "with sufficient fullness the number, position, shape, substance, connection with other parts, use and function of each part of the human body."[3] This was clearly an approach conceived along the lines of geometry and mechanics, an approach that had to be phrased in its most extreme form first by the one whose main ambition was to formulate the new and definitive form of physics. In fact, the laws of Cartesian physics were proposed as equally powerful in dealing with non-living or living phenomena. Descartes' *Treatise of Man* forms accordingly a part of his synthesis of the physical sciences where vitalism is banned in any form whatsoever. In the same vein, the concluding section of *Discourse on the Method* speaks of Descartes' determination to devote the remainder of his life to the development of a science that will enable man "to arrive at rules for medicine more assured than those which have as yet been attained."

The little details of Descartes' private communications are, however, no less expressive in this connection than the grandiose formal statements of his major works. Thus, in 1630, he tried to comfort the ailing Mersenne by a reference to a project that seemed to occupy all his attention at the time: a "medical theory based on infallible demonstrations."[4] Clearly, what he envisaged was a cure-all in line with his unbounded trust in the effectiveness of geometry, which alone, in his view, possessed definitive and errorless conclusions. If medicine was to be squeezed out of geometry, then in ultimate analysis the bodies, too, healthy or sick, were to be patterned along the figures of geometry, just like machines: an ingenious mixture of curved and straight lines, as embodied in all mechanical contrivances of which the chief paradigm was the clock. For as Descartes put it, the body of man was a "machine made of earth," and all the functions "follow naturally in this machine simply from the arrangement of its parts, no more and no less than do the movements of a clock or other automata."[5] This was a sweeping dictum by any count, but Descartes conformed to it, even to the most minute details of his style. After defining man's body as a machine, he simply kept referring to it throughout his *Treatise of Man* as *this machine*. The most he was willing to concede concerned the incomparable perfection of the body machine as contrasted to man-made contrivances.[6] To forestall any temptation to correlate the various "spirits" operating in the body to vitalistic, non-mechanical principles, he went to great lengths to show "in a beautiful comparison," as he called it, how all those "spirits" and their functions are but a perfect version of the air that moves through the intricate pipe system of an elaborate church organ.[7]

Such an attitude was just as complete a declaration of triumph as any type of mechanism could hope to gain over vitalism, and Descartes clearly enjoyed sketching the details of it. He saw in the return of swallows in the spring the action of clockwork and said that "all that honeybees do is of the same nature."[8] With obvious relish he drew a detailed parallel between the actions of the human body and the workings of the elaborate fountains reproducing mythological scenes with moving figures. There the nerves corresponded to pipes, muscles and tendons to engines, body fluids to running water, the heart to the mainspring, and the cavities of the brain to the outlets of water.[9] Picturesque as was this comparison, it did little justice to the rules of scientific reasoning for which not every flight of fancy can pass as a respectable explanation. Yet, the proponent of a mechanistic philosophy to which there could be no secrets, no puzzles, no paradoxes, no perplexing aspects of nature, could hardly have had any misgivings, in taking so much liberty with the physical world. As for the realm of life, machines could perform any feat according to the mechanistic creed, and Descartes stretched this sacred tenet of mechanism so far as to speculate on how machines could beget machines. This he did in an essay entitled *First Thoughts on the Generation of Animals*, which is long on conjectures and exceedingly short on "second thoughts." A little honest reflection could have showed him that on the subject he could say no more than Galileo could when trying to explain by physics the physiology of vision. On the things pertaining to this question, wrote Galileo, "I pretend to understand but little; and since even a long time would not suffice to explain that trifle, or even to hint at an explanation, I pass this over in silence."[10]

Not that Galileo as a physicist had always kept a prudent silence about physiological topics. How could he help it after stating that only the quantitative or dimensional aspects of things are rooted in reality? From there it took but a step to offer unhesitating explanations of physiological questions, such as the origin of tastes that Galileo categorically ascribed "to the various shapes, numbers and speeds of particles of matter," which reached the upper surface of the tongue.[11] In general Galileo believed, even in the absence of rigorous proofs, that to experience the sensations of tastes, odors, and sounds "nothing is required in the external bodies except shapes, numbers, and slow or rapid movements . . . ; if ears, tongues and noses were removed, shapes and numbers and motions would remain, but not odors or tastes or sounds."[12] Yet, for him the tongues, ears, and noses were ultimately but constructs of shapes, and clearly it should have appeared to him as puzzling, to say the least, how the contact of two realms — man's senses and the outer world — equally composed of shapes and motions, could produce something distinctly different in the perceiving man. To pause briefly and ponder problems of elementary logic, however, was never the forte of too enthusiastic scientific creeds.

Physiology thus began to be viewed by some as a branch of study wholly subject to mechanics, and it should be recalled that such an approach was not without some signal success in the early seventeenth century. The chief of these was the discovery of the circulation of blood by Harvey, to whom the harmonious and rhythmical motion of auricles and ventricles evoked the image of a machine in which "with one wheel moving another, all seems to be moving at once."[13] In a truly mechanistic vein, Harvey lets his reader share his fascination with the mechanical details. The working of the heart should remind us, he states, of that "mechanical device fitted to fire-arms in which, on pressure to a trigger, a flint falls and strikes and advances the steel, a spark is evoked and falls upon the powder, the powder is fired and the flame leaps inside and spreads, and the ball flies out and enters the target; all these movements, because of their rapidity, seeming to happen at once in the wink of an eye."[14] To describe a mechanical process with such attention was no doubt the mark of a "mechanical philosopher," and Harvey was of the sounder types making up this group. As such he made resolute efforts to go beyond mere analogies between bodies and machines. Of course he needed to start with analogies. As Boyle reported the aging Harvey's words, it was the presence of "valves in the veins of so many parts of the body"[15] that convinced him of the correctness of the mechanical approach to the problem of the circulation of blood. It was clear to him, however, that quantitative evidence alone could assure lasting triumph to his conclusions. He therefore proved by a series of measurements that even if the smallest possible value is adopted for the quantity of blood sent through the heart and lungs with each pulsation, the amount of blood pumped into the arteries in half an hour would be equal to the total amount of blood present in the body of a man, or sheep, or an ox. Clearly the heart or any other organ could not generate so much new blood in such a short time. Yet, this was precisely what traditional views implied. The contrast between the old and new views could hardly be greater and Harvey himself admitted that it took great courage to come forth with his findings. To hear him say it, they were "so novel and hitherto unmentioned that, in speaking of them, I not only fear that I may suffer from the ill-will of a few, but dread lest all men turn against me."[16]

As time went on, however, the mechanical philosophers needed less courage to proclaim the supreme competence of physics in physiology. With carefree abandon, they began to see mechanisms everywhere they looked. Peering through his microscope, Hooke felt entitled to state that in minerals nature begins to "geometrize and practice as 'twere, the first principles of Mechanicks." Passing from minerals to plants, nature proceeds, Hooke contended, merely to the level of advanced mechanics "by adding multitudes of curious Mechanick contrivances in their structures." The supreme skill of nature in mechanics is displayed, according to Hooke, in the animals

— in the "most stupendous mechanisms and contrivances."[17] In plant seeds Hooke saw "little automatons or Engines"[18] put in motion by the sun's heat, and to explain the corruption of certain substances, he could conceive of only one "intelligible" approach: an analogy with a clock. This he did in a way that illustrates better than anything the incomparable assuredness of the mechanical philosopher in his art. For, as Hooke put it, a clock too can be corrupted by losing some of its parts, and he believed, although he assuredly never tried the experiment, that on further shakings the remaining components might fall into a new arrangement with the result that the clock suddenly will start again and keep working but in an entirely new fashion.[19] Being a faith, the mechanical creed had to claim miracles of its own.

Here then was the semidivinity of mechanical physics, the clock, that in Borelli's *De motu animalium* performed its most astounding feat: the "explanation" of the fertilization of an ovum by a male sperm. The process, according to Borelli, is like the working of a timepiece made of cogwheels, endowed with a motive force through an affixed machine or a weight.[20] Clearly, only an enormous scientific faith could find mental satisfaction with this sort of explanation even in 1685. On such faith hinged ultimately this alleged supremacy of physics over physiology, and it was with such illusions of the geometrical spirit that Borelli attacked every conceivable problem of physiology. He had no other choice. Everything in an animal's body, he declared in the preface of his work, must be subject to mathematical analysis. The motion, structure, and composition of animals are rooted, according to him, in mechanical causes, means, and reasons (*rationes*), among which he lists the balance, bars, rods, blocks, pulleys, drumwheels, wedges, cones, and so forth. Such mechanical "reasons" represented for him the acme of intelligibility, and as a result, he readily brought the supreme intelligence into the picture. "In the construction of the organs of animals," he noted with obvious satisfaction, "God resorts to geometry . . . which is the only and proper science that makes legible and intelligible the Divine Code inscribed in the body of animals."[21]

Although a confident recourse to a divine code could hardly settle the issue between mechanism and vitalism, the procedure was highly expressive of the state of mind for which "real" was practically synonymous with "mechanical," living things not excepted. Few of the "mechanical philosophers" could remain completely immune to this attitude. Leibniz himself flatly declared that "whatever is performed in the body of man, and of every animal, is no less mechanical, than what is performed in a watch."[22] Accordingly, he saw the basis for the study of medicine "in the science of motion which is the key to physics."[23] It made little dent on such a thorough subordination of biology to physics that in Leibniz' view "animals are never formed naturally of a non-organic mass"[24] and that he decried the opinion of those "who transform or degrade animals into pure machines."[25] The "immense distance" and the "difference of kind," not only of degree that he stated to exist between machines and animals[26] merely served the purpose of extolling the perfection of a divine contrivance based on the same mechanical principles.

Newton himself referred in both the *Principia* and the *Opticks* to the ethereal medium that makes possible gravitational attraction, electrical effects as well as sensory perception, propagation of nerve impulses, and the functioning of animal instincts.[27] That such suggestions of Newton, which were preceded by his great feats in celestial mechanics, proved to be very influential is understandable. Thus S. Hales introduced his experiments on plant physiology with a reference to the "great philosopher of our age" (Newton), who by measuring and numbering discovered the gravitational system of planets.[28] Although Sir Isaac could hardly be considered an authority on plants, his success in physics seemed to guarantee the validity of the mechanical philosophy in every field of study, physiology not excepted. W. Wotton in his *Reflections upon Ancient and Modern Learning* went to great lengths in contrasting the traditional views in physiology to the marvelous intelligibility of the modern methods, by which he meant of course the application of mathematics and mechanics in physiology. Thanks to these methods, he wrote, "the general Consent in Physiological Matter . . . is an almost infallible Sign of Truth."[29]

In those years of general enthrallment with the new science, it took some intellectual courage to note that such a consensus could also be the sign of very fallible shortsightedness. One of the few physiologists who refused to bow to the dictates of a rigid mechanistic philosophy was the London physician, G. Cheyne. Experiments performed by him indicated, so he contended, that vegetables and animals could not be produced from matter and motion alone.[30] Such and similar examples greatly encouraged those who on theological or humanistic grounds were unable to accept the reduction of living organisms to machines. Like Cheyne, they too set great store on the futility of the mechanistic explanation of the generation of animals. In an article reviewing "the modern theories of generation," the Scottish divine, George Garden, voiced in 1691 the antimechanistic view with a scathing reference to Descartes: "All the Laws of Motion which are as yet discovered can give but a very lame account of the forming of a Plant or Animal. We see how wretchedly *Des Cartes* came off when he be-

gan to apply them to this subject; they are formed by Laws yet unknown to Mankind."[31] Two decades later, in the November 22, 1712,
issue of the *Spectator*, Addison, that persuasive spokesman of humanistic viewpoints in his day, went a step further and pointed to the basic
role which chance played in the mechanistic explanation of life. This
he contrasted to the purposefulness of living bodies and emphasized
that the formation of such bodies could not be ascribed "to the fortuitous concourse of matter."

Throughout the latter half of the eighteenth century, the opinions
varied over a broad spectrum about how competent physics was to
explain the phenomena of organic life. This was an age in physics that
was fascinated with the basic phenomena of static electricity, and
speculations ran rampant as to the identity of electricity and the
ether. As the allegedly finest brand of electric fluids, the ether began
to be taken as the ultimate vehicle of living phenomena as well. John
Wesley no doubt expressed the fashionable belief of his day when he
wrote in 1760 that this fluid is "subtle and active enough, not only to
be, under the Great Cause, the secondary cause of motion, but to
produce and sustain life throughout all nature, as well in animals as
in vegetables."[32] Ultimately such a view was more vitalistic than
mechanistic in the sense that it readily submerged any difference
between physics and physiology in a flood of hazy words about the
ether.

Such a state of affairs could hardly escape the truly perceptive
figures of eighteenth-century physiology. While reviewing the various conjectures about the seat of the soul in the brain, Albert Haller
gently poked fun at those who regarded "the ether, this invisible and
experimentally unverified element, as the source of all phenomena
whose cause was not apparent."[33] In addition, he emphasized that
the introduction of mechanics and hydraulics into physiology was a
mixed blessing. "There are many things in the animal machine," he
warned, "that are most alien to the common laws of mechanics."
Therefore, "those laws should never be applied to the machinery of
animal bodies unless experiments warrant such procedure." It was his
warning to students of physiology that they should guard against being swayed by the false promises of the endless use of geometry in
matters physiological. This he noted with an eye on those whom he
called "iatromathematicians," and he contrasted in particular the
geometrical fantasies of Borelli to Hales's preference for experiments.[34]

People whose main achievements lay in physics or in mathematics
were on the whole not especially verbose on the subject. Few of them
were as versatile as Maupertuis, whose *Venus physique* (1745)[35]

shows not only a genuine interest in problems such as fecundation, development of the foetus, and varieties within the human species, but also displays a solid information on current research in biology. Maupertuis, for one, tried to chart a middle course between mechanism and vitalism, strongly stressing the irreducibility to mechanism of the purposefulness manifested in the functioning and development of living organisms. This was, however, only one of the various attitudes adopted by men of science who discussed in those days the relevance of physics to biology. Actually Maupertuis had to defend himself against the criticism of Diderot, who resolutely pleaded the case of strict mechanism.[36]

Men of science whose main interests lay outside physics but who tried to emulate its exactness displayed of course a distinct preference for the mechanistic view. The most distinguished of these scientists was Lavoisier whose connections with biology lay in his study of the respiratory process. He had shown in fact that the ratio of heat to carbon dioxide given out either by animals or by candle flames was in both cases of the same order of magnitude. The inference was therefore unavoidable that the chemical energy of animals must have its source in the combustion of their food. Evidently in studies of this type the strictly quantitative method could leave no place for vitalistic interpretations of any sort. Another notable figure of eighteenth-century science who deserves to be mentioned in this connection was Buffon. Primarily a naturalist who devoted more than half of his *Natural History* to the description of a long list of animals, Buffon preferred nevertheless to be considered a physicist. He spoke of physics as "the primary science"[37] and envisioned a future stage where all phenomena of nature would be explained by physics. Of course for him physics meant "rational mechanics," to which he assigned without any restriction the role of investigating the powers of nature.[38] According to him, these powers originate in one single mechanical source, determining everything in inanimate matter, and "from this force, combined with that of heat, originate those living particles which give rise to, and support all the various effects of organized bodies."[39] As Buffon stated, "The living animated nature, instead of composing a metaphysical degree of beings, is a physical property, common to all matter."[40] For him, however, matter was strictly mechanical in its behavior, and consequently he exclaimed: "What variety of springs, what forces and what mechanical motions are enclosed in this small part of matter which composes the body of an animal!"[41] Clearly, then, one would look in vain for anything basically vitalistic behind those statements of Buffon in which he speaks of the constancy of the amount of living matter and of the strict permanence of living forms or molds on which individuals and species are fashioned.[42]

Firm as Buffon was in his belief in the general validity of mechanism, he was not reluctant to confess his ignorance about many of the puzzles presented by physiology. Indeed he had an eye for the foibles present in many specious arguments advanced in his day in the name of exact sciences and remarked wryly that "people in general reason only from their sensation and natural philosophers determine from their prejudices."[43] No trace of such an awareness, however, was evident in the manner in which the principal eighteenth-century propagandists of the mechanical creed declared the problems of physiology solved by physics. This class of "intellectuals," as always, had ideas of their own, which in accordance with the acoustics of hollow bodies excelled mainly in loudness. Those who earned the most fame along these lines were Holbach and La Mettrie. The latter's *L'Homme machine* (1748) is undoubtedly the classic example of that baffling degree of disproportion between evidences and conclusions that are presented at times in the disguise of the "latest findings of science."

It was precisely the awareness of such disproportion between the claims and proofs of mechanism in the field of physiology that determined ultimately the organization of various branches of biology in the eighteenth century along non-mechanical principles. Such was the case with embryology, where Harvey's revival of the Aristotelian epigenesis had to yield for a while to a mechanistic vision which could easily turn at times into sheer illusion. Thus Hartsoeker professed to have seen *homunculi* in the male sperm through his microscope and Swammerdam spoke in the best mechanistic vein about the egg as containing in itself all future generations in miniature as a set of boxes encasing one another. After all, was not this principle of *emboîtement* a basic tenet of the mechanical world picture in which physical reality had the same features on all scales, however large or small?

Although not every student of embryology went so far as Hartsoeker, preformationist embryologists were divided only on the point whether to place these unborn generations in the male sperm or in the female ovum. When it became evident, however, that neither man nor animal was ready-made in the sperm and ovum, vitalism reasserted itself. C. F. Wolff's *Theory of Generation* (1759) exposed not only the weaknesses of mechanical theories but also argued the existence of a vital force directing the development of homogenous organic matter. This non-mechanistic view received a further boost when Spallanzani showed in 1767 that Needham's experiments on spontaneous generation were basically defective. Developments like these could of course but hasten the emergence in embryology of the various recapitulation theories, distinctly vitalistic in outlook. What happened in embryology was paralleled in the classification of plants and animals. Here, too, one may notice a change from a preponderantly mechanical preference in the late seventeenth century to an approach that in the second half of the eighteenth century stressed the primacy of an intuitive method looking for archetypes and plans of unity embodied in any organism. This trend reached its peak in the work of Cuvier, who in 1817 wrote in his *Animal Kingdom* that the only persistent element in the organic world was the form that could multiply and perpetuate itself in a mysterious operation called generation.[44]

Clearly, this was an idiom revealing not even a faint affinity with the concept of mechanics. The investigation of living matter seemed to necessitate a language and method different from those of physics, and no one emphasized this more than Xavier Bichat, the great pioneer of tissue research. Actually, he devoted a short but vibrant section of his *Recherches physiologiques sur la vie et la mort* to this problem.[45] There he dwelt on the contrast between the fixed, invariable laws of physics and the perpetual variability of vital phenomena. The latter, he noted, may pass in an instant from an almost inert condition to the highest degree of excitation and can also take on a thousand various modifications under the impulse of the slightest change. To the uniformity and stability of physical forces, he opposed the irregularity of vital processes and called attention to the difference in the behavior of inorganic fluids and of fluids active in living systems. In Bichat's eyes chemical analysis of living fluids could merely issue in animal chemistry, which he called, somewhat contemptuously, the cadaverous anatomy of fluids. Physiological chemistry, or the analysis of fluids in a living body, was, in his view, so complex a problem as to be forever beyond the reach of mathematical analysis. Powerful as physical science had proved itself in calculating the speed of projectiles, the velocity of water in canals, or the return of comets, it had, according to him, no place in physiology. Attempts such as Borelli's to calculate the strength of a muscle, or Keil's to calculate the velocity of blood in the veins, or Lavoisier's to determine the quantity of air entering the lungs, were in Bichat's view nothing short of building "on quicksand an edifice solid in itself, but which collapses soon for lack of secured foundation."

In the instability of vital forces, Bichat saw the stumbling block that made a shambles of the calculations of physicist-physicians of the eighteenth century. The same frustration was in store, he contended, for the chemist-physicians of the century to follow. Consequently he pleaded in physiology for a language free of the concepts of physics. If a discussion of gravitation, hydrodynamics, and crystallography in physiological terms was obviously preposterous, no less ridiculous, he argued, had to be the introduction of gravity and chemical affinity into physiology where these could have, according to him, but a most obscure role. "Physiology," he declared, "would have made progress even if nobody had brought into it ideas borrowed from so-called auxiliary sciences." Closely related as physics and chemistry were to one another, an immense interval separated both, so Bichat claimed, from physiology, owing to the enormous difference that existed between the laws of physics and the laws of life. "To say," he concluded, "that physiology is the physics of animals, is to form about it a very inexact idea; by the same token one might say as well that astronomy is the physiology of stars."

That Bichat in his attack on mechanism singled out pieces of research the merits of which he could hardly deny was not without some paradoxical significance. Much as the vitalists could decry the intrusion of physics into biology, they had to realize that if a vitalistic theory was to achieve the status of science, it had to go beyond mere description and establish invariable connections or laws for "vital" phenomena. A vitalistic law, as Cuvier put it, was supposed to have a validity equaling that of the laws of metaphysics and mathematics. According to him, such a law was to explain, for instance, how the form of an animal's tooth determines the form of the jaw, or how the shoulder bone does the same for the shape of the claws, with no less accuracy, to be sure, than the equation of a curve determines all its properties.[46] In his cell theory, Schwann saw the manifestation of a similar "vitalistic law," peculiar to biology, a law through which the same blind necessity operates that obtains in the laws of physics.[47] Schwann considered the cell a unit equivalent to the atoms or molecules of physics and believed that the function of cells was determined by forces, one of which, the attractive force, was operative both in organic and inorganic matter, while the other, the metabolic force, was a feature of living matter alone.

Although the vitalists did not like to admit it, whenever such a vitalistic law was formulated with enough concreteness, the presence of quantitative considerations implicit in it was unmistakable. It also became more and more apparent that meaningful advances in the various branches of physiology always implied the isolation of a problem from the *whole* and its subjection to the quantitative procedures of chemistry and physics. It was therefore inevitable that alongside the radical vitalists and the radical mechanists a middle-road group should emerge which, while recognizing the paramount importance of the methods of physics and chemistry in biological research, remained fully aware of the limitations of this approach in biology mainly owing to the immense complexity present in life phenomena.

The most consummate synthesis of this position is Claude Bernard's *Introduction to the Study of Experimental Medicine* (1865), which contains all that was enlightening about the relevance of physics in biology. Not only did Bernard acknowledge the "powerful support of physico-chemical sciences,"[48] but he also proposed time and again the exactness of physics as a goal to be emulated by physiological research.[49] On the other hand, he deplored the rashness of some physicists and chemists who advocated, as Bernard put it, a "false simplicity"[50] by declaring that life phenomena have been completely explained by physics and chemistry. Bernard warned that "in the present state of science it would be impossible to establish any relation between the vital properties of bodies and their chemical composition."[51] By this he meant, and in this respect he was willing to concur with the vitalists, that at present "the manifestations of life cannot be wholly elucidated by the physico-chemical phenomena known in inorganic nature."[52] Still, he was emphatic in considering living organisms as wonderful machines, as complicated physico-chemical systems and was convinced that "every phenomenon called vital today, must sooner or later be reduced to definite properties of organized or organic matter."[53]

All this, however, did not prevent him from stressing the immense complexity of living beings, which, according to him, imposed on physiology an empirical character as contrasted to the exactness of physics. As a result, he advocated the acceptance of anatomical or histological units[54] as the provisionally ultimate or fundamental blocks with which any branch of physiology has to work. These units he called "vital" and considered them to be strictly provisional, because, as he put it, "we call properties vital which we have not yet been able to reduce to physico-chemical terms."[55] It is important to note, however, that in Bernard's view these vital units were not the end products of calculations and measurements. They had to be recognized as such by a mental process in which intuition, imagination, or analogous reasoning played a decisive role. Thus he insisted that these units could not be manipulated and interpreted by the methods of the physicist. Their meaningful handling demanded, according to Bernard, the approach peculiar to a biologist. This is why he could say that while "biology must borrow the experimental method of physico-chemical sciences, [it must] keep its special phenomena and its own laws."[56] In other words, the life sciences had to adopt a methodology which, according to Bernard, should differ markedly in many ways from the methods of physics. For much as Bernard appreciated the contributions of physics to physiology, he had a keen sense of the immensity of problems still unresolved in his field, and it was inconceivable to him that in those problems one could make any headway by leaning on mathematics and physics alone. Owing to his keen awareness of the enormous complexity of life phenomena, Bernard was wary of an extensive application of statistical methods in physiology and pleaded for the application of the simplest instruments of physics in physiological research. On these two points developments hardly followed the pattern laid down by him. Nor was he wholly successful in making clear how he conceived the relation of the physico-chemical determinism and the purposefulness, wholeness, and entelechy that he unreservedly recognized in living beings.[57]

Wholesome and balanced as was Bernard's recognition of the complementary aspects of the method demanded by the problems of physiology, his program could not have been attractive to the older generation of vitalists, who even tried to brush Pasteur aside as a "mere chemist." Many of them were also practicing physicians, and their attitudes hardly seemed to have changed since Laplace proposed their admittance to the Academy as a means of exposing them to the spirit of science. In Laplace's motion, however, there was more than the often vaunted superiority of physics over biology. After all, Laplace was not completely foreign to physiological research, as witnessed by his experiments on energy exchange in digestion. What is more, the nineteenth century saw a number of physicians who left an indelible mark on physics and who would also testify through their own research the important role physics could play in physiological research with no detriment at all to the peculiar characteristics of physiology. Physiological and physical research in fact went hand in hand in the scientific pursuit of T. Young, the founder of the modern wave theory of light. The first to announce the principle of the mechanical equivalence of heat, R. Mayer, was also a physician by training. Speaking of the chemical force (energy) evident in the functions of plants, Mayer emphatically stressed that this energy could not come from vital forces without the expenditures of an equivalent amount of energy that could be measured with exactness. A vital force that created chemical energy out of nothing was, as Mayer put it, an abdication of scientific inquiry and could lead in physiology only to a chaos of unbridled fantasy. Consequently he stated as an axiomatic truth that "during vital processes a conversion only of matter, as well as force, occurs and that creation of either the one or the other never takes place."[58]

The rejection of vitalism on the ground of the conservation of energy was not to be construed, as Mayer took pains to emphasize, as proof of a purely mechanistic interpretation of life. Far from trying to subject physiology to the exclusive domination of quantitative methods, Mayer spoke eloquently of the purpose and beauty that prevailed in the realm of the living. Between animate and inanimate matter, he said, "figures are the boundary mark. In physics numbers are everything; in physiology, they are little; in metaphysics, they are nothing."[59]

When Mayer uttered these words in 1869, sharing the platform with him was none other than Helmholtz, who more than any of his contemporaries was qualified to speak on the role of physics in physiology. One of the undisputed leaders of nineteenth-century physics, Helmholtz was a physiologist by training, and it was from physiology that he derived some of his most influential insights in physics. Dissatisfaction with the vitalistic interpretation of the animal heat provided him with the stimulus that led to the enunciation of the principle of the conservation of energy. Preoccupation with some aspects of hearing prompted him to treat the propagation of sound as the advance of waves in an incompressible fluid. This in turn presented him with the problem of solving certain equations in hydrodynamics, a work that served from the late 1860's on as the foundation of the theory of vortex atom, a most avidly discussed topic in late nineteenth-century physics. Although some of his contemporaries grudgingly noted that he introduced the cosine into physiology, Helmholtz could rightly insist in his address following Mayer's that physiology first had to shake off the yoke of vitalism and adopt the method of physical science before its unimpeded development could be secured.[60] There was hardly an exaggeration in his claim that the ensuing progress that took place in physiology in the last three decades was greater than all the advance during the two preceding centuries. Nor could his contribution to that progress be easily exaggerated. In the study of hearing, vision, and nerve functions, to mention only the principal fields of his physiological work, his investigations were trailblazing. The ophthalmoscope, which he invented in 1851, unfolded an entirely new world. As a result, ophthalmology achieved in a short time a degree of perfection that, in Helmholtz's words, compared favorably with the accuracy of astronomy and stood as a model of exactness for other branches of medicine as well.[61]

The successes of the physical method in physiology by 1869 were almost impossible to count. Physics could time and again cut through the barriers declared by the vitalists as insurmountable, as was, for instance, the case of the measurement by Helmholtz in 1850 of the

velocity of nerve impulses. Still, Helmholtz, who equated an animal
with a steam engine as regards metabolism, was not, however, equat-
ing life with machinery. He was willing to recognize the possible
presence in living beings of factors other than the forces operating in
the inorganic world. What he stipulated was merely that inasmuch
as these factors produced chemical or mechanical effects, "their effects must be ruled by necessity, and must always be the same when
acting under the same conditions; and so there cannot exist any arbitrary choice in the direction of their actions." [62]

This was a carefully worded position on a complicated problem,
and it must be noted that no prominent man of physics in the nineteenth century lent support to the uncritical identification of life with
mechanical processes. As Maxwell warned, the utilization of physics
in biology could be just as misleading as fruitful. He deplored the
manner in which some of the cultivators of biological sciences "have
cut and pared at the facts in order to bring the phenomena within the
range of their dynamics." This, he added, "has tended to throw discredit on all attempts to apply dynamical methods to biology." [63] One
of these discredited attempts concerned the explanation of the origin
of organic life, and on this point a long array of nineteenth-century
physicists stressed the inadequacy of the methods of physics. Among
them were Ampère, Faraday, Helmholtz, Stokes, Lodge, Larmor, and
Poynting, to name a few. Faraday, for example, felt justified in searching for the electrical properties of the nervous system only because
he was convinced that "the direct principle of life" was not touched
upon by such investigations. [64] Kelvin, for one, declared emphatically
that "the origin of life anywhere in the universe seems absolutely to
imply creative power." [65] Maxwell, too, shared this view and called
attention to the fact that the smallest units of life visible under the
microscope can hardly contain more than a million molecules. For
Maxwell, however, such a number seemed inadequate to embody, in
a code as it were, all the structural characteristics and complexities of
a full-grown animal. Much less could he find it reasonable to assume
that the same number of molecules could contain in capsule form all
the varieties, organic and structural, displayed by the members of a
long ancestral tree. Molecular science, as he put it, entailed a rigorous
confrontation of certain physiological puzzles and theories and forbade the physiologist to indulge in speculations along the lines of
biological pangenesis. [66]

Physicists, who always sought to discern in complex reality some
very simple ideal case to make possible the formulation of a law,
could hardly help being taken aback by the immense complexity of
the molecular configuration present, say, in a cell. How indeed could

a late nineteenth-century physicist venture to say anything about the structure of protein molecules, for instance, when such purely physical problems as interatomic or intermolecular forces could only be vaguely speculated on at best. Pasteur's negative conclusions about the possibility of spontaneous generation going on at present were also taken by and large as an indication of an unbridgeable gap between living and non-living matter. The semantics of the evolutionists, so hopelessly imprecise when compared with the equations of physics, did not help matters either. The supporters of the mechanical explanation of life had to fall back on vague expressions and lengthy circumlocutions, which in the last analysis, as Huxley admitted in a rare candid moment, demanded nothing less than "acts of philosophical faith." [67]

Nor could the physicist find much light in the "mechanism" that no mechanistic explanation of life could ultimately avoid: the mechanism of random chance. Not that random processes were simply beyond understanding. They offered indeed some enormous possibilities through statistical methods, as evidenced by the sudden development of the kinetic theory of gases. It is well to remember, however, that behind this sort of statistics there lay a well-defined set of physical properties and laws specifying the motion and interactions of molecules. The chance the evolutionists had to rely upon in the last analysis, however, was a horse of another color. Rather it had no color, or underlying physical specifications, and when fed into the elementary theory of random events, it provided the most potent weapon, for a while at least, against Darwin and his followers. The final factor shaping the growing controversy between physicists and evolutionists was the early exploitation of the Darwinian theory in support of agnostic and materialistic tenets, a circumstance that could hardly have endeared Darwinism to many outstanding men of physics, noted for their deep attachment to Christian beliefs.

These factors had to lead almost inevitably to a conflict that for several decades seemed to pit physics against biological science. For biological thought came to be dominated and reshaped by evolutionary theories, which in turn seemed to stand or fall with a geology assuring a very long time span for the earth. The controversy that ensued is highly instructive, for it brought into sharp focus the point that the same physicists who on the one hand were so eager to recognize the shortcoming of the methods of physics could on the other hand remain strangely insensitive to it. The geological timetable interpreted according to the specifications of organic evolution was clearly based on arguments which although not vitalistic were nevertheless strongly intuitive and had nothing in common with mathematical and dynamical analysis. This intuitive, non-mathematical, at times simply poetic aspect of British geology of the 1830's and 1840's, on which Darwin leaned so heavily, was sharply criticized by Liebig, who in a letter written to Faraday shortly after his visit to England in 1844 deplored the fact that "without a thorough knowledge of physics and chemistry, even without mineralogy, a man may be a great geologist in England."[68]

What Liebig could notice in a few months could hardly be missed two decades later by British physicists who, with Kelvin in the van, tried to discredit Darwinism mainly through their attack on its geological basis. They saw, to recall Stokes's words, a "scientific romancing" in the unquestionable acceptance by many a man of science of Darwin's theory.[69] Although for a while the physicists scored heavily, they failed to realize that in spite of its sadly insufficient explanations on many particular points, Darwinism offered a sweeping unitary vision of the phenomena of life in both space and time. It conjured up the image of an irresistible march of living forms, changing, progressing, and declining, but always blending into a great cosmic unity. Physicists failed to remember that physics, too, had achieved some of its greatest advances by resorting to great visions of simplicity and unity that at the moment of their formulations had hardly an answer to many momentous objections. This was the case of the Copernican revolution, of the various conservation principles, and of the never-ending efforts to find some connection among *all* the forces acting in nature.

Kelvin and his colleagues, when combating Darwinism, saw only the other side of the coin, namely, that it was of the very essence of physics that concepts, theories, and visions must have at least an indirect relation to the world of observation and must be subject to calculations. Setting off theories against experiments and measurements was both the chief pride and safety device of physics and rested on the belief that any physical phenomenon is, in principle, measurable. The unfeasibility of measurement in complicated cases was never attributed to the lack of distinct laws underlying it. To base the origin, the course, the fortunes of the variations of life on blind chance, that is, on randomness operating without any known connection to physical and chemical laws, was utterly repugnant to the spirit of classical physics. This spirit was the omniscient mind described by Laplace as able to calculate any future and past event from the exhaustive knowledge of all parameters specifying the present configuration of things and forces. But even lacking such knowledge, the law of statistical averages indicated clearly enough that randomly moving atoms and molecules would tend toward a state in which particular features, or individual traits, characteristic of the realm of the living, would be helplessly obliterated into undifferentiated lumps of matter. There was complete unanimity among the leading physicists and chemists that the laws of physics in a sense work diametrically opposite to the immense richness of forms displayed by living organisms. Reminiscing on the first few years following the publication of *On the Origin of Species*, Kelvin wrote in the London *Times*, of May 2, 1903: "Forty years ago I asked Liebig walking somewhere in the country, if he believed that the grass and flowers which we saw around us grew by mere chemical forces; he answered, 'No, no more than I could believe that a book of botany describing them grew by mere chemical force.'"[70]

Not that Darwin was eager to base his theory on blind chance, blind laws, and blind forces. He believed that there ought to be a basic biological mechanism governing the hereditary changes, but about its nature he did not conceal his profound ignorance.[71] When, however, the nature of any salient factor eludes a researcher, it is always natural psychologically to emphasize the imperceptibly small rate at which this factor might operate. To let extremely small effects slowly accumulate to an obvious magnitude implied, however, great amounts of time, and Darwin had no choice but to postulate geological time in almost unlimited measure. On this score uniformitarian geology had already established a climate receptive to his contention. Many decades earlier Hutton closed his *Theory of the Earth* by declaring that in geological history "we find no vestige of a beginning — no prospect of an end."[72] One of the reasons Hutton felt so reassured in this respect came precisely from physics, and Playfair, who was so instrumental in promoting the Huttonian theory, took great pains to emphasize the point. He quoted the work of Lagrange and Laplace on the stability of the solar system and found strict parallelism and connection between unchangeable geological and celestial mechanisms. In both, he said, changes are confined within relatively small limits and "in both a provision is made for duration of unlimited extent, and the lapse of time has no effect to wear out or destroy a machine, constructed with so much wisdom. Where the movements are all so perfect, their beginning and end must be alike invisible."[73] Decay and rejuvenation on earth therefore followed one another in interminable sequence and as a result, the succession of animals and plants, too, like the motion of planets, was taken as a process in which "we discern neither a beginning nor an end."[74] Lyell's *Principles of Geology* struck an identical note as a reviewer put it characteristically: "The concession of an unlimited period for the working of the existing powers of nature has permitted us . . . to solve every difficulty in the path of the speculating geologist." [75]

In securing this cure-all for geology, physics appeared a most reliable partner for the time being, and Darwin, in all likelihood, had physics in mind when he wrote to Asa Grey outlining his theory: "We have almost unlimited time; no one but a practical geologist can fully appreciate this." [76] In the same breath he conjured up the millions of millions of generations of shells that might have existed in the glacial period alone. Time was clearly the magic device on which the evolutionist had to fall back of necessity, and Darwin warned in *On the Origin of Species* that anyone who "does not admit how vast have been the past periods of time, may at once close this volume." [77] So confident was Darwin on this score that in the first edition of the work he even tried to calculate the time needed for the denudation of the Weald. The result was 306,662,400 years, which he modestly rounded off to a plain 300 million. [78]

No doubt, it must have been with heavy heart and perplexed mind that Darwin eliminated this little exercise in mathematical physics from later editions. In doing so, he seemed to admit that physics suddenly made chimerical not only his thousand million years, which his theory badly needed, but also the apparently modest requirement of three hundred million years. Physics, the once powerful ally of geological evolution, turned almost overnight into an implacable foe of organic evolution as well, and the vague conjectures of Lyell and Darwin on geological time span had no chance whatever against the strict quantitative formalism of the precise data and conclusions of mathematical physics. Not that Darwin's theory was lacking in experimental or quantitative data. His book was almost bursting with them. But those quantities could hardly be poured into the molds of mathematical functions. To make sense, they needed above all the bold imagination and intuitive generalization of a genius. Darwin's theory was a construct that covered quantities physics could not utilize but could not ignore either without trapping itself in its own vaunted quantitative precision. The chief spokesman of this narrow approach to a complex problem with several "quantitative" features of its own was Kelvin, who, according to his own admission, was, from the very start of his long career as a physicist, keenly aware of the direct bearing of physics on geological theories. For eighteen years, he said in 1862, "it has pressed on my mind" that the contentions of the uniformitarianists went against the principles of thermodynamics. [79] The very same year saw two of Kelvin's three main arguments expounded.

The first of these, entitled "On the Age of the Sun's Heat," called attention to the fact that gravitational contraction, as proposed by Helmholtz, could keep the sun's radiation on the present level for twenty million years only if its temperature remained approximately constant. Kelvin, however, easily showed that the sun's temperature had to be much higher in the past, and this suggested a substantial reduction in the age of the sun as estimated by Helmholtz.[80] But even if the sun had not been warmer in the past, Kelvin deemed it *most probable* that the sun has not illuminated the earth for one hundred million years and *almost certain* that it has not done so for five hundred million years. He stated with *equal certainty* that the inhabitants of the earth shall not have at their disposal the necessary heat for many million years longer "unless sources unknown to us are prepared in the great storehouse of creation."[81] The clause of reservation, however, was pretty much an exercise in rhetoric. Not only did Kelvin not set much store on it, but not even his opponents were willing to seek support in sheer possibilities. What is more, the sun was far away, and physicists alone controlled the narrow bridge to the interpretation of the available experimental data about its properties.

The March issue of *Macmillan's Magazine* that carried Kelvin's foregoing arguments had hardly reached all its subscribers when on April 28 Kelvin read a paper "On the Secular Cooling of the Earth" before the Royal Society of Edinburgh.[82] In it Kelvin exploited Fourier's equation on the flow of heat due to temperature gradient. He "exploited" even more the *value* of the temperature gradient in the uppermost layer of the earth's crust. This value, derived from measurements taken in deep mines, he assumed to be the same at even greater depths. Thus the equation readily yielded the time necessary to reach the presently observed temperature gradient from the solidification of the earth's crust, since the temperature gradient had to decrease with the cooling of the earth. For a melting temperature of 10,000° F, the time calculated was 200 million years, for 7,000° F, 98 million years. True, Kelvin admitted that "very wide limits" must be allowed in this type of estimate. Nevertheless he believed that it could be said "with much probability"[83] that the consolidation of the earth had taken place between 20 million and 400 million years ago. For the moment nothing was asked about the reliability of the calculations squeezed out of the experimental data that covered less than one two-thousandth of a physical parameter, the temperature gradient of the earth's interior. Apparently then as now experimental data were at times too confidently extrapolated and even venerated as sacred cows with hardly any profit for the cause of science.

The evolutionists began to beat a retreat with Kelvin relentlessly in pursuit. In 1865 he carried the battle to the evolutionists' grounds and argued that the principles of uniformitarianism lead to impossible results when taken as a basis upon which to infer heat conditions on the earth millions of years ago.[84] Still, Kelvin was not satisfied. His was a mind bent on routing the evolutionists, and he canvassed constantly the field of physics and astronomy for further arguments. Thus he stumbled on the studies of Adams and Delaunay, who proved that Laplace ignored tidal friction when calculating the value of the secular acceleration of the moon's mean motion. By 1866, together with Adams and Tait, Kelvin concluded a long series of calculations and found that the corresponding retardation in the earth's rotation amounts to twenty-two seconds in a century. Upon this result Kelvin built his third argument, which he presented in a paper entitled "On Geological Time"[85] to the Geological Society of Glasgow, on February 27, 1868. There Kelvin established a connection between the centrifugal force due to rotation and the deformation caused by it on a sphere of liquid. Kelvin's calculations showed that a hundred million years ago the centrifugal force would have been 3 per cent greater than its present value — the higher figure being due to the faster rotation of the earth — and that such was the maximum force permissible to give the present oblateness to the molten mass of earth upon solidification. Kelvin also pointed out that the shape of the earth would have become markedly different had its solidification taken place ten thousand million years ago. Such, however, was the amount of time, or even more, that would have best suited the theories of the evolutionists and their allies among geologists. Consequently Kelvin could state that geology and evolution stood in contradiction to natural philosophy (physics) regardless of whether the twenty-two seconds as calculated will ultimately prove exactly correct or not. His judgment was peremptory: "There cannot be uniformity. The Earth is filled with evidence that it has not been going on for ever in the present state, and that there is a progress of events towards a state infinitely different from the present."[86]

That both geologists and evolutionists reacted vehemently was only natural. The geologists, as one of them put it candidly, had long been accustomed "to deal with time as an infinite quantity at their disposal."[87] As for the evolutionists, Darwin himself referred to Kelvin as "an odious spectre."[88] Darwin was, in his own admission, "greatly troubled at the short duration of the world according to Sir W. Thomson,"[89] and the later editions of the *Origin* show clearly how Darwin was forced into inconsistencies in his desperate efforts to accommodate somewhat his theory to Kelvin's conclusions. The reaction of Huxley, chief spokesman of the evolutionists, was probably the most characteristic. In an address entitled "Geological Reform" (1869), he tried to discover "vagueness and conjecturality" in Kelvin's arguments and insisted that the twenty-two seconds could not be looked upon as an absolutely certain result.[90] Huxley called attention to the fact that mathematics, like a mill, will yield only what is fed into it: "As the grandest mill in the world will not extract wheat flour from peascods, so pages of formulae will not get a definite result from loose data."[91]

Clever as Huxley's footwork was, and meaningful as was his remark on the limitations of the methods of mathematical physics, he obviously tried to avoid the main issue: Was it scientific to uphold rigidly for the time being natural selection as *the* mechanism of evolution, in the face of the arguments offered by physics, although these latter were only very probable at best? Even if Kelvin was guilty of stressing his points too boldly, did this justify Huxley's tactics of *ignoring* physics completely? For this is what he did essentially when he declared in reference to the arguments of the physicists that most of the evolutionists "are Gallios who care for none of those things."[92] In stating this, Huxley merely provided a classic case of one's dismissing as irrelevant the conclusions of one branch of science and sticking to the vision of another with the stubbornness of "scientific faith." This "faith" the advocates of natural selection needed in generous measure, that is, if they wanted to hold their own against the harsh implications of Kelvin's conclusion: "The limitation of geological periods, imposed by physical science, cannot of course, disprove the hypothesis of transmutation of species; but it does seem sufficient to disprove the doctrine that the transmutation has taken place through 'descent with modification by natural selection.'"[93]

For several years the developments brought only support for those views expressed by Kelvin, who followed the question closely even after his retirement. His authority completely dominated the question, and no physicist of any standing saw it attractive to challenge his conclusions, the validity of which seemed to be beyond any doubt. If someone with no established name took exception to the consensus, his contention was readily dismissed as can be seen from the avalanche of letters that appeared in the columns of *Nature* between January and April of 1895. It was started by a communication from J. Perry who claimed that the accepted value of the age of the earth should be increased by a factor of two to six, because Lord Kelvin had mistakenly derived a wrong value for the conductivity of the upper crust of the earth.[94] Perry, who could claim the privately expressed agreement with his views of such people as Fitzgerald, Reynolds, Larmor, Lodge, Heaviside, and some others, submitted his calculations with the utmost diffidence. With a pointed reference to the established consensus, he found it almost impossible to assume that Kelvin could have made an error and that it could have gone undetected for so many years. Again when Tait rebuffed him, Perry confessed to have been "so accustomed to look up to you [Tait] and Lord Kelvin that I think I must be more or less of an idiot to doubt when you and he were so cocksure."[95] They could indeed sound very cocksure at times, and in his next letter to Tait, Perry put his finger to the heart of the matter by asking: "But surely the real question now is not so much what geologists care about as — Had Lord Kelvin a right to fix 10^8 years, or even 4×10^8 years as the greatest possible age of the earth?"[96]

At this point Kelvin could no longer remain silent, and his letter to Perry struck a genuinely authoritarian note. While admitting some merit in Perry's remarks on the constant of conductivity, Kelvin quoted authorities like Helmholtz and Newcomb who, as he put it, "are inexorable in refusing sunlight for more than a score or a very few scores of million years of past time." Lastly, he chided Perry on his taking lightly the argument of the twenty-two seconds. "Don't despise," he instructed Perry, "the secular diminution of the earth's moment of momentum. The thing is too obvious to every one who understands dynamics."[97]

This was a rather high-handed manner and evolutionists lost no time to decry it. Thus A. Geikie, the principal figure in British geology at that time, called the attitude of physicists "insatiable" and "inexorable." They were, in his words, like Lear's remorseless daughters in their relentless drive to bring the age of the earth down to a few million years. The physicists were guilty, as Geikie put it, of ignoring a "multitude of facts which become hopelessly unintelligible unless sufficient time is admitted for the evolution of geological history."[98] This was unquestionably true. But the physicists could have very well said the same of the evolutionists. Actually neither side displayed enough scientific maturity to recognize that far-reaching generalizations should be supplemented with a generous dose of caution and reserve.

For a few more years physicists enjoyed their triumph undisturbed. None of them could suspect in 1895 how precarious it was. In fact, in the very same year Kelvin could refer with no small satisfaction to the recent data of the American physicist. C. King, on the melting temperature of rocks,[99] from which he inferred an age of twenty-five million years for the solid crust of the earth and stated that this result "throws the burden of proof upon those who hold to the vaguely vast age, derived from sedimentary geology."[100] Just as disconcerting for the evolutionists must have been the fact that results coming in from other areas of physics were strangely coinciding with Kelvin's figures. Thus J. Joly estimated in 1899 the "geological age" of the earth's surface at eighty-nine million years, taking this amount of time as equal to the one needed to bring the sodium content of ocean water to its present level.[101] Evolutionists could have found this result all the more disconcerting, as in Joly's theory the effect was attributed to denudation by solution of land surface going on at an essentially *uniform* rate. Once more uniformitarianism appeared to be upset by its own basic principle. That Joly later revised this figure to three hundred and thirty million years could be of little consolation to them. The figure was still badly inadequate for their purposes.

Physics was not the only source of trouble for the evolutionists. Mathematics was equally inimical to their early constructs. Darwin, as already mentioned, made no secret of the fact that nothing was known about the cause or mechanism of the small variations that set apart offspring from parents. It was, however, perfectly clear to him that unless the benefits of such variability could be retained and accumulated through successive generations, they could not play a basic role in the process of evolution. Darwin himself set great store in the effectiveness of selective livestock breeding well in evidence in the middle nineteenth century, particularly in England, but the limitations of such a process were largely ignored by him. Before long, however, in 1867, a Scotch engineer, F. Jenkin, called attention to the major weakness inherent in such an explanation of the mechanism of evolution.[102] For one thing, the possibilities of developing a desirable quality in a particular breed were very limited, and this limit was approached very rapidly. For another, the great majority of the offspring would not show improved qualities but would rather tend to revert toward an average degree of the quality in question in spite of the most careful selective breeding. It then required only simple mathematics for Jenkin to show that in nature where selective breeding was not operative, the individuals with mediocre qualities would so outnumber the few with superior qualities that by the sheer force of great numbers the survival of the race would depend on the mediocre strain alone. It was also pointed out by Jenkin that the chance of survival was clearly dependent on a large number of physical qualities, all of which, or at least the great majority of which, should be concentrated in a few individuals and accumulated in many successive generations to bring about with some plausibility the emergence of a distinctly different strain. A little mutation of only one specific physical quality could secure only an infinitesimally better chance for survival.

It was now clear why the evolutionists needed a tremendous amount of time. For if each favorable physical variation carried forward the process of evolution to only such a small degree, and if a favorable step could be canceled by a slight numerical inferiority, the game of chance had to be given an enormous time to operate through an almost countless sequence of breedings and thereby work its way from the amoeba to the mammals. Truly, such a process takes time beyond measure, and it was natural for Jenkin to recall Kelvin's figures for the age of the earth. But independently of physics he was right to state that "the vague use of an imperfectly understood doctrine of chance has led Darwinian supporters . . . to imagine that a very slight balance in favor of some individual sport must lead to its perpetuation."[103] What Darwin really demanded was nothing less than a chain of "successive creations," and this was diametrically opposite to the main inspiration of *On the Origin of Species*. At any rate, Jenkin struck a seemingly mortal blow at the "mechanism" of natural selection, which Darwin liked to call "the keystone in the arch of evolution." As a result, the sixth edition of the *Origin* acknowledged that Jenkin's remarks could not be disputed. In fact, *the Descent of Man* (1871) no longer viewed natural selection as a major factor in evolutionary process. By a strange twist of history, it was not given to Darwin to learn that experiments and simple mathematics in the hands of Mendel had already provided his vision with a solid foundation. What is more, Mendel could not have been more explicit in emphasizing the crucial role of statistical analysis in arriving at reliable ideas about the appearance of new variations.

Meanwhile, however, Jenkin's point received further corroboration when A. W. Bennett made a highly entertaining mathematical analysis of the impossibility of accounting for evolution by the perpetuation of favorable mutations through chance matings.[104] The force of Bennett's reasoning can be felt only by studying its details, which would take too long to reproduce here. It was in part owing to Jenkin's and Bennett's arguments that Darwin felt utterly crushed and retreated to the Lamarckian position of the inheritance of acquired characteristics. Bennett, an evolutionist himself, was less pessimistic. Instead of trying to rehash the discredited Lamarckism in some disguised form, he gave voice to a robust faith in evolution and rested his hope on some as yet undefinable discoveries that may come in the future. The true mechanism of evolution would then replace, he hoped, the clearly inadequate natural selection both as the initiating and sustaining agent of the evolutionary process.[105] All this, however, was pure speculation. The evidences, or perhaps lack of them, pointed in the opposite direction, and the apparent triumph of physics and mathematics over the Darwinian theory could not fail to strengthen also the view that in nature there is a clear line of demarcation between the living and the non-living and that the range of physics ends where life begins.

The low fortunes of evolutionary theory brought barren years for speculations on spontaneous generation as well. As M. Calvin remarked, there had been no serious discussion of the origin of life between 1870 when A. Winchell published his *Sketches of Creation* and J. B. S. Haldane's study dating from 1928.[106] Others, like M. Florkin, made essentially the same appraisal by assigning the turning point to A. I. Oparin's booklet of 1923.[107] Whatever the true limits of this unproductive epoch, it was not short on fervent declarations about the reality of spontaneous generation. Thus Darwin, who admitted that "it is mere rubbish thinking at present of the origin of life," expressed at the same time his regret of having truckled to public opinion by using the "Pentateuchal term of creation." What he meant by it, he explained, was that life "appeared by some wholly unknown process."[108] To seek refuge in the unknown was not, however, an attitude Darwin could live with long. Intense as his interest was in the problem, he could not get too far. At most, it became evident to him that even "if (and oh! what a big if!) we could conceive in some warm little pond, with all sorts of ammonia and phosphoric salts, light, heat, electricity etc. present," the process of spontaneous generation would still elude us. Owing to the all-pervading presence of living organisms, protein compounds, formed by chance in such a little pond, would be instantly devoured and absorbed.[109]

All of this, of course, was more of a dialectical strategy than an experimental science. The theory of spontaneous generation remained essentially what it had been — a matter of faith — although the reluctant recognition of this was too often couched in scientific phraseology giving the opposite impression. Thus T. H. Huxley, after admitting that the proponents of biogenesis had been victorious all along, claimed that spontaneous generation at some time or other must have occurred because it was a "necessary corollary from Darwin's views if legitimately carried out."[110] All that was necessary, he said, was "to look beyond the abyss of geologically recorded time . . . to be a witness of the evolution of living protoplasms from not living matter."[111] Herbert Spencer for his part put his trust in vague processes and geological periods lost in the immensely distant past, "when the temperature of the Earth's surface was much higher than at present, and other physical conditions were unlike those we know, inorganic matter through successive complications gave origin to organic matter."[112]

Candid admissions about the true status of spontaneous generation were few and far between. Just as rare was the number of those
who could rise to a higher level of scientific sanity as did Pasteur, who
protested against drawing inferences either theistic or materialistic
from his experiments on spontaneous generation. In fact, he warned
both sides in an impassioned discussion of the topic at the March 8,
1875, meeting of the Academy of Medicine in Paris that whatever the
final dictum of science on the matter, "all the worse for those whose
doctrines or systems do not agree with the truth of the facts of Nature."[113] Some monists tried of course to camouflage their materialistic exploitation of the alleged fact of spontaneous generation with a
propaganda long on scientific words but very short on science. Thus
Haeckel boldly informed the German Association in 1877 that once
the chemical components of a cell — carbon, hydrogen, nitrogen, and
sulphur — are properly united, they "produce the soul and body of the
animated world, and suitably nursed, become man." In all this of
course there was only as little of "exact" physical science as there was
logic in his concluding sentence: "With this single argument the
mystery of the universe is explained, the Deity annulled and a new
era of infinite knowledge ushered in."[114]

This was scientific propaganda of rather shady character that had
nothing to do with legitimate assumptions of any sort whatsoever. Of
this the general public remained largely unaware. Sensationalism on
the part of some men of science was no less attractive then than it is
now, and organs of publicity were in general far less receptive to sober
appraisal of facts, such as the one made in 1898 by the prominent
physiologist J. S. Haldane. Reviewing the progress of the previous
fifty years, he warned that science is farther than ever from the
physico-chemical explanation of life. Discoveries, he said, merely
brought out the immense proportions of the problem inherent in the
idea of spontaneous generation, and he felt that "with more refined
methods of physical and chemical investigation they will only appear
more and more difficult to surmount."[115]

To ignore obvious difficulties always takes strong convictions,
which in the case of spontaneous generation were in part derived from
the principle of the constancy of physical laws in general and from the
principle of the conservation of energy in particular. Thus A. Weismann based his belief in the spontaneous generation "as a logical
necessity" on the fact that organic matter could completely be
resolved into inorganic matter.[116] More explicit was the reference
to the laws of physics made by C. von Nägeli, another fervent advocate of spontaneous generation. He went so far as to state that in the
first place the question is not that "of experience and experiment, but

a fact deduced from the law of constancy of matter and force."[117] At most he would have been entitled to say "postulate," "hypothesis," "assumption," and the like. But prejudice took hold of his logic and forced him to write "fact," which in this case lacked only the hallmark by which science recognizes facts: observability.

Statements like Nägeli's were echoed by only a very few physicists, none of whom could be credited with the formulation of any of the great generalizations of nineteenth-century physics. In his highly controversial Belfast address of 1874, Tyndall sided with the cause of spontaneous generation in phrases that proved nothing but which were at least very revealing of the type of reasoning that so readily takes possibilities for facts. For to declare the reality of spontaneous generation, Tyndall had, as he admitted, to "cross the boundary of experimental evidence," to "supplement authoritatively the vision of the eye by the vision of mind," and to appeal to the "unknown powers of matter."[118] But most of the outstanding figures of nineteenth-century physics and chemistry could hardly be sympathetic to "latent powers," "successive complications," and especially to an attitude so ready to minimize the weight of experimental evidence that persistently failed to support these contentions.

Their misgivings were rather understandable. Clearly, an attitude like Tyndall's came dangerously close to the pretentious pose struck by some disreputable popularizers of science. The great German chemist, Justus von Liebig, described them as men who "after some leisurely walk on the periphery of natural science claim for themselves the right to inform the uneducated people and the incredulous masses about the origin of the world and life."[119] Their noisy chatter had more often than not drowned out the voice of the truly qualified spokesmen of physical science, who took, as already mentioned, a rather dim view of the capability of physics in dealing with the origin and evolution of life. For besides the general recognition of the immense difficulties facing a physico-chemical explanation of life, the conclusions of thermodynamics also seemed to raise an impenetrable barrier around the origin of life. To admit spontaneous generation seemed to go against the law of entropy and to entail the possibility of the reversal of life processes. To illustrate what this actually would imply Kelvin described in vivid colors the patently impossible case of boulders recovering from the mud the materials required to rebuild them into their previous jagged forms and reuniting to the mountain peak from which they had formerly broken away. Were it not for thermodynamics, warned Kelvin, "living creatures would go backwards with conscious knowledge of the future, but with no memory of the past and would become again unborn."[120] No wonder that Kelvin termed speculations along these lines "utterly unprofitable." In his view "the real phenomena of life infinitely transcended human science,"[121] and as proof he referred to "mechanical reasoning" (thermodynamics) and geological history, which by establishing a fiery origin for the earth show that all living forms must ultimately owe their origin to the will of a Creator.[122] Yet, insisting as he did in too general terms on the impossibility of understanding "either the beginning or the continuance of life without an overruling, creative power,"[123] he defined too narrowly the ability of physics to illuminate the puzzle of life phenomena: "I need scarcely say that the beginning and the maintenance of life in earth is absolutely and infinitely beyond the range of all sound speculation in dynamical science. The only contribution of dynamics to theoretical biology is absolute negation of automatic commencement or automatic maintenance of life."[124]

Presenting the conclusions of physics too rigidly and setting its limits in a peremptory way proved, however, to be a very precarious policy. The way in which the results of physics were weighted reflected a state of mind that was singularly insensitive to the difference that exists between the solid and conjectural parts in a physical theory. Nor was there sufficient consideration of the extent of what might still be unknown about the physical reality. The second half of the nineteenth century was an age in physics that viewed the advances in physical science as a coherent sequence of triumphs bringing man ever closer to the final, all-embracing scientific synthesis. That discoveries unfolded at the same time unsuspected aspects of physical reality was lost on most men of science who could not perceive the supreme irony in treating the ether as a reality and deriding in the same breath the imponderable fluids of the physics of yesteryear. Even more disastrous for the intellectual climate was the lack of apprehension about the true significance of grave question marks crowding almost every branch of physics. That Kelvin could see only two clouds on the horizon of the physics of the 1890's was one of the most incredible statements ever uttered by a prominent physicist. Beside the difficulties encountered by the principle of equipartition, and the stubborn inability of detecting the ether drift, there were major problems in magnetism, in electrical conductivity, in the behavior of high-vacuum tubes that could have warranted at least a very strong note of caution. As to spectroscopy, which was to yield the clue to the world of the atom, the experimental data were without any explanation at all. The idea that solutions to these problems might come from areas whose existence and nature physics could not even suspect was hardly given a serious thought, which could very well have tempered the cocksure style of the Victorian physicists. True, Kelvin made a few lame allusions to the possibility of a still unknown energy source at work in stars and planets, but neither did he sound convincing nor was he taken so by any of his colleagues. Such reluctant references to the unknown proved, however, a most welcome panacea in the hands of the younger generation of physicists. It was by recalling these references that they tried to soothe the feelings of the old guard that watched with incredulous eyes the new world of radioactivity.

For, as if by magic, the horizons of physics sprang wide open. In the same year when Joly derived from the salinity of the ocean an age of roughly one hundred million years for the earth, T. C. Chamberlin, the geologist, raised the crucial question: "Is present knowledge relative to the behavior of matter under such extraordinary conditions as obtain in the interior of the sun sufficiently exhaustive to warrant the assertion that no unrecognized sources of heat reside there?" As Chamberlin warned, no careful chemist could ignore the fact that the internal constitution of atoms was at that time still an open question for science. Consequently no one could rule out that, as he put it, there might be locked up in the atoms "energies of the first order of magnitude." The contraction theory of the sun's heat, he noted, "takes no cognizance of latent and occluded energies of an atomic or ultra-atomic nature."[125] No remarks could indeed be more prophetic about what was to follow in physics. A year later N. Lockyer speculated on a subatomic change going on in the stars to maintain their internal energy.[126] In 1904 Jeans put forward the idea that in the enormous stellar heat positive hydrogen ions and electrons might coalesce, neutralize one another, go out of existence, and emit thereby an electric wave of enormous power.[127] Such ideas were of course received by and large as vague conjectures, but when Curie and Laborde measured the total rate of heat production from radium,[128] and Rutherford did the same for the radium emanation a little later,[129] it became immediately evident that a heat source of truly unsuspected magnitude is operating in the earth itself and permits a much longer age for the earth as an abode of organic life. To some, like Kelvin, the new results appeared almost embarrassing; to others, positively liberating. In fact, as Rutherford recalled his lecture given on the subject at the Royal Institute in 1904, Kelvin, who was in the audience, cocked a "baleful glance" at him on hearing the new estimate for the age of the earth. At that moment, continues Rutherford's account, "a sudden inspiration came and I said that Lord Kelvin had limited the age of the Earth *provided no new source* [of heat] was discovered. That prophetic utterance refers to what we are considering tonight, radium. Behold! the old boy beamed upon me!"[130]

Clearly the panacea became handy, but if anyone had cause to cheer it was the geologist, the evolutionist, and the biologist. In 1905, Rutherford told them that radioactivity assures them of at least five billion years for the age of the earth.[131] They were also informed that radioactive decay constants furnish a fairly exact method for estimating the age of the oldest rocks on earth and the minimum age of the earth required to account for the present ratio of radioactive elements. Here, too, the trail was blazed by Rutherford who, while still at McGill University, showed a piece of pitchblende to the professor of geology and inquired about the age of the earth. To the answer of one hundred million years, he simply remarked to his astounded colleague: "I know that this piece of pitchblende is 700 million years old."[132]

The basic intuition of geologists and evolutionists about a process a billion years long shaping the organic and inorganic world alike became almost overnight a position no physicist dared to poke fun at any longer. The publication of Joly's monograph *Radioactivity and Geology* in 1909 symbolized indeed that the conflict between physics and evolutionary theory was a thing of the past. What is more, with the aid of various radioactive methods it became possible to chart the chronology of organic evolution with an ever more impressive accuracy. By such methods it was proved recently that advanced forms of life, as clamlike brachiopods, existed 720 million years ago, that is, prior to the beginning of the Cambrian period. About evidences concerning the most primitive forms of life, it is now believed, following the findings of E. S. Barghoorn and others, that they might indicate the presence of biological processes as early as three billion years ago. In connection with the problem of the evolution of man, the potassium-argon method has proved particularly useful, permitting as it does the determination of the age of a fossil at fifty million years with a probable error of 2 per cent. Thus it became possible to show that the *Zinjanthropus*, a possible ancestor of man, lived about 1,750,000 years ago. Yet, what undoubtedly would have most delighted Darwin was the recent experiment based on radioactive tracer methods that showed that the closer two species are on the evolutionary ladder the more readily do their DNA strands combine with one another.[133] Physics indeed has transformed itself from the antagonist of evolution to its most powerful ally.

Actually all branches of physiology began to draw almost overnight immense profits from the revolutionary discoveries of modern physics.

Only a few months after Röntgen first reported his findings in late 1895, the first medical application of X-ray photography had taken place, and as early as September, 1896, Sir Joseph Lister mentioned the possibility of treating internal tissues by X-rays. By 1910 the radium that had already shown its fateful properties on physicists experimenting with it was being used to cure cancerous tissues, symbolizing perhaps the paradoxically opposite roles physics was to play in human life in the coming decades. On the brighter side it is sufficient to recall the widespread use of radioisotopes in medicine, or the possibility of investigating by radioactive tracer methods such so far hidden processes as the metabolism in cells, to illustrate the full engagement of physics in basic biological research.

How intimately physics and biology came to be interconnected is perhaps best demonstrated by the newly developed bio-batteries in which electricity is produced through the catalytic functions of certain bacteria or enzymes from a wide variety of otherwise unusable fuels and oxidants. Among these are such materials as certain types of potatoes and of all things, raw sewage, and substances that are dissolved in limitless quantities in the oceans. These bio-batteries are already generating enough power to operate devices such as signal lights, and it appears to be within the reach of technology to develop bio-batteries powerful enough to propel ships and provide cities with electricity.

But living organisms can do far more than deliver energy to new devices of physics. In the interplay between physics and biology the stage has already been reached where the extremely refined sensory mechanisms of certain animals and plants are now under intensive study to improve man-made mechanisms. Nature, it becomes more and more evident, can produce devices of both extreme sensitivity and complexity. To wrest such secrets from nature is the goal of one of the youngest branches of science, bionics, the name being forged from *bio*logy and electro*nics*. It was found, for instance, that chlorophyll captures light energy by exactly the same layer method by which the recently developed solar batteries operate. Only nature has anticipated man's inventiveness by a billion years or so. Nature was also first, as evidenced by the skill of bats, to exploit supersonic vibrations to measure distances. She also developed perhaps a billion years before man the first effective counter sonar as indicated by the evasive action of certain moths against oncoming bats. Clearly, ingenuity of this kind can but prompt efforts to duplicate it. The study of the eyes of beetles has already been put to use in constructing instruments that measure the ground speed of airplanes at landing. Again, an electronic replica of the frog's eye might provide a system that could distinguish between "meaningful" and "non-meaningful" flying targets and report, for instance, only those aircraft and missiles that are potentially dangerous. Other spectacular performances of nature are still to be duplicated by man. For instance, the heat-detecting devices by which rattlesnakes locate their prey respond to a temperature change of one-thousandth of a degree. A certain tropical fish can detect currents of two one-hundred billionths of an ampere per square centimeter on its body. Such a delicate sensitivity permits the fish to discriminate between glass rods that differ in diameter by less than one-tenth of an inch.

Yet, much as the frog's eye, or that of a beetle, might show the features of a feedback mechanism, they work only as part of an organism that, for all we know, transcends the conceptual framework of physics. Therefore their study and similar studies should keep alive the awareness of the usefulness of methods that are not strictly those of physical science. This is perhaps the most important benefit physics might derive from its growing involvement in the analysis of living organisms. In essence, it is the same benefit that should have been drawn from the fact that not so long ago physicists had to come around to the position of "dreamers" concerning the disputed point of the age of the earth. What happened then should remain a warning to all those who still believe so uncritically in the absolute value and effectiveness of mathematical handling of problems and at the same time so readily overlook the shortcomings of their data and assumptions. The progress of science — physics not being an exception — needed and still needs vision, imagination, and bold unifying ideas as much as it needs the cold edge of numbers, measurements, and mathematical formulas.

In the interplay between physics and biology, the exactness demanded by the physical method should not cease to protect biology and biologists from the failure to distinguish visions and hypotheses from experimentally established facts. This is a highly beneficial effect and is well illustrated in the role played in this century by mathematico-physical considerations in the evaluation of the scientific status of the concept of spontaneous generation. To such a chain of ideas, the road was opened by Boltzmann's interpretation of entropy as a function of probability and permitted for the first time the comparison between living and non-living systems or units on a strictly quantitative basis. While such an approach made it evident that the emergence of life through random chance was not an absolute impossibility, it had to become just as clear that such a prospect was improbable beyond imagination. Thus the application in biology of the exactness of mathematical physics helped once more to bring into sharp focus that it is one thing to have great intuitive ideas and quite another to try to disguise the lack of reliable information by a furtive recourse to the cure-all of random chance.

The gist of the argument was first intimated by Pasteur, who through his research on paratartaric acid was led to notice the characteristic presence of dissymmetric molecules in organic matter. As the years went by, he came to regard this dissymmetry as the only well-marked line of distinction between the chemistry of dead matter and that of living nature. Although he believed that symmetrical molecular configurations could not be transformed into dissymmetrical ones under the impact of symmetrical forces, he took pains to emphasize that such differences between the living and the non-living represented to him a de facto barrier between the two realms and not necessarily an absolute and impassable one.[134]

With Pasteur this interesting topic remained at the stage of relatively simple molecules with mirror images that could not be superimposed on the original. With the development of the kinetic theory of gases, the topic, however, lent itself to a much broader analysis done in a series of papers by the Swiss chemist, Charles E. Guye.[135] He pointed out that at the level of large-scale molecular arrangements a symmetrical configuration means that the different kinds of atoms composing the unit are rather evenly mixed, whereas a dissymmetrical arrangement implies the clustering of a given type of atoms in a particular sector of the whole unit. Different molecules moving at random in a given volume of space would of course tend toward symmetrical arrangements, or even mixing, which is also the distribution approaching equilibrium. As a result, the more dissymmetrical a molecular configuration, the higher is the improbability of its occurring through random chance. As Guye's calculations showed, the measure of improbability is exceedingly high even for rather simplified models of large molecules. In Guye's model of the molecule, only two types of atoms occur, while even in the simplest protein molecule there are at least four kinds of atoms — carbon, hydrogen, nitrogen, and oxygen — and in most cases several others, such as copper and iron. Clearly, the situation for random chance becomes even more desperate when one thinks of giant molecules, the molecular weight of which is not two thousand as is the case with Guye's hypothetical protein, but can be as high as a hundred thousand or a million. An even greater difficulty for random chance is presented by the specific structure of proteins, which are essentially chains of about one hundred, or sometimes two hundred links. The links are amino acids of about twenty different types. The number of different proteins that can in principle be imagined is clearly great enough to surpass even the customary astronomical figures. At the same time, the number of different proteins actually occurring is relatively small. Yet, as J. B. Leathes pointed out in 1926 in his presidential address to the Physiology Section of the British Association, hardly anybody stopped to reflect on this peculiar fact in the excitement following the elucidation of the chemical structure of proteins.[136]

The utter inability of a mechanism based on random chance to produce specifically demanded patterns comes to its peak in the explanation of the reproduction of such giant molecules as the genes. Quantitative evidence of this was provided by the calculations of D. M. Wrinch who investigated the number of possible patterns that may occur under the following conditions: the number of constituents is limited to six types, five-sevenths of them belonging to one type, as is the case with protamine, while the whole protein pattern is extended to a distance of forty microns. The result is $10^{50,000}$, a truly inconceivable figure.[137] Such exponents can speak for themselves, and this is why unqualified references to chance production of life had to be abandoned at least in reputable scientific circles. How indeed could one support a contention about which A. I. Oparin remarked that the spontaneous occurrence of a microorganism is far less probable than, say, the sudden emergence of an entire factory from a heap of inorganic matter through a volcanic eruption.[138]

The utter impossibility of explaining life on the basis of random chance should certainly make one wary of some sophisticated theories that still keep chance the backbone of life-producing mechanisms. Although a table could suddenly rise into air owing to a chance coordination of the motion of the great majority of atoms in it, the occurrence is, however, of such infinitely small probability that it could never serve as the basis for scientific discussion of the motion of the table. This point is still being missed at times, and it is clearly begging the question to say that "since the origin of life belongs in the category of at-least-once phenomena, time is on its side."[139] This is to assume, however, that life *is* a product of chance — which is still to be demonstrated. Nor has the contention been proved that one or two billion years are enough to cope with the compounded improbability of the successive chemical steps implied in the "chance" production of life. Of very dubious merit also is the assertion that in a problem like the spontaneous generation of life "we have no way of assessing the probabilities beforehand, or even of deciding what we mean by a trial."[140] Readily available estimates show the contrary. It is only speculation that can make two billion years the "hero of the plot" of spontaneous generation. The backbone of such an approach is a rather tenuous logic that can categorically assert: "Given so much time, the 'impossible' becomes possible, the possible probable, and the probable virtually certain. One has only to wait: time itself performs the miracles."[141]

A logic that can gallop with such abandon through the immense distance separating the extremely improbable from the virtually certain can achieve only one thing: reintroducing miracles into science through the back door. Such reasoning betrays the same philosophical poverty that one could rightly deplore in the attitude of many a biologist who felt that once the actual occurrence of spontaneous generation had been disproved, they were left with nothing.[142] Even if this were so, should scientific method fall back on the "cure-all" chance? The trouble, to be sure, is not with chance, but with the role assigned to it in particular cases. Not that one should postulate, as Lecomte du Nouy and others did, an external agency to account for the emergence of life in view of the inability of chance to bring about spontaneous generation. But none should either, and certainly not in the name of science, expect chance to do what it cannot, as far as the situation is known today.

One of the signal contributions of modern physics to the problem of life might consist precisely in pointing out the true importance of the role chance can play in various molecular processes. This chance, however, is no longer the know-it-all chance but a chance playing its part within limits established by quantum mechanics, free energy, and entropy. Such a chance is of course of limited power, but at the same time is also a reliable guide. Once the pitfall of omnipotent chance is recognized, the various recent theories on spontaneous generation will clearly display their fundamental shortcomings. Such a distinguished expert on these theories as H. Blum had to admit in 1956 that none of them has yet succeeded in avoiding some sort of circularity, that is, introducing at an intermediate stage a process that can at present only be performed by an already living unit of matter.[143]

Yet, whatever advances have been made in edging closer to the most minute units and processes of life, modern physics shares in them a major credit. Although a century ago leading figures of classical physics saw an impenetrable wall between the living and the non-living, modern physics contributed tools of power and precision that could have defied even the boldest imagination of biologists only two generations ago. With X-rays an access opened to the systematic study of genetic mutations and to the all-important program of mapping the structure of chromosomes and genes. The mechanism of evolution turned out to be a problem in radiation and heat-energy vibrations. Cosmic rays and radioactivity came to be recognized as ever-operating factors in the production of sudden mutations. Furthermore, an analysis of the energy spectrum and density of cosmic radiation and of natural radioactivity and, most of all, a precise knowledge of energies involved in the chance fluctuations of molecular vibrations could yield a quantitatively satisfactory answer to the question why spontaneous mutations are such a relatively rare event.

What is even more important, quantum mechanics proved capable of accounting for the permanency of such changes. This represented a sharp contrast to the inability of classical physics to explain the stability of large molecules in view of the disordering tendency of heat motion. Actually the stability of large molecules posed as recently as forty years ago a seemingly insoluble dilemma for the physicists and compelled them to disclaim any competency in matters of organic life. With the application of quantum mechanics by Heitler and London to the problem of molecular forces, the situation became completely different. There was physics at its best working at the very core of biology. Its conclusions sounded almost embarrassingly quantitative: "We believe," wrote Schrödinger, "a gene, or perhaps the whole chromosome fibre to be an aperiodic solid."[144] This was, however, not to be taken as a relapse into crude mechanistic thoughts. The chromosomes could be taken as "cogs of the organic machine," but none of them was "of coarse human make." They were "the finest masterpiece ever achieved along the lines of the Lord's quantum mechanics,"[145] in which, one may add, there is no place for "machines" in the ordinary sense.

It was modern physics with its X-ray diffraction method that revealed the helical structure of polypeptide chains. The photoelectric effect in its turn unlocked the door to the understanding of photosynthesis in plants. Again, it was certainly not an "experiment in biology" in which S. L. Miller in 1952 exposed water, ammonia, methane, and hydrogen to ultraviolet light produced by electric discharge and found amino acids in the solution at the end of a week. This was a major step in the effort to reproduce artificially essential constituents of the polypeptide chains that in turn form the protein molecule. Today biochemists are confident that they understand the nature of all chemical reactions that play roles in the step-by-step building up of polynucleotides, and W. M. Stanley even felt encouraged to outline the steps through which chemistry might in the foreseeable future synthesize a small polynucleotide possessing genetic continuity.[146] His scheme is not the only ambitious project advanced recently, one of the most highly regarded of which is due to M. Calvin. In this theory the all-important point is centered in the property of certain organic substances that combine by piling up face to face or plane to plane.[147] Such a process, unlike crystallization, does not require a concentrated warm soup but can take place in dilute concentrations as well. It is now believed by some that DNA, the backbone of the genes, is exactly the type of molecule that could form this way and in turn could influence neighboring molecules to unite in exactly the same way.

Valuable as such speculations are, they are still far from solving the puzzle of life. Calvin himself warned that processes like these are not sudden occurrences, and it is doubtful, as he put it, "that we will ever be able to put all the chemicals in a pot and place it in a radiation field, and go away and leave it for a while and come back and find nucleic acids."[148] Although it can be argued that either in the Urey-Miller scheme or in Calvin's theory, single phases of the process are not unlikely to occur in a test tube, it still does not follow that when the time scale is long enough the improbability resulting from compounded probability becomes equivalent to an inevitable process. In spite of all the recent advances in biochemistry, it still reflects beliefs, not facts, to say that life "in all its complexity is nothing more than one of the innumerable properties of the compounds of carbon."[149] It may very well be, but it is still far from being proved, and if anywhere then in scientific parlance, hopes, expectations, hypotheses, and probabilities must not be expressed in terms reserved for well-established facts. It is distressing that only a small minority of those writing today about the physico-chemical explanation of life point out in a straightforward manner the basic inadequacy of all theories offered so far. One of them is N. W. Pirie, very much devoted to the idea of spontaneous generation but who in utter candor quoted H. Belloc to convey his impressions: "Oh! let us never, never doubt what nobody is sure about."[150] This was in 1956. A year later, at the International Symposium on the Origin of Life on Earth, held in Moscow, spontaneous generation was still an enigma. In Pirie's words the symposium "demonstrated that there is no basis for dogmatism about the processes involved in the origin of life." He felt compelled, however, to go even beyond that. "My contention," he stated, "is that there is not yet even a basis for dogmatism about the materials undergoing these processes."[151] This meant that the situation looked more complicated than ever. To illustrate this, Pirie could do no better than to agree with the poet, Louis MacNeice, that more than we would think the world is crazy, sudden, and incorrigibly plural.

It is indeed time to recognize that there is nothing to gain in trying to ignore great difficulties and perplexities by too hopeful statements. It is hardly a sign of healthy intellectual atmosphere that sanguine appraisals of the prospects of producing life in a test tube turn up with rugged regularity. Thus in early 1966 a prominent nuclear physicist predicted the synthesis of life within five years. Some caution would probably have stood him in good stead. Perhaps he should have remembered that in 1958 it was claimed by a high authority in the field that the synthesis of nucleic acids is the remaining essential step in closing the gap between living and non-living matter. Four years later the synthesis was in fact achieved by M. Schramm of the Max Planck Institute for Virus Research in Tübingen. Having exposed simple sugars, amino acids, and nucleotides to moderate heat, pressure, and other influences that probably were present in the primitive ocean, Schramm found in the solution simple nucleic acids. They showed under a powerful electron microscope even the twisted ladder-like structure peculiar to nucleic acids occurring in living matter.

Impressive as such advances are, they have not closed the gap between the living and the non-living. The physicalist claim that life is but the complexity of matter is still a hypothesis and not a demonstrated fact. Curiously enough, the principal frustration of the view that equates life with molecular complexity lies in the successive complexities that biochemical research keeps unfolding in its way of inching closer to the hypothetical unit of life. Perhaps foremost of these baffling complexities is that at a very early stage of the biochemical evolution a reproductive mechanism should emerge to secure the perpetuation of patterns already produced. In other words, certain relatively simple molecular configurations must have the ability to create in their immediate neighborhood a field of force that would group surrounding free molecules into the very same configuration. The gist of the problem is to argue the existence of relatively simple molecules formed by "chance," whose field of force has the mathematically demonstrated ability to catalyze in their immediate vicinity the synthesis of exact replicas. Since Miller's epoch-making experiment, it is evident that such molecules as amino acids can be produced by "chance" under suitable conditions. The further multiplication of amino acids is, however, an entirely different problem.

The first to attempt an explanation of self-reproduction on a quantum mechanical basis was P. Jordan.[152] He referred to the quantum mechanical resonance phenomenon in which the stability of an aggregate of two identical molecules is greater than that of two similar but non-identical molecules. What is supposed to happen is that owing to some special stabilizing force acting between two identical molecules, one molecule creates a field of force around itself that can lead to the production of another exactly similar molecule. It was shown, however, by L. Pauling and M. Delbrück[153] that since the difference between these two degrees of stability is negligibly small, the resonance phenomenon could not be effective enough to account for the reproduction of molecules. They rather argued that the duplication must be a two-step process using a complementary molecule as a "negative" image or template of the first molecule. Pauling also called attention to the complexity of the forces taking part in such a process. Beside the so-called chemical bonds there would be the forces of the Van der Waals attraction, electrostatic attraction between charged groups and the hydrogen bonds. An exact solution of such a complex problem is of course well-nigh impossible.

The real crux of the question seems to be that genes, cells, and units of living matter in general are reproducing themselves with almost absolute reliability and that somehow this lack of malfunctioning must be accounted for to obtain a satisfactory scientific explanation of the origin of life. J. von Neumann, in his work on the construction of self-duplicating machines, admitted[154] that his model cannot be applied to biological systems. Another model of self-reproduction proposed by F. H. C. Crick and J. D. Watson is based on classical concepts, and the possibility of its malfunctioning was not investigated.[155] This is, however, a crucial issue and once again it must be emphasized that chance is wholly powerless to prevent malfunctioning. As W. M. Elsasser warned, this point is of such paramount importance that every theory should be scrutinized with care whether it does not assume implicitly the mistaken principle that "mechanism can generate information out of random elements."[156] An extremely complicated array of information is at work, however, in the reproduction of any living form, and as Elsasser pointed out, several grave difficulties stand in the way of ever imitating the function of chromosomes and genes by an electronic system. While in living units the information is stored in soft tissues and is exposed to a high level of heat-energy disturbance owing in part to metabolism, electronic engineering cannot dispense with hard parts and with a very low level of noise. Again, while the amino acid components of a protein molecule are exchanged within a few months, an electronic system operates on the basis that parts are not continually and automatically exchanged or replaced.[157] Such and similar considerations prompted Elsasser to postulate the presence in the living organism of biotonic laws that cannot be expressed in mechanistic terms and that are the source of the morphological stability of the individual organism and of the species.[158] Structures that produce other structures identical to them appear to be a "miracle from the point of view of the physicist," to quote E. P. Wigner, who studied the problem from the angle of quantum mechanical probability.[159] Starting from the assumption that the laws of reproduction are practically absolute, he concluded that "according to standard quantum mechanical theory, the probability is zero for the existence of self-reproducing states."[160] Although Wigner added with exemplary caution that his arguments claimed to be only suggestive and not conclusive, it seems that physics operates in connection with the problem of life too in its persistently baffling way: solving problems and creating new ones alike.

But independently of physics, life phenomena keep revealing new puzzling aspects and too often at the very moment when final elucidation of a basic problem appears to be at hand. It should suffice here to refer to the rapidly accumulating evidence that the genetic code, paradoxically enough, may not be genetically determined. For contrary to the confident expectations of most geneticists, recent findings indicate that the evolutionary gap between two widely different organisms does not manifest itself in their respective codons, or genetic code-words.[161] Nor does biochemistry have an unequivocal success with its vaunted motto, "DNA is the secret of life." A series of considerations suggest rather that DNA depends essentially on the complexity of the cell and this implies, as B. Commoner put it succinctly, that "biology might be more wisely guided by the aphorism, Life is the secret of DNA."[162]

No less paradoxical should appear another chief problem of the biochemistry of our day, the understanding of how the different RNA's are instructed by the DNA to carry their specific amino acid "bricks" in the proper sequence. The process is in part at least teleological in character and as such is hardly accessible to physical analysis. In essence, the question boils down to how the DNA present in the fertilized egg in a great variety of chemical patterns can determine that the egg would develop into a mouse or an elephant, or taking the case of man, into a Newton or an idiot. What is already known is that each type of DNA produces a specific type of RNA carrying specific instructions to produce one of the thousands of enzymes, which in turn should carry out in proper sequence the production of an immense variety of proteins characteristic of each tissue and organ of all forms of life. In all this an extremely complicated code is implied, the breaking of which, although expected by some to be achieved before the end of 1962, still did not materialize.

Deciphering the whole genetic code presupposes of course the exact mapping of genes in the chromosomes and the position of every atom in the genes themselves. Highly improved forms of the electron microscope might perhaps lead to "seeing" each atom in the gene if its resolving power can really be pushed to distances of two to three Angstrom units. The so-called breaking of the genetic code, however, will not answer what makes a particular cell develop in a specific direction. While all the cells in a human body contain the same hereditary information, the body itself is made up not only of one kind of tissue, say epidermis, but of nerve, blood, bone cells, and so forth. The enormous complexity of the mechanism of such differentiation is only beginning to unfold and will likely prove an even more formidable problem in biochemistry than any hitherto encountered. Actually research has not even reached yet the stage of establishing a vocabulary of differentiation.

Clearly, it is one thing to talk in generalities about biochemical processes in the evolutionary framework and another to ascertain with precision the molecular steps making up the whole process, especially if one considers the biophysical or biochemical stages through which new species appear. The uncounted steps "from atom to Adam" are not as straightforward as it has been claimed recently. Nor does nature comply of necessity with some sweeping predictions based largely on physics and biochemistry. The face of Mars turned out to be shockingly different from the one depicted in the report of April, 1965, of a panel of the National Academy of Sciences on the existence of living organisms on that planet. In general it is difficult to avoid being struck by the difference between the way of thinking of the physicist and that of the biologist appraising the same problem, using the very same tools of research and relying on the same general concept of evolution. For the physicist, the concept of unitary physical laws is paramount, and he is too ready to believe that evolutionary processes follow the same basic lines everywhere in the universe. As F. Hoyle, of steady-state theory fame, recently insisted, life forms existing on the countless planetary systems should look "similar, remarkably similar in their ground plan."[163] It is this type of thinking that underlies the present-day conviction of the possibility of establishing radiocontact with manlike beings on other planetary systems. To what extent such a way of reasoning might differ from the reflections of a biologist can be gathered from the remarks of H. J. Muller, fully aware of the incalculably many turns evolution might take. To suppose, he said, "that humans have evolved there is about as ridiculous as to imagine that they speak English."[164] It seems that the difference between a physicist's and a biologist's appraisal of the same question can easily be as enormous again as in the days of Kelvin.

That such a contrast can arise between the physicist's and the biologist's view is owing in no small measure to the discrepancy that exists between the understanding of physics of biochemical mechanisms on the molecular level and its severe limitations to duplicate them. Understanding of atomic processes, however fundamental, does not necessarily imply the technical ability to manipulate or reproduce them with sufficient control. An example is the process of photosynthesis that has largely been clarified by Calvin's work and about which such details are known that it can be triggered even by electrons as is the case with certain bacteria living in the soil cut off from sunlight and oxygen. Still, trying to copy the intriguing chemical factory of nature, as is photosynthesis, in a perfectly controlled way with the available highly advanced tools of physics would be nothing short of trying to make a modern watch with Stone Age instruments. Similarly, it is one thing to predict confidently that the next century will find mankind able to control the hereditary traits of plants, animals, and humans, and another to cope with the complexities of such a task.

It is also well to ponder that however extraordinary the available tools might be in the future for biochemistry and biophysics, quantitative precision will not fail to turn up deep problems through its sheer use. Furthermore, if the merits of the quantitative method will be appraised without taking into account its limitations, it may hamper again as it did in the not so distant past a fruitful cooperation between physics and biology. The ever closer connection of these two branches of research has in recent decades brought out with particular emphasis the unity of scientific knowledge and was a principal factor in making biology a branch of the "exact" sciences.

Unity of science as evidenced by an ever more "exact" biology should not, however, let one forget the *wholeness* of science which demands an uncompromising respect for every facet of the scientific method. One could indeed never be sufficiently aware of the danger of the temptation of ignoring everything but the quantitative component in scientific research. It is therefore reassuring to see that a good many physicists and biologists are not willing to be stampeded into a wholesale quantization of biology so attractive today in various quarters. Actually even those who plead for an outright reduction of biology to physics admit on occasion that such a task is beyond the reach of the tools and concepts of modern physics. A complete explanation of biology in terms of physics might very well make necessary, as N. Rashevsky noted, the formulation of physical laws that would surpass as much the conceptual framework of quantum physics as this does classical physics.[165] When and how such a new physics will be formulated is, however, anybody's guess and guesswork was never supposed to provide a reliable basis for a serious scientific stance. Furthermore, as noted some time ago by a prominent physicist, it is not impossible that new discoveries in biology would render imperative the reformulation of the laws of physics to the effect that physics, instead of gobbling up life sciences, will find itself reduced to biology.[166]

Apart from these future possibilities, the forced reduction of biology to physics is already producing a biology which, as pointed out quite recently by René Dubos, has almost completely become irrelevant to the understanding of man within the framework of the humanities. The persistent emphasis on the mechanical aspects of man's nature led, in his view, to the gross neglect by biologists of problems that arise from civilized man's responses to his rapidly changing cultural environment. Clearly, the "humanistic biology" for which he pleaded implies more than the analysis of the lifeless bits of man's body with the tools and concepts of physics.[167]

Yet, even if the biologist enchanted by the sweeping generalizations of the methods of physics tends to frown on the humanistic concerns implied in his art, he cannot ignore, in his capacity of pure scientist, that physics is a science of homogenous classes while biology is that of inhomogenous ones. As opposed to physics where one works with items and data that in principle can always be homogenized and their behavior therefore strictly predicted, in biology one has to deal with a universe of inhomogenous classes or units that have features peculiar to them: they develop in time, and by impressing order and organization on their surroundings manifest a distinct individuality. A biological unit is a "repetitive production of ordered heterogeneity," and its proper understanding demands the acceptance of all four of the concepts expressed in this definition.[168]

Thus the thinking of the physicist, as E. Mayr put it some years ago, is bound to remain on the whole "exceedingly different from that of the biologist who works in the evolutionary field." [169] As reasons for this Mayr referred to an almost unlimited number of phenomena that for the purposes of the biologist are to be taken as unique events. It is this uniqueness that is alien to the method of physics and to the thinking of the physicist. The reliance of the physicist on the concept of universal laws predisposes him to appreciate only the typological aspects of phenomena. Furthermore, as another distinguished student of life phenomena, P. B. Medawar, noted some years ago, there is nothing "empirical" about concepts such as heredity, instinct, adaptive activity, organic growth, and the like. It is, however, around such concepts that the findings of biology are organized and consequently biology will represent a conceptual organization different from the conceptual framework of physics. This is why Medawar could characterize talks about the eventual reduction of biology to physics as "the outcome of confused thinking." [170]

In the light of this, one can all the more appreciate the wisdom and insight of such a giant of modern physics as Bohr, who viewed the attempt to quantize everything in biology as a reappearance of the iatrochemists of old disguised as iatroquantists, none the better for science. Assumptions, ideas, and visions, nay, beliefs, of non-quantitative nature can never be supplanted by the wizardry of numbers, however powerful these latter might be. The successive stages of the interaction between physics and biology that has been traced in the preceding pages indicate this lesson all too clearly. Apart from the lessons of history of which one can never be sufficiently aware, an impressive reason to this effect is provided, as Bohr indicated, by atomic physics itself. Observations of atomic structures cannot be done without disturbing the system appreciably, and in the case of a living unit of molecular dimensions, such disturbance is equivalent to its destruction. The portentous consequences of this were formulated by Bohr in a classical passage:

On this view the existence of life must be considered as an elementary fact, that cannot be explained but must be taken as a starting point in biology, in a similar way as the quantum of action, which appears as an irrational element from the point of view of classical mechanical physics, taken together with the existence of the elementary particles, forms the foundation of atomic physics. The asserted impossibility of a physical or chemical explanation of the function peculiar to life would in this sense be analogous to the insufficiency of the mechanical analysis for the understanding of the stability of atoms.[171]

It seems therefore that the very principles operating on the atomic level set a limit to the extent to which physics can penetrate the puzzle of living matter. Furthermore, although an absolutely sharp distinction between animate and inanimate matter cannot be made and although quantum theory proved a marvelous tool in understanding the functions of molecules in living matter, differences still remain, that cannot be ignored. While, for instance, in a crystal one encounters a stability of form, living matter displays a stability of process or function. Owing to such discrepancies, Heisenberg stated, the analogies between molecular mechanism in animate and inanimate matter do "not prove that physics and chemistry will together with the concept of evolution, someday offer a complete description of living organism."[172] In the same vein, Bohr called attention to the fact that while in physics the amount of matter under consideration always remains constant, in the sense that the same atoms and molecules remain in a closed system, in a living unit they are in constant replacement by metabolic processes. Again, while in physics the borderline between areas where classical mechanics and quantum mechanics apply respectively is fairly clearly drawn, in biology this cannot be done with the same accuracy. Also the teleological concepts, clearly alien to the methods of physics, remain indispensable in biology and are no more irrational, as Bohr stressed the point, than the quantum is in physics and might be just as unfathomable.

Impressive as were the advances in biochemistry and biophysics during the following thirty years, Bohr found no reason to part with such conclusions. As he stated it in 1962 in an address delivered not long before his death: "Life will always be a wonder." The only change produced by scientific progress with regard to the puzzle of life concerned, according to him, the balance "between the feeling of wonder and the courage to understand."[173] The range of physics with regard to life phenomena seems therefore to have been set by a variant of the principle of complementarity. On the one hand, terms of physics and chemistry alone can describe unambiguously the results of what is quantitative in biological observations and experiments, whereas on the other hand, the existence of life is to be taken as an irreducible fact. To emphasize this latter point is, however, to repeat merely the wisdom of Aristotle's remark: "It is to live that makes living things to exist."[174] This basic fact of life is hardly a subject for physics, nor is it an easy puzzle for biology, but it certainly ought to be an object of reverence for all.

CHAPTER EIGHT

Physics and Metaphysics

"A METAPHYSICIAN," wrote Maxwell, "is nothing but a physicist disarmed of all his weapons, — a disembodied spirit trying to measure distances in terms of his own cubit, to form a chronology in which intervals of time are measured by the number of thoughts which they include, and to evolve a standard pound out of his own self-consciousness."[1] These are not flattering words, to be sure, but neither are they the most acid remarks penned by physicists about the metaphysician and his trade. Denunciation of metaphysics is, in fact, one of the threads that has run unbroken through the history of physics since its spectacular upsurge in the seventeenth century. As physics advanced by leaps and bounds, the distaste of physicists for anything savoring of metaphysics also grew stronger. The process was mirrored even in the changes in semantics. "Natural philosophy" came to be viewed more and more as the only respectable branch of philosophy and before long physicists found themselves talking about their own trade simply as "philosophy." Galileo was already eager to secure for himself the title of "philosopher," in addition to that of "mathematician" at the Tuscan court, for as he put it in referring to his own accomplishments: "I may claim to have studied more years in philosophy than months in pure mathematics."[2] By philosophy he meant physics, of course. In the same way, Newton, disgusted by the criticism provoked by his early investigations on light, wrote to Oldenburg: "I intend to be no further solicitous about matters of Philosophy."[3] He, too, meant physics and nothing else. Like most other men of science of his time, Newton was hardly concerned that the rest of philosophy, in particular the search for causes in terms of the nature of things, was becoming more and more identified with the obscure domain of occult qualities. The style of some truly great books of post-Newtonian science well illustrates this process. Thus Lavoisier's *Elements of Chemistry*, in which the ill-fated chemist tried to put chemistry on a Newtonian basis, contrasts what is scientific, or well demonstrated, to the metaphysical.[4]

The disdain of metaphysics was just as obvious and widespread among physicists when classical physics reached its zenith in the latter half of the nineteenth century. A few details will suffice to throw sharp light on what could be described as the typical state of mind of physicists two or three generations ago. One of Kelvin's favorite classroom asides was that "mathematics is the only true metaphysics,"[5] and he readily classified anything as "metaphysics" that did not fit promptly into the patterns of classical mechanics. In this respect his attitude had not changed during a very long career. While he grudgingly admitted in 1860 that "the mystery which hangs around the ultimate nature of matter . . . is a question that must be answered by the metaphysician," he reserved, in the same breath, "facts and phenomena" to the domain of natural philosophy.[6] In doing so, he only provided an unwitting instance of the often repeated pattern that dismissing metaphysics will lead to dismissing facts. For late in his career, when physics had to face almost unbelievable facts, Kelvin actually came close to dismissing them. Thus, when Balfour gave an account of the revolutionary aspects of radioactivity at the 1904 meeting of the British Association, Kelvin's comment was: "Arthur Balfour is just in the arms of metaphysics."[7] Such reaction was, of course, to some extent in keeping with Kelvin's temperament, but even Helmholtz' more reserved attitude could turn into pugnacity when it came to metaphysics. Helmholtz heaped praises on Faraday — whether rightly or wrongly, is of no concern here — for striving "to express in his new conceptions only facts." This Helmholtz considered real progress, "destined to purify science from the last remnants of metaphysics."[8] Even in the 1870's, when German *Naturphilosophie* and Hegelianism were largely out of fashion, Helmholtz still poured scorn and irony on metaphysics. According to his capsule formulation, metaphysics stood in its relation to philosophy as astrology to astronomy.[9] No more flattering for metaphysics was the verdict of the mathematician, C. S. Peirce, according to whom "metaphysics has always been the ape of mathematics."[10]

In coining desultory aphorisms on philosophy in general and metaphysics in particular, physicists merely imitated some philosophers. To give only one example, Kant, in *The Only Possible Ground for a Demonstration of God's Existence* (1762), spoke of metaphysics as a "bottomless abyss" and a "dark ocean without shore and without lighthouses."[11] It was this state of affairs with metaphysics that he tried to remedy in his own way. Yet, he succeeded only in throwing out the baby with the bath water, as later developments were to show. By turning metaphysics into a handmaid of science, he effectively prepared the way for those scientist-philosophers and philosophers of science (these two categories do not always coincide) who in the latter half of the nineteenth century were busy pruning physics of its last, hidden metaphysical ingredients. They proceeded about this task with a rather uncommon devotion. To see this, one need only open Mach's famous analysis of the history of mechanics. There the very first sentence of the preface declares boldly that the chief characteristic of the work is "enlightening, or to put it more clearly, anti-metaphysical."[12] The antithesis between enlightening and metaphysical is worth noting if one wants to feel the breadth and depth of mistrust and misconception of metaphysics that could be entertained without a second thought. Mach's English counterpart, K. Pearson, struck exactly the same tone in *The Grammar of Science*. According to him, the difference between science and metaphysics consists in the fact that "the laws of logical thought" are valid in the former and do not obtain in the latter.[13] Consequently, "we must conclude," advised Pearson, "that metaphysics are built either on air or quicksands — either they start from no foundation in facts at all, or the superstructure has been raised before a basis has been found in the accurate classification of facts."[14] This is not the place to ponder the "scientific" merits of such evaluations of metaphysics. Their undisciplined wording should not fail, however, to illustrate the color of the glasses through which the closing decades of classical physics looked at philosophy in general and at metaphysics in particular.

One might expect, of course, that the demise of classical physics as the last word in the physical explanation might have made physicists more aware not only of the shortcomings of a rigid positivistic interpretation of the scientific method, but also of the clearly metaphysical nature of certain problems that kept emerging as twentieth-century physics forged ahead. This indeed took place to some extent, but metaphysics to many physicists still remained something that could be dismissed without first being carefully weighed. Thus the word metaphysics was used on occasion by Einstein as synonymous with "empty talk,"[15] although he had indeed a metaphysical bent of mind. Einstein, of course, was not alone among the greats of modern physics who occasionally poked fun at metaphysics. Rutherford, for one, grandly rejected the prompting of the metaphysician, Samuel Alexander, to look into the philosophical foundations of science. "Well, what have you been talking all your life, Alexander? Just hot air! Nothing but hot air." Characteristic words indeed for a great experimental physicist who was not reluctant to tell a joke or two even about theoretical physicists. No less suggestive, however, is the remark of C. P. Snow, who added to his account of the story, that "scientists have sneaking sympathy for Rutherford."[16] They have a great deal indeed and are not shy about revealing it. One comes across this deep-seated distrust of metaphysics in the writings of contemporary physicists at the most unexpected moments. Thus Max Born's *Atomic Physics* concludes with a contrast between the world of experience and that of philosophy. The former, he states, is "wide, rich enough in changing hues and patterns," the latter is the realm of the "dry tracts of metaphysics."[17] Dirac also made the point that physicists are quite happy if they know how to calculate results and compare them with experiments, that is, if they can handle what he calls the class two difficulties of quantum mechanics. Physicists, he says, do not worry about the class one difficulties, which include questions like: how can one form a consistent picture behind the rules that govern quantum theory?[18]

If prominent figures of modern physics evaluate the respective merits of physics and philosophy in such a manner, what will the scientist of lesser stature and vision do? Obviously much the same, but with one difference. He will use an even more unrestrained style in echoing his masters. One such example is worth quoting:

The truth to a scientist is not the vague metaphysical concept about which philosophers talk and write so much and know so little. To him truth is that body of statements and conclusions about any set of features and phenomena in nature which represents most accurately all the best observations he can make and which conform most closely to all findings in adjacent or related phases and fields of investigations. He may and does often wonder what the so-called ultimate truth may be, but he does not worry about it. He knows that a priori pure thinking will never reveal it so far as knowledge of the physical universe is concerned, and that observation and deduction alone in the manner of science will ever do it.[19]

This strange mixture of truths, half-truths, and illusions shows, however, only one thing: a quarrel with metaphysics usually goes hand in hand with the erroneous belief that science consists only of observations and deductions.[20]

There are many reasons for this deeply ingrained distaste of physicists for metaphysics. One is the heritage of the violent reaction of nineteenth-century physics against Schelling, Hegel, and their followers, who tried to derive the whole of physics not from observations and measurements, but from their own fancies, which they identified with metaphysics. Their opposition to research based on carefully planned experiments could hardly be more obstinate and this physicists were eager to recall. Thus Planck quoted as a horrible aberration that according to Schelling "the method of experimental physics is worthless and futile, false in principle and is an eternal and unsurmountable source of error."[21] A position like Schelling's of course evoked a most passionate rebuttal by the physicists. Speaking of the mathematical incompetence of Schelling, Hegel, and their followers, Gauss could not help asking in his letter to the astronomer, Schumacher: "Don't they make your hair stand on end with their definitions?"[22] Clearly, the theories of German idealism made any reasonable dialogue between science and philosophy outright impossible, but there were even more deplorable aspects to this strained situation. Not rarely the devotees of Naturphilosophie did their best to make life difficult for the scientists given to experimental research, as shown by various incidents in Ohm's life, for instance. One can well imagine that the reaction of the men of science did not limit itself to the abstract academic level, and the result was a clash where passions often got the best of reason. The complete mutual alienation was best summed up in 1862 by Helmholtz:

The philosophers accused the scientific men of narrowness; the scientific men retorted that the philosophers were crazy. And so it came about that men of science began to lay some stress on the banishment of all philosophic influences from their work; while some of them, including men of the greatest acuteness, went so far as to condemn philosophy altogether, not merely as useless but as mischievous dreaming. Thus it must be confessed, not only were the illegitimate pretensions of the Hegelian system to subordinate to itself all other studies rejected, but no regard was paid to the rightful claims of philosophy, that is, the criterion of the sources of cognition, and the definition of the functions of the intellect.[23]

Perhaps the majority of scientists did not become "enemies" of philosophy, but as Helmholtz himself admitted in a letter of 1857, the greater part of them grew indifferent about philosophical matters, and Helmholtz laid the blame for this on Schelling and Hegel, "two writers who have been taken to represent all philosophy."[24] It was to this state of affairs that Maxwell referred when he spoke with biting sarcasm of "the den of the metaphysician, strewed with the remains of former explorers, and abhorred by every man of science."[25]

Not only could the physicists be rightly indignant of the bunglings of the philosophers in physical science, but they could also hardly remain unaware of the long chain of prestigious philosophical systems discredited in part or in whole by the discoveries of physics. The devastating arguments brought by physics against many conclusions and assumptions of the Aristotelian philosophy in general and cosmology in particular are too well known to review here. Descartes' system, which tried to replace that of Aristotle, could not cope with the advances of physics and had to yield eventually to Newtonian conceptions even in France. In general, philosophers, eager to borrow "definitive" data from physics, had bad luck oftener than not. Physics was no respecter of philosophers and its "definitive" data changed at times with bewildering rapidity. When Berkeley argued against the Newtonian concept of space and tried to connect his theory of relative motion with the supposition of fixed stars in his *De motu*,[26] Halley had already showed rather plausibly that certain stars, Sirius, Arcturus, and Aldebaran, move relative to the solar system. Kant, whose great ambition was to write a philosophy sanctioning the unconditional validity of Newtonian physics, was no more fortunate either, although he did not live long enough to see the handwriting on the wall. His table of categories, tailored to fit exactly the contours and structure of the Newtonian universe, began to lose its firm moorings in physics with the construction of non-Euclidean geometries. As Gauss himself pointed out in 1844, Kant's distinction "between analytic and synthetic propositions is one of those things that either run out on triviality or are false."[27] The list of philosophical statements discredited by the discoveries of physics is long indeed and shows a pattern that was best summed up by Maxwell: "Taking metaphysicians singly, we find again that as is their physics, so is their metaphysics."[28] As can be seen from the admission of the distinguished philosopher and historian of philosophy, E. Gilson, things had not changed in the hundred or so years following Maxwell: "All philosophers perish by their science."[29] A philosopher could indeed have hardly more completely concurred with a physicist. In fact, the last hundred years witnessed at an accelerated rate the iconoclastic effect of physics on philosophy, due mainly to the ever more rapid changes operating within physics proper. It was precisely in this latter circumstance that Eddington found an excuse for physicists who are unconcerned about the philosophical implications of their theories and discoveries.[30]

That future discoveries and attainments should make short shrift of many a speculation is of course not a weakness peculiar to philosophy, nor is it necessarily an indication of lack of talent on the part of the philosopher. It is, however, less pardonable when a philosopher who attempts to codify a specific type or phase of physics into philosophy is found wanting in understanding the physics and mathematics of his time. Unfortunately such has been the majority of cases ever since the birth of the new physics in the seventeenth century. Hobbes, for instance, boldly claimed that every branch of study — scientific, moral, and political — derives from the concept of mechanical motion. He asserted a continuity between physics and psychology and flatly stated that ratiocination means computation. Yet, his many mistakes in mathematics provided more than enough material for the *Elenchus geometriae hobbinae* composed by Wallis, the outstanding mathematician of the time. Clearly, to utilize judiciously the data of science, it was not enough to be a friend of Mersenne and to hear from Galileo's lips that the concepts of body and motion alone yield an adequate explanation of the universe. Not that Hobbes by knowing more science would likely have noticed the fallacy in the exclusive intelligibility attributed to the mechanistic world view. Yet, greater familiarity with science might have inspired in him more appreciation for the role of experiments in science and in all probability would have kept him from ending up in doubt about the applicability of scientific propositions to reality.

Far more moderate and cautious than Hobbes, Locke had an eye for the enormous deficiencies of the science of his day. He was aware of the fact that the sciences of his age were "ignorant of the several powers, efficacies and ways of operation whereby the effects which we daily see are produced."[31] The magnitude of this ignorance led him to question whether physics would ever reach the stage of definitive conclusions, or what he called the "scientifical knowledge."[32] Still, by fashioning metaphysics on physics (D'Alembert and Voltaire considered him the Newton of metaphysics and Benjamin Franklin called him the Newton of the microcosmos), he added the decisive impetus to British empiricism that step by step pushed man and the universe farther and farther apart. Thus the trend that claimed to be a quest for reality had to culminate on a note of despair as to what reality was. By applying relentlessly the dissecting mechanistic and atomistic patterns of physics to the processes of human cogitation, Hume was forced to postulate a *belief* to justify the continuing existence of bodies; as to the explanation of the perceiving and thinking mind, he had to claim finally "the privilege of a skeptic."[33] In doing so Hume claimed of course to be on genuinely scientific grounds. Yet, the paradoxical thing was that the men of eighteenth-century science were kept from such extremes by the mental demands of scientific work. For how could it make sense to observe, measure, and compute if the relation of such work to reality had to be considered highly questionable if not meaningless? One may indeed speak of an existentialist impact of scientific research on the mind of the scientist, and it seems that it is this influence that keeps men of science from accompanying certain "scientific" philosophers on the road to wholesale doubting. The divergence between philosophical convictions animating scientific research and the analysis of the conclusions of science by empiricist philosophers not versed in scientific work could hardly have been lost on the classical physicist, nor could it have endeared philosophy to him.

The attempt of idealist philosophy to revitalize metaphysics on the basis of the Newtonian physics was no more successful either. Of the difficulty of the task he had set for himself, Kant was fully aware. "Metaphysics," he wrote early in his career, "is without doubt the most difficult of all human studies." This difficulty lay, according to him, in the fact that metaphysics had to be created anew, for as he viewed all earlier attempts, "No metaphysics has as yet been written." [34] In stating this, he had in mind not only the scholastics but most of all the empiricist school. The latter, strangely enough, would have concurred with his contention that the "genuine method of metaphysics is fundamentally of the same kind as that which Newton introduced into natural science and which was there so fruitful." [35] Little did Kant suspect how fatal such an assumption was for metaphysics. Metaphysics was far more than an abstract mirror image of the concepts of quantitative sciences; yet what Kant wanted to secure for metaphysics was, as he put it, "standard weight and measure to distinguish sound knowledge from shallow talk." Thus, in spite of all his labors and genius, Kant could in a sense only perpetuate what he deplored in the metaphysics he so severely criticized: "Its followers having melted away, we do not find men confident of their ability to shine in other sciences venturing their reputation here, where everybody, however ignorant in other matters, may deliver a final verdict." [36] For it was Kant who inspired ultimately that era of "final verdicts" in metaphysics that wanted to legislate for physics as well. Yet, apart from the preposterous contentions of Schelling and Hegel in matters of physics, how could a metaphysics secure the respect of the physicist when it was physics that undermined in the long run the "scientific" foundation of the idealistic metaphysics?

Nor could much trust be generated for philosophy in the physicist when he saw how the same physical theory could be eagerly seized upon as confirmation of philosophical systems that had little or nothing in common. In times past, Newtonian gravitation was claimed as proof by so divergent trends of thought as neoplatonism and relationalism. In our times, Einstein's theory of relativity has been described as an operational system by the positivists and as an idealistic construct by the exponents of the neo-Kantian school. Physicists could also see that philosophers were quick to endow their great generalizations with a metaphysical halo, which in a brief period of time often turned into a garment of ridicule. One of the laws of physics that fascinated philosophers excessively was the principle of conservation of energy. In it H. Taine found the "immutable ground of being" and the "permanent substance" of the universe.[37] For Nietzsche it served as the main source of inspiration to describe the universe as a "metallic quantity of force" in his posthumous *Wille zur Macht*.[38]

A few years later, however, when the enormous energies carried away by alpha, beta, and gamma particles from the interior of the atom took the world of science by surprise, some philosophers were again quick to draw far-reaching conclusions. G. Le Bon, for instance, viewed the momentary inability of physicists to verify the law of conservation of energy in radioactive processes as the wholesale debacle of "sacred dogmas" of classical physics. To him this meant that "nothing in the world is eternal," and everything, including this "great divinity of science" (the law of the conservation of energy), must submit "to that invariable cycle which rules things — birth, growth, decline and death."[39] Le Bon's handling of the philosophical implications of major discoveries in physics is highly characteristic on several counts. His two works, *L'Evolution de la matière* (1905) and *L'Evolution des forces* (1907),[40] show a thinker familiar with much of the physical research of his time; but they also show him incapable of spotting several of the truly seminal aspects of current research. Foremost of all, they show him as a philosopher eager to draw something definitive from still highly fluid theories and findings.

Le Bon's familiarity with physics was no doubt far above that of most philosophers who discussed physics at that time. In fact, he could refer with pride to the ten years of experimental research that had preceded the main conclusions reached in his book of 1905, namely, that matter is not indestructible. He often refers to the works of Rutherford, J. J. Thomson, Becquerel, Poincaré, and many other prominent contemporary physicists but at the same time belittles the bearing of radioactivity on the electron theory of matter. In Le Bon's words: "From all matter we can extract electricity and heat; but there is no more reason to say that matter is composed of particles of electricity than to assert that it is composed of particles of heat."[41] He saw in the electron theory only a lure of the "old craving for simplicity with which nature is doubtless not acquainted."[42] Le Bon speaks a great deal of the ether, but keeps silent on the problems posed by the failure of all experiments designed to detect motion relative to the ether. He also seems to be unaware of the problems posed by the black-body radiation. What he says of spectroscopy again shows the same curious insensitivity to what is really important in current research. A similar, almost self-defeating ambivalence characterizes his appraisal of the great changes that were going on in physics in the first years of this century. In view of the drastic transformations in scientific knowledge, he admits that "it now appears pretty clearly that we know very little of the universe,"[43] and again, that "we now feel ourselves surrounded by gigantic forces of which we can only get a glimpse, and which obey laws unknown to us."[44] Still, he enumerates eight guiding principles, the validity of which, according to him, cannot be questioned. Actually these principles betray a strange mixture of correct and incorrect guesses, and above all, exhibit the luckless philosopher who trusted too much the finality and apparent irreducibility of certain phenomena discovered by physics and his own incautious interpretation of them.

The first of these principles states that "matter, hitherto deemed indestructible, slowly vanishes" due to radioactivity. The rest can be summarized as follows. The product of the "vanishing of the matter," or the dematerialization of matter, is a substance occupying an intermediate place between ordinary matter and the ether. The process of the dematerialization shows that matter is far from being inert and that it is an enormous reservoir of energy. From this interatomic energy derive not only solar and stellar energy, but also electricity and heat and most of the forces in the universe. (How electricity derives from inter-atomic energy, Le Bon does not specify, of course.) He states the identity of force and matter, or rather the existence of matter, under two forms: stable (ordinary matter) and unstable (energy, heat, electricity, and the like). The continuous transformation of matter into energy argues further the instability and evolution of all chemical elements and finally the most fundamental precept, namely, that "energy is no more indestructible than the matter from which it emanates."[45]

It is easy to see how far some of these statements were from the actual thinking of most of the physicists. Hardly any of them was willing to consider seriously the possibility of giving up the principle of the conservation of energy, and this deep-seated unwillingness was in fact the reason that led, decades later, to the assumption of the existence of the neutrino. Also the way Le Bon speaks of the transformation of matter into energy shows the striking departure from the simple, technical expression used by physicists stating their great findings. The concluding words of Einstein's classic paper of 1905 might serve as an example: "The mass of a body is a measure of its energy content."[46] Einstein's paper also could have shown Le Bon that the equivalence of matter and energy was derived from the principle of the conservation of momentum, regardless of the reference system in which the momentum of the body is measured. Le Bon also could have easily learned that the transformation of matter into energy proved rather than disproved for any sound physicist the principle of the conservation of energy, which Le Bon classed quickly with other "dethroned dogmas" of classical physics. But even apart from Einstein's paper, there was nothing in the papers of Kaufmann and Abraham on the increase of mass with velocity to suggest that the principle of the conservation of energy was to be abandoned. A conclusion like the one reached by Le Bon was indeed much more the result of rash thinking than the fruit of painstaking, cautious information.

In general philosophers who wanted to find confirmation in modern physics of one or several of their cherished tenets fared as poorly as Le Bon. To the monists, for instance, the transformation of mass into energy apparently came as a handy principle and they tried to make the most of it. As early as 1906 a German exposition of the monistic world view declared: "The mass of a body is synonymous with its energy."[47] This was not physics of course but philosophy, and a very dubious one at that. The monists, however, were not alone in their inept exploitations of modern physics. Indeed there was hardly a branch of twentieth-century philosophers that failed to seek justification in one way or another in the findings and theories of modern physics. Their successes, if any, were ephemeral. This is, however, hardly surprising. Long gone is the day when a Newton could assure a Bentley or a Locke that both the scientific and the philosophical content of physics could be mastered without mathematics. No philosopher can any longer become "Master of all the Physics" by following the procedure of Locke, who contented himself with the assurance of Huygens that the mathematics of Newton's *Principia* is reliable indeed. While Locke could become in Newton's estimate, as Desaguliers informs us, a "Newtonian philosopher without the help of geometry,"[48] no one can repeat today this performance and become a philosopher à la Heisenberg or à la Lee and Yang, without first mastering advanced quantum mechanics. Jeans was very much to the point when he stated that "no one except a mathematician need ever hope fully to understand those branches of science which try to unravel the fundamental nature of the universe — the theory of relativity, the theory of quanta and the wave mechanics."[49]

One should note, however, that even in the eighteenth century it was already risky for a philosopher lacking a thorough training in physics to speak about physics or utilize for philosophy some of the findings of physics. Kant's case shows this all too well. In fact, it was precisely his self-assuredness about his scientific competence that proved to be his undoing. When he tried to atone in his *Critique of*

Pure Reason for the wild conjectures of his youthful work, the *Universal Natural History and Theory of Heavens,* he was led by a desire to do better justice to Newtonian physics, which he believed himself to have finally fully understood. He was hardly up to it.[50] In his *Metaphysical Foundations of Natural Science,* he succeeded only in petrifying the basic concepts of Newtonian physics into hard-set, immutable categories for which no less a physicist than Helmholtz criticized him severely a hundred years later, in his address of 1870 "On the Origin and Significance of Geometrical Axioms."[51]

The persistent mishandling of the results of physics was not the only symptom that could strike a physicist taking a long look at philosophy and philosophers. Just as appalling for him was the contrast between the almost complete disagreement among the numerous schools of philosophy and the apparently firm consensus to which men of science are led, at least as their research progresses. Such evaluation of the respective merits of physics and philosophy was no doubt a view to which most physicists were all too ready to subscribe. Galileo had already contrasted the humanities, "wherein there is neither truth nor falsehood," to the "true and necessary" conclusions of natural science.[52] After Newton's death, this contention became a climate of opinion, which received its literary seal in Voltaire's *Micromégas* (1752). This short essay, or rather science fiction, culminates in a discussion between Micromégas, an eight-league-tall giant from one of the planets of Sirius and a group of philosophers from the earth. As one of the latter put it succinctly: "We are in accord on two or three points which we understand, and we disagree on two or three thousand which we don't understand."[53] Accordingly, when the giant asked them about the distance of the moon from the earth, or about the angular separation of certain stars, the earthlings answered in unison. Micromégas could not keep himself from commenting: "Since you know so well what is outside you, no doubt you know even better what is inside you. Tell me therefore what is your soul and how do you form your ideas."[54] Again the earthlings spoke simultaneously, but not in unison. No two of them had the same answer. Philosophy could hardly have been belittled in a more entertaining fashion.

A brilliant piece of scientific propaganda, the *Micromégas* shared, however, the basic fault of all such writings: it permitted only a one-way vision. Viewed from the opposite direction, the physicists looked no better than the philosophers. Almost exactly two hundred years later, when it had already become abundantly clear that at times even the physicists are hard pressed to sound in unison, Gilson wrote with a balance Voltaire never had: "Nothing equals the ignorance of modern philosophers in matters of science, except the ignorance of modern scientists in matters of philosophy."[55] What Gilson said of the moderns, however, applies to the earlier generations of scientists as well. Yet, ignorance is just the surface phenomenon. Below it there is the often frustrating limitation of the human mind, which only very rarely unites in one and the same person what Pascal once called the *esprit de finesse* and the *esprit géométrique*. Leibniz too noted the fact that "most of those who like mathematics have no taste for metaphysical reflections and find light in the one and darkness in the other."[56] Voltaire, who in spite of all his pretensions about understanding Newtonian physics remained a humanist at heart, also remarked that "mathematical studies leave the intellect as they find it."[57] Yet, in an age that resounded with statements alleging the total subordination of metaphysics to physics, one could hardly expect people like Voltaire or D'Alembert to include metaphysics among those occupations of the human mind that they tried to save from being gobbled up by scientific preoccupations. The most they did along these lines was to plead for the cultivation of literature. D'Alembert, for one, scorned those who wanted to make a choice between Newton and Corneille.[58] Of philosophy proper, D'Alembert was far less appreciative. He stated in fact that the "nature of motion is a riddle only for the philosopher"[59] and dismissed the metaphysical analysis of the concepts of space and time as absolutely irrelevant and useless for the purposes of natural science.[60] He was therefore hardly consistent though right in noting the curious phenomenon that great geometers (physicists) are far from being excellent metaphysicians, although one could naturally expect them to be so, at least in connection with concepts they are constantly dealing with. As D'Alembert put it:

The logic of some of them is confined to their formulae and does not extend beyond. They can be likened to a man who has the sense of sight contrary to that of touch, or in whom the latter of these senses can be improved only at the expense of the former. These bad metaphysicians in a science in which it is so easy not to be so, would infallibly be much worse, as experience proves, in matters in which they will not have the calculus for a guide. Thus the geometry which measures the bodies, can serve in certain cases to measure the minds as well.[61]

This is indeed a harsh indictment, coming as it does from no mean geometer, and applies even to Sir Isaac himself. Genius as he was in mathematics and physics, as a philosopher he was "uncritical, sketchy, inconsistent, even second-rate."[62] So goes the rather reliable verdict of E. A. Burtt, and it is no more flattering to hear its sequel that concerns the philosophical judiciousness of Newton's followers. Believing that Newton freed them from the shackles of metaphysics, they became the unsuspecting captives of a very definite, though at times almost naïve, metaphysics implicit in Newton's physics itself. What is worse, they could hardly care less. Apparently mechanics could make its progress without getting bogged down in analyzing the concept of space, a question which with other similar topics was declared unimportant in the *Encyclopédie* of Diderot and D'Alembert.[63]

That such a solution was far too simple came to be recognized, though unwittingly, by D'Alembert himself. For it was such a simplistic approach to intricate puzzles that led to the situation that D'Alembert could not ignore. That in his day certain persons tried to erect a nebulous philosophy on the concept of infinitesimals was in D'Alembert's estimate due to the fact that the great inventors of calculus neglected to clear up the difficulties in the basic concepts of the calculus. His comment could not be more dejected: "After having misused the method of geometry in metaphysics, it remained only to misuse metaphysics in geometry, and this is what they are doing."[64]

What seems to be most reprehensible, or unscientific, in such a flippant attitude toward matters of philosophy is that it accepts unfamiliarity with the history of a specific topic as a scholarly attitude. For even if one despairs of finding any solution to a metaphysical problem, knowing the history of the meanings and explanations of a term remains a *conditio sine qua non* for anyone, even the best of "geometers," to claim an air of reasonableness for his clearly metaphysical statements. Apologies, however sincere, will not supplement the lack of information. One can perhaps feel sympathy for Riemann, who asked in his famous paper of 1854 for "indulgent criticism" for not being "practised in such undertakings of a philosophical nature where the difficulty lies more in the notions themselves than in the construction."[65] Yet, this does not change the fact that the fundamental problems of physics will always raise metaphysical questions. Actually, not even the positivists' disclaimers of all metaphysics could so far dispense with postulates that are genuinely "beyond" physics. Thus even the most radical positivist method in physics will have to suppose the universe to be such as to fit the method and vice versa. Otherwise such physics will not be about the physical universe. Or, as Einstein put it, "every true theorist is a kind of tamed metaphysicist, no matter how pure a 'positivist' he may fancy himself."[66] What is more, very often the same reasoning that claims the elimination of all metaphysics serves as the foundation of a new but rather spurious metaphysics. True, in much of his detail work the physicist can ignore metaphysics without his results being any the worse. When it comes to the basic questions of physical theory, however, even the physicist forms no exception to the rule phrased by Burtt with unique forcefulness: "The only way to avoid becoming a metaphysician is to say nothing."[67] And if the testimony of a professional philosopher, such as Burtt, is not above suspicion for the average scientist permeated by antimetaphysical prejudices, one can recall what Whitehead said in rebuttal of the concept of science proposed once by J. S. Mill: "Once you tamper with your basic concepts, philosophy is merely the marshalling of one main source of evidence and cannot be neglected."[68]

If this is so even with physics proper, it is all the more indispensable that the physicist take a serious stab at philosophy before discussing the philosophical implications of physics. Humble protestations about one's insufficiency in philosophical matters does not remedy one's incompetence. Nor can a respectable measure of competence be gained by taking, as Jeans did, "the proverbially advantageous position of the mere onlooker."[69] Viewed at close range, or from a comfortable distance, a play will still reveal vastly different things to a competent critic than to a chance theatergoer. The fact remains that all Jeans wrote about the impact of physics on philosophy, or about philosophy as such, bears out only what he himself admitted: "I am not a philosopher either by training or by inclination,"[70] or again, "My acquaintance with philosophy is simply that of an intruder."[71] To claim that a non-technical account of the deeper layers of physics is of necessity imprecise shows only that inexact language can never be the carrier of exact thought. Clearly, when a physicist engages in such a venture he still cannot abdicate his obligation of making clear to the reader as much as possible the measure of inexactness present in his account. Such a duty is, however, almost impossible to discharge when one wades deep in the waters of philosophy without serious preparation. Instead of a clarification of ideas, the result will be, as Whitehead once put it, the reaffirmation "of chance philosophic prejudices imbibed from a nurse or schoolmaster or current modes of expression."[72]

Such a state of affairs will only perpetuate the validity of the proverbial saying, which Einstein himself once quoted approvingly: "The man of science is a poor philosopher."[73] Furthermore, the gap between physics and philosophy will remain wide, which as De Broglie admitted, has been "harmful to both philosophers and scientists."[74] There are a number of reasons why this is so, and all of them point to a fact of which both philosophers and physicists begin to grow more aware, namely, that the nature of both disciplines is such that they cannot ignore one another's work any longer. The physicist has had to realize that the positivist program was an illusion as regards the origin of basic concepts and assumptions with which physics has to work. No one voiced the recognition of this fact more forcefully than Einstein. It was indeed the leading idea of all his reflections connected with the method and nature of theoretical physics. "The axiomatic basis of theoretical physics cannot be extracted from experience but must be freely invented," he stated, and he considered "doomed to failure" all attempts to derive them from elementary experiments.[75] According to him, "the basic philosophical error of so many investigators in the nineteenth century" consisted precisely in not recognizing the fact that "there is no inductive method which could lead to the fundamental concepts of physics."[76]

To enable the physicist to reach such concepts, Einstein urged nothing less than complete freedom from the epistemological prohibitions legislated by positivism. Such freedom of thought he defended from the charge of being "fanciful" and urged the physicist "to give free rein to his fancy, for there is no other way to the goal."[77] This was not only a far cry from Comte's or Mach's precepts, but was evidently a procedure that had on it the hallmark of genuine metaphysics. From a positivist point of view, however, it could be nothing short of a blind jump into the darkness. To justify such a step, there was only one expedient to resort to, which could be designated by that most unscientific word in the eyes of Mach and his followers: faith. Faith in the intelligibility of nature, faith in the possibility of reaching the outer world with our concepts, faith in the existence of a universe structured according to definite patterns and revealing orderly arrangement in its entirety and it its details. As Einstein expressed it: "Belief in an external world, independent of the perceiving subject, is the basis of all natural science."[78] To him the nature of such faith was so genuinely metaphysical that he did not refrain from putting it in the sphere of religious beliefs. As he emphatically argued the point, the man of science needed no less than "profound faith" to secure for himself the assurance that "the regulations valid for the world of existence are rational, that is, comprehensible to reason." As a matter of fact, a scientist without that faith was simply beyond his comprehension. Clearly, such a disclosure reflected convictions the depth of which can better be gauged in that aphorism that put the crowning touch on his consideration of the subject: "Science without religion is lame, religion without science is blind."[79]

This clarion call for the recognition of faith as a major foundation of scientific work could come as a shock only to those swayed by an extreme form of positivism. For, contrary to claims of some positivist historians and philosophers of science, from Comte down to our days, this faith has always been a mainspring of scientific work. It was a firm trust in something that clearly was not physical but metaphysical that conjured up for Copernicus an entirely new vision of the universe. Indeed, three generations later Galileo could still find no words of praise for Copernicus who, as Galileo put it, believed so much in the mathematical simplicity of the universe that he was willing rather to commit rape on his senses.[80] That such testimonials to the role of faith in science erupt most spontaneously in times when science is in the throes of a rebirth is only natural. At such a stage of the course of science, the experimental evidence is still too sparse to hide the true nature and origin of basic assumptions. Later, when both experimental evidence and familiarity with principles increases, the temptation to regard the principles as results of observations becomes almost irresistible. With a sharp turn in the course of science, such illusion is, however, quickly dispelled and the recognition of the role of faith again comes to the forefront. In a sense, the more important one's contribution in revolutionizing scientific thought, the firmer is one's appreciation of this "scientific faith." Planck in particular esteemed it dearly: "Science demands also the believing spirit. Anybody who has been seriously engaged in scientific work of any kind realizes that over the entrance to the gates of the temple of science are written the words: *Ye must have faith.* It is a quality which the scientist cannot dispense with."[81]

Heisenberg, whose indeterminacy principle was no less a break with patterns of the past than Planck's quantum theory, also found it important to stress the role of this philosophical faith in scentific work. "What is and always has been our mainspring is faith. . . . I believe in order that I may act; I act in order that I may understand. This saying is relevant not only to the first great voyages, but to the whole of Western science."[82] The types of metaphysics, implicit or explicit, which this faith bespeaks, vary of course from physicist to physicist. Nevertheless, some common element is clearly noticeable. This holds true even of persons like H. Weyl, who, as he expressed it, became more hesitant as the years passed about the metaphysical implications of science. Yet, he readily acknowledged that "science would perish without a supporting transcendental faith in truth and reality."[83] In a similar vein, two prominent experts on the theory of the nucleus, R. G. Sachs and E. P. Wigner, saw in the marvelous correspondence between intricate mathematical laws and the facts of nature a factor which alone can provide the indispensable encouragement to the physicist's labors.[84] That such a factor is rooted in metaphysics was unwittingly acknowledged by them when they called the recognition of this marvelous correspondence "an article of faith" for the theoretical physicist.

A metaphysician can register satisfaction that physicists, long strangers to and distrustful of metaphysics, have to fall back on terms like article of faith, belief, and the like, when trying to account for the intelligibility of nature. In doing so physicists bear witness to the indispensability of a metaphysics that steers clear of the shallows of both naïve realism and of dreamy idealism. For this is precisely the point intimated in a passage of Einstein and Infeld, who emphasized that "without the belief that it is possible to grasp the reality with our theoretical constructions, without the belief in the inner harmony of our world, there could be no science. This belief is and always will remain the fundamental motive for all scientific creation."[85]

It is not difficult to single out some of the reasons why statements of this kind appear in increasing numbers in the writings of physicists of our times. As physics forges ahead, the physical world not only reveals more and more of its grandiose unity of plan, but it also gives us a closer glimpse of the staggering dimensions of its complexities. Past successes of science, of course, give strong support to the view that the human intellect will be able to cope with these complexities. Still, men of science feel today that in addition to this they also need the support of an intuitive assurance of the intelligibility of nature, if they are to carry on confidently with their often frustrating search for the simple laws lying behind enormous complexities. Such a feeling is especially noticeable today among those in the forefront of physical research. "We cannot make much progress," said one of them — Oppenheimer — in discussing the prospects of fundamental particle research, "without a faith that in this bewildering field of human experience . . . there is a unique and necessary order."[86] Next to the complexity of nature, it is the increasing awareness of the revisability of the laws of physics that prompts references on the part of the physicist to the need of faith in the meaningfulness of scientific research. For a typical illustration of this, one need only read H. Margenau's scientific creed composed of six articles of faith. Of these, four assert the confidence of the scientist in his endeavor, while admitting that science is a never-ending quest and that its particular laws are forever subject to revision.[87]

Clearly, the faith needed by the scientist as such is a most serious, nay, sacred matter. As K. Lonsdale, a leading expert on crystal structure, put it a few years ago: "The scientist, as well as the man of religion lives by faith and not by certainty."[88] There are of course physicists who speak, write, and work without paying any attention to the metaphysical roots of their methods and basic concepts. Dame

Lonsdale had them in mind when she noted in the same context: "It is so easy for a scientist to become narrow-minded and even to forget some of the basic assumptions upon which his own science is founded: to forget indeed that there are basic assumptions." If such a warning appears too pedagogical to antiphilosophical men of science they might well recall what the logical outcome is when someone like Russell takes the view that "order, unity and continuity are human inventions just as truly as are catalogues and encyclopaedias." For the one who said that "of unity, however vague, however tenuous," he can see "no evidence in modern science considered as a metaphysic" had also to say, to remain consistent, that the "universe is all spots and jumps, without unity, without continuity, without coherence or orderliness or any of the other properties that governesses love." [89] Yet, different as the world of governesses and that of scientists may be, the work of both of them becomes hopeless if certain basic assumptions about the realm of reality are creations of the imagination.

The principal one of these basic assumptions is the intelligibility of nature. If this is mentioned by scientists only on occasion, it is only because we don't talk too much about the air we breathe, unless the situation becomes critical. There are other assumptions, too, often and explicitly used by physicists, without being any the less metaphysical for that. The first to mention is simplicity. It is a principle particularly effective in giving assurance to the physicist about the correctness of his theories and conclusions. As Galileo noted in a moving passage of the *Dialogue*, it was in the "wonderful simplicity" of his bold new vision of the arrangement of planets that Copernicus "found peace of mind." [90] The importance that Newton accorded to the principle of simplicity can best be seen from the fact that it is the first of the four rules of reasoning that introduce the third book of the *Principia*. Moreover, the rule explicitly refers to the universal agreement on this point among those investigating nature. Countless indeed are the passages in which one or several references to the simplicity of nature, which "does nothing in vain," occur in the writings of Descartes, Galileo, Boyle, Huygens, and Leibniz. Simplicity was a hallmark of scientific truth in Laplace's time as well. In his eyes, Huygens' law of double refraction was raised to the status of rigorous law following its derivation from the simplicity of the law of action. Previous to that, Laplace contended, Huygens' law was no more than "the result of observation approximating the truth within the limits of error to which the most exact experiments are subject." [91]

Of Ohm's law Maxwell spoke in a similar vein. His report of 1876 to the British Association on the experimental testing of Ohm's law is revealing on several counts. It shows the classical physicist to whom a law, however accurate, is only "experimental" until it has been "deduced from the fundamental principles of dynamics." But at the same time, it also shows the physicist, classical or not, to whom the accuracy of the law, verified over a greater range than ever before, proves that "the simplicity of an empirical law may be an argument for its exactness even when we are not able to show that the law is a consequence of elementary dynamical principles."[92] Dalton's atomic theory also reflected an effort toward achieving the greatest possible simplicity in explanation. Gay-Lussac, too, was led by the conviction that gases always combine in the simplest proportions by volume. So was Avogadro's rule also a simplicity principle. When Gibbs was awarded the Rumford Medal for his work "On the Equilibrium of Heterogenous Substances," he wrote, no doubt with awareness of his own deepest motivation in scientific work: "One of the principal objects of theoretical research in any department of knowledge is to find the point of view from which the subject appears in its greatest simplicity."[93]

Convictions of the value of the principle of simplicity only became strengthened when in the course of scientific development a particular pattern of simplicity had to yield to a more inclusive one. It was Fresnel's awareness of this pattern in scientific history that gave him confidence in the correctness of his wave theory of diffraction. In his famous memoir of 1819, Fresnel explicitly referred to the repeated triumphs of simple assumptions in the field of classical mechanics. Furthermore, Fresnel accorded the principle of simplicity a genuinely metaphysical sense. That the mathematics of the corpuscular theory was far simpler he readily admitted. But he warned in the same breath that it is not the simplicity of mathematical formalism that should have the final word when a choice among various hypotheses was to be made. The decision, he insisted, should rest rather on the conceptual simplicity. Therefore a hypothesis aiming at acceptance should respect not so much the simplicity of mathematics but rather the simplicity of nature, which, as he noted, was not conditioned on mathematical difficulties. Nature always achieved a maximum of effects with a minimum of causes, and therefore a hypothesis based on a minimum of assumptions was, in Fresnel's eyes, always the most productive step toward sighting the true simplicity of nature.[94]

Faith in the simplicity of nature was not weakened when toward the end of the nineteenth century it became evident that there was no simple criterion as to what is the simplest of the various "simple" laws of physics. Physicists agreed, as H. Hertz noted, that in physics all phenomena should be reduced to the "simple laws of mechanics." This consensus rested on the assumption that the forces evidenced by those laws are "of simple nature and possess simple properties." Yet, Hertz himself had to acknowledge that "we have here no certainty as to what is simple and permissible and what is not: it is just here that we no longer find any general agreement."[95] As he knew only too well, to many of his colleagues the removal of this uncertainty seemed to be utterly superfluous if not meaningless. Physics, according to them, had long since renounced such metaphysical preoccupations. To Hertz, however, such a position was wholly gratuitous and superficial. "A doubt," he warned, "which makes an impression on our mind cannot be removed by calling it metaphysical."[96] The metaphysical overtones of the principle of simplicity were just as clearly voiced by Einstein. He not only rejected the positivist allegation which branded this principle a "miracle creed" but also pointed to its intimate connection with what he called a "tamed" metaphysical interpretation of our comprehension of nature.[97]

The principle of simplicity assumes of course that an all-pervading uniformity underlies the manifold complexity of natural phenomena as they appear to the ordinary observer. That atoms of the same substance or element should be absolutely alike was an article of faith for physicists long before physics could learn anything concrete about the atoms. In the same way, the principle of uniformity constituted the backbone of the most seminal speculations about the large-scale patterns of the universe in the era that antedated the telescopes of Herschel. The assumption that the universe should look the same to an observer situated anywhere in space was the idea that led Kant to his conception of the universe as a system of galaxies.[98] Clustering of stars into galaxies supposed, however, a regularly repeated pattern in the scattering of stars across space. This was again a subtle variation of the principle of uniformity. Or as the originator of the "grindstone theory" of the Milky Way, Thomas Wright, put it, one had to assume "all the Stars scattered promiscuously, but at such an adjusted Distance from one another, as to fill up the whole Medium with a kind of regular Irregularity of Objects."[99] When a generation or so later Herschel decided to see for himself the true pattern of the scattering of the stars, he was led by an assumption identical to that of the "regular irregularity" of Wright. To Herschel, the principle also involved a loosely uniform pattern for the size of the stars. He postulated, in a picturesque style that could not conceal the metaphysical vein, that the distribution between fixed limits in size, which is evident in the case of oak trees and men, should be analogously valid for the stars as well.[100] Herschel's use of the analogous application of the uniformity principle was far from being the last of such instances in the speculations of physicists. Uniformity of physical laws in all reference systems is the basic postulate of the Special Theory of Relativity. Uniformity in space is pivotal in all world models based on the General Theory of Relativity. Again, uniformity in time is one of the leading ideas of the steady-state theory. The uniformity assumed in such contexts is of course a large-scale one. In this regard, there is no difference between the "regular irregularity" proposed by Thomas Wright and the little "wrinkles" created by the presence of material bodies in the space-time curvature of General Relativity. In both, the intuitive element appears to be the same and also the metaphysical component it contains.

Not a bit different is the case of the principle of symmetry. There is enough asymmetry in the world, especially in the microcosmos, to make one hesitate about accepting symmetry as a fundamental pattern in nature. Still, as a guiding principle in physical research, it indeed worked wonders. The conviction that action between different forces of nature must be reciprocal, that is, symmetrical, led to the discovery of magnetic induction. Faraday's research was in fact based on his holding "an opinion almost amounting to conviction"[101] that since electricity can produce magnetism, magnetism too must be able to produce electricity. In connection with his research in thermo-electricity he jotted in his notebook a remark that conveys no less forcefully the same conviction: "*Surely* the converse of thermo-electricity *ought to be obtained* experimentally. Pass current through a circuit of antimony and bismuth, or through the compound instrument of Melloni"[102] (italics mine).

In modern physics also, the role played by symmetry is nothing short of spectacular. Einstein's formulation of his Principle of Relativity was motivated by his effort to remove the asymmetries to which Maxwell's electrodynamics — as understood at that time — led when applied to moving bodies. Dissatisfaction with the lack of symmetry in the relativistic form of Schrödinger's equation led Dirac to a solution that contained not only mathematical symmetry but also the first indications of the greatest discovery made on the basis of symmetry in the history of science: the world of antimatter. Dirac's symmetrical equation was the starting point of an experimental work unique in the history of physics. It stimulated the construction of the huge bevatrons and cosmotrons, the principal objective of which was the detection of various antiparticles. It regaled physics not only with a puzzling set of units of matter but also with the concepts of the thus far hypothetical antiatoms, antistars and antigalaxies, all composed of antimatter, ruled by antigravitation and perhaps by a backward running time. No wonder then that physicists clung fondly to a principle having such effectiveness and were ready, after the overthrow

of parity in weak interactions, to search for a law in which the lack of
parity in weak interactions fits into a more fundamental law of symmetry.

Followers of Mach would indeed only be puzzled by an attitude
that is clearly inspired in this instance not by "facts" (for the facts
supporting this deeper law of symmetry are still being looked for)
but rather by what Mach decried as "mystical inclinations."[103] By
this expression he meant metaphysical "vagaries," and he was determined to show their fallacy even in the particular case of the symmetry principle. He alluded to the well-known fact that dynamic
equilibrium can be upset even in the perfect symmetrical configuration that obtains when a magnetic needle lies parallel both to the
magnetic meridian and a wire conducting a current. As Mach claimed,
this could have served as a shock to the "believers" in symmetry, but
Mach himself would have been shocked even more had he realized
that the symmetry principle holds even in this case. One need only
assume, as Weyl pointed out "that a reflection with respect to the
plane in which current and needle lie, maps the current into itself,
but interchanges the north and south poles of the magnet."[104] This is
a perfectly legitimate procedure from the point of view of physics,
since positive and negative magnetism are of the same nature and can
only appear together.

It is of entirely secondary importance, however, whether in a particular problem of physics the assumption of symmetry, or uniformity,
as applied in a given configuration of data, leads to a result or not.
The point at issue is that physicists cannot dispense with these assumptions. The well-known fact that the seventeenth-century giants
of physics believed in the law of the inverse square for gravitation,
and in several other laws, long before they had observational evidence
for them, cannot be explained away by remarking that they were still
far away from the golden age of "positivist maturity." Belief in metaphysical principles is still indispensable in scientific work whether
some physicists like it or not, or whether they admit it or not, or
whether they are fond of the word metaphysical or not. It is the new
return to such assumptions that keeps physics moving and enables the
physicist to look at old facts in a new way. At times the newness thus
achieved is bewildering, as was the case with ideas propounded by
Copernicus or Einstein. Yet, neither of them had new facts at his disposal. They both had to look at old facts from a radically new angle,
or in other words, they had to rely on a mental process which could
appear utterly irrational to some antiphilosophical theoreticians of
science. Their precepts as to how the scientific mind does or should

work stand at complete variance with the analysis of the genesis of scientific laws as given by no less a man of science than Planck:

The man who handles a bulk of results obtained from an experimental process must have an imaginative picture of the law he is pursuing. He must embody this in an imaginary hypothesis. The reasoning faculties alone will not help him forward a step, for no order can emerge from that chaos of elements unless there is the constructive quality of mind which builds up the order by a process of elimination and choice. Again and again the imaginary plan on which one attempts to build up that order breaks down and then we must try another. This imaginative vision and faith in the ultimate success are indispensable. The pure rationalist has no place here.[105]

The pure rationalist, the extreme positivist, the antiphilosophical physicist are clearly at a loss when coming face to face with the task of accounting for the constant use of assumptions, the nature of which they try to ignore. Yet, without recourse to philosophy, one is indeed hard put to find justification for the fact without which physicists simply cannot do their work: the unavoidable necessity of translating mathematical formalism into a conceptual system based on sense perception. The problem is extremely acute today as physics is more and more exposed to a one-sided trend of abstraction. In this trend, as the Nobel-laureate physicist, H. Yukawa, noted a few years ago, theoretical physics is being reduced to the mathematics of complex functions of complex variables. "The probability amplitude," he charged, "which is a complex number and which does not correspond directly to any intuitive image, has become something like the almighty dollar."[106] What Yukawa called an intuitive image has of course a close connection with the realm of commonsense thinking. It is to the vigorous reliance on this intuitive imagination that Yukawa ascribed the brilliance of early twentieth-century physics and its great power of prediction. Neglect of this factor led, according to him, to the inability of mid-twentieth-century physicists to handle the richness of subatomic physics.

Acute as the problem of relating some details of mathematical physics to the physical world is today, it is not a new problem for physicists. A hundred or so years ago, Faraday, who in spite of his total lack of expertise in higher mathematics could grow into a giant of physics, discussed it most instructively in a letter to Maxwell. There he complained of physicists who failed to translate "their hieroglyphics," as he referred to their mathematics, into common language. Interestingly enough, he found that Maxwell, perhaps the most mathematically minded among Faraday's younger colleagues, always succeeded in giving him a "perfectly clear idea of conclusions, which though they may give me no full understanding of the steps of your process, gave me the results neither above nor below the truth, and so clear in character that I can think and work from them."[107]

Maxwell responded in a most sympathetic way. Years later, in his article on Faraday in the *Encyclopaedia Britannica*, he even poked fun at those mathematicians "who have rejected Faraday's method of stating his law as unworthy of the precision of their science." For all their mathematics, Maxwell noted, they "have never succeeded in devising any essentially different formula which shall fully express the phenomena without introducing hypotheses about the mutual actions of things which have no physical existence, such as elements of currents which flow out of nothing, then along a wire, and finally sink into nothing again." By contrast, Faraday, with no recourse to mathematics, came up with one of the most seminal laws in physics. For Maxwell this was not without the deepest significance, because Faraday's original statement was a model of perfection. It remains to this day, wrote Maxwell, "the only one which asserts no more than can be verified by experiment, and the only one by which the theory of phenomena can be expressed in a manner which is exactly and numerically accurate, and at the same time within the range of elementary methods of explanation."[108] There is indeed much food for thought in the fact that Maxwell deemed it very fortunate for physics that Faraday was not a "professed mathematician" and was thus left at leisure to "coordinate his ideas with his facts and to express them in natural, untechnical language."[109] Helmholtz, too, found it "astonishing in the highest degree" how Faraday could find a "large number of general theorems . . . by a kind of intuition, with the security of instinct, without the help of a single mathematical formula."[110]

What Faraday so excelled in was far more than some subtle ability of visualizing theories. It was more recondite and fundamental than that. He could indeed "smell the truth," as Kohlrausch once put it,[111] for his was an intuition that, on the one hand, gave him a sixth sense about physical reality, and on the other hand, kept him convinced that physics ultimately is and must be about reality, however abstract the mathematical formalism might appear at times. This is why he insisted that physicists, whatever their mathematical prowess, should try their best to establish ties between mathematical formalism and physical reality. This program, so indispensable if physics is to retain its claims of being about *physis*, nature, involves reasoning and judgments which are distinctly *meta*-physical, implying, as they ought to do, the recognition of things, entities, on which relations, mathematical or otherwise, can be predicated.

Faraday's apprehension was prophetic. In an age of excessive model-making, concerned so much with *things* in physics, some mathematical physicists perhaps felt justified in considering Faraday's request too pedestrian. Before long, however, the course of physics took a turn that made it impossible to ignore the bearing on physics of problems implicit in the ever-widening gap between mathematical formalism and the concepts needed to translate the reality handled by mathematical symbols into the ordinary spoken word, the ultimate vehicle of understanding. The difficulty experienced by late nineteenth-century physicists in paraphrasing Maxwell's equations was merely a foretaste of what was to come. If Boltzmann could already claim at the turn of the century that "the time has now come for the alliance" between physics and philosophy,[112] the situation of physics a decade or two later made the need for this all too obvious. In 1920, Eddington felt impelled to write that "the mind is not content to leave scientific Truth in a dry husk of mathematical symbols" and that the mathematician "who handles x so lightly, may fairly be asked to state . . . the meaning which x conveys to *him*."[113] Yet, all the questions raised by the various x's, y's, and z's in mathematical physics were only a mild anticipation of the puzzles occasioned five years later by the appearance of the famous ψ in Schrödinger's equation. By that time, physics saw introduced quantum "jumps," in which there was no jumping but merely the endpoints of a jump, and a tantalizing series of other paradoxes that were on the borderline of being self-contradictory.

Physicists began to speak of the curvature of space, insisting in the same breath that not only is space *not* curved, but that there is no such "thing" as space. Later physicists announced a long series of "fundamental" particles, which not only failed to prove themselves "fundamental," but could hardly be called "particles" in the common acceptance of the word. They rather gave the impression of being mere mental constructs, or a "state of affairs," defined by the Schrödinger or Dirac wave equations. Most interpreters agreed wholly with the definition of a "particle" given by Eddington: "By a particle I mean not a classical particle, but a conceptual entity whose probability distribution is specified by a wave function."[114] It was in view of this paradoxical situation that one could recently suggest that instead of particles science should speak of manifestations.[115] One advantage of this suggestion is undeniable. The word manifestation should force on the mind of the scientist the question so laden with metaphysics: "The manifestations of what?"

Regardless of whether such suggestion would prove popular, it should appear amusing to an outsider to see physicists trying heroically to pin down the elusive neutrinos which they refer to both as little particles and also as entities having neither mass nor electric charge, but to which they attribute nevertheless a sharply defined spin. Neutrinos, so physicists are constrained to say in a full-blooded metaphysical idiom, when away from their blackboards, are created inside the stars, quickly escape into empty space, and carry their energy away with them. (A great feat indeed for anything that possesses not even "thingness," though it possesses a perfect mirror image, the anti-neutrino.) In all this exercise in paradoxes, there runs an unquenchable metaphysical vein that brings to the surface a constant stream of references to things and entities. No doubt, certain types of talk about things that are supposedly behind things (meta-physics) are meaningless. Yet, is not the fact highly remarkable that ordinary language and physicists, who must needs use language, keep referring incorrigibly to things in the very sense in which Aristotle discussed them in his *Metaphysics*? It is so often forgotten by physicists that this sense of things physical is called metaphysical only because it refers to a meaning that cannot be grasped by the methods of physics. Such a contention is not the desperate rearguard action of almost vanquished metaphysicians as some would have it. As a matter of fact, this contention found one of its most telling formulations in the words of a physicist of no less stature than Planck, who wrote:

As there is a material object behind every sensation, so there is a metaphysical reality behind every thing that human experience shows to be real . . . the word "behind" must not be interpreted in an external or spatial sense. Instead of "behind" we could just as well say, "in" or "within." Metaphysical reality does not stand spatially behind what is given in experience, but lies fully within it. Nature is neither core nor shell: she is everything at once.[116]

Contrary to the widely shared misconception, metaphysics is not the branch of study set up for determining the maximum number of angels that can be accommodated on a pinhead. Nor is justice given to the whole history of metaphysics by the position taken by R. Carnap, who called metaphysical "all those propositions which claim to present knowledge about something which is over and above all experience." He was hardly on objective ground when stating that in general metaphysicians "are compelled to cut all connection between their propositions and experience; and precisely by this procedure they deprive them of any sense."[117] Of course, in the resolutely positivist outlook of Carnap, no other appraisal was possible. Yet, to remain consistent with the evidence, a reappraisal on Carnap's part proved just as impossible to avoid. The program of "Unified Science," aiming at a thoroughly non-metaphysical interpretation and systematization of scientific ideas, ran into difficulties that could be resolved only by acknowledging the indispensability of considerations patently metaphysical in character. The road covered by Wittgenstein from his *Tractatus* to the posthumous *Philosophical Investigations* illustrates the same point just as impressively.

What is particularly important to keep in mind in this respect is that notwithstanding the insistence of some, metaphysics, while being a step beyond science, is not a step beyond nature. Metaphysics is metascience but not metanature or a study independent of nature. Of this, Andronicus, who edited Aristotle's works, was well aware, and when he called certain books of Aristotle *Metaphysics*, he meant simply those works that inquired beyond the reach of the science of physics without ignoring the physical. Such inquiry is not merely a pastime of bygone ages, but rather a perennial feature of the quest of human understanding. If fresh proof is needed that this is so, one need only recall the rather desperate attempts of some modern physicists to try to bridge that ever-widening gap of which Einstein spoke in 1933 "between basic concepts and laws on the one side and the consequences to be correlated with our experience on the other."[118] Efforts to bridge this gap can, of course, degenerate into a sort of metaphysics having insufficiently strong ties with sensory experiences, and in this sense metaphysics is surely no more than empty talk. But the physicist, too, can get lost in empty talk when trying to translate the almost outlandish mathematical formalism of modern physics into common talk, if he ignores that aspect of reality that is beyond the science of physics but rooted in the physical.

When a physicist falls into this error, it is because he erects unawares a method into metaphysics. The result is a position which at times exhibits a glaring lack of logic. Clearly, the oddity of such a situation cannot be redeemed by condescending rhetoric. Eddington, for instance, was quite right in stating that there is a world of difference between the commonsense and the scientific notion of a table or of an elephant sliding down a grassy hillside. The table of commonsense judgment is a thing with properties, and so is the bouncing elephant, although in the scientific or "true" viewpoint both table and elephant shrink to a set of pointer readings correlated by mathematical functions. It was Eddington's claim that modern physics assured us "by delicate test and remorseless logic" that the scientific table "is the only one which is really there."[119] But is such a claim really sound from the viewpoint of science? Why is it that according to Eddington's own admission "modern physics will never succeed in exorcising that first table — strange compound of external, mental imagery and inherited prejudice"[120] which lies visible to our eyes and tangible to our grasp? Is not there, in this stubborn resistance of commonsense thought to the pretensions of some men of science, something far more than "mental imagery and inherited prejudice"? Or should man perhaps capitulate to a certain type of physical theory so preposterously identified with physics and renounce his own nature as a ghost just because it tells him consistently and irresistibly of aspects of the external world which are beyond the reach of physics, or to use that discredited technical term, are simply metaphysical? Did not Eddington himself assure us, as if it needed his assurance, that "that overweening phase when it was almost necessary to ask the permission of physics to call one's soul one's own, is past"?[121] Did not he acknowledge that "the process by which the external world of physics is transformed into a world of familiar acquaintance in human consciousness is outside the scope of physics"?[122] If this process is outside the scope of physics, however, has it thereby ceased to be a part of reality or is it simply severed forever from reality? After all, did physics ever succeed, or can physics ever logically succeed, even within its own narrow range, in reducing all its pointer needles, which are commonsense objects, to pointer readings or abstract data? Must not the last or ultimate set of pointer readings rest on the functioning of a real pointer needle? Clearly man cannot think, speculate, reason, or build specious theories even within physics itself, let alone outside it, except in a way in which metaphysical aspects of physical things and phenomena are inextricably present. Is there at the bottom of this persistent pattern of human thought a perennial fallacy or rather a perennial reality accessible only to a perennial philosophy?

True, physics in a sense owes its origin to a resolutely critical attitude toward what may be called the world of common sense. That there was and still is plenty to criticize in the data and judgments of common sense no one would deny. Yet, one should not forget either that all that physics can touch in the world of common sense concerns strictly the quantitatively evaluable relations among univocal terms. The richness of nature is far from being exhausted by univocal terms or univocal classes of things. To see this one should merely recall some of the conclusions reached by philosophical attempts which, swayed by the success of physics, recognize the legitimacy only of univocal terms. On such a ground, even the continuing identity of a thing through time becomes incomprehensible, and one is led to assert with Russell that each of us is no more than a succession of "momentary men."[123] The consequences of such a position are disastrous not only for ethics but equally so for the meaning and interpretation of scientific laws, since these must aim at discovering persistent patterns in things and among things. This is why when one tries to answer in a meaningful way the problem of continuity in time, one has to steer a middle course between the monism of Parmenides and Spinoza and the strict pluralism of logical atomism espoused by not a few modern philosophers. Over such rigid extremes common sense ultimately prevails and with common sense comes inevitably the recognition of the manifold analogies pervading the realm of being. The most obvious and most fundamental of these analogous realizations of existence is the permanence through time, and for this reason alone, exact physical science cannot rely exclusively on strictly univocal terms, for it cannot eliminate time as one of its chief parameters.

Faith in the existence of the external world, faith in the existence of a typical order in nature, assumptions used by the scientific method to detect this order, the necessity of referring the mathematical formalism to commonsense terms, the indispensability of non-univocal, analogous concepts even within the framework of a science characterized by emphasis on using univocal terms whenever possible, all these are not the only instances of the unavoidable involvement of physics in metaphysics. The physicist must also believe that things behave in the same way whether they are being observed or not. He has to assume that his memory is basically trustworthy. He has to have a large measure of confidence in the reports of fellow scientists especially when he is not in a position to test their results personally. He has to use terms such as I, you, it, is, same, different, unity, diversity, and a host of others which by their very use raise questions that are pregnant with metaphysics. All this takes on an added metaphysical significance for the modern physicist, who must always be aware that the role of the observer forms an integral part of many of his experiments.

No less metaphysical is the task of the physicist when he tries to fulfill his cultural role by discussing and interpreting the relevance of his findings to the other sectors and to the wholeness of human experience. In essence this task boils down to the recognition that physical science deals with only one segment of reality. To take a specific example, the task consists of admitting that while it is legitimate for physics to treat colors quantitatively, it is presumptuous for some physicists to reduce the whole reality of colors to wave numbers. Such a remark is motivated not only by a desire to secure the "legitimacy" of the art of painting or to justify the cultivation of poetry. It is also prompted by a consideration that pertains to physics proper. For as Whitehead pointed out in speaking of the pseudo-metaphysics erected by some seventeenth-century philosopher-scientists about the primary and secondary qualities, ". . . Any reasons which remove color from the reality of nature should also operate to remove inertia."[124] To deny any sort of reality to colors was clearly a metaphysical judgment and when this judgment was made originally in the seventeenth century, one could little guess that ultimately it would lead physics into a framework of thought, narrowly rigid, and deprived of liberating insights. Whitehead called this dead end the "nineteenth-century dogmatism" in physical science. Its collapse, as he put it, should remain a "warning that the special sciences require that the imaginations of men be stored with imaginative possibilities as yet unutilized in the service of scientific explanations."[125]

It should indeed give some food for "metaphysical" thought that the same error that erected a distinction between two aspects of nature into a radical bifurcation in nature had to prove in the long run both misleading in physics and deadly divisive in modern culture, which finds itself split in two. Both sides of this split, of course, keep arguing and reasoning and explaining, which means that all are busy tying together causes and effects. Actually both sides use the term cause in its full sense; yet one side, the scientific, is all too often reluctant to recognize the full metaphysical connotation of the term cause. This in spite of the fact that the principle of indeterminacy could have given men of science many opportunities to enrich their concept of causality by some timely considerations. As is well known, this principle has raised weighty questions about the concept of causality both as an explanation in physics and as an axiom in epistemology and ontology. What emerged in the ensuing controversies was the recognition that the notion of causality is ultimately only an aspect of the principle of sufficient reason, which plays a central role in any scientific proof or demonstration. This principle is just as indispensable when one argues within the framework of the mechanistic determinism of classical physics or within the probabilistic formalism of quantum mechanics. Of course one may twist the concept of causality to satisfy even the most extravagant ways of sophistication. Yet, as long as there is any sort of science, it will remain true what Russell observed so pointedly: "The discovery of causal laws is the essence of science, and therefore there can be no doubt that scientific men do right to look for them. If there is any region where there are no causal laws, that region is inaccessible to science. But the maxim that men of science should seek causal laws is as obvious as the maxim that mushroom-gatherers should seek mushrooms."[126]

A face-saving escape from the dilemma offered itself, however, if one was willing to declare that the principle of indeterminacy presents the problem of causality with an unresolvable enigma that must remain forever so. But before such an expedient had become widely adopted, especially among the representatives of the Copenhagen interpretation of quantum mechanics, farfetched conjectures were proposed, which clearly illustrate the difficulties of physicists distrustful of metaphysics. In this respect, it should be instructive to recall the theory proposed in 1924 by N. Bohr, H. A. Kramers, and J. C. Slater on the interaction between matter and radiation. Their paper raised the question "whether the detailed interpretation of the interaction between matter and radiation can be given at all in terms of a causal description in space and time of the kind hitherto used for the interpretation of natural phenomena."[127] What their theory really proposed was that in the interaction between matter and radiation, energy and momentum are conserved only statistically. In other words, while the radiation is continually scattered by all electrons in an atom exposed to radiation, only occasionally does this process concentrate on one electron with the result that a so-called recoil electron is emitted. On such a view, there should be as a rule a considerable time lag between the detection of the scattered ray and the recoil electron. It was precisely the all too ready assumption of such a long time lag that indicated the lack of proper concern for the philosophical implications of the conjecture. Yet, even on the level of physics, this theory was at sharp variance with the explanation of the Compton effect, in which one has to assume that both energy and momentum are strictly conserved in such processes. A long time lag (and by inference a causal disconnectedness) in such an explanation is clearly inconceivable.

The experimental test for the respective merits of the two theories was quickly designed and carried out by W. Bothe and H. Geiger. The results showed that the scattered gamma rays and the recoil electrons activate the two monitoring Geiger counters simultaneously to within 10^{-2} second. Furthermore, the number of such coincidences for a given gamma radiation and for a given amount of time was such that three hundred years would have been required to obtain them if the Bohr-Kramers-Slater theory was correct. The result cast serious doubts, therefore, on the idea of the statistical conservation of energy and momentum, which fell into even greater disrepute because of the experiments carried out between 1936 and 1954 by Shankland, Hofstadter, McIntyre, and Voelker. Their results verified the coincidence of detection to within 10^{-7} seconds and the respective angular direction of the photon and the electron, as required by the conservation of momentum and energy, to within ± 2 degrees.[128]

The theory of Bohr, Kramers, and Slater was based on a mixture of concepts borrowed from both classical and modern physics. As such, it could hardly yield, even if verified, the basis for a wide-ranging theoretical speculation. It was otherwise, however, when quantum mechanics became formalized and shortly afterward when Heisenberg enunciated his indeterminacy principle. As a law of physics, it rested both on consistent theoretical grounds and on indirect observational evidence.[129] It was therefore all the more tempting to erect it into a law of philosophy. The first determined steps in that direction were made by Heisenberg himself in the concluding remarks of his now classic paper. Heisenberg emphasized man's inability to predict the future with precision and dismissed as sterile and meaningless all speculations about the existence of a real, causally determined world behind the statistically ruled realm of quantum mechanics. For Heisenberg, the true situation could be characterized in only one way: "Since all experiments are subjected to the laws of quantum mechanics and thereby to the equation $[p_1\, q_1 \sim h]$ the invalidity of the law of causality is definitely proved by quantum mechanics."[130]

With this Heisenberg stirred up, probably without meaning to, an extraordinary dust storm of loose talk. The responsibility for this rests in part with him and some of his eminent colleagues. One of them, J. von Neumann, went as far as to state that "there is at present no occasion and no reason to speak of causality in nature — because no experiment indicates its presence, since the macroscopic experiments are unsuitable in principle, and the only known theory which is compatible with our experiences relative to elementary processes, quantum mechanics, contradicts it."[131] Such a conclusion, already strange in itself, appears even more so, if one recalls that a few lines earlier the outstanding mathematician and theoretical physicist declared that one should not think that because of its contradiction to quantum mechanics, "causality has thereby been done away with." What is more, he also spoke of the "several serious lacunae" present in quantum mechanics, hinting thereby at the possibility that quantum mechanics might one day be superseded by a theory built on different principles. Much of the uneasiness that one might feel in the face of this inconsistent arguing could have been prevented if leading physicists like Von Neumann had made it clear that whatever the weakness of the notion of mechanical causality, this does not exhaust the whole spectrum of the concept of causality. Nor does its elimination in physics entail the overthrow of the principle of sufficient reason, which is the fundamental aspect of causality. Von Neumann dismissed causality in general, however, as merely "an age old way of thinking of all mankind." Yet, Heisenberg, Von Neumann, and others who blithely dismissed causality as such on the basis of quantum mechanics, had to resort to a long chain of causal reasoning to prove the absence of mechanical causality in the realm of quantum mechanics. In this connection, one cannot help recalling Whitehead's famous remark about those who spend their lives with the purpose of proving that life is purposeless. Such efforts constitute, to say the least, an interesting subject of study.

At any rate, it is indeed baffling how such careless philosophizing could appear on the pages of a book that even today is still regarded as perhaps the deepest and most rigorous probing into the mathematical foundations of quantum mechanics. Yet, it was through the superficial and at times simply naïve use by physicists of the words determined, caused, and the like, that the principle of indeterminacy began to mislead many a heedless mind. It remains a perennial documentation of the incompetency of many a physicist in basic philosophical matters that so few of them could recognize what J. E. Turner put so concisely in 1930 in the pages of *Nature*: "Every argument that, since some change cannot be 'determined' in the sense of 'ascertained', it is therefore not 'determined' in the absolutely different sense of 'caused', is a fallacy of equivocation." [132]

The pattern of fallacy could not be clearer. The state of mind of physicists, so preoccupied with univocal terms, had to be trapped in elementary equivocation. Being unaccustomed to the realm of analogous terms, their thinking led them and many others into what was nothing short of confusion. As the theoretical physicist, H. Margenau, described this era of orgy in reasoning: "No simple slogan save 'violation of causal reasoning' was deemed sufficiently dramatic to describe the revolutionary qualities of the new knowledge." [133] One may indeed say that never before had a slogan originating in physical science been adopted with more shocking gullibility by all those who wrote and spoke much and thought little. From the speeches of politicians and even some clergymen to the conferences of sociologists and psychologists, the stunning references to the alleged demise of causality were repeated to no end. The imprecise style of many physicists who happened to amplify the subject just kept the dust whirling merrily.

When the dust began to settle, points of importance began to appear. First, more attention was paid to the limitations of the content of the indeterminacy principle. It was recognized that the principle cannot of itself disprove that atomic particles really have definite positions and velocities. The principle has such bearing only when taken jointly with the methodological assumption that only what can be observed in a laboratory is endowed with reality. Whatever the limits of such a contention, it is this methodological principle taken without any limitations that lies at the root of the wholesale rejection of the principle of causality based on quantum mechanics. This can be seen clearly, for instance, in a revealing passage of a technically excellent text on quantum chemistry. There it is stated that the principle of causality is still adhered to because only in this way can we "believe literally what our gross senses tell us about the world around us." To this naïve belief, the same passage contrasts the enlightenment of most scientists "who know only too well that our gross sensations are unreliable and that common sense is fallible." Therefore, concludes the author, "most scientists see little reason for clinging to a dogmatic position," such as the admission of causality.[134]

The only merit of such contention is that it reveals, though unwittingly, the dogmatic reason behind the rejection of the "dogma of causality." What also distinguishes the passage is that it is replete with omissions. It fails to mention that scientific observation too is not only fallible but that it has to fall back ultimately on commonsense observation, which is also fallible (but not always), for otherwise science itself would be impossible. The passage also should have mentioned, for the sake of scientific objectivity, that many outstanding physicists refused to conclude the overthrow of causality from the principle of indeterminacy. Among them was Max Born, one of the chief architects of quantum theory, who categorically stated in the best study written by a physicist on the subject: "The statement, frequently made, that modern physics has given up causality is entirely unfounded."[135]

In the polemics following the enunciation of the uncertainty principle by Heisenberg, more light was also shed on the fact that the concept of causality is far richer than was believed by most moderns who equated it with the possibility of mechanistic prediction of future events. The meaning of causality, however, is deeper than that. It is inextricably present in the domain of psychological motivation and in the realm of logical inferences. Mathematical intuition and reasoning, generalizations from sense perception, fall back on it invariably. However great the disagreements might be among philosophers of all ages about the shades and depths of its meaning, all the professional and amateur wrestling with the concept of causality brings out forcefully what Helmholtz once said of it: "It is nothing but the demand to understand everything."[136] As long as science exists, ample use will be made of causality, if not openly, at least furtively. For science is not a sophisticated tinkering with highly refined instruments. It is not a profit-motivated search for new commodities, or new tools of warfare. It is a search not for gadgets but for understanding; its principal aim is not to invent but to comprehend. This urge to comprehend — or the eternal curiosity of man — is a quest for clues to account for the observed phenomena. Man will forever be looking for clues, for he is constituted in a way that impressions made on him will keep him in unceasing wonder. It was this wonder that Plato once identified with the most genuine philosophical attitude, and for Aristotle, it was the mainspring that first prompted man to philosophize.[137]

Physicists may one day succeed in eliminating the word cause from their vocabulary. It may even come true, though it has not yet, what B. Russell forecast in 1914: "In a sufficiently advanced science, the word 'cause' will not occur in any statement of invariable laws."[138] The word itself may prove, as Russell contended half a century ago, "a relic of bygone age."[139] Yet, the absence of the word cause will never mean that physics ceased to look for causes in the conviction simply that there are no causes. What physics looks for is the interconnectedness and coordination of nature and not merely some purely mental constructs having no relation to nature. Yet, the more successful the mind is in finding order and unity in complexity, the more amazed it will be by its ability to correlate the jumble of sensory data in such a way as to obtain laws that work independently of the perceiving mind. The amazement felt in this connection by a truly perceptive mind is but a tribute to the miraculous role of causality in the process of understanding. It was in this sense that Einstein noted: "The very fact that the totality of our sense experiences is such that by means of thinking . . . it can be put in order, this fact is one that leaves us in awe."[140] For Einstein, the comprehensibility of the world was nothing short of a miracle upon which rested the possibility of science.[141] De Broglie, too, gave expression to his wonderment at the progress of science, which "has revealed to us a certain agreement between our thought and things, a certain possibility of grasping, with the assistance of the resources of our intelligence and the rules of reason, the profound relations existing between phenomena. We are not sufficiently astonished by the fact that any science may be possible."[142]

If the ability of an individual to know is already a miracle, what should one say about the fact that different world pictures of different individuals can be communicated and correlated? There can probably be only one answer, which reads in Schrödinger's words as "the greatest and absolutely inexplicable marvel."[143] Marveling, however, is fundamentally a metaphysical attitude. A pseudometaphysician like a Kantian, or an antimetaphysician like a positivist, to whom the intellect is not a disciple but an absolute master of nature, will look puzzled at anyone's marveling at the ability of mind to find coherent answers to the phenomena of nature. But for modern physics the mind in relation to nature is both a master and a disciple. The mind had to find out that there is far more in the world than what can be suggested by "common sense" or by "a priori categories" or even by the most recondite mathematical theorems. It is no less amazing, however, that regardless of how uncommon are the data of the physical world revealed by experiments, the mind rooted in common sense still has answers to the puzzles raised.

The physicist wary of metaphysics would also do well to remember that philosophy has some heuristic ability highly valuable for the purposes of physics as well. It was from philosophy that physics borrowed the concept of the atom, and modern physicists, who speak more and more often of the atomicity of time, and of its units *chronons*, had long been anticipated in this by philosophers like Maimonides and Descartes. Maimonides even went so far as to indicate the number of time atoms in a second as being, to use the modern notation, 60^{10} or more.[144] Modern physical cosmology, so strongly dominated by the discovery of the expansion of the universe, might also recall that it was Locke who devoted a chapter in his *Essay Concerning Human Understanding* to a joint consideration of space and time and concluded: "Expansion and duration do mutually embrace and comprehend each other; every part of space being in every part of duration, and every part of duration in every part of expansion."[145] It was a philosopher, E. Boutroux, who, years before modern physics was born, pointed out that the laws of physical science do not mean either an absolutely necessary or a complete one-to-one correspondence with reality.[146] To what extent the philosopher Bergson could anticipate insights of time and motion that are now part and parcel of modern physics was shown by no less a physicist than De Broglie.[147] Another distinguished physicist, W. M. Elsasser, had in his turn to admit "his indebtedness to Bergson for the idea that the deviation of organisms from mechanistic behavior ought to be studied in terms of the failure of information storage."[148]

These examples of course should not be taken even as an indirect suggestion that metaphysics can ever establish a physical law that should necessarily stand up to the test of any subsequent research in physics. It is in this sense that Maritain denied the heuristic role to metaphysics.[149] The main heuristic role in the physico-philosophical dialogue on nature will remain forever that of physics. The richness of new philosophical viewpoints, problems, and questions raised by physical research can most vividly be sensed when physics enters a new era. This was clearly the case in the seventeenth century and was no less so more recently. Harnack had precisely this in mind when he once said, in the conference room of the University of Berlin: "People complain that our generation has no philosophers. Quite unjustly: it is merely that today's philosophers sit in another department, their names are Planck and Einstein."[150] One might equally well recall the advice Heisenberg gave the young Weizsäcker, who was planning a career in philosophy: "One can't get anywhere in philosophy nowadays without knowing something about modern physics. But you will have to start on physics pretty soon if you don't want to be too late."[151] Truly, there is no exaggeration in the words of H. Margenau, who referred to the "enormous metaphysical wealth reposing largely untapped in modern physical theory."[152]

Just as abundant in modern physics, however, are the metaphysical questions that are to be answered not only out of metaphysical curiosity but even more out of a concern that connotes both a cultural and a scientific responsibility. As to the cultural responsibility implied in the proper and careful philosophical interpretation of science, Sir Cyril Hinshelwood, president of the Royal Society, made a remark in an address delivered at its tercentenary meeting which is worth recalling. Sir Cyril warned "the more serious and responsible men of science" that if they "remain aloof from the task of relating their subject to its philosophical background, the less serious and responsible will do it for them, and will startle the uninitiated with fantastic paradoxes which may impress the layman but do science real harm."[153] The plight of the laymen bewildered by those questionable interpretations of science will be discussed later in this book. But for the harm done to science, or physics to be specific, one need only note that the farther physics gets in exploring the vast reaches of space, or the minute details of matter, the more numerous and obvious will be its contact points with metaphysics. It should be highly significant that as physics advances through the centuries, the intricacies of such basic concepts of physics as space, mass, and time, for instance, are growing in number instead of diminishing. Statements to the contrary reveal only a grossly superficial acquaintance with the real situation. Physicists and writers about physics may grandly declare, for instance, that action at a distance is an absurdity, or that it has been definitively discarded as a hypothesis. A thorough review of the history of the problem, however, casts a very different light on this question. Such a survey shows that if physics is unable to decide whether bodies act at a distance, it is only so because the question "is not so much a problem of particular theories as of metaphysical framework."[154] Quite similar is the pattern revealed in the conclusions of the most exhaustive modern monographs available on the history of such basic concepts of physics as space, mass, and time. These conclusions attest the "historical, heuristic and logical importance for physics of ideas and assumptions commonly called metaphysical."[155] This is why it was possible to state without exaggeration that "a society which is uninterested in metaphysics will have no theoretical science."[156]

The most immediate issue concerning the relation of physics to metaphysics is to give a fresh and unbiased look to the question whether it would not be advantageous for physics to pay constant and explicit attention to metaphysics. Perhaps an affirmative answer would entail some physicists giving up their long-ingrained revulsion to metaphysics. Perhaps it would appear imperative too that at least basic information about metaphysics should be included in the training of physicists. This might appear highly repulsive to those who never tire of recalling the old clichés that metaphysics, unlike physics, has no record of progress, is incapable of securing universal assent to its basic propositions, and can offer no effective criterion for eliminating false theories. It might be just as difficult to pick the most reliable type from among the numerous varieties of metaphysics engaged in a relentless intramural fight. Still, however numerous the divergences of philosophical systems might be, philosophy remains indispensable in the process of scientific understanding. As H. Weyl warned: "In spite of the fact that the views of philosophy sway from one system to another, we cannot dispense with it unless we are to convert knowledge into a meaningless chaos."[157]

For great as the number of metaphysical systems might be, far greater is the number of those non-quantitative concepts that are indispensable for a scientifically productive investigation of things physical. The conscious or unconscious use of these concepts constitutes, however, a recourse to metaphysics on the part of the physicist whether he admits it or not. He might choose not to reflect on what he is actually doing. He might refuse to enter upon a thorough analysis of the conceptual foundations of his work. He might disdain to wax philosophical. But in doing so he will not advance the cause of science. He will merely help science degenerate, as Whitehead once put it, "into a medley of *ad hoc* hypotheses."[158] The physicist would merely deceive himself by thinking that a lackadaisical attitude toward philosophy is justified by the scandal of the multiplicity of metaphysical systems, often contradicting one another. Instead of proving that metaphysics is impossible, this scandal only illustrates that metaphysics, or the study of the non-quantitative aspects of things and processes, is a difficult subject. It is very likely that a physicist accustomed to the straightforward handling of univocal concepts will more likely than not feel mental discomfort in trying to cut his way through the countless shades of analogous terms to which the non-quantitative aspects of existence give rise. He might find, as Helmholtz did a hundred years ago, that, as contrasted with the clear contours of the world of experiments and mathematics, "much philosophizing eventually led to a certain demoralization and made one's thoughts lax and vague." [159] With Helmholtz, however, he should guard against concluding that philosophy is therefore useless for the physicist.

Whatever in philosophy might appear to the physicist scandalous and demoralizing is due only in part to the defects of philosophy. And in any case, a subject, however lucid, will very easily appear demoralizing to one not prepared for it. Still, even a physicist with more than cursory training in philosophy might find philosophy demoralizing. The reason for this is very human and is evident in all cases where man shifts his line of investigation from a simple terrain to a very complicated one. For physics investigates the external world by focusing on the simplicity that lies behind its complicatedness. Philosophy deals, however, with the inner world where even the clearest contours of conceptual refinement are vague at best when set off against the sharp outlines that are the hallmark of quantities and things physical. It is the haziness inextricably present in that inner world of mentation that will almost of necessity produce in its investigator that feeling of a vague and lax mind of which Helmholtz spoke. Yet, uncomfortable as the condition of a vague and lax state of mind might be, it will be the inevitable share of anyone with a healthy respect for the whole truth. Conversely, it is the wholeness of truth that will suffer when a physicist engages, as is not rarely the case, in a patently metaphysical discourse rich in loud disclaimers about metaphysics. The final outcome of such an attitude is bound to be embarrassing, to say the least, and will only provide one more illustration of the truth of Gilson's remark: "If one does not want to listen to the answers of metaphysics, one should not address questions to it." [160]

Physics will have to raise questions, however, as long as physics is to remain an investigation of nature, and many of these questions will be metaphysical in character. If these are answered by the methods of metaphysics, a strict logic will prevail in the answers which at times might appear mysterious to minds sensitive only to quantities. If, however, scientists try to answer such questions by the methods of science, not only will the answers look mysterious, but that aspect of the universe which it is the business of the scientist to make lucid will appear equally mysterious.

One should not too readily presume of course that the words of philosophers and historians of science, pleading for more appreciation of metaphysics by scientists, will be willingly heard. Let it be remembered therefore that it was a leading mathematician and a mathematical physicist, H. Weyl, who insisted that the mathematical physicist "should be reminded from time to time that the origins of things lie in greater depths than those to which his methods enable him to descend. Beyond the knowledge gained from the individual sciences, there remains the task of comprehending."[161] Let it also be remembered that even an Einstein, who seemed to be so much in agreement with the school of logical empiricism, decried on occasion the "fateful fear of metaphysics . . . which has become a malady of contemporary empiricistic philosophizing."[162] Again, it was in essence for the cultivation of metaphysics that Einstein pleaded when he defined the whole of science as the refinement of everyday thinking. He noted that for this very reason the physicist cannot proceed without critically considering a problem which in his eyes surpassed in difficulty the speculations of physics proper: the problem of analyzing the nature of everyday thinking.[163] Similar concern prompted De Broglie to deplore the separation between philosophers and scientists as harmful to both sides. But apart from considerations of gain and loss, one should above all keep in mind that metaphysics and physics originate in the very same source. A hundred years ago Maxwell called this source freshness of mind and challenged those who kept asserting that physical science made metaphysical speculation a thing of the past. The discussion of the categories of existence, he warned, continues to be "as fascinating to every fresh mind as it was in the days of Thales."[164]

Thus, as long as the fresh insights of fresh minds remain the indispensable means of advancing physics, physicists are bound to turn up new questions. Many of these will point beyond the boundaries of physics. For, as the physicist-author of one of the most incisive commentaries on how physics works, put it at the time when modern physics was coming of age: "At some stage in our inquiry we must stop and accept judgments without argument; is it certain that these judgments will not be found to be metaphysical? Or again, are we sure that the process of reasoning by which we develop our conclusions from these fundamental judgments does not depend on the acceptance of doctrines that are distinctively metaphysical?"[165]

CHAPTER NINE

Physics and Ethics

AT THE CAMBRIDGE MEETING of the British Association in 1938, Robert John Strutt, the fourth baron of Rayleigh, devoted the second part of his presidential address to a firm rejection of the charge that major scientific discoveries were motivated by a desire to provide tools for warfare. He could indeed point out that the poisonous gases, the basic forms of explosives, or the art of flying were not invented with an eye on possible military application. He was also wholly right in calling attention to the fact that "the application of fundamental discoveries in science is altogether too remote for it to be possible to control such discoveries at the source." [1]

For all the soundness and value of such remarks they are hardly the ones that make his address attractive reading even today. The outstanding feature of the address lies rather in its involuntary omissions. Being a man of his time, the former professor of physics at the Imperial College of Science and Technology could not, with most of his countrymen, bring himself to believe that a war was in the making. Prognostication of the political future, even of the immediate one, is of course a thankless task. Not much easier is the prediction of what is in store for science in the very near future. If this is possible to some meager extent, it is only so by keeping in mind the pattern of past events. Perhaps Lord Rayleigh should have recalled that no sooner had his father, John William Strutt, put the finishing touch on his classic investigations of sound, than the first steps were taken to make ultrasonic waves a highly prized tool in submarine warfare. Some application to warfare could just as conceivably develop out of the research on natural radioactivity in which the younger Strutt made his name as a physicist. Yet, even

371

by contemplating such a possibility, he could hardly guess in 1938, that the very next war was to be won, in part at least, by a tightly organized army of physicists. He could hardly foresee that their task would be to transform with the aid of artificial radioactivity that gem of theoretical physics, the equivalence of matter and energy, into a deadly bomb. Reading his speech from the distance of a quarter of a century, one cannot help noticing how limited the scientific mind can be in charting, however tentatively, the course to be taken by scientific work in the immediate future. On the other hand, Lord Rayleigh's short survey of the past misuses of scientific discoveries for purposes of destruction shows that the ethical concern about the use of the products of science is as old as science itself.

That the scholars of classical Greece opted by and large for an organismic type of physics, for instance, was in no small measure due to ethical considerations. Plato, for one, proposed harsh punishments for those who followed "brash physicists" in denying the essential difference between the substance of heavenly bodies and that composing the earth. According to him, such deviation from traditional tenets was an inadmissible departure from beliefs that alone could secure the foundations of society.[2] Later Epicurus and his followers refused to admit that the laws of physics were valid for the heavenly regions as well for fear that this might subject the gods, providence, and the free will of man to an ironclad physical determinism. The tension between ethical convictions and scientific outlook grew especially acute during the long centuries when magic and alchemy became synonymous in popular thought with scientific investigation. Although the advent of mechanism in the seventeenth century proved very effective in dissipating the suspicion that science was ultimately a sort of consorting with evil spirits, the basic tenets of the new physics influenced in two diametrically opposite directions the further development of ethical philosophy.

One of these directions had a strongly optimistic overtone. Descartes, for instance, saw in the complete knowledge of mechanistic sciences the indispensable steppingstones that should lead to "the last degree of wisdom, the highest and most perfect moral science."[3] That geometry, which stood as the symbol of the perfection of mechanistic philosophy, might revitalize the study of what is just and unjust was a fervent hope with Leibniz,[4] who also thought that rewards follow good deeds "by mechanical ways through their relations to bodies," just as sins "by the order of nature and perforce even of the mechanical structure of things" are punished automatically.[5] How could there be anything wrong with this fusion of ethics and mechanical science if, as Leibniz put it, "God as architect satisfies in every respect God the legislator." [6] On the ground that "God is the same Yesterday, Today and Forever" did G. Cheyne, the London physician and an admirer of Newton, predict that general laws, analogous to those of mechanics, would be formulated and recognized one day in the realm of morality. He envisaged in ethical science the same final codification to obtain once and for all in every respect, which he believed to have already been realized in mechanics, whose laws, as he wrote shortly after Sir Isaac's death, "are pretty well known and adjusted." [7] In Cheyne's reasoning the typical pattern was all too obvious: a wishful belief in the finality of the physical science of his day prompting bold conjectures about reducing other fields of thought, such as ethics, to the "definitive" pattern of physics.

Other admirers of Newton did much the same. Locke, in the third and fourth books of his *Essay Concerning Human Understanding* proposed a "rationalistic" ideal of ethics, in the conviction that morality is capable of demonstration as much as mathematics. It was in espousing the views of Locke that Berkeley spoke of a "method affording the same use in Morals etc., that this [algebra] doth in mathematics." [8] Yet, the part of Berkeley's *Principles* that was to have dealt with ethics was never written. His optimism about the feasibility of writing an ethics equivalent in exactness to "mixt Mathematics" was dampened by his realization that a universal agreement about the meaning of ethical terms presupposed far more than did a consensus about the meaning of algebraic terms. This, however, was far from being the only disconcerting sign about the possible use of mechanical science in ethical philosophy.

Among the other signs was the specter of the possible misuse of technical inventions, at which Bacon hinted in his *New Atlantis*. At Solomon's House regular consultations determined which of the new inventions should and should not be made public. What is more, in this scientific citadel of a Utopian society, a compulsory oath of secrecy made sure that certain scientific discoveries would always remain the exclusive property of a small circle of top scientists and never get into the hands of the political and military rulers.[9] In all probability this section of the *New Atlantis* appeared as unreal to Bacon's contemporaries as did much of the rest of his flight of fancy. An age possessed of unbounded admiration for mathematics and mechanics could hardly be expected to worry about the possible misuses of the tools mechanical science might devise. After all, by the seventeenth century, mankind for over two hundred years had already learned to live with gunpowder, and little did anyone expect a similar discovery to come in the future.

Yet, Bacon the man of affairs as well as of ideas proved prophetic. But long before secrecy had become a basic feature of scientific research organized and supported by the body politic, the conflict between mechanistic and ethical philosophy had to come into the open. In fact, Bruno already felt impelled in the First Dialogue of his *On the Infinite Universe and Worlds* to vindicate the equal legitimacy of the two apparently contradictory concepts: the freedom of human choice and the determinism of physical events.[10] Descartes, too, saw the need for emphasizing that the universal mechanistic conception of the world should not undermine the truth of man's moral freedom. Man's awareness of his freedom was for him a primary datum that logically preceded even the axiom of the *cogito ergo sum*.[11] Leibniz' doctrine of preestablished harmony aimed, too, at the defense of the moral world to which he subordinated mechanical causality.[12]

In 1714, when Leibniz concisely stated these points in his *Monadology*, mechanistic physics had already found systematic interpretations which allowed no room for human freedom in the universe. Hobbes, for one, wanted moral philosophers to treat the motions of the mind — appetite, volition, and the like — as Galileo treated the motion of bodies. While such ethics would banish the freedom of will, it certainly could not achieve what it was supposed to do — eliminate war and secure stable peace.[13] Along with Hobbes there was Spinoza, whose writings exemplified what results one could expect when the Cartesian view of extension as supreme tenet in philosophy was carried to its logical extremes. In his *Ethics*, emulating the lifeless rigor of Euclidean geometry, Spinoza had no choice but to deny the freedom of will.[14] The basic distinction between good and evil, too, had to disappear in Spinoza's ethical system, in spite of his desperate efforts to uphold such a distinction in a way logically consistent with his basic assumptions.[15]

Against such extrapolations of the principles of mechanistic science some men of science did not fail to lodge firm protests. One need only recall Boyle's attitude, but neither he nor philosophers like the Cambridge Platonists or Shaftesbury could much influence the direction toward which human thought was gravitating in the age of mechanism. Below the surface glitter of the concept of an ethics as exact as geometry, there was developing an ethical philosophy in which the price of "mechanical intelligibility" was the abandonment of transcendental ethical aims and of the traditional concept of moral freedom. Hume's writings on ethics illustrated in all seriousness both the full breadth and the reluctant acceptance of the consequences in ethics of an exclusively mechanistic philosophy: the dilution of the distinction between good and evil, the absence of eternal laws of justice, and the erosion of the concept of free will.

From the mid-eighteenth century on, no thinker of any standing could avoid comment on the apparent bearing of physics on ethics in one of the many tones that ranged from the sarcastic to the outright tragic. At one end of the gamut was Voltaire, who after his conversion to Newtonian physics penned remarks on the freedom of will that brought to light only his flippant superficiality. On the principle that nothing is without a cause, Voltaire presented the voluntary choice as completely determined by the judgment immediately preceding it, and since judgments, good or bad, were in his view necessary effects, so was the human choice following them. It would be strange indeed, he said, "that all nature, all the planets, should obey eternal laws, and that there should be a little animal, five feet high, who in contempt of these laws, could act as he pleased, solely according to his caprice. He would act at random, but we know that randomness means nothing. We have merely invented the word to denote the known effect of all unknown causes." [16] In resisting passions, man is just as little free as he is in succumbing to them. In both cases, man follows irresistibly the last idea that has already been formed by necessity.

Still Voltaire tried to save the freedom of will at least in a nominal way. Necessary as man's inclination is to do this or that, man, as Voltaire believed, was capable in some measure to implement his wishes. This ability of man Voltaire identified with freedom of will. That humans could hardly be proud of such "measure" of freedom, Voltaire readily admitted. To comfort his fellowmen, he merely noted that the stars had not even that little in the way of freedom.[17] This was indeed a meager if not a desperate solace, but so was Voltaire's effort not to go all the way down the road of the mechanistic world view. His *Ignorant Philosopher* clung to the notion that the concepts of the just and unjust are grafted on human nature by God and saw in the universal moral conscience the best proof of God's existence.[18]

The manner in which Voltaire treats free will, physical determinism, the concept of good and evil, and moral law shows that he tried to learn from Maupertuis not only the rudiments of Newtonian physics but also some of Maupertuis' preoccupation with a "scientific" approach to the problems of ethics. But Voltaire failed to come close to the originality of his master who nowhere had shown so much of his zest for bold enterprises as he did in his attempt to quantize ethics.

In his *Essai de philosophie morale*, Maupertuis identified the good with the sum of happy moments and the evil with the sum of unhappy times[19] and believed himself to have found in this a quantitative scale on which the various ethical systems could be appraised with mathematical precision. As the happy and unhappy moments have both duration and intensity, all that was to be done, according to him, was to take the product of the intensity and duration of each feeling and sum up the happy ones with a positive sign and the unhappy ones with a negative sign.[20] Maupertuis' view was that in an ordinary life the sum of unhappiness usually exceeds the amount of happiness, but this was not his main point. Having equated the ethical good with what is pleasant in a truly noble sense, Maupertuis felt that his calculus could show convincingly the superiority of Christian ethics over the system of the Stoics.

Though speaking of the Stoics of old, Maupertuis had in mind the Stoics of his time, the rising leaders of the Enlightenment, who like D'Alembert, tried frantically to oppose the exploitation of the "scientific approach" in support of a crude, ethically lawless materialism and atheism. Voltaire himself, though bent on destroying the church, carefully kept on the throne of mechanism the pale supreme being of deism and was eager not to contaminate the populace with atheism, lest they overthrow with church and throne the sinecure of the *philosophes* as well. In fact, he advised some of his atheistic friends to keep their "refined music" from the uneducated. He feared that a disillusioned populace might one day grab "the instruments" of his friends and break them on their heads.

But the genuinely "non-mechanical," gospel-like zeal of the materialistic version of the mechanistic creed could hardly be contained by advices of caution. It blared forth without any restraint in De la Mettrie's *L'homme machine* (1749) where the author went to great lengths to disprove the belief that there exists in the human "soul" a non-material principle, or natural law, that dictates ultimately the choice between good and evil. De la Mettrie argued passionately that a dog's repentance differs only in degree, not in kind, from that shown by a man. The slight difference depended, so De la Mettrie contended, on the degree of the mechanical complexity and refinement of the structure of the brain: "A few more wheels, a few more springs than in the most perfect animals, the brain proportionately nearer the heart and for this very reason receiving more blood — any one of a number of unknown causes might always produce this delicate conscience so easily wounded, this remorse which is no more foreign to matter than to thought, and in a word all the differences that are supposed to exist there."[21] Truly, could this be otherwise when the

soul was defined as an "enlightened machine"?[22] Nebulous as such
a definition may at first appear, it must be admitted that De la Mettrie
left none of his readers in doubt about what he had in mind. To por-
tray the reduction of ethics to mechanics could have hardly been
done in a more explicit manner. To be a good man and deserve
thereby the confidence and gratitude of society consisted from then
on in the observance of the laws of mechanics.[23] Once man had identified himself with a machine, he would be as tranquil as a smoothly
running piece of machinery and would have a full measure of reverence, gratitude, and affection for the whole and every part of the big
machine, nature. In a machine-like man, so De la Mettrie promised,
there would be no hatred but only compassion toward his enemies
and the wicked, for in them he would see only the faulty parts of a
most worthy machinery. Nor would the man-machine maltreat fellow
components, for, as befits a useful machine part, "he will not wish to
do to others what he would not wish them to do to him."[24] Scamps
and hypocrites were all those, so De la Mettrie called them, who
sought for ethical conduct a foundation outside the realm of pulleys,
cogwheels, and levers. This of course had to be so if one was willing
to go along with De la Mettrie's basic "ethical" tenet: "Given the least
principle of motion, animated bodies will have all that is necessary
for moving, feeling, thinking, repenting, or, in a word, for conducting
themselves in the physical realm, and in the moral realm which depends upon it."[25]

Least principle of motion, universal attraction, inverse square law,
and the like were now being debased into the role of a fountainhead
for a pseudo-ethics, and Holbach's *Système de la nature* proudly bore
the ambitious subtitle: "The Laws of the Physical World and the
Laws of the Moral World." Not that he intended a distinction between the two. As a result, his was a book that had to be short on
physics and long on fantasies.[26] Its first chapter was characteristically
devoted to the contention that there is no difference whatever between the physical man and the moral individual. The former, says
Holbach, "is a man acting under the impulse of causes which our
senses make known to us," whereas the moral person "is a man acting
under the influence of causes which our prejudices prevent us to
recognize."[27] All that man can do therefore is to submit himself unreservedly to the inexorable laws of physics, in the conviction that
"all the errors of man are errors in physics."[28]

Holbach wrote at the dawn of an age that was seized by an uncon-
trollable urge to give concrete forms to sweeping "scientific" reasonings. Everything was to be given a new structure, a new outlook, that
had to mirror as much as possible the patterns of mathematics and

physics. In this unbridled program of the reorganization of society, such ethical matters as judiciary procedures received of course high priority. Condorcet was one of those who tried to put the problem of securing fair judgments in the courts on the only supposedly rational foundation: the formulas of mathematics. That the injustices of the tribunals of the *ancien régime* were glaring, no honest mind could deny. Yet, was one to find a more reliable source of justice in the premise that opened Condorcet's *Essay* of 1785, namely, that moral doctrines are susceptible to the precision and certitude that characterizes the physical sciences? Was not there a deadly dangerous illusion hidden in what Condorcet called "the consoling hope that mankind will of necessity make a progress towards happiness and perfection inasmuch as it recognizes the truth"?[29] The fallacy of this promising statement was, however, not difficult to detect. For the truth which Condorcet had in mind was merely the truth of mathematics. While admitting that not every question of social life can be clarified by calculus, Condorcet insisted that one must try to extend the range of applicability of calculus as much as possible because this was the only safe road toward a more enlightened life.[30] Thus for him the problem of securing just tribunals had to be a matter of mathematical analysis. Trusting to his unquestionable expertise in the calculus of probabilities, Condorcet felt he could safely ignore all aspects involved in the organization of a just tribunal that were not mathematical. His basic view was that unjust judgments can be tolerated in the same measure in which society and individuals are willing to take such risks as implied, for instance, in taking a mailboat in good weather from Calais to Dover. He contended that one should be willing to accept the miscarriage of justice when the chances for it are not greater than are the chances for any adult between the age of thirty-seven and forty-seven to die of a sickness that lasts less than a week. His tables of statistics showed that the chance of this was approximately $1/150,000$. He could then readily prove that a defendant's chances of being unjustly condemned would be about the same if he were to be tried by a bench of thirty judges, each of whom would make only one wrong judgment out of ten and if the consent of nineteen judges would be required to pass a valid judgment against him.

This was hardly a legal or moral philosophy, but was rather a perfect mathematics and a bold program befitting an age so oblivious to limitations inherent in ideas, things, and human beings. Condorcet himself was so fascinated by the prospects of his approach that he continued working on the subject, as he hoped to secure with "more precise proofs" the fruits of the revolution. His mathematical analysis of the composition of "enlightened tribunals" may have advanced the cause of the probability calculus but hardly helped human justice. Nor was it given him to see the publication of the second and improved edition of his work in 1805. Eleven years earlier the revolutionary ethics failed to give him what would have saved his great mathematical talents for further scientific contribution: human treatment and a fair hearing.

The tribunals of "reason" had to run their gory course before Laplace could voice the view that judiciary decisions have motivations that simply cannot be submitted to the calculus of probabilities. Not that Laplace or any other prominent man of science could have put an end by his warnings to this sad aberration of the human mind that kept dreaming about a "scientific" ethics molded ultimately on the study of mathematics and physics. Saint-Simon's wild conjectures about an ethics reduced to the law of gravity[31] were perhaps the most grotesque but certainly not the last of such attempts. The obstinacy with which such leanings asserted themselves could of course easily adapt itself to the changing nomenclature of science. Thus on Holbach's, De la Mettrie's, and Condorcet's statements latter-day addicts of such views offered only variations but no essential reformulations. Perhaps the most expressive of these came from the pen of B. Russell, and oddly enough at the time when the first big cracks in the world picture of classical physics began to appear. Russell, however, was still solemnly asserting the perennial validity of space-time concepts on which classical physics rested.[32] Had he not adopted mechanics as the basis of his "scientific" ethical manifesto, perhaps he could more readily have recognized the handwriting on the walls of the hallowed edifice of classical physics. In the billiard-ball universe of colliding atoms, however, he saw the ultimate reality, and he exploited this premise with the religious zeal of the wholly dedicated materialist:

. . . man is a product of causes which had no prevision of the end they were achieving; that his origin, his growth, his hopes and fears, his loves and beliefs, are but the accidental collocations of atoms; that no fire, no heroism, no intensity of thought and feeling, can preserve an individual life beyond the grave; that all the labors of the ages, all the devotion, all the inspiration, all the noonday brightness of human genius, are destined to extinction in the vast death of the solar system, and that the whole temple of Man's achievement must inevitably be buried beneath the debris of a universe in ruins — all these things, if not quite beyond dispute, are yet so nearly certain, that no philosophy which rejects them can hope to stand. Only within the scaffolding of these truths, only on the firm foundation of unyielding despair, can the soul's habitation henceforth be safely built.[33]

How safely, two great world wars must have brought home sharply to anyone unmindful of the limits within which the viewpoint of mechanistic physics can safely be applied. It was not a scholarly expert on transcendental ethical theory but a prominent physicist who located in the irresponsible extrapolations of mechanistic ideas the source of the ethical erosion that led to two wholesale cultural debacles. "In the decline of ethical standards," wrote W. Heitler in 1949, "which the history of the past fifteen years exhibited, it is not difficult to trace the influence of mechanistic and deterministic concepts which have unconsciously, but deeply, crept into human minds."[34] While the large public inhaled unawares the fallacies of the mechanistic creed, its propagandists moved, however, with firm purpose to make their beliefs prevail. On the ethical level their concerted efforts were clearly an exercise in destruction, and unfortunately some men of science had their share in this work of distinctly non-scientific character. Those who find such an indictment of some scientists harsh would better ponder the words of Poincaré, uttered in reference to the exploitation of mechanistic physics in support of pseudoethics:

One should be fearful of incomplete science only; for that is what deceives and lures us by vain appearances and engages us thereby in destroying what we would like to reconstruct later, when we are better informed, and when it is too late. There are people who fall for an idea, not because it is correct, but because it is new, because it is fashionable; they are indeed frightful destroyers, but they are not . . . I almost said they are not scientists, but I recognize that many among them rendered great services to science; therefore they are scientists, only they are such not because of this, but rather in spite of it.[35]

Outstanding scientist as M. Berthelot was in his day, he had no scientific justification in stating that the scientific conception of the world extends its "fatal determinism even to the realm of morality."[36] The gratuitousness of statements of this sort could of course be surpassed only by their destructiveness. The decades closing the last century and opening the present witnessed indeed an intellectual orgy of exploitation of classical physics in support of "ethical" systems that tried to make short shrift of anything perennial and non-material in the ethical code. Unfortunately, the prophets of mechanistic ethics, parading in the robes of science, all too often were paid more credence by the public than were persons opposing them by some "not-so-exact" humanistic methods. Of the latter was Dostoevski, who was particularly appalled by the individual and social convulsion that was emerging from an intellectual atmosphere poisoned by a distinctly unethical "scientific" ethics.[37]

His warnings went largely unheeded, as did the persistent protests of the great figures of classical physics against views that claimed competence for physics in matters of ethics. To Maxwell, the possible bearing of physical determinism on the freedom of will was a topic that had to force itself on anyone in whom, "the cares of this world do not utterly choke the metaphysical anxieties." In a posthumously published paper which he read before a select circle of friends on February 11, 1873, he dealt at length with the question, "Does the progress of physical science tend to give any advantage to the opinion of Necessity (or Determinism) over that of Contingency of Events and the Freedom of the Will?"[38] As Maxwell pointed out, the main criterion of determinism — the accurate predictability of future physical events — is not rigorously applicable even in statistical (molecular) physics, or in phenomena that are chain reactions resulting from changes in an originally unstable configuration. While he willingly admitted that free will is not so absolutely free as at times it is believed to be (its range, he said, is in a sense "infinitesimal," and it cannot do anything, just as a leopard cannot change its spots), on the other hand, physics was not in his view so extensively deterministic as it was often thought to be.[39]

The occasion of publicly discussing the alleged bearing of physical determinism on the freedom of will came to Maxwell when in 1879 he reviewed in the pages of *Nature* a book entitled *Paradoxical Philosophy*.[40] The book, which purportedly pleaded the cause of spiritual values, tried in effect to utilize the new science of psychophysics in support of spiritualistic speculations. Actually it represented a thinly disguised materialism defining the human mind as a subtle kind of matter. In the same vein, it also claimed that the freedom of will could not form an exception to the universal validity of the "principle of continuity" by which the book meant the conservation principles of physics. Were this principle overthrown, the book stated, there would be no ground to consider real such entities as the human mind and will. Clearly, this sort of reasoning was hardly a defense of non-material values or entities. Maxwell, for one, could easily point out that the position of the book implied a general collapse of every intellect or soul in the universe, since a genuinely free will would have meant a break in the chain of universal material causation on which the existence of any "material" soul and will had to depend. The alleged scientific moorings of such materialism Maxwell could of course brush aside with ease. He tersely recalled the progress made by science "in clearing away the haze of materialism which clung so long to men's notions about the soul."[41] This state of affairs he contrasted to the speculations and predictions made by certain men of science "in their non-scientific intervals," who contended that man shall soon have to confess "that the soul is nothing else than a function of certain complex material systems."[42]

More important in his criticism was the point concerning the limits of scientific method. As Maxwell stated, "one of the severest tests of a scientific mind is to discern the limits of the legitimate application of scientific methods," and he voiced his classic warning, "there are many things in heaven and earth which, by the selection of our scientific methods, have been excluded from our philosophy."[43] One of the subjects barred to the incisive tools of natural sciences was, according to Maxwell, the human personality: "As soon as we plunge into the abysmal depth of human personality we get beyond the limits of science."[44] For, as Maxwell saw it, the progress of physical science, far from unfolding the mystery of personality and the innermost recesses of human decisions, had rather tended "to show that this personality with respect to its nature as well as to its destiny, lies quite beyond the range of science."[45]

Kelvin was no less emphatic in rejecting attempts that tried to bring spiritual or ethical problems under the competency of theoretical physics. Of his many utterances on this point, it will suffice to quote his concise remark of 1903: "The mystery of radium, no doubt, we shall solve it one day, but the freedom of the will, that is a mystery of another kind."[46] Fifty years earlier Gauss spoke of problems to whose solutions he attached "an infinitely greater importance than to those of mathematics." Among them he enumerated the questions "touching on ethics," "our relation to God, or concerning our destiny and our future." Their solution, he stated categorically, lies "completely outside the province of science."[47] About the same time, when Vogt and Moleschott tried to shore up their brand of materialism by "scientific" arguments, Helmholtz firmly disclaimed any affiliation with their bent of thought in his letter of March 4, 1857, to his father. Not only did he point out that both these gentlemen still had to learn through more scientific work "the respect for facts" and "the prudence in reaching conclusions," but also made an unequivocal statement about the incompetency of physical science on certain questions: "A prudent investigator knows very well that the fact of his having penetrated a little way into the intricate process of nature gives him no more right, not a scintilla more, than any other man to pronounce dogmatically on the nature of the soul."[48]

Among the principal figures who were the most instrumental in laying the foundations of modern physics, Planck in particular gave much attention to the relation between ethical questions and the laws of physics. Planck, as is well known, never could accept the probabilistic interpretation of atomic phenomena as required by quantum mechanics. The universal validity of physical determinism in the domain of material processes had always been for him an indispensable principle of correct reasoning in physical science. Yet, he was adamant in upholding the special character of human will and moral responsibility as being wholly outside the range of physical determinism. To this point Planck returned time and again during the second half of his scientific career. His address of 1914 on the "Dynamical Laws and Statistical Laws" concludes with a warning that "he who considers it [moral freedom] logically irreconcilable with absolute determinism in all spheres of philosophy, makes a great mistake of the same nature as that made by the physicist . . . who does not take adequate precautions to eliminate errors in his observations."[49] Physical science, as Planck insisted, is by its method fixing "for itself its own inviolable boundaries"[50] and is unable to furnish answers to questions like "What am I to do?"

Planck's longest address of a non-technical nature was devoted to the defense of the freedom of the will against the claims of the Monistic League, that tried to sell the public on a world outlook (Weltanschauung) based on purely scientific grounds.[51] For Planck, the basic shortcomings of attempts of this type consisted in ignoring the fact that the determinism of the scientific method is inapplicable when it comes to the innermost core of the personality where human decisions have their ultimate roots. He repeated this conviction seven years later in a lecture that dealt with the concept of external reality as conceived by positivism.[52] In doing so, he made no digression from the subject of his lecture. For, as Planck had clearly seen, in both monism and positivism, taken as general philosophical explanations, freedom of the will, moral responsibility, ethical autonomy, and the like are "meaningless questions." To the dangers and consequences of such a position, Planck called attention in a most resolute manner: "It is a dangerous act of self-delusion if one attempts to get rid of an unpleasant moral obligation by claiming that human action is the inevitable result of an inexorable law of nature."[53] Freedom of the will, or individual responsibility, constituted in Planck's eyes a point "where science acknowledges the boundary beyond which it may not pass, while it points to farther regions which lie outside the sphere of its activities."[54] To recognize in such a manner the limits of the range of physical science was in Planck's view anything but a partial renunciation of the rights of scientific method. On the contrary, he stressed that such an acknowledgment of the limitations of science "gives us all the more the confidence in its message when it speaks of those results that belong properly to its own field."[55]

In another major address of Planck, given in 1937, on "Religion and Natural Science," the familiar theme came back again with undiminished force: "Questions of ethics are entirely outside of its [physical science's] realm."[56] The events of the year of 1937, when this statement was made by Planck, could have brought home to anyone in Planck's situation whether it was meaningful to look upon science as the source of moral fiber. In that year the Nazi government forced Planck to resign the post as permanent secretary of the Prussian Academy of Sciences. This was an ill-boding sequence to his futile attempt in 1933 to convince Hitler of the propriety of supporting German scientists of non-Aryan extraction. Of this confrontation with Hitler, Planck in later years would speak only in his innermost family circle and even there "with horror." Subsequent fateful events of history more than substantiated his lifelong scientific conviction that recognition and fulfillment of moral responsibility were not a derivative of any or all the laws of mathematical physics.

Einstein, who, like Planck, could see at close range the convulsion produced by the negation of universally valid ethical standards, was also careful not to submerge the realm of human values into the narrow category of the exact sciences. However much he believed in the universal harmony of nature, as expressed most beautifully in the laws of mathematics, he never lost sight of the fact that personal and social harmony stem from principles of distinctly other nature. Science, he noted, "can only ascertain what *is*, but not what *should be*, and outside of its domain value judgments of all kinds must remain necessarily"[57] (italics mine). He branded as "fatal error" the attempt often made by representatives of science who tried "to arrive at fundamental judgments with respect to values and ends on the basis of scientific method."[58] In his usual candor he made no secret of his own experience in this respect. Toward the end of a most distinguished scientific career, he remarked to a visitor: "I have never obtained any ethical values from my scientific work."[59] In Einstein's view, it was totally wrong to ask questions of the type: "What hopes and fears does the scientific method imply for mankind?" Whether such method would bring blessing or destruction depended, according to him, on the type of goals animating mankind. "Once these goals exist," he emphasized, "the scientific method furnishes means to realize them. Yet, it cannot furnish the very goals."[60]

Why physics or scientific method in general can claim no competency in ethical matters was most concisely formulated perhaps by Poincaré.[61] For purely grammatical reasons, as he pointed out, a so-called scientific morality is just as illusory as is the prospect of an "immoral" science. As is well known, the premises of a scientific proof are always in the indicative form and so is of necessity the conclusion. The statements of moral philosophy, or rather of an ethical code, aim on the other hand, at imposing an obligation and require therefore the use of the imperative form. In order to have a conclusion of this type, at least one of the premises must be in the imperative. One would look, however, in vain for an imperative among either the axioms of geometry and mathematics or the countless observations and measurements of experimental physics. They are statements of factual character, and no dialectician, however skilled, would ever be able to derive from them a conclusion that would not be in the indicative. The dialectician and the scientist "will never obtain a proposition which says do this or don't do that; that is to say a proposition which either confirms or contradicts ethics."[62] This was the reason Poincaré could reassure the readers of his *Value of Science* that the possibility of a basic conflict between moral truth and scientific truth is simply illusory. "Ethics and science," he emphasized, "have their own domain which touch but do not interpenetrate. The one shows us to what goal we should aspire, the other, given the goal, teaches us how to attain it. So they can never conflict since they can never meet. There can no more be immoral science than there can be scientific morals."[63]

The reasoning is elementary and almost embarrassingly obvious.[64] Its real value and novelty lay in the concise formulation of a basic idea that was, for centuries before Poincaré, a commonly shared conviction of critically minded students of ethics and scientific method. It was then all the more surprising that many physicists hailed the advent of quantum mechanics, and in particular the enunciation of the principle of indeterminacy, as the factors that assured for the first time on a scientific basis a place for the freedom of will in the universe. As a result, at the centenary meeting of the British Association in 1931, General Smuts spoke in his presidential address "of the liberation of life and spirit from the iron rule of necessity." Nature, as he rather naïvely announced the good tidings to his festive audience, turned out to be "not a closed physical circle, but has left the door open to the emergence of life and mind and the development of human personality."[65] He uttered words of course that merely echoed an interpretation of the indeterminacy principle already fashionable for several years.

As early as 1926, Jeans saw modern physics in this light. Presenting the gold medal of the Royal Society to Einstein, he said: "There is now for the first time since Newton room in the universe for something besides predestined forces."[66] Truly astonishing words. In all honesty, did classical physics ever afford a proof, however tentative, to the effect that there are only "predestined forces" in the universe? Did classical physicists ever demonstrate that fear, hope, joy, decision, responsibility, and remorse of conscience are either the products of those forces or that they do not exist at all? Or was not Jeans's contention the reflection of a sadly uncritical state of mind rather than the critical appraisal of scientific results? Did humanity have to wait for the advent of modern physics to learn through the "comforting" words of Jeans that after all "humanity may not have been mistaken in thinking itself free to choose between good and evil, to decide its direction of development, and within limits to carve out its own future"?[67] What is even stranger, Jeans put the blame on Victorian science, which according to him challenged the belief of the average man in his moral freedom. The record, however, shows the very opposite. The giants of Victorian physics staunchly defended the existence of the freedom of will and drew a sharp line of demarcation between the material and the non-material realm. In their view, the determinism of classical physics had nothing to do with the exercise of free will. How could it still happen that a generation or so later Victorian physics could be used as a dark background against which the "liberating truths" of modern physics would shine forth all the more powerfully? Does not such a strange turn in the evaluation of certain stages of the history of physics indicate strongly that evaluations of this type have little or nothing to do with the "coldly" objective scientific method or with the careful study of the historical record? For is there, in all truth, any "scientific" merit in another similar passage of Jeans, which deserves to be quoted in full:

The classical physics seemed to bolt and bar the door leading to any sort of freedom of the will; the new physics hardly does this; it almost seems to suggest that the door may be unlocked — if only we could find the handle. The old physics showed us a universe which looked more like a prison than a dwelling place. The new physics shows us a universe which looks as though it might conceivably form a suitable dwelling place for free men, and not a mere shelter for brutes — a home in which it may at least be possible for us to mould events to our desires and live lives of endeavor and achievement.[68]

Just as gratuitous is Eddington's statement assuring us that following the formulation of the indeterminacy principle, "science thereby withdraws its moral opposition to free will."[69] But can the conclusions of science issue in a "moral opposition," or in any moral position whatsoever? Or is it along the intellectual mischiefs of some nineteenth-century popularizers of science that one should evaluate the true position of classical physics in its relation to the question of free will? Is it not rather a very poor reflection on the judiciousness of any physicist, however eminent, who insists with P. Jordan that the gaps in the causal chains open the way to moral freedom?[70]

Those, however, who were lured into believing that modern physics restored to them their ethical freedom, should not have any illusion about the margin of freedom thus recovered. For as Eddington estimated its magnitude: "Our new-found freedom is like that of the mass of 0.001 mg, which is only allowed to stray 1/5000 mm in a thousand years. . . . The physical results do not spontaneously suggest any higher degree of freedom than this."[71] This is not much to boast of, Eddington admitted, and in the whole paragraph this remark is perhaps the only one with any "scientific" merit. For in what follows, there blares forth in full strength the utter disregard of the incompetence of physics to arbitrate problems of ethics. And to crown the comedy, after admitting that arguments based on philosophy, psychology, and common sense demand a larger margin of freedom than this, Eddington raises the question that betrays as little scientific as common sense: "How can this be done without violence to physics?"[72]

It is appalling indeed that so keen an intellect as Eddington's could entangle itself in such glaring inconsistencies. How could he forget so readily his own statements that openly admitted the inability of physics to decide what is good and what is evil? Was it not he himself who wrote that "there is not much hope of guidance from it [physics] as to ethical orientation"[73] and that "science cannot tell whether the world-spirit is good or evil"?[74] As a matter of fact, he eloquently pointed out: "Life would be stunted and narrow if we could feel no significance in the world around us beyond that which can be weighed and measured with the tools of the physicist or described by the metrical symbols of the mathematician."[75] Contrasted to such sound utterances, his reflections on the link between free will and the uncertainty principle seem to be wholly out of joint. No one expressed this more frankly than the distinguished mathematical physicist, R. C. Tolman:

> I must caution you . . . that the opinion of one good physicist that the uncertainty principle brings free will and moral responsibility back into the world can hardly be regarded as sensible. As far as I know, moral responsibility has never left the world and, indeed, could hardly be helped by a principle which makes physical happenings, to the extent that they are not determined, take place in accordance with the laws of pure chance.[76]

These are harsh words indeed but wholly justified. The balances of science can weigh only what has physical weight, but are insensible

to factors, principles, and facts that carry the real weight in the full
context of human existence. Certainty is not restricted to the appraisal
of quantities like weight, and there is many an imponderable component of human reality with an evidence far surpassing the data of
measuring rods and balances. The deepest aspect of the relation between free will and the laws of physics lies precisely in this consideration, which received its most forceful statement in the words of a
Nobel physicist, Arthur H. Compton. For it was the recognition of
two distinct sorts of "weights" or evidences that led Compton to conclude that if there were a real conflict between physical laws and free
will, the burden of proof should rest with physics as possessing the
lesser degree of evidence: "One's ability to move his hand at will is
much more directly and certainly known than are even the well-tested
laws of Newton, and . . . if these laws deny one's ability to move his
hand at will, the preferable conclusion is that Newton's laws require
modification."[77] What is true in this respect of Newtonian physics is
true also of modern physics operating mainly with statistical considerations. As Schrödinger once remarked with obvious scorn for this
alleged contribution of modern physics to the dignity of free will and
ethics, such a contribution should consist in nothing more than in
helping us realize "how inadequate a basis physical haphazard provides for ethics."[78]

The problem of the freedom of will was perhaps the most outstanding but by no means the only ethical question caught in an apparent conflict with physical science. Another much agitated issue
concerned the problem of the purposefulness of human existence as
related to man's changing concepts of the physical universe. Here,
too, the conflict was in large part artificial, created mostly by an utter
disregard of the nature of basic assumptions of both ethics and
physics. The Aristotelian physics, tailored to save an anthropomorphic purpose in every nook and cranny of the cosmos, was in fact just
as much without solid foundation, as was the fear that human existence might have lost all purpose and direction owing to the drastically
changed world picture emerging from the Copernican revolution.
True, some cherished views and traditions proved to be wishful thinking. Yet, the purposes and goals that had been discredited by scientific revolutions were always the fanciful embellishment, never the
genuine core of ethics. Geocentricism and celestial spheres even before Copernicus' time were merely an incidental device to make man
feel assured of the protection of divine Providence. Again, when it became popular to speak of an infinite number of worlds, the only ones
who took the view that the idea of Providence became therefore
meaningless were those who never cared much for it anyhow. The

contribution of such thinkers to the relation of ethics and physical sciences consisted mainly in raising phony issues, creating a false atmosphere, and popularizing half-baked views that were only effective in compounding an already growing confusion.

It is indeed highly characteristic that it was in discussing *Uranie*, a book by the widely influential popularizer of nineteenth-century astronomy, Flammarion, that Anatole France was prompted to speak of the impact of the rapidly expanding world of astronomical science on traditional human values and standards. Adopting the view that scientific progress only makes man aware of an ever-increasing universe, which is basically unknowable, France wrote: "The Unknowable grips and envelops us. During the last two centuries it has terribly increased. Physical astronomy has not shown us the objective reality of things, but it has changed all our illusions; that is to say, our very soul. In that it has worked such a revolution in men's ideals that it is impossible for the old beliefs to subsist without transformation."[79]

The superficiality of the passage speaks for itself to anyone whose judgment is not easily swayed by the beauty of exquisite prose. Encumbered as man's vision might be by a host of illusions, man's inner world is not lacking in convictions that remain the same while the world of illusions continually changes. But science, which claims rightly or wrongly a relevance to ethical considerations and principles, changes too. France would probably be surprised to see how great those changes might become even within the lifespan of one generation. It was only a few decades ago that Jeans asked his reader to ponder the implications of life's being restricted to a small corner of the universe.[80] In our day, however, another distinguished astronomer, H. Shapley, insists on the need of an ethical outlook based on the recognition of man's awareness of the presence of manlike beings on milliards of other planets.[81] Of a true awareness of our insignificance there can of course never be enough, since such an awareness can effectively generate a deeper ethical attitude. It is highly doubtful, however, whether man will become through "interplanetary" considerations more possessed of ethical strivings, unless he has already been permeated by ethical convictions that were operative and imperative long before science began to suspect with some factual probability that there might be civilizations beyond our own planet. There is indeed not much point in being dissatisfied with the alleged narrowness of what Shapley called a one-planet ethics or deity until the countless opportunities for ethical conduct have already been seized on earth.

In all frankness, physical astronomy seems to have business other than to determine through its conclusions *and* conjectures ethical thought, conviction, and conduct. Jeans himself recognized that while science can unveil for man entirely new cosmic situations that might add new dimensions to traditional world views, "it is not for the cosmogonist to suggest answers to these wide questions, or even to the more limited questions directly raised by the sequence of events which is his own special study."[82] He even indicated one good reason why a cosmogonist — or any scientist as such, we might add — should be reluctant to attempt to trace out the possible ethical implications of a particular scientific world picture. The cosmogonist, he warned, should be aware how incomplete his system is, "how dimly most of the sequence of events can be seen, and how much of it cannot be seen at all."[83]

There are other reasons as well. If scientific method restricts itself to counting, measuring, and weighing, then the scientific world view, as Schrödinger once warned, "contains of itself no ethical values."[84] Science is not about persons, personal commitments, human destiny, or responsibility. Strictly speaking, the world picture of science does not contain even as much as a bright color, a soft touch, or a pleasant taste. This remains so, even if, like Schrödinger, one is inclined to think that of all items of knowledge the one obtained by the scientific method is the safest and the least controvertible. Still, science is not the source of decisions, principles, directives, or attitudes that can make or break not only the individual, but mankind as a whole. To hear a Nobel physicist restate it: "Science, in itself, is not the source of the ethical standards, the moral insight, the wisdom, that is needed to make value judgments; though it is an important ingredient in the making of value judgments. Social, political and military decisions are made on grounds other than those in which science is authoritative."[85] How hopeless science, or scientific method, is indeed in the face of fateful decisions that cannot be avoided and that at the same time might lead to total annihilation has been evidenced day by day for the past twenty-five years. One need only remind himself of the agonizing arguments that torment the scientific community about atom bombs, hydrogen bombs, nuclear stockpiling, neutron bombs, Laser rays, fallout, disposal of radioactive waste, and a host of other questions created by the results of the scientific method.

That the products of scientific discovery can be tools of evil as well as of good was recognized centuries ago. Still, as classical physics made its triumphal advances for almost three centuries, there was hardly a single case when a physical scientist would have considered suppressing a discovery for fear of its being eventually misused. Most of the time such concerns could hardly occur to physicists as they were inching toward their discoveries. How could Volta see in his primitive pile the tool that a dozen or so years later served to detonate explosive mines across the Seine, during the occupation of Paris in the last phase of the war against Napoleon? How could Maxwell, Hertz, and Marconi foresee that radio waves would one day be the most potent carrier of global propaganda campaigns aimed at poisoning men's hearts and minds? How could Urey suspect in 1931 that heavy water, the investigation of which earned him the Nobel Prize, would render possible the construction of the atomic bomb and not the development of neon signs as he had originally thought? Or how could the inventors of the coherent light beam (Laser) guess that three years later the prototype of a Laser rifle would be delivered to the armed forces? Leonardo, as a man of science, had in fact set a rather lonely example by destroying his plans for a submarine for fear that it might be used for the destruction of civilization. Again, no participant of the 1832 meeting of the British Association is known to have agreed with that dignitary of the University of Oxford who, upon seeing the demonstration of Faraday's newly invented method of eliciting a spark from a magnet, sadly repeated: "I am sorry for it; it is putting new arms into the hands of the incendiary."[86]

For the most part, men of science welcomed the new inventions and applications of physics with unreserved optimism. More comfort, longer life, and increased control over the forces of nature seemed to accompany the chain of discoveries. It was clearly tempting to believe that science surpasses any other factor in making man happier and "more able," as Priestley once put it, "to communicate happiness to others." There appeared to be more than a daydreamer's wish in Priestley's vision of the future, which promised that "whatever was the beginning of this world, the end will be paradisiacal beyond what our imaginations can now conceive."[87] Not that all men of science would have fully subscribed to Priestley's hopes. Still, the conviction that the results of science would of themselves assure their proper use was rather commonly shared by people interested in the scientific foundation for the progress of mankind. Did not physics seem to unveil in nature that lawfulness, stability, and effectiveness that could be taken as an enviable pattern even in the non-physical areas of human aspirations? Did not Emerson speak the mind of a whole age in declaring that "the laws of moral nature answer those of matter as face to face in a glass"? In his view, so great was the correspondence between the two realms that, as he wrote, "the axioms of physics translate the laws of ethics."[88] What is more, the physical law seemed to lack the capriciousness that so often thwarted human planning in securing its goals.

Thus, when the discovery of radium ushered in a new era in physics and in human history as well, the enthusiastic reports about its beneficial potentialities found far more credence than the scattered references to a gigantic new means of destruction hidden in it. Such indeed was the case when the *St. Louis Post-Dispatch* informed its readers on Sunday, October 4, 1903, in two articles [89] about the four grains of radium to be exhibited at the World's Fair in 1904. Of the two write-ups, one came from the pen of a staff reporter, the other from a scientist, David T. Day, chief of the Department of Mining and Metallurgy of the World's Fair. Still, the two reports were remarkably similar in tone. Both stressed what was pleasanter to believe: the apparently unlimited power of radium to illuminate rooms, streets, and entire cities. Much less credible or realistic were the few references to the inconceivably great destructive power of a future device based on a possible tapping of the energy content of the radium. After all, the fair itself was intended to be an unsurpassed documentation of the benefits and promises of applied science. A year or two later, the widely read French sociologist and scientist, Gustave le Bon, conjured up the image of an unlimited supply of energy placed at the disposal of man in radium, which, as he hoped, could put an end to arduous labor. The willingness to hear the ring of Utopian possibilities was unmistakable: "The poor would then be on a level with the rich and there would be an end to all social questions."[90]

The reference to social problems was an apt one. Most social movements and programs looked upon science as the chief justification of the contention that a better life for the wide masses is indeed feasible. With such "social faith" in science, how could one seriously entertain the idea that science itself could bring about the end of civilization in the not-so-distant future? How could the delegates of the Independent Labour Party, meeting at Aberdeen on November 17, 1915, fancy that in a few decades another world war would be brought to an end by a fantastic explosive, the vision of which was outlined to them by no less an expert on radioactivity than F. Soddy? For they were told not only of the fact that one can extract in theory at least as much energy from one pound of radium as from 150 tons of coal, but they also were warned that one pound of radium could be made to do the work of 150 tons of dynamite as well. "Imagine, if you can," Soddy told his startled audience, "what the present war would be like if such an explosive had actually been discovered instead of being still in the keeping of the future."[91] And in one of the most accurate prophecies in the history of science, Soddy outlined the availability of such a device "for the use or the destruction . . . of the next generation."[92]

Soddy's was a lonely voice. It could cause hardly more than a ripple on the mental inertia of scientific and general public alike. After all, were not the men of science almost without exception simply incredulous that such a disastrous technical feat could ever be achieved? In 1930 R. Millikan referred in fact to a "new scientific evidence," when trying to reassure those "living in fear lest some bad boy among the scientists may some day touch off the fuse and blow this comfortable earth of ours to star dust." They may go home, he noted in a tone of supreme confidence, and may "henceforth sleep in peace with the consciousness that the creator has put some fool-proof elements into his handiwork and that man is powerless to do it any titanic physical damage, anyway."[93] During the September, 1933, meeting of the British Association, Rutherford voiced the view that atomic processes are a "very poor and inefficient way of producing energy and anyone who looks for a source of power in the transformation of atom is talking moonshine."[94] In 1937 Eddington stated at the tercentenary celebration of Harvard University that "the release of subatomic energy on a practical scale is, and seems likely to remain, a Utopian dream."[95] Toward the end of 1938, Einstein spoke graphic words to William L. Laurence, science reporter of *The New York Times*, about the infinitesimal chances of ever unlocking the energy of the atom: "No. We are poor marksmen, shooting at birds in the dark in a country where there are very few birds." A few months later, in February, 1939, Fermi told the same reporter that an atomic bomb, even if it proved theoretically feasible, was at least twenty-five to fifty years away.[96] Perhaps Fermi deliberately overestimated the time to allay the anxiety of laymen. Yet, scientists talking to one another could strike just as emphatically the note of skepticism in this regard. Not long after Fermi made this remark, Niels Bohr listed for E. P. Wigner fifteen reasons against the practicability of drawing energy on a large scale from the process of fission.

Undoubtedly, the technical difficulties that in 1939 stood in the way of the utilization of atomic energy necessarily appeared gigantic. Still, the skepticism of most physicists in this regard had other motivations as well. In the 1930's physicists were more dazzled perhaps than ever by the great discoveries of physics and by its theoretical advances. It could therefore hardly occur to them that their triumphs might bring anything but good. A strong note of optimism was struck in Rutherford's words, which described the first quarter of the century as the heroic age of science.[97] If the first two decades of the century were, as he put it, "periods of intense activity in physics," the 1920's and the 1930's were nothing short of feverish. As the German physicist, P. Jordan, described the exalted state of mind prevailing among his colleagues: "Everyone was filled with such tension that it almost took their breath away. . . . The ice had been broken."[98]

Beneath the ice, however, there lay not only magic keys to the secrets of matter, but also the vision of terror. Almost overnight there emerged the spectacle of helpless physicists involved in a work raising the gravest moral questions, to which answers were not to be found in mathematical formulas and measurements. But before this became abundantly clear, leading physicists found themselves enveloped in a shroud of secrecy that put an end to what seemed to be the most valuable ethical aspect of scientific research: the free flow of ideas. Scientists suddenly urged secrecy on their colleagues and smuggled all-important data across heavily guarded borders at the risk of their lives. Discoveries could no longer be considered the property of all humanity. It was impossible to carry on the hallowed attitude exemplified by the Curies, who made public property of what they worked out with incredible labor — the process of separating radium from pitchblende. In its effects, however, radium was but a weak daughter of uranium. The parent element had such potentialities of destructiveness that Marie Curie's reasoning, "physicists always publish their researches completely,"[99] could no longer be the highest rule of research. The gigantic proportions of evil emerged in their naked reality and the line between those who could and those who should not share the results of the uranium research suddenly had to be drawn along lines that were clearly ethical in essence, the borderlines of democracy.

It was not without the greatest reluctance that men of science acquiesced in such a policy of secrecy, although it meant to forestall a global abuse of their findings. Yet, it was precisely the possibility of abuse on a gigantic scale which they could no longer disregard. It was impossible to consider any longer the laboratory as a workshop automatically assuring the beneficial effects of its products. Scientific research appeared rather suddenly as what it is and has always been: a human act, as human as any other, having immediate links with the categories of good and evil. True, science had been greatly engaged in World War I, but not through its basic, theoretical branch. Some first-rank theoretical physicists like K. Schwarzschild and H. Moseley then discharged their patriotic duty, if they had to, by dying on the battlefield. In 1939 the case was wholly different. Theoretical physics found itself mobilized on both sides and physicists had to busy themselves daily with the details of a device capable of destruction surpassing any yet in history. All this was a far cry from even the most bitter years of the Napoleonic wars, when Humphry Davy as a scientist could travel with a French passport throughout France and Italy, although both France and England had interned British and French citizens, respectively. How deadly ironic could Davy's once proud words sound by 1940: "If two countries are at war, the men of Science are not. That would indeed be a civil war of the worst description."[100] Yet, by 1940 a global war was already engulfing mankind, and it fell to men of science to make the war deadly beyond description. That the study of the nucleus should first issue in an awesome bomb must have appeared to most of them, as Lise Meitner put it, a blasphemy. Kapitza surely expressed the thinking of most of his colleagues in saying that to speak about atomic energy in terms of the atomic bomb was comparable with speaking about electricity in terms of the electric chair. Still, there was no ignoring the closeness and magnitude of evil implied in the possible success of a uranium explosive.

This is not the place to retell the well-known story of the heatedly discussed pros and cons that suddenly filled the air as the production of the first atomic bomb neared its completion. In fact, no sooner had the project got into high gear than physicists discovered the principle of the hydrogen bomb, and the question arose whether the tremendous heat developed by an atomic bomb could not set off a hydrogen-fusion chain reaction in the ocean, or a nitrogen chain reaction in the atmosphere. The top scientific coordinator of the atomic bomb project, the late A. H. Compton, recalled this fateful possibility and the pending choice in words worth quoting: "This would be the ultimate catastrophe. Better to accept the slavery of the Nazis than to run a chance of drawing the final curtain on mankind. We agreed there could be only one answer. Oppenheimer's team must go ahead with their calculations. Unless they came up with a firm and reliable conclusion that our atomic bombs could not explode the air or the sea, these bombs must never be made."[101]

This was only one of the many agonizing moments physicists had to learn to live with after 1939. The almost unbearable weight of moral considerations steadily diminished the number of those, who, like Fermi, could at times brush aside objections against the explosion of the finished bomb with the flippant words: "Don't bother me with your conscientious scruples. After all, the thing is superb physics."[102] But on the fateful morning of July 16, 1945, even Fermi's vision became sensitive to shades of truth hitherto hidden to him. For the first time in his life, he asked a friend to sit at the wheel and drive him back from the test site to Los Alamos. As he told his wife next morning, it seemed to him "as if the car were jumping from curve to curve, skipping the straight stretches in between."[103] The most revealing of all the comments, however, was that of K. Bainbridge, Oppenheimer's principal assistant. As he accepted the congratulations of his boss following the spectacular outburst, he just looked into Oppenheimer's eyes and said softly: "Now we're all sons-of-bitches."[104] All this, of course, signaled the beginning of a most excruciating ethical reappraisal. With the use of the bomb over Nagasaki and Hiroshima, the burden of ethical considerations came to the surface in full force. The scientific community became the scene of mutual accusations charged with high emotions. The responsibility for the bomb suddenly proved unbearable. As Oppenheimer summed it up poignantly: "In some sort of crude sense which no vulgarity, no humor, no overstatement can quite extinguish, the physicists have known sin; and this is a knowledge which they cannot lose."[105]

It was only natural that scientists connected with the construction of the bomb began to apologize and disclaim at least part of the responsibility. Einstein himself felt it necessary to play down his decisive role in alerting President Roosevelt's attention to the necessity of going ahead with the bomb project. "I really only acted," he wrote to his biographer, Antonina Vallentin, "as a mail box. They [Szilard and Wigner] brought me a finished letter and I simply signed it."[106] Others, analyzing the fateful years in retrospect tried to fancy, as Heisenberg did, that a firm refusal on the part of twelve leading physicists could have prevented the construction of the atomic bomb. Weizsäcker, for instance, thought for a moment after the war that physicists perhaps should have formed a sort of international monastic order with disciplinary power over its members. Or should perhaps all of them have followed Kapitza's daring and risked their lives by refusing to participate in atomic weapon projects?

To hope that by any such means the chain reaction of discovery can be stopped is, however, simply ignoring the inner dynamism of scientific knowledge. As Soddy remarked in 1915, in referring to the concept of an atomic explosive: "The discovery . . . it is pretty certain, will be made by science sooner or later."[107] And as the rate of growth of scientific discoveries shows, it is safer to say "sooner" if one cannot resist the temptation to make prophecies. To think that a potentially dangerous scientific discovery could be nipped in the bud would be illusory, even if all the members of the scientific community were possessed of the most acute ethical awareness. Nor is the suggestion made by the Bishop of Ripon to the British Association meeting in Leeds in 1927 practical: that men of science should take an extended holiday to allow time for the minds of men to catch up with them and assimilate the moral consequences of their actions.

The disconcerting fact, however, is that the gigantic responsibilities of the atomic age caught many a physicist mentally unprepared. In 1915 Soddy's was a lonely voice warning that the world had fooled too long in the past with the achievements of science. Hardly anyone felt a sense of urgency in reading his prophetic words: "The cold logic of science shows without the possibility of escape, that this question if not faced now, can have only one miserable end."[108] The state of mind of the scientific community was unaccustomed to ponder such warnings. Thus when the scientists were suddenly faced with the choice to make or not to make the bomb, what they regarded as the imperative necessity of securing this superweapon for the democratic societies dwarfed any other consideration about the morality of the enterprise. The scientists at Los Alamos felt, as H. Bethe later recalled those years, that first the job had to be done. "Only later," said Bethe, "when our labors were finally completed, when the bomb was dropped on Japan, only then or a little bit before then maybe, did we start thinking about the moral implications."[109]

Few human actions are as ethically commendable as rallying to the defense of democratic societies. Refusal to produce the bomb would no doubt have jeopardized the very possibility of democracy. And this no free man had the right to let happen. As E. P. Wigner so aptly noted in connection with the twentieth anniversary of the destruction of Hiroshima: "The scientist, like any other civilian, should not act in such a way as to make democracy impossible."[110] Those who made the bomb could rightly refer to the fact that by the admission of the most reliable Japanese sources, the use of the bomb had prevented the continuation of the war and saved millions of lives, both civilian and military.[111]

All these considerations should not draw attention from another factor that served as a most effective inducement to scientists to work wholeheartedly on the construction of the bomb. This factor is closely connected with the emotional components of scientific research and can be described as the utter fascination of the challenging, or to use Oppenheimer's words, the "technically sweet" aspect of research.[112] It is equivalent in essence to what Fermi called "superb physics," and its high attractiveness was clearly evident in the deliberations leading to both the atomic and hydrogen bombs. How this fascination with a possible major discovery could force into the background moral considerations was stated with a highly respectable candor by Oppenheimer himself: ". . . it is my judgment in these things that when you see something that is *technically sweet*, you go ahead and do it and you argue about what to do about it only after you have had your technical success. That is the way it was with the atomic bomb. I do not think anybody opposed making it; there were some debates about what to do with it after it was made"[113] (italics mine).

Hardly anything can compete with the expressiveness of the words "technically sweet" in characterizing the state of mind of many a scientist at this particular juncture. Nor can anything be farther removed from the realm of keen ethical awareness about the possible consequences of one's work than a mind preoccupied with what is "technically sweet." Undeniably this state of mind can act as a powerful barbiturate and help one ignore at a crucial moment the ethical components of a situation and make him do what James Franck, another pioneer in atomic research, described so graphically: "So we took the easiest way out and hid in our ivory tower. We felt that neither the good nor the evil applications were our responsibility."[114] When, however, the hour of awakening arrived, there was no more place for such sophistication. The demands of the times suddenly made everybody, even scientists, appear in his true size, and as Von Neumann admitted, that size was puny: "I know that neither of us were adolescents at that time, but of course we were little children with respect to a situation which had developed, namely that we suddenly were dealing with something with which one could blow up the world."[115]

True, scientists as a body can no longer withdraw into an ivory tower. The course of history makes this type of evasive tactic simply impossible. What is more, scientists face even greater ethical problems than those involved in the construction of the atomic bomb, if gradation in this respect is meaningful. If the very existence of a hydrogen bomb and the knowledge of its construction could be declared wholly evil by some prominent physicists,[116] what about the neutron bomb, let alone the genetic effects of the radioactive fallout of all these devices? The grave ethical problems involved in space exploration are only beginning to emerge. What if the large amounts of fuel burned by space rockets passing through the earth's atmosphere should produce fatal changes in its chemical composition? What if space probes returning from the moon or from any of the planets should bring back strains of bacteria against which our biosphere would be defenseless? Did not the establishment of a contact between two isolated biospheres, Europe and America, result in widespread disease and countless deaths on both sides of the Atlantic? Again, what if successful radio contacts with the inhabitants of other planetary systems make a potentially hostile race aware of our existence? Was it not precisely the preoccupation with such a possibility that prompted the Nobel physicist, C. N. Yang, to say that we should not try to answer an eventual radio message coming from another planetary system? Or should mankind rather adopt the advice of the German astronomer, S. von Hoerner, who saw in such radio contacts man's last hope of avoiding global self-destruction?[117] In his view, several advanced extraterrestrial civilizations had already gone through cycles of possible self-destruction, and man might very well profit from learning their solutions to such problems. Truly, man today is pondering problems that only a few years ago still belonged on the pages of science fiction. As science keeps forging ahead, many other fearful question marks and danger signals will emerge. At present, mankind has already reached the stage where a decision has to be made whether nuclear explosives will be used for "overkill" or for digging canals and unearthing rich mineral deposits.

Ignoring these and similar problems can only court disaster. But can science, or the so-called scientific method, alone lead to solutions in such dilemmas? Is not the paramount lesson of the opening decades of the atomic age that science of itself can give no clues as to how to use its products? Should it not be abundantly clear that in the intricate structure of the modern political state scientists are unable to retain exclusive control of the use of their inventions? Should it not be evident that the more far-reaching and ramifying are the effects of the use of a tool, the more futile becomes the attempt to tabulate and compare the "good" and "bad" effects and establish the morality of the tool on the basis of a so-called calculative ethics? Should it not be obvious that the scientific method, so intent on searching out its limited truth and this truth alone, cannot provide the enlightenment, ethical earnestness, and good will required to reach a firm consensus on the ethical problems raised by the very same scientific method? Should not the principal human lesson of the atomic turmoil consist in the realization of how ridiculous is the contention that familiarity with science is the most effective way to produce good citizens, firm characters, and men of good will?

Age-old fallacies, however, are much more difficult to uproot than is generally believed. The world has been served, ever since the French *philosophe* movement, huge doses of the tenet that the safest road to virtue lies in the study of the sciences. Repeated a million times over, this "truth" had to sink sooner or later into the innermost recesses of minds and there was little exaggeration in the words of the famous chemist, M. Berthelot, who stated in 1897 in an address on "Science and Popular Education": "People begin to understand that in the modern civilization, every social utility derives from science, because modern science embraces the entire domain of the human mind: the intellectual, moral, political, artistic domain as well as the practical and industrial."[118]

These were the times that saw the publication of important works devoted to the analysis of scientific method that, like K. Pearson's *Grammar of Science*, opened with an ambitious chapter on "Science and Citizenship." True, Pearson emphasized that men of science are not necessarily good citizens and that the judgments of natural scientists are not necessarily sound judgments in problems relating to "Socialism, Home Rule, or Biblical Criticism."[119] True, he warned that judgments made by scientists on such matters would be dependable only if the scientist carried his scientific habits into the fields in question. Still, Pearson believed that only one specific method is and can be scientific or reliable, and it was only in such a method that he could find a firm basis for the unity of science and human knowledge in general: "The unity of all science consists alone in its method, not in its material."[120] Again, it was this method alone that in his view could produce a state of mind which, as he put it, was "an essential of good citizenship."[121] The method, the source of everything good and desirable, was of course best exemplified by that used in physics and mathematics. Little could Pearson guess that half a century later the analysis of modern physics would force no less a physicist than Bridgman to deny the existence of the so-called scientific method: "I am not one of those who hold that there is a scientific method as such. The scientific method, as far as it is a method, is nothing more than doing one's damndest with one's mind, no holds barred. What primarily distinguishes science from other intellectual enterprises in which the right answer has to be obtained is not the method but the subject matter."[122]

In the meantime, however, Pearson's views on science and citizenship enjoyed an immense popularity and were constantly embellished with sanguine variations. Many of these appeared in textbooks on science or in papers devoted to scientific education. In reading these passages, one is tempted to think that ethics was merely an extension of analytical mechanics, introductory quantum mechanics, or physical chemistry. The claim is made with sustained regularity that the study of science will teach man "how to think straight, how to avoid deceit, and how to benefit mankind most by honoring the authority of Truth."[123] Such is not merely the conviction of isolated individuals. Policy-making bodies of powerful educational organizations can be seen just as eager to submit the sweeping proposal that all fields and facets of education be infused "with the spirit of science." It is claimed that this is the only way to bring about the "development of persons whose approach to life as a whole is that of a person who thinks – a rational person."[124]

Knowledge is of course a foremost source of benefits. Yet, not every component of useful and indispensable knowledge is "scientific" in the ordinary meaning of this term. Nor will one's familiarity with science issue, of necessity, in the recognition of ethical norms, let alone in their faithful observance. It is rather frightening that the disaster of Hiroshima and the ominous proliferation of nuclear weapons were simply not effective enough to demolish in some the faith in science as the main source of ethical standards and ethical inspirations. True, there were not lacking voices, like J. B. Conant's, resolutely rejecting as a "very dubious educational hypothesis" the claim "that the study of science is the best education for young men who aspire to become impartial analysts of human affairs."[125] He had to acknowledge that even today there are still "brave men" around who keep insisting that the habits of thought, the methods, and the viewpoints of scientists are applicable to the solution of almost all problems of the modern world. One can still hear scientists expound the view that peace and sanity in this world will be secured only by the widespread application of scientific method. Thus Heisenberg presented science in the aftermath of the war as the basic means of international understanding.[126] According to him, in science, as contrasted to political thought, there will always be a right and a wrong. Furthermore, as he put it, this "right," or the truths of theoretical science, are so obvious and convey such force of conviction that "scientists of the most diverse peoples and races accepted it as the undoubted basis of all further thought and cognition."[127] (Fortunately for mankind, a great number of scientists do not derive all their "further thought and cognition" from the propositions of theoretical science.) In a much similar vein the British mathematician and lecturer, J. Bronowski, claims, for instance, that it is science that provides the two primary values: independence of thinking and tolerance in attitude. Once these are secured, there follows step by step the full spectrum of values, or rather virtues: "dissent, freedom of thought and speech, justice, honor, human dignity and self-respect."[128] At the basis of such a claim, however, one finds neither the factual analysis of the situation nor the consensus of scientists. The true support of such a claim is not reason but a vision bordering on a dream: that alone could inspire the phrase introducing the foregoing quotation: "The society of scientists is simple," Bronowski tells us, "because it has a directing purpose: to explore the truth."[129]

Implicit and at times explicit in such assertions is the misguided belief that truth is a monopoly of science, that anything that does not pass through the filter of "scientific method" is irrational or undemonstrable at least. Such was the contention of B. Russell, who argued that "whatever knowledge is attainable, must be attained by scientific methods." Consequently he could say that "what science cannot discover, mankind cannot know," and if science cannot resolve problems of values, it is only because such problems "lie outside the realm of truth and falsehood."[130] More recently a similar note was struck by E. Rabinowitch, a scientist-publicist gravely concerned about the responsibility of men of science in shaping the society of the new scientific age. According to him, the responsibility of scientists as regards the future of mankind consists of two points. First, they have to secure a wider understanding of the dynamical and irreversible impact of scientific discoveries on society; second, they have to convince the world about the role of scientific method in problems bearing on the future of mankind. While admitting that many "non-scientific" factors are also to be considered, Rabinowitch classes all these as "irrational." To hear Rabinowitch name them, they are the "political and national animosities, established traditions, ideological, racial and religious fanaticisms."[131] Apart from the gratuitous assertion that whatever is not within the ken of science is simply irrational, many scientists would refuse to go along with Rabinowitch's "cure-all" proposition, which assigns to the scientists the role of finding a "reasonable compromise between these forces and the arguments of reason."[132]

The reasons for this refusal are not difficult to fathom. First, one cannot advocate with logical consistency a "reasonable" compromise between what are, by definition, "rational" and "irrational" attitudes. Second, are scientists pure "rational" beings, free from any of those "irrational" motivations? Would it not exemplify a far greater degree of scientific objectivity to acknowledge with Conant that as "human beings scientific investigators are statistically distributed over the whole spectrum of human folly and wisdom much as other men"?[133] Not only can scientists claim no special virtues and wisdom when faced with non-scientific problems but even in evaluating current scientific problems they often show a surprising lack of unanimity. One need only recall the heated controversies concerning the proper interpretation of the first television pictures obtained of the surface of the moon, or the question of the alleged presence of microorganisms in some meteorites. Actually, some scientists involved in the latter dispute felt impelled to appeal to an "impartial panel" to assure an "objective evaluation" of the data and the arguments. Not that any of the parties involved had tried to falsify their findings in any sense. Such brazen disregard for facts is very rare in the more recent annals of science. The experience of P. Jordan, the noted German physicist, is typical in this respect. Although closely associated with several branches of science for forty some years, he could recall only three cases of scientists deliberately submitting false data.[134]

For all that, the scientist is just as subject as any other intellectual to the disturbing influences of subjective preferences, commitments, and wishful thinking. Furthermore, when it comes to some not strictly metric aspects of his work, such as the recommendation of a research project to be supported by public funds, the scientist can display, as any other human being, an appreciable disregard for strict ethical norms. In self-defense, such a scientist will always point to the fluctuations and uncertainties of scientific opinions, but in doing so he unwittingly serves notice that positions still in a flux should not be advocated with a patent disregard to other possibilities. Yet, such narrow-minded advocacy of half-truths or shades of truth is not infrequently undertaken by men of science. Thus atomic scientists offered widely differing views on a postwar policy concerning atomic disarmament. The magnitude of their differences prompted in effect the comment attributed to the late Secretary of State, James F. Byrnes: "In this age it appears every man must have his own physicist."[135] What was a statement of slight scorn in 1945 soon became a practical precept, however, and political leaders used it to secure for their views the desired scientific "authority," with many physicists eagerly offering their services. Expeditious as this procedure could be at times, it could also easily turn into something ridiculous. The alleged conclusion of physicists, for instance, that a nuclear explosion might knock the axis of the earth out of kilter was referred to in 1956 by a presidential candidate favoring a complete ban on nuclear testing.

True, one can always be misled by a single consultant, but safe guidance is not necessarily assured when hundreds of scientists volunteer their views on a matter still incompletely understood or investigated. Have we not witnessed recently the spectacle of two groups of scientists heatedly contradicting one another about the possible interference with astronomical observation by metallic needles placed into orbit?[136] Again, which section of scientists is to be believed, the innocent layman might wonder, when it comes to the appraisal of the genetic dangers of radioactive fallout? Did not C. P. Snow, himself well trained in science and a devoted advocate of science, warn passionately that in matters of this type there is no such thing as an "authoritative opinion" that could be handed down by any scientist?[137] What Snow said of a single scientific "overlord" holds as well for the body scientific: it disposes of no foolproof means of forging infallible conclusions concerning the uses of the products of science. To try to sell the non-scientist intellectual and political leaders on the promise that "scientific method" can do this would be, to say the least, preposterously non-scientific, but probably even worse: it is simply courting disaster.

Another cliché in support of a "scientifically" argued ethics is that owing to their familiarity with the scientific method, scientists, as compared with other professional groups, display a "greater humility in the face of the unknown, greater tolerance of each other's points of view, and greater respect for mutual difficulties of nations and societies."[138] One could only wish this were so. Of genuine intellectual humility one can never have enough. But is intellectual humility possible for any scientist to whom the limitations of scientific method are almost non-existent? For no matter which aspect of the interplay between physics and ethics is considered, one is invariably led back to the root of the question: What is the competence of physics in matters that are not part of it but only related to it, at times but remotely? And lest one be tempted to believe that the physicist's familiarity with the scientific method can transform him into an oracle of truth in general and the apostle of good in particular, the following disclaimer is one to be pondered, coming as it does from the distinguished scientist, A. V. Hill, a former president of the British Association. Speaking at the 1952 meeting of the Association on the "Ethical Dilemma of Science," he warned that as regards morality and ethics, "There is no such thing as 'the scientific mind.' Scientists for the most part are quite ordinary folk. In their particular scientific jobs they have developed a habit of critical examination, but this does not save them from wishful thinking in ordinary affairs, or sometimes even from misrepresentation and falsehood when their emotions and prejudices are strongly enough moved."[139]

Far from being a source of ethical values and moral rectitude, the scientific method presupposes the presence of moral qualities in the researcher if a well-balanced, reliable, and objective evaluation of the observations is to be secured. One of the most articulate writers on the structure and conditions of the experimental method, Claude Bernard, stated long ago: "The sound experimental criticism of a hypothesis is subordinated to certain moral conditions; in order to estimate correctly the agreement of a physical theory with the facts, it is not enough to be a good mathematician and skillful experimenter; one must also be an impartial and faithful judge."[140] Far from generating an upright character, scientific work will contribute to the strengthening of ethical convictions in the scientist only if he is already possessed of ethical values and practices. Such an indirect ethical inspiration, deriving from research, has often been noted by men of science. From the advancement of natural philosophy, Newton expected the "Bounds of Moral Philosophy" to enlarge.[141] Humphry Davy thought that scientific work would guard the scientist "against the delusions of fancy," make him "reason with greater reverence concerning beings possessing life," put him on guard against "turbulence and hasty innovation," and exhibit him "as the friend of tranquillity and order."[142] It was this noble ethical atmosphere that prompted Maxwell to describe the past generations of the Royal Society as "the company of those men who aspiring to noble ends . . . have risen above the regions of storms into a clearer atmosphere, where there is no misrepresentation of opinion, nor ambiguity of expression, but where one mind comes into close contact with another at the point where both approach nearest to the truth."[143]

Victorian rhetoric thrived of course on fashionable exaggeration and was only too willing to gloss over obvious non sequiturs in science as well as in other matters. The Royal Society had, like other societies, both serene and stormy days and hardly deserved the ethical apotheosis implied in Maxwell's words. To hear the ethical inspiration of science described in its true and moderate proportions, one must turn to Poincaré.[144] Science, he said, puts the individual constantly in contact with something grandiose, lifts him above himself into an objective sphere, and offsets egotistical tendencies. Scientific work inspires the love of truth, demands unconditional sincerity, and forms intellectual habits that in their turn influence moral decisions. As a collective work, science makes the individual aware of his social role and responsibilities and might even incite in the scientific worker an increased respect toward mankind and a deeper appreciation of human problems. Yet, as Poincaré pointed out, only a genuine scientific attitude will issue in such ethical benefits. The genuine scientific attitude he defined as the willingness to be mindful of all the possible aspects of a question to which science might have some measure of relevance. In effect, Poincaré contrasted full science to half-science, as he called it, and the latter he denounced as a grave intellectual danger. It was for him the half-science (*demi-science*) that produced the narrow-minded thinking sensitive only to what is quantitative in the whole texture of human existence and experience. He described the psychological effect of half-science as a source of unethical attitudes and protested against giving science an exclusive role in education. It was at the door of half-science that Poincaré laid the blame for hasty generalizations, for the rejection of everything traditional, for the "scientific snobbism that can be duped so easily by novelties."[145]

Poincaré clearly realized that in moral action and attitude there is an irreducible sector impervious to the analysis by quantitative laws. In his view morality had an existentialist character, to use a modern term, that led him to say that the experimental study of morality is no more morality itself than the study of the physiology of digestion is a good dinner. The independence of ethics from the laws of physics he asserted even for the hypothetical case of absolute determinism being demonstrated on the psychological level. Thus he could rightly state: "Of a science animated by a real experimental spirit ethics has nothing to fear; . . . the best remedy against a half-science is more science."[146]

It is indeed to more science, or in other words to the fullest possible consultation of all available information, that ethics must resort in order to reach a clear and objective appraisal of a given situation. For the moral good must always be a reasonable good, that is, something in accordance with all the aspects of all the legitimate claims of human nature. In a culture ever more dependent on the use of the products of technology, the reasonableness of such a use can often be established only by a most accurate scientific analysis of the situation and circumstances. The classical cases concern the use of drugs and foods, but there are innumerable others. To permit their use, society or government clearly must adjudge the bearing of all scientific data relevant to such items. Statistics, graphs, and formulas can prove a most useful means in helping to clarify ethical problems provided they are not elevated from the status of method into an ethical philosophy. For as Quetelet pointed out in 1829, in his analysis of the first Belgian census, certain aspects of moral phenomena can be analyzed by mathematics without usurping morality itself. Quetelet's views on the matter found a memorable echo in the thinking of Florence Nightingale, one of his students and also distinguished in many other aspects of human excellence. She felt that one could be greatly helped in understanding God's thought by studying statistics, the data of which she took as the measure of God's purpose.[147] There is indeed room in Condorcet's program to make use of the calculus of probabilities in analyzing human affairs provided one remains aware of the warning of A. Cournot, one of the great nineteenth-century authorities on the application of probability calculus in social statistics: "The acts of living, intelligent and moral beings are by no means explained, in the present state of our knowledge, and there is good reason to believe that they never will be explained by mechanics and mathematics."[148]

If ethics must have recourse to science and statistics out of respect for the *totality* of pertinent viewpoints in a question, it is again the insistence on a complete view of science that will show that scientific laws and theories as such are not basic principles or guidelines for human action. That W. Ostwald could go so far as to lecture widely on an "ethics" based on the second law of thermodynamics was clearly due to the fact that his Comtean convictions gave him a very deficient view of the nature of scientific law in particular and of science in general.[149] His motto, the "energetical imperative" had as little to do with genuine thermodynamics as it had with trustworthy ethics and at best can be remembered today as a symbol of the failure that is in store for efforts of this type.

Not that such a failure would have effectively discouraged similar attempts. Maupertuis' leading idea is still luring away many scientific minds, though their connection with physics is only indirect. Some of them are the latter-day advocates of the utilitarian ethics of Jeremy Bentham, an avowed admirer of Maupertuis, who promote with evangelical zeal the alleged effectiveness of the "felicific calculus." Others are the representatives of the so-called evolutionary or Darwinian ethics, but the biological nomenclature they use only partially hides the fact that the laws of physics are supposed to support ultimately such ethical theories. Their forerunner, H. Spencer, was at least very explicit on this point. In his *Data of Ethics* he stated with complete clarity that the roots of ethical conduct lie in human activities that "in common with all expenditures of energy, conform to the law of the persistence of energy: moral principles must conform to physical necessities."[150] In a chapter entitled "The Physical View," he stated that there is "entire correspondence between moral evolution and evolution as physically defined."[151] Thus he merely paraphrased a law of physics when he defined moral life as "one in which this maintenance of the moving equilibrium reaches completeness, or approaches most nearly to completeness."[152] Similarly, he could offer only variations of the same concept borrowed from physics when he went on to deal with the biological, psychological, and sociological aspects of ethics. For Spencer's view on ethics was distinctly quantitative, as the chapter on the "Relativity of Pains and Pleasures" proves, and there he merely echoed Maupertuis' thought in the sophisticated prose of his evolutionary philosophy.

Among those who espoused Spencer's views on an evolutionary ethics, the correlation between morals and the laws of physics was perhaps not so elaborate but still clearly discernible. John Dewey, for one, stressed that morals based upon the study of human nature would find the facts of man continuous with those of the rest of the universe "and would thereby ally ethics with physics and biology."[153] He, too, identified what was morally right with the multitude of concrete physical demands which man had to take into account if he were to survive. C. H. Waddington's *Science and Ethics* went along the same lines and Chauncey D. Leake in his "Ethicogenesis" looked for a basic ethical principle as naturally operative as the principle of the conservation of energy. The essence of ethics he therefore placed in the recognition of the fact that "the probability of survival of a relationship between individual humans or groups of humans increases with the extent to which that relationship is mutually satis-fying."[154] One might of course give such a theory the euphemistic name of "harmony theory," but in essence it is physics in disguise, and in a very crude one at that. Instead of leading to harmony, it speaks rather of the deadly extinction of all human harmony in that remorseless equalization of all personal talents, distinctions, heroic innovations, and original insights that kept mankind on the road of material and spiritual progress. Indeed, a little reflection on the na-ture of the physical laws of energy conservation and dissipation might indicate the fallacy of any attempt to hang a "moral flavor," as T. H. Huxley remarked long ago, around the Darwinian phrase of the sur-vival of the fittest.[155]

The laws of physics, the laws of the conservation and dissipation of energy, to be specific, are not only incapable of serving as a foun-dation for ethics, but it is highly doubtful whether they imply value principles even in a general sense as argued by the physicist, R. B. Lindsay.[156] True, the activity of any living organism aiming at self-preservation is in a sense a struggle for a greater degree of order within the boundaries of its life sphere. The same holds of any cul-tural endeavor on the part of intellectual beings. The construction of machines, the creation of language and science, the development of systems of communications, the establishment of political order, the repression of law-breakers, or *disorderly* persons, is no doubt an effort in the direction of greater order. Although the degrees of the order thus attained can be evaluated to some extent, ethical choice is only in an indirect sense an option for the least probable, or most ordered state. The analogy between the trend imposed by ethics and the de-crease of entropy is very remote and will lead to contradictions if pushed too far. For one, the second law of thermodynamics will never issue in a moral imperative. Furthermore, moral obligation at times demands actions, such as self-sacrifice, which if carried out, will not only enhance the ethical order, but will, as physical acts, diminish physical orderliness and increase the amount of entropy.

Just as the law of entropy is incapable of leading to value conclu-sions, no genuine value concept is contained in the physical law of least action. Although it is possible to see some goal-seeking tendency expressed in this law,[157] the "goal" of a physical process and the pur-pose of an ethical act remain two basically different notions. Again, one should be somewhat wary about the alleged ethical content of the principle of complementarity. Undoubtedly this principle forces one to recognize the equal importance of irreducible aspects of physi-cal reality. It may therefore be suggested that such a recognition might foster an attitude of tolerance when it comes to the reconcilia-

tion of conflicting political, economical, religious, and cultural aspirations.[158] Yet, our lot would hardly improve in any sensible degree
if, to grow in the spirit of tolerance, we were to fall back on medita-
tions on laws of physics. Nor can the various similarities that may
exist between the derivation of a physical and an ethical law justify
the contention that a universally valid ethics can be established only
on a "scientific basis." A noted physicist who discussed this view with
obvious sympathy is H. Margenau. Yet, even he felt impelled to warn
that for all we know the ethical "ought" cannot be derived from
empirical observations.[159]

It is doubtful that anyone will ever derive ethics from physics. Such
an effort, it should be noted, is not only futile; it is also a "most dangerous" one, as Max Born, an outstanding physicist of our century,
remarked in an address delivered on June 24, 1964, before a gathering
of Nobel laureates.[160] The dangers implied in such an effort should
help one recognize that ethics is not sociology, that is, a sociology
limited to summarizing the statistical averages of human behavior.
For instance, informative as quantitative accounts of the sociological
pattern of suicide might be, they can hardly contain an ethical insight
into the tragedy of such acts. Much less is ethics an ennobled form
of the so-called social physics, which aims at establishing laws stating,
for instance, that the speed of propagation of rumors between two
cities is directly proportional to their populations and inversely to
the distance that separates them. It is puzzling that Margenau found
such points instructive for a scientific understanding of ethics. Yet,
it was on the basis of that tenuous parallelism between the methods
of social physics and ethical studies that he expected to confer "upon
moral philosophy a measure of universal authority and attractiveness
now enjoyed by science."[161] That such attempts at explaining ethics
invariably bog down in the morass of inconsistencies constitutes but
a new chapter in an old story, the details of which are not necessary
to repeat. Actually, did not Margenau himself see that pitfall when
he emphasized in the same context that his approach to ethics should
not be taken as an example of the "naturalistic fallacy of reducing
ethics to science"?[162]

That some men of physical science fascinated with the spectacular
sweep and power of physics are tempted to make it a basis for other
fields of study is understandable of course. Physics will hardly appear
in a more glorious light, however, by being used for purposes that
by definition it cannot serve. No positive contribution to society can
be achieved either by forgetting that the advances of physical science
are bought at the price of methodically eliminating from its subject
matter aspects of reality that cannot be measured, though they are

of crucial importance in the whole texture of human existence. Only the inveterate worshipers of science are apt to forget this. They will continue talking about "scientific ethics," and will state, if they dare to be thoroughly consistent with their basic assumptions, that all forms of social activity or human achievement are ultimately reducible to electron-proton interactions.[163] Statements of such crudeness serve only one beneficial purpose. They reveal the hopeless vista of a physicalist ethics in its true nature. As is usually the case, such vistas emerge from a lack of proper information about physics. Those who are really familiar with the development of physical science know only too well that, to quote Heisenberg, "as facts and knowledge accumulate, the claim of the scientist to an understanding of the world in a certain sense diminishes."[164]

That science has had to come to recognize the steady narrowing of its range of explanations might be discomforting to those to whom more science is the only answer to the basic irrelevance of science in some fields. It was, however, more than discomforting, in fact outright tragic, that in not a few men of science awareness of this limitation of scientific answers considerably decreased. Actually the terrible upheavals of recent history had to come to make some of them realize that there is such a thing as "sin" outside the laboratory, that science can see only one of the many dimensions in which human beings and society develop. In particular did not the fact fade away in many a scientific mind that the deepest of human dilemmas is still with us — that no knowledge, not even the knowledge of the atom will of itself beget good will and active conformity to what ought to be done? Did not a whole scientific culture espouse the misleading syllogism of Confucius, who anticipated by more than two millennia the fundamental article in the creed of the physicalist ethics: "When things are investigated, then true knowledge is achieved, when true knowledge is achieved, then the will becomes sincere." Herein lies, however, the momentous non-sequitur, although one could only wish that Confucius had been right, for then the rest of his aphorism would readily follow and the cause of universal peace would be steadily gaining ground.[165] Still, to what degree did our culture succeed in achieving peace even though it had an incomparably greater share of scientific information than any earlier phase of human history? Was it not precisely the horrible experience of atomic devastation that prompted Einstein to confess that "the real problem is in the minds and hearts of men"?[166]

Had the elementary truth of the existential difference between knowledge and goodness been remembered, there would not be heard well-justified warnings today about the professional trespassings and temptations of the "scientific priesthood." There would not be grave questions raised about certain contentions of cybernetics and "social engineering." It would not be necessary to ask the most pertinent and ominous question of our times: What or who will engineer those who claim the scientific skill needed to engineer the rest of mankind? For however great are the successes science derives from limiting itself to the problem of how things happen, the question why certain actions are taken, or whether they should or should not be taken, remains central in human endeavor. The inordinate emphasis laid on the *how* and the simultaneous neglect of the *why*, the gradual encroachment of the quantitative method on the investigation of the non-quantitative aspects of human existence, manifest a cultural myopia for which an all too heavy price has already been paid in recent historical upheavals. Scientific progress, however spectacular, will never obliterate the difference between means and goals. Belittling this difference can only deepen that tragic contradiction to which Einstein once referred in a moving remark: "It is easier to denature plutonium than it is to denature the evil spirit of man." [167]

Had this elementary difference been kept sharply in mind, men would not have looked in vain for scientific panaceas against cruelty, envy, greed, hatred, and kindred sources of evil. Had the fact been clearly kept in mind that there is no scientific substitute for hope, for love, for compassion, a man of science like B. Russell would not have had to "discover" that it is simply Christian love that the world really needs today. It took him fifty years and the prospect of atomic annihilation to learn that much. His plea for Christian love at Columbia University in 1950 was indeed a far cry from his "scientific" ethical manifesto that recognized only the meaningless whirl of atoms half a century earlier. It was no doubt a difficult change of mind to make and candidly to admit. He could probably see the ridicule that his "capitulation" might touch off among wise cynics whom he had known only too well. He also felt the need to apologize. To whom, he did not say; perhaps he meant those who had valued above everything else a certain outlook on life which he so lightly dismissed before. He should have felt reassured that those sincerely possessed of it could feel no selfish triumph but only sincere joy upon hearing his words:

The root of the matter is a very simple and old-fashioned thing, a thing so simple that I am almost ashamed to mention it, for fear of the derisive smile with which wise cynics will greet my words. The thing I mean — please forgive me for mentioning it, is love, Christian love or compassion. If you feel this, you have a motive for existence, a guide in action, a reason for courage, an imperative necessity for intellectual honesty.[168]

CHAPTER TEN

Physics and Theology

Since the conclusions of physics are an attempt to understand the universe, they naturally lend themselves to confrontation with cosmological views originating in religious beliefs. Such confrontation, as history shows, can take many forms ranging from the domination of physics by religion to a full-scale exploitation of physics by militant atheism. To a large extent, it was religious motivation that helped decide the course of physics for almost two thousand years when Aristotle, following Socrates' and Plato's lead, made physics serve the "living and divine principles" that were believed to rule the cosmos. Aristotelian physics, as already pointed out, had a well-defined role in the campaign waged against the materialistic leanings of the presocratic *physikoi*. In popular parlance, too, the difference between the two kinds of physics amounted to a choice with distinctly religious overtones. One may well recall from Aristophanes' *Clouds* the indignant reaction of old Strepsiades about the "scientific" explanations that tried to reduce atmospheric phenomena to vortices at the expense of Zeus, the god of thunderbolts: "What! Vortex? That is something I own. I knew not before that Zeus was no more, but Vortex was placed on his throne!"[1]

The mental turmoil that came to such a climax in late fifth-century Greece is graphically described in Plutarch's *Life of Nicias*, in connection with the military disaster suffered by the distinguished general at Syracuse because of his superstitious dread of eclipses. Eclipses were believed to be a sign of the wrath of the gods, and against the deadweight of ingrained superstitions, the sober reasoning of Anaxagoras could hardly prevail. The book of Anaxagoras, "the first man to put in writing the clearest and boldest of all doctrines about the changing phases of the moon,"[2] in fact had to be circulated in secret. For, as Plutarch tells us, natural philosophers, or "visionaries," as their antagonists called them, were suspected of substituting blind, "necessary" causes for hallowed spiritual agencies. Science came under suspicion, its representatives were thrown into jail or banished, and it was only much later, as Plutarch comments, "that the radiant repute of Plato . . . took away the obloquy of such doctrines as these, and gave their science free course among all men."[3] The liberation of science in Plato's hands, however, was a paradoxical procedure, since Plato achieved this feat, to quote Plutarch, by subjecting "the compulsions of the physical world to divine and more sovereign principles."[4]

The return in physics to "divine principles" of course did not mean a relapse into explaining the physical world by the Homeric tales about the gods. The trend initiated by Plato was in essence just as opposed to crude superstitions as was the tack taken by the atomists. The day was still far away when Lucretius would spell out bluntly the "iconoclast" effect of scientific investigations in general.[5] For the time being, the attempt was made to accommodate the "spiritual" core of the mythological tradition in the new, theologically oriented approach to nature championed by Plato. In this respect, his writings are full of references that are worth considering in some detail. In the *Laws*, for instance, Plato chastised time and again the Ionian *physikoi* who held, as he put it, that the divine entities of the sun, moon, and stars were merely stones, "which can have no care at all of human affairs," and that "all religion is a cooking up of words and make-believe."[6] Plato also referred to the suspicion widely shared among the populace that the cultivation of the sciences, and astronomy in particular, made men godless, by developing in them the pernicious inclination "to see, as far as they can see, things happening by necessity and not by an intelligent will accomplishing good."[7] Such studies, continued Plato, "gave rise to much atheism and perplexity, and the poets took occasion to be abusive, — comparing the philosophers [the Ionians and the atomists] to she-dogs uttering vain howlings and talking other nonsense of the same sort. But now, as I said, the case is reversed."[8]

That a distinct change in the intellectual climate began to assert itself by the time the *Laws* was written is very likely true. But it was not until Aristotle came to discourse on topics of natural science that the scientific codification of the change was actually accomplished. Whereas in the Ionian and in the atomistic world views there is hardly place or admitted need for a divine Power, the Aristotelian physics presented the world as a well-designed, tightly interlocking hierarchical staircase. Its uppermost echelon was nothing else but the realm or embodiment of "the foremost and highest divinity."[9] Next to the divinity, or Prime Mover, was the sphere of stars, described invariably as having a divine nature, or made up at least of some "divine" substance. For, as befitted a student of Plato, Aristotle could never bring himself to doubt the divinity of stars, which was an article of faith in the Platonic creed. How closely Aristotle aligned himself with the trend that aimed at the restoration of the traditional and religious interpretation of the cosmos can best be seen when he argued from the belief in the gods to the existence of the ether as the substance constituting the stars and heavens.[10] To prove that the ether was immortal and eternal, he stated that "we may well feel assured that those ancient beliefs are true, which belong to our own native tradition, and according to which there exists something immortal and divine in the class of things in motion, but whose motion is such that there is no limit to it."[11]

According to Aristotle, the purest occurrence of this substance in the cosmos was the first heaven, which was hardly distinguishable from the Prime Mover. At any rate, in the passages that indicate some distinction, the Prime Mover was said to move the first heaven by inspiring love and desire. This in turn implied for Aristotle that the first heaven, too, had a soul. Once soul and motion became intimately connected, however, there was no end to the multiplication of "souls" all across the starry spaces. Aristotle clearly stated that the other celestial spheres and bodies also had their own souls or "intelligences" whereby they moved. Consequently the motion of stars became a chapter in theology. For as Aristotle tells us, it was desire and love in the stars that produced their physical motion, inasmuch as they strove to imitate a Prime Mover whose life was a continuous unchanging spiritual act. The function of the spheres was to reproduce this in an analogous way by performing the only perfectly continuous physical movement, motion in a circle.

Being a part of the cosmos, the first cause, or the divine principle, also had to become a part of physics and not without affecting the very core of physics as a science. Religious motivation was basic to the Aristotelian choice of an organismic conception of physics, which divided the universe into two completely distinct regions made up of different matter and ruled by different laws or "natural motions." These views not only discredited to a large degree the quantitative approach in kinematics and dynamics, but also gave further support to the age-old superstition that stars have a decisive effect on any physical, personal, and social change that may take place on the earth.

The welding of a somewhat ambiguous deity, the first cause of Aristotle, into the physical description and explanation of the world posed grave problems to the same antimechanistic or "religious" attitude that inspired it. This became apparent a generation or two later in the teaching of Epicurus. By that time, the Stoics had carried to logical extremes the pantheistic aspects of the Aristotelian interpretation of the cosmos, with the result that the all-pervading purposefulness in the cosmos, which Socrates and Plato felt their mission to defend, came to be absorbed in an eternal chain of causality, hardly distinguishable from fate. From this great dilemma Epicurus saw only one way out: the deity was to be denied any role in either originating or implementing physical laws. What was to be preserved above all was man's peace of mind, and Epicurus decided that "it were better to follow the myths about the gods than to become a slave to the destiny of the natural philosophers: for the former suggests a hope of placating the gods by worship, whereas the latter involves the necessity which knows no placation."[12] This position, however, implied the rejection of anything in the physical method that savored of a unitary principle, because in Epicurus' view the acceptance of all-inclusive laws would subject the realm of the divine to compulsion also. Thus Epicurus had no choice but to part with the concept of a universally valid law or explanation in physics. Speaking of eclipses and other celestial phenomena, such as the rising and setting of the stars, Epicurus warned against choosing one possible explanation over others. Physical phenomena, he said, admitted more than one interpretation, all of which should be kept on equal footing. With a strange twist of reasoning, clearly expressive of a deeply troubled mind, he even added that preferring one explanation to another would be leaving the path of scientific inquiry and reverting to myths. It would be, warned Epicurus, falling prey to irrational beliefs and groundless fantasies, but as his famous injunction stated, "we must live free from trouble."[13] No less important for Epicurus was it to secure this trouble-free life for the gods, whom he relegated to the realms of eternal bliss, where they were no longer taxed with any responsibility for anything that went on in the cosmos: "And do not let the divine nature be introduced at any point into these considerations but let it be preserved free from burdensome duties and in entire blessedness."[14]

Such a solution was, of course, as S. Sambursky put it, nothing short of "scientific bankruptcy."[15] It may be said with no less justification, moreover, that the whole range of the mutual involvement of Greek physics and theological thought represented a hopeless labyrinth from which classical Greek thought could offer no escape. When the deity was extended into the world, as in the system of Aristotle, physics was deprived of its proper method. When the deity was negated, as in atomism, the realm of human values came to be undermined.

When the deity was absolutely severed from the cosmos, as in Epicurus' teaching, the concept of a generally valid physical law had to fall.

The story was markedly different when scientific thinking encountered views on the cosmos deriving from Judaeo-Christian monotheism. Although a full-scale confrontation of Christian thought and Greek science had to wait until the late Middle Ages, the isolated case of the Christian philosopher, Philoponus, in the sixth century, well indicates what was to be expected in such an event.[16] Although star worshiping was resolutely rejected by Jewish and Christian theologians alike, it was Philoponus who first clearly recognized that the idea of a Creator is hardly compatible with the implicit identification of the deity with the uppermost heavens. This type of reasoning, however, necessarily cast grave doubt on the cornerstone of the Aristotelian physics, the dichotomy between the celestial and terrestrial regions in the cosmos, on which this implicit identification ultimately rested. Led by the Christian view of a sharp difference between Creator and creature, Philoponus asserted that for the purposes of science, too, the whole cosmos, created in all its parts by God, and therefore essentially different from Him, was rather to be viewed as composed of the same type of matter and governed by the same laws.

To support this general principle, Philoponus insisted on rigorous observations, thereby exemplifying for the first time the purifying effects of theological tenets in physical reasoning. It was on such grounds that he pointed out that Aristotle's contentions about the dichotomy between the celestial and terrestrial regions in the universe were far from conclusive. According to Philoponus, the various colors of the light of celestial bodies strongly indicated the fiery state of many of them as opposed to the sublimely cold Aristotelian ether. Nor could the radiant color, as Philoponus argued, be an exclusive attribute of the heavenly bodies, since the same color was often displayed by glowworms and the heads and scales of various fish. The stars could hardly be more degraded from their exalted position than by being compared to worms, however glowing these could be at times.

To show further that all the properties of the skies were duplicated on earth, Philoponus referred to the transparency of air, glass, water, and various stones. From transparency and visibility, he then argued in a sense diametrically opposite to the Aristotelian view. To Philoponus whatever was visible, was also tangible, vested with the ordinary qualities of hardness, softness, roughness, dryness, humidity, and heat to which all the others were reduced. Thus Philoponus concluded in effect that the common features of terrestrial matter were present in the matter composing the heavenly regions. Again, whereas for Aristotle the apparent lack of changes in the heavens proved the eternity of the stars, for Philoponus such a reasoning was far from being watertight. He referred pointedly to Mount Olympus, which showed no more evidence of growing or diminishing than the stars did, without therefore being taken as eternal. The invariability of the stars, he argued, could be explained far better in analogy to a phenomenon generally valid on earth: objects with huge mass displayed a permanence and unchangeability in direct proportion to their bulk. To undermine the special status of the heavens in every possible way, Philoponus set great store on the deviation of the planets' path from a perfect circle. This, he said, could not be taken lightly, as had Aristotle, who also tried to explain away the problem presented by the oppositely directed motion of the stars and of the planets. Philoponus, however, insisted that precisely this difficulty made meaningless the ether itself, the existence of which was postulated by Aristotle to account for a uniquely directed circular motion.

Stressing the uniformity of the physical world was not the only contribution of monotheistic beliefs to scientific thought. Even more important was another consequence of the belief in a Creator. It opened the way to an autonomous scientific world picture in which nature neither usurped the attributes of God nor was itself exposed to incessant intervention by other-worldly powers. What monotheism helped Philoponus arrive at was a conception of the world in which God, having accomplished the act of creation, "hands over to nature the generation of the elements one out of another and the generation of the rest out of the elements."[17] These words are from Simplicius, Philoponus' great and bitter antagonist, who was unable to understand how the laws of nature could operate without unceasing recourse to a deity constantly modifying and rearranging the processes of nature. To the heathen Simplicius, the approach of Philoponus to the physical world was "an extraordinary way of a stupid person to inquire into truth." Yet, it turned out to be not so stupid after all.

When Greek science and the full Aristotelian corpus in particular were rediscovered in the Middle Ages, the doctrine of creation did not fail to assert itself anew. Not that the Aristotelian physics had not been admired by such prominent exponents of Jewish and Christian thought as Maimonides and Thomas Aquinas. In fact, both tried their best to interpret with a benevolent attitude those parts of the Aristotelian physics that seemed to contradict their belief in a Creator. Yet, mainly owing to the contentions of the Averroists, those "radical" followers of Aristotle, an unequivocal position had to be taken on some propositions involving the very essence of Aristotelian physics. One of these stated that there could be but one world, and the other postulated that heavenly bodies or the heavens in general could not be given a rectilinear translational movement. To admit the unconditional validity of such tenets seemed to result in limiting the power of the Creator, an evident absurdity for Christian reasoning. The solemn condemnation of these propositions by the Bishop of Paris in 1277 was therefore more than a defense of Christian dogma. It was also an implicit affirmation of the view that the laws of the universe were not necessarily identical with inferences based on a priori philosophical speculations. The decision was momentous indeed, encouraging as it did efforts that aimed at both keeping scientific research within its own range and securing to it a proper measure of autonomy. For no less a historian of science than Duhem, the decision of 1277 marked the birth of modern science.[18]

Although three centuries had yet to pass before the basic laws of classical physics came to be formulated, medieval thinkers interested in science, such as Oresme and Buridan, did not fail to notice the theological advantages of a world picture based on mechanistic considerations. The mechanistic view of the world, denounced by so many from Plato to Simplicius as sheer materialism, was regaining credit in the monotheistic context as the manifestation of the Creator's power and intelligence. The benefits to Christian theology were obvious. Since clockworks needed only mechanical components, the celestial intelligences assigned to drive the spheres, stars, and planets could be dispensed with. Harmony with the Bible, which spoke of no such spirits, thus became more satisfactory. The discredit brought to the "intelligences" also meant a heavy blow to astrology, which was blocking with particular effectiveness the ascendency of more reasonable notions about the physical universe.[19] In a conceptual framework rid of astrological illusions, all that was necessary to assume besides the first impetus given to celestial bodies at the moment of Creation was, to quote Buridan, "the general influence whereby [God] concurs as a coagent in all things which take place."[20] But it was precisely this type of action on the part of the Creator that had no direct bearing on the formulation of physical laws.

To understand this was not only a major step forward, but also a very daring one. For, had it not been for the Christian idea of creation, it would have easily led to a wholesale doubting whether the world was really rational in all its parts. Granting the supreme rationality of a personal Creator, as conceived in the Christian context, his handiwork too had to be supremely rational. Such a view furthermore was not merely a hesitating play with premises and conclusions, but as Whitehead once put it, "an inexpugnable belief that every detailed occurrence can be correlated with its antecedents in a perfectly definite manner, exemplifying general principles." For science in the modern sense to rise at all, it was not enough to have available the findings and insights of a few extraordinary minds devoted to scientific speculations. A widely shared belief, an "unquestioned faith of centuries," an "instinctive tone of thought" were necessary, for without such a belief, such a climate of thinking, so Whitehead argued, "the incredible labors of scientists would be without hope." Highly unorthodox as may be one's interpretation of the merits of certain theological tenets, the historical fact remains hardly controvertible: "The faith in the possibility of science, generated antecedently to the development of modern scientific theory, is an unconscious derivative from medieval theology."[21]

Such a conclusion necessarily startled most of those attending Whitehead's Lowell Lectures of 1925. At that time, not only the general public but even professional historians were largely unaware of the medieval roots of modern science. A generation later the situation was notably different. As a result, more attention has been paid recently to the influence that theological tenets may have had on the furthering or stalling of scientific thought in various great cultures. An example is the monumental study by J. Needham on the development of science in China from the earliest times. The most crucial question of such a study, of course, is the problem of why the medieval Chinese, so proficient in technical inventions, failed to formulate scientific laws. It should be highly instructive that Needham, no particular friend of theology, found only a theological answer to this puzzle. As he put it, among the Chinese, long without a faith in a supremely rational and personal Creator, "there was no conviction that rational personal beings would be able to spell out in their lesser earthly languages the divine code of laws which he [the Creator] had decreed aforetime."[22]

Similar studies of Mayan, Hindu, Persian, Egyptian, and other great cultures would no doubt be highly enlightening. As to the Greeks, it is now coming more and more to be recognized that their theological outlook is in part responsible for the fact that science, after its marvelous birth in Greece, came to a standstill there. Viewing the whole world as a supreme living organism, and themselves tiny organs in it, could the Greeks muster a sustained confidence of ever deciphering, let alone mastering, all the regular and capricious movements of the cosmic Leviathan? Was not far better prepared the state of mind of those whose faith reminded them daily to look upon the world as the product of a most rational and reasonable Creator and to look upon themselves as the stewards of their Father's handiwork?

Such reflections on some crucial problems of the scientific past in the Western world do not claim originality. They were first put forward in essence by Lord Verulam, who diagnosed more clearly than all his contemporaries that science was opening a new era in history. His writings in particular evidence a high degree of awareness of the different ways in which Greek and Christian religious thought affected the course of science. Bacon laid the blame for the thwarting of the progress of science on the "heathen opinion" that "supposed the world to be the image of God, and man to be an extract or compendious image of the world." On such premises, Bacon argued, it was only too tempting to take the world as one huge pantheistic organism whose strivings duplicated human volitions. As a result, an excessive concentration on final causes had taken the upper hand in scientific investigations. The consequences for science were nearly disastrous, for as Bacon tells us, engrossment with final causes "intercepted the severe and diligent inquiry of all real and physical causes" and brought about a "great arrest and prejudice of further discovery."[23] The inordinate quest for final causes prompted hasty generalizations: facts contrary to theories were subtly glossed over and, to quote Bacon's words, "this flying off to the highest generalities ruined all."[24] Actually, not even God's role in the universe could be safeguarded in a physics fashioned to final causes. "Aristotle," remarked Bacon, "when he had made nature pregnant with final causes, laying it down that 'Nature does nothing in vain' and always effects her will when free from impediments, and many other things of the same kind, had no further need of God."[25] With remarkable insight, Bacon pointed out that natural philosophers like Democritus, "who removed God and Mind from the structure of things," got a more genuine glimpse of the processes of nature than did Aristotle and Plato, for the very simple reason of not mixing final causes into the study of physics. Aristotle, who did so, said Bacon, lost out not only on physics, but lost sight of God too by substituting nature for God.[26]

In contrast to the easy "explanations" of nature offered by the antique views steeped in pantheism, the world as the work of an infinitely superior Creator had to appear to created intellects anything but simple. Time and again Bacon recalled that "the subtlety of nature is far beyond that of sense or of understanding."[27] For him, the universe was as complicated as a labyrinth, presenting countless ambiguities, "deceitful resemblances of objects and signs, natures so irregular in their lines and so knitted and entangled."[28] Consequently, physics "carrieth men in narrow and restrained ways, subject to many accidents of impediments, imitating the ordinary flexuous courses of nature."[29] In such a complexity, it is hard enough to decipher some mechanical connections, let alone final causes. The latter are not barred from metaphysics, but only from physics. Although the heavens declared the glory of God, they were not ordained to manifest the will of God, Bacon remarked,[30] and to the idols of the human mind he opposed the infinitely superior divine ideas.[31] The summary law of the cosmos, which for Bacon meant the all-embracing, exhaustive knowledge of nature, or the "vertical point of the pyramid of knowledge," was judged by him as being very likely beyond human capabilities.[32] What little could be learned about the intricacies of nature was to be done therefore in a laborious way through long, patient experimentation. It is highly significant that none of the seventeenth-century men of physics failed to refer to the infinite richness of possibilities implicit in the infinite power of the Creator when they emphasized the indispensability of experiments in science.

Although the concept of divine creation sets science free and at the same time prunes it of exaggerated ambitions, science, according to Bacon, may exert no less beneficial effects on the interpretation of revealed religion. Science for him was an antidote against superstitions and errors,[33] and as the sole source of knowledge about the physical world, science was also a safeguard against misguided exploitation of the Scriptures on behalf of views concerning the physical universe. All this was very wholesome, but far less so was the sharp dichotomy Bacon came to erect between the two sources of truth, nature and the Bible. Undoubtedly, as Bacon stated correctly, these two sources related to different levels as regards human assent and therefore were to be carefully distinguished.[34] But he went too far in stressing the imperviousness of the mysteries of faith to rational analysis. In contrasting the little boat of human knowledge to the great ship of the Church, Bacon said that the latter, symbolizing theology too, could be guided only by a divine "nautical needle" (compass) and not by the stars of philosophy. Had he added that when watching the former one should keep an eye on those stars as well, he would have done better justice to the manner in which the mainstream of Christian tradition viewed the relation of faith and reason. Of their interconnectedness Bacon saw little, and their distinctness he raised into a "holy" but in effect disastrous opposition.[35]

Comments on the relation between science and Christian faith had of necessity to keep turning up in the writings of all the major and minor figures of the emerging classical physics. There was a program, a creed for science that had to be adhered to in a Christian framework. Thus the first day in the *Dialogue concerning the Two Chief World Systems* closes with Salviati's remarks on the shortcomings of human knowledge as contrasted to the perfection of divine intelligence: "The way in which God knows the infinite propositions of which we know some few is exceedingly more excellent than ours"; while man's intellect makes its advances "laboriously and step by step," all truth is always present to the divine Mind. As a result, God's knowledge infinitely surpasses ours, "as well in the manner as in the number of things understood."[36] The study of the world therefore could not claim exhaustive knowledge, all-embracing final conclusions. It had to be rather content with a slow penetration into the infinity of unknown facts and hidden laws of nature without ever being sure of fully achieving its aim. "Those who know very little of philosophy are numerous," wrote Galileo. "Few indeed are they who really know some part of it, and only One knows all." The more science grows and "partakes of perfection, the smaller the number of propositions it will promise to teach, and fewer yet will it conclusively prove."[37]

Like Bacon, Galileo stressed that both nature and the Bible have a common source of truth, the divine Word, and that God is not "any less excellently revealed in nature's actions than in the sacred statements of the Bible."[38] Their purpose, however, is as distinct as their mode of reasoning. While the parlance of the Scriptures is tailored to a particular cultural setting, nature's way of speaking is, as Galileo emphasized, "inexorable and immutable." Consequently, in matters of natural science, no one should be permitted "to usurp scriptural texts and force them in some way to maintain any physical conclusions to be true when at some future time the senses and demonstrative or necessary reasons may show the contrary."[39] On the other hand, the well-established conclusions of science can and should be used in eliminating interpretations of the Bible that rest only on ignorance or inherited errors. "Certainties in physics," wrote Galileo, "are the most appropriate aids in the true exposition of the Bible," for "God who has endowed us with senses, reason and intellect . . . would not require us to deny senses and reason in physical matters which are set before our eyes and minds by direct experiences or necessary demonstrations."[40]

The way Galileo spoke of certainties, direct experiences, and necessary demonstrations in physics shows that much as he tried to pay both theology and science their due, his balance was, unknown to him, badly tilted. No doubt, nature as a source of truth possessed in principle at least a precision that was an exclusive feature of hers. But was this precision of nature self-evident or easily demonstrable in all her phenomena? Again, could the laws of physics, or the description of phenomena, represent at a given stage of physics the ultimate word in the matter? Actually physics could provide no absolute assurance about its propositions, and the realization of this seized Galileo's mind at times, though he failed to sense the full import of this as regards the more distant future of the relation between physics and theology. While stating that only Holy Writ could decide such problems as the freedom of will, predestination, and even the finitude of the world, he unwittingly set too much store on the clarity of nature as contrasted to the haziness the Bible offers on many points. By doing so he undoubtedly contributed to a trend, which two centuries later strongly insisted on appraising by natural and rationalistic means even the redemptive message of the Bible. To foresee such a development in Galileo's day would have been well nigh impossible, and as a devout son of the Church, Galileo would certainly have disavowed it.

Much less could he be aware of the fact that for the correct interpretation of the Bible as a source of truth it was far more important to ascertain the literary form of its books than to know physics. Still, it must be admitted that Galileo's insight was exceedingly good in choosing the best from theological tradition as regards the principles of interpreting the Scriptures. As a result, his quoting and discussing the weighty texts of Tertullian, Jerome, and Augustine produced for seventeenth-century Catholic theology a little gem known as the *Letter to the Grand Duchess Christina*. If the guess of Bellarmine's foremost contemporary biographer, J. Brodrick, is right, the saintly cardinal probably did not read Galileo's theological masterpiece. Had he done so, muses Father Brodrick, how different a course history could perhaps have taken.[41] Speculations aside, it remains true that for a physicist to best a doctor of the Church on a particular theological problem was no mean achievement. For this success, however, a large share of credit is due to the suggestions given to Galileo by progressive theological minds among his friends in the clergy. One of these was Galileo's most outstanding pupil, Benedetto Castelli, who was, however, forbidden by his abbot to speak up in his master's behalf when the hour of the infamous trial struck.

This is not the place to retell a well-known story. It must suffice here to say that the weight of appearances, the routine of customary thinking, the heavy bulges of tradition, corporate interests of established academic circles, and personal animosities are almost always impossible to overcome, even when taken singly let alone together. Witness the case of Oresme, the learned bishop of Lisieux and a man of very open mind, who found it impossible to depart from the traditional meaning given to the words of the Scriptures indicating an immobile earth and a moving sun.[42] Similar was the case of Tycho Brahe, who found no way of reconciling Copernicus' vision with Holy Writ. Brahe, in fact, heaped scorn time and again on the "absurd discoherency" of Copernicus, who in his fancy ascribed to the "opaque, gross and fat body of the Earth" a threefold motion notwithstanding "all the truth of physics and the authority of the Sacred Scriptures to the contrary which ought to be supreme," as he put it, in such matters.[43] Furthermore, Brahe did not want to incur possible ecclesiastical censures.

Censures of this sort had their short-lived triumph when Galileo dismally failed in his well-meaning but reckless effort to prevent his Church from sanctioning a scientific error. Unlucky as he was in charting his strategy, he could take some comfort in the fact that Rome stopped short (some would say through the help of Providence, others, through mere luck or shrewdness) of an irrevocable, infallible pronouncement. This was abundantly clear to his contemporaries, such as Descartes, who nevertheless remained silent for fear of reprisals by Church authorities. Two generations later, the Protestant Leibniz, argued in his letters to Landgrave Ernest of Hesse that no principle of Catholic theology stood in the way of removing the strictures both from Copernicus and Galileo.[44] When this at long last happened in 1822, it was not without some touch of irony: The demonstration of stellar parallax — this most sought-after proof for the motion of the earth on which Galileo's critics and judges had set so great a store — was still several years away.

In their own way, both the critics and the supporters of Galileo sought harmony between physics and theology. Neither side could see, however, that in order to achieve this far more was needed than an isolated adjustment of the interpretation of some biblical passages to the findings of science. The source of the conflict between science and theology resided rather in the reluctance on both sides to reflect with utmost intellectual detachment on the limitations of both theological and scientific statements alike. That various ages, peoples, cultures, and groups of believers could build on basic Christian principles intellectual and cultural superstructures that were Christian in name only is too well known to be discussed here. Much less is it remembered how quickly far-fetched generalizations start cropping up when word is received about some new discovery in science. The fallacy inherent in such procedures is not always immediately obvious, and thus an intellectual climate may easily develop in which spurious proofs are readily taken for granted. Essayists, poets, and philosophers will then do the rest by dramatically portraying the collapse of the traditional set of values and beliefs. Yet, a deeper understanding of human nature and a more thorough familiarity with physics would have showed them that the discoveries of physics leave untouched the deepest dimensions of man's existence. But it is there that man's religious orientation is anchored and not in the perishable Aristotelian spheres or in the evanescent Newtonian space or in the expansion of the universe that might be followed, who knows, by its contraction.

To see in the seventeenth century how crucial this point was, if a balanced relation between physics and theology were to be obtained, it required nothing less than a personality in which an outstanding scientist and a genuine mystic were united. This personality was Pascal's. His main scientific achievements ranged from the theory of conic sections, written when he was only sixteen, through the invention of the calculating machine and the discussion of the vacuum and air pressure to the analysis of probability theory, of which he was one of the principal originators. To the field of religion, he brought not only a deep erudition but also the experience of a Christian mystic. Furthermore, his was a discerning mind to which the foibles and anxieties of his age could not remain hidden. The key for the correct interpretation of the *Pensées* lies indeed in Pascal's effort to show how shaky were the tenets forged by the seventeenth-century *libertin* from the scientific discoveries of his day. To hear Pascal say it, the basic fallacy of the "scientific" materialism and atheism was that it ignored much of what it took to make a whole human person. "Man is a whole," he warned and asked in the same breath, "If we dissect him, will he be the head, the heart, the stomach, the vein, the blood, each humor of the blood?" [45] If one was to understand man, argued Pascal, this wholeness had to be of overriding concern.

Not that Pascal wanted, by putting the emphasis on man, to draw attention from the twofold infinites that science was just beginning to open up before the truly bewildered man of his day. In fact, hardly anyone wrote more impressively of those frightening infinities. But he was quick to point out that only the *libertin* ought to lose hope and cry out in despair, "The eternal silence of those infinite spaces strikes me with terror!" [46] This terror was a heavy price the *libertin* had to pay if he wanted to be consistent. Such a *libertin* had no choice but to ignore the fact that immense spaces are merely quantitative relations of matter, and whatever their numerical magnitude, they fall far short of the greatness of a thinking man, who belongs to a higher level of existence where considerations, other than quantitative, master the solutions of outstanding problems:

Man is but a reed, the weakest in nature; but he is a thinking reed. It does not take the whole Universe in arms to crush him. A vapour or a drop of water is enough to kill him. But if the Universe were to crush him, man would still be nobler than his killer. For he knows that he is dying, and that the Universe has the advantage over him; the Universe knows nothing of this. Our whole dignity consists, therefore, in thought. By thought we must raise ourselves, not by space and time, which we have no means of filling. Let us endeavour, then, to think well; this is the beginning of morality.[47]

The physical universe was not therefore to be taken as the source of the intellect in man. Much less could the material order of things be viewed on the same level with the realm of moral values. "All bodies together, the firmament, the stars, the earth and its kingdom, are not equal to the lowest mind. For it knows them all and itself, and bodies know nothing. All bodies together, and all minds together, and all their products, are not worth the least prompting of charity."[48]

To speak of charity and practice it, one of course did not have to be a scientist. As a scientist, Pascal could, however, effectively remind the *libertin* that although science speaks only of bodies and is therefore "materialistic," materialism, if it is to be "scientific," should say things that are perfectly evident. But, as Pascal remarks, "it is not perfectly evident that the soul is material."[49] What is more, for a scientific atheism to obtain, there should also be the assurance that the findings of science are final at some particular stage. Science, however, presented its results and problems to an unprejudiced observer under an entirely different light:

We see that all the sciences are infinite in the extent of their field of inquiry. For who can doubt that mathematics, for example, has an infinite infinity of problems to be solved? These are equally infinite in the multitude and the subtlety of their premises. For anyone can see that those which are put forward as ultimate are not independent, and that they rest on others which, resting on yet others, permit of no finality.[50]

If one were to look for the main reason for such lack of finality in the conclusions of science, Pascal would have pointed to the disproportion between man's mind and the vastness of the physical world. Powerful as man's imagination might be, "it will exhaust its powers of thinking long before nature ceases to supply it with material for thought."[51]

To sum up Pascal's view, then: if unnecessary clashes between science and religion were to be avoided, one had to be keenly aware of the limitations of the human intellect and of the difference that exists between the findings of science and their alleged bearing on problems essentially impervious to the methods of science. This, however, was a hard injunction for those who wrote about science, and it was as vain to expect them to adopt it as it was to expect theologians to be cautious in matters clearly subject to experimental verification, in theory if not in practice. The *Pensées*, to be sure, did not lack persuasiveness. This was recognized to such an extent that in 1776, on Voltaire's advice, Condorcet published a garbled version of it to dim somewhat a beacon of light that was uncomfortably too brilliant for the much advertised brightness of the Enlightenment. In fact no one was more haunted by the penetrating light of the *Pensées* than Voltaire had been all through his long literary career. But even if the *Pensées* had enjoyed a wider circulation a hundred years earlier, it is still doubtful whether Boyle's and Newton's age would have taken Pascal's warnings more to heart. For he pleaded in essence for a recognition of the respective limits of science and theology, and this was going against the grain of a generation that was absolutist not only in politics but in other more abstract fields of human endeavor as well. Religious eagerness very frequently in those days had the upper hand over the objective appraisal of arguments, and the theological evaluation of science was no exception to this rule.

Thus it happened that in the seventeenth century a carefully balanced relation between theology and physics could not materialize in spite of the clear enunciation of wholesome principles. The laws of physics that at most could underscore the orderliness and contingency of nature were drawn rather into a role that in the long run would bring only disillusionment with natural theology. Religious preoccupations weighed heavily on men's minds and turned physics into a handmaid of theology, but as later events proved, a not-very-submissive handmaid. For a while, however, theological considerations had a strong influence in physics. Gassendi, for instance, who brought about the revival of atomism, was careful to point out that the self-movement of atoms was through *Dei gratia*.[52] After all it was only about twenty years earlier, in 1624, that the Parliament of Paris, no doubt under the pressure of some overzealous divines, threatened with the death penalty anyone maintaining or teaching atomism, or any doctrine contrary to Aristotle. Another early atomist, W. Charleton, also mixed theology with physics in contending that the rejection of atomism would be tantamount to imputing a lack of proper reasoning on the part of the Creator.[53] Mersenne, who never could bring himself completely to the Copernican view, admitted that he would accept the motion of the earth provided he could convince himself that "God always did things in the shortest and easiest way."[54] In a similar vein, Descartes anchored the clinching proof of his law concerning the conservation of motion in the universe not in experiments or mathematics, but in God's immutability.[55] Leibniz, too, maintained that "one has to derive everything in Physics" from the ultimate causes, that is, from the consideration of the Supreme Being acting with wisdom.[56] The precept was hardly a reliable one. For one thing, it led oftener than not to contradicting the views of other physicists. Thus Leibniz argued from the wisdom of God against the existence of vacuum and atoms.[57] For another, it frequently aroused the indignation of the theologians. They could indeed hardly go along with Leibniz, who thought that by identifying the material body with motion, as physics allegedly did, the problems implicit in the dogma of transubstantiation could be readily solved.[58]

The eager willingness of the seventeenth-century scientist to connect some particular law of physics or an interesting numerical relation with God's wisdom could be illustrated by many instances. Let it suffice here to mention Huygens who by his discovery of Titan in 1655 brought the number of planets and satellites to twelve and commented that this number "can be considered as predetermined by the counsel of the Supreme Artifex." [59] It was still to be learned that God's mind could not be charted so easily. At any rate, scientists in the seventeenth century viewed their work as giving a direct look at the Creator's supreme blueprint. This attitude was described in Leibniz's letter of 1716 to Peter the Great, where he explained why he valued scientific studies above jurisprudence and political administration, which brought him so much recognition and fame: "It is especially in sciences . . . that we see the wonders of God, his power, wisdom and goodness; . . . that is why, since my youth, I have given myself to the sciences that I loved." [60] In so doing, Leibniz only followed in the footsteps of his elders in science, one of whom, Boyle, spoke of "Telescopes, Microscopes, Anatomical Knives, Chymical Furnaces" as irresistible proddings to sing the Lord's praises with the psalmist.[61]

That these men of science reacted vigorously whenever science was exploited in support of materialistic tenets is only natural. In this regard, it made no difference whether the scientist was a deist like Huygens, or a devout Christian like Boyle. The former, who once admitted to his fear of being poisoned if his disbelief in Calvinism were ever to come to light, spoke in his *Cosmotheoros* with contempt of those who "are spreading false opinions, such as attributing the origin of the earth to the accidental union of the atoms, or of the earth being without a beginning and without a creator." [62] Boyle, on his part, was incensed by Hobbes' attempt to harness mechanistic physics in the cause of a thinly veiled materialism. Not only did Boyle write a book on *The Wisdom of God Manifested in the Works of Creation*, but he also endowed an annual lectureship to be devoted to the defense of religion against indifferentism and atheism. The first lecturer to deliver the eight-sermon series was R. Bentley, chaplain to the Bishop of Worcester. His seventh and eighth sermons dealt with proofs of God's existence based on conclusions in Newton's *Principia* about the solar system. In 1692 this was all the more surprising, as the first edition of the *Principia* was conspicuous by its silence about God. By not mentioning God, Newton meant of course no more than to keep physics and supernatural causes apart. But he was all for the utilization of the results of physics in natural theology. Thus, when Bentley sought out Newton's approval before publishing his sermons, he obtained it in the most generous terms.

"When I wrote my treatise about our own system," reads Newton's first letter to Bentley, "I had an eye upon such principles as might work with considering men for the belief of a Deity; and nothing can rejoice me more than to find it useful for that purpose."[63] Among the particular aspects of the solar system that cannot be explained without direct intervention by the Creator, Newton lists first the uniqueness of the sun as a heat- and light-giving body in the solar system. Of this fact he says, "I know no reason, but because the author of the system thought it convenient."[64] Another phenomenon that according to Newton also postulates divine intervention is the regular motion of planets in the same plane as opposed to the comets, which follow all sorts of paths with respect to the sun. In the same way, the proper adjustment of the velocities, masses, and distances of planets "argues that cause to be not blind and fortuitous, but very skilled in mechanics and geometry."[65] For Newton the fact that the planets with larger masses are farther from the sun than the smaller planets also points to the action of a Creator who prevents thereby a "considerable disturbance in the solar system."[66] In a similarly unexplainable category Newton classed the inclination of the earth's axis and the rotation of the sun and the planets.

In his second letter to Bentley, Newton called attention to the fact that for planets to move in definite paths in a gravitational field, a transverse impulse of precise quantity must be imparted to them. In his view, however, it was impossible to designate any known "power in nature which would cause this transverse motion without the divine arm."[67] Concerning the cosmological aspects of gravity, which he dealt with in his third and fourth letters, Newton declared that only divine Power can account for the fact that the members of the solar system are not aggregated in one huge central lump of mass.

To him it was inconsistent with his system to try to derive the present "frame of the world by mechanical principles from matter evenly spread to the heavens." [68] Only a supernatural power, he insisted, could reconcile this supposedly original state of matter with the effects of gravity, for "that which can never be hereafter without a supernatural power, could never be heretofore without the same power." [69]

What was more concretely in Newton's mind, when stressing the necessity of the continued intervention of God in the universe, can best be seen in the queries appended to his *Opticks*. There he discussed the effects on the planetary system of two physical factors, which should upset in the long run the regular motion of the planets unless God intervened from time to time. The first of these factors was the resistance encountered by the planets moving through the ether, the density of which he put at $1/700,000$ that of the air. [70] Although he estimated that the cumulative effect of such resistance would even after thousands of years still be unobservable, he warned that the long-term effect could not be ignored. The other factor, demanding the eventual "reformation" of the solar system derived, according to him, from the mutual disturbances of comets and planets. [71]

The publication of Bentley's sermons in 1693 was chiefly instrumental in publicizing the apparent advantages afforded to natural theology by a physical system such as that presented in the *Principia*, which had remained until then a closed book for educated circles. What followed was an unbridled exercise in natural theology that witnessed the publication of books with most alluring titles, such as W. Derham's *Physico-Theology* (1713) and a host of others. Derham's work, like that of Bentley, grew out of the sixteen sermons preached as the annual Boyle Lectures in 1711 and 1712. It showed natural theology on the rampage, for Derham not only "proved" God's existence from the existence of an atmosphere in general, but he also forged a distinct proof from the winds, from the clouds and rain, from light, and of course from gravity. All this, however, covered only the first of eleven long-winded dissertations, which surveyed the earth, man's body, the kingdom of four-footed animals, birds, insects, reptiles, and vegetables with a view to finding material useful for natural theology. For Derham and many others, the vigorously growing body of emerging science appeared indeed a goldmine of proofs pointing to the existence, goodness, and power of a Creator. Unfortunately, it took a long time to realize that many of the shiny bits were only fool's gold. In the meantime theologians and scientists were busy sealing that "holy alliance" [72] between science and religion that extended at least in England well into the nineteenth century. The

Continent followed suit, for a while at least, and saw the appearance of books that boasted of such titles as "theology of stones" and "theology of insects,"[73] to give only a little detail of a literature which bore witness not so much to God as to a pathetic absence of sobriety of mind.

Theologians, however, are to be blamed only in part. Scientists were no less enthusiastic in preparing fantastic mixtures of physics and sacred history. Halley, in a paper read before the Royal Society in 1694, explained the deluge by huge tidal waves that followed a near collision between a comet and the earth.[74] Newton himself was not safe from indulging in and condoning similar extravaganzas. When W. Whiston, his successor at Cambridge, published his *Theory of the Earth* in 1696, Newton's praise of it was not long in coming, a really startling fact in view of the fantastic details of this book. While Whiston connected the spinning of the earth on its axis with the eating of the forbidden fruit by Adam and Eve, J. Craig, another professor of mathematics at Cambridge, fixed the date of the end of the world at A.D. 3150 on the basis of calculations that claimed to have utilized a version of the inverse square law of Newton. The title of Craig's work, *Theologiae christianae principia mathematica,* indicated only too well the extent to which Newton's immortal opus could dominate the speculations of some of his contemporaries even in theological matters.

Whiston's and Craig's writings clearly bore the marks of odd reasoning, to say the least, and they should now be remembered only as indicative of the intellectual tastes of a bygone age. Yet, even around 1700 they failed to command a lasting interest. It was otherwise with the proofs of God's existence based on the physics of the *Principia,* which Newton himself explicitly summarized in the General Scholium, appended to the second edition of his celebrated work. These arguments touched off a serious and heated debate that characteristically enough never questioned the direct bearing of the conclusions of physics on the existence of God. For some time yet this seemed to be beyond any doubt. The controversy revolved rather around the point whether a perfect clock, or a clock in need of periodic repairs, should serve as the basis of a scientific demonstration of God's existence. Samuel Clarke, who delved into the controversy in Newton's name, argued that the concept of a perfect machine excluded God from the universe and lent support to materialism and absolute fate. Leibniz, for one, could not have objected more strongly to the concept of a God "who must consequently be so much the more unskilful a workman, as he is often obliged to mend his work and to set it right."[75]

Not that Leibniz envisaged a physics or a scientific world system disregarding God. Far from it. He pleaded in his *True Method in Philosophy and Theology* for the introduction of the method of geometry into theology to make theological reasonings more cogent and effective.[76] In poking fun at a "God" capable of constructing only a deficient clock, Leibniz merely tried to discredit proofs of God's existence that were based on gaps in our scientific knowledge. In this he not only took a sound position, but also anticipated the tastes of the deism of eighteenth-century Enlightenment. Signs of this were not long in coming, and, as often happened, the declaration of preferences well preceded the clinching of scientific evidences. Thus in Kant's *Universal Natural History and Theory of Heavens*[77] little if any proof is offered for the point that those phenomena of the solar system, which according to Newton demanded the intervention of God, can be explained on a general mechanical basis. This, however, could hardly disturb Kant, who resolutely abandoned that part of the Newtonian approach that based God's existence on proofs erected on gaps. On the other hand, he reaffirmed wholeheartedly that part of the Newtonian natural theology that argued God's existence from the order and lawfulness manifested in nature. This he stressed strongly in his work, part of which he republished several years later as the only possible proof of the existence of God.

This was clearly the style the age of Voltaire would relish. If, however, there was a point where Voltaire did not adopt the views of Newton, which he made to prevail in France even at the price of being thrown into jail, it concerned precisely the problem of the perfection of the world machine. On the one hand, Voltaire proclaimed that Newton's philosophy "leads necessarily to the knowledge of a superior being who has created everything, arranged everything freely." He even deprecated Cartesianism as a system that led many people "to admit no other God than the immensity of things" and claimed that he saw "no Newtonian who was not a theist in the strictest sense."[78] On the other hand, Voltaire's God had nothing to do with the celestial clockwork once it had been wound up. For God, as defined by Voltaire, was the unsurpassed "eternal machinist."[79] Similar preferences animated the youthful Le Sage, the author of *Lucrèce newtonien*, who claimed to have achieved in respect to the Newtonian system what Lucretius had once done for the cosmologies of his day — free them from the continual intervention of the gods. What is more, the task of the physicist, in Le Sage's view, was "to put himself in the place of the Being who produced the phenomena and who, having clearly grasped in advance all the consequences of the various intensities with which the affections of bodies might be endowed, would have chosen that intensity which was the most suitable to fulfil his aims, and who would have implemented it without any previous trial."[80] Such a program was presumptuous, if not simply arrogant, and happily for physics, Le Sage received no encouragement to carry it out. Yet, it clearly showed the thorough dissatisfaction with the concept of an imperfect world machine, a dissatisfaction that culminated, on a scientific level, in the work of Lagrange and Laplace.

As is well known, these two found that the perturbations within the solar system never exceed a certain amount and repeat themselves over a period of about two million years. Therefore, contrary to Newton's stipulations, it no longer required divine intervention to keep the planets in the plane of the solar system and their distances from the sun essentially constant. To base atheism on Laplace's equations and calculations, however, was as unsafe as trying to notice with Newton the Deity in every dark corner of the solar system. The story of Laplace's boastful words to Napoleon, "Sire, je n'ai pas eu besoin de cette hypothèse," might of course sound blasphemous to some pious ears, but whether authentic or not, those words reflect a thoroughly sound scientific attitude. For Laplace saw clearly that the dispute between Newton and Leibniz on the role of the Creator in his world machine had ended in a stalemate. To Leibniz's criticism that a God repairing his universe must be short on wisdom, Newton could answer that Leibniz's preestablished harmony was nothing short of a perpetual miracle. Was it not much simpler to ask with Laplace: "But may not this disposition of the planets be itself an effect of the laws of motion: and may not the Supreme Intelligence, to whose intervention Newton had recourse, have made this orderly disposition dependent on a phenomenon of a more general character?"[81]

Clearly, for the purposes of science, a simple reference to a more general physical law, though unspecified, was perfectly sufficient and, in all honesty, not at all detrimental to the prerogatives of Deity. Such a stance was eminently honest from the viewpoint of physics, which simply had no criteria for singling out any of its laws and observations as representing the very limits of scientific search. This was equally true of all the assumptions science could imagine about any hypothetical "phenomenon of a more general character." None of them could be vested in an aura of finality. Thus theologians could keep pointing out that postponing the final explanation was not giving it. If nebular hypotheses and kindred assumptions, as W. Whewell noted, "are put forwards as a disclosure of the ultimate cause of that which occurs, and as superseding the necessity of looking further or higher . . . their pretensions will not bear a moment's examination."[82] Laplace, for one, could not be charged with such pretensions. For him the nebula out of which the solar system was supposed to have formed was but a conjecture, and he submitted it "with all the distrust which everything that is not a result of observation or calculation ought to inspire."[83]

The recognition that the gaps in scientific knowledge need and should not be filled with obsequious references to Deity was a signal step forward. Its impact was felt almost immediately. Thus W. Paley admitted as early as 1803, in his widely acclaimed *Natural Theology*, that "astronomy is not the best medium through which to prove the agency of an intelligent Creator."[84] As he put it in a most revealing passage, only subjects sufficiently complex were good "for this species of argument." Clearly, it was easier to find "gaps" in a complex subject than in one with the high degree of simplicity that characterized the Laplacian account of the solar system.

One should not suppose, however, that after Laplace all physicists stopped basing God's existence on "gaps." Those gaps referred at times not only to gaps in scientific information but also to "gaps" in some sectors of the universe itself. Thus, for instance, J. E. Bode, the German astronomer, when discussing in 1802 Piazzi's discovery of Ceres, noted with unconcealed satisfaction that he had written as early as 1772: "Now, however, there comes a gap in this regular progression. From Mars outward there follows a space of $4 + 24 = 28$ units in which, up to now, no planet has been seen. Can we believe that the Creator of the world has left this space empty? Certainly not!"[85] This was a strange mixture of astronomy and theology but regardless of Bode's personal gratification, the filling of the "gap" between Mars and Jupiter could hardly have had any bearing on any proof concerned with God's existence.

Bode at least was fortunate. Physical discovery verified what he predicted. Other physicists, baffled by other "gaps," were rebuffed by nature. She proved to be far more resourceful than some physicists imagined, and in the long run she invariably accounted for puzzles that prompted some physicists to invoke God too hastily. Thus an unexplainable fact that postulated, according to Kelvin, the direct action of a Creator, was the permanence of atoms.[86] Such a view was all the more strange on Kelvin's part, since his was not a mind to abdicate problems too easily. "Science," he once said, "is bound by the everlasting law of honour, to face fearlessly every problem which can fairly be presented to it. If a probable solution consistent with the ordinary course of nature can be found we must not invoke an abnormal act of Creative Power."[87] But believing as he did that atoms are truly something that cannot be split, Kelvin had no choice but to refer the case to God. In this reasoning, he was not alone. The puzzle of the stability of atoms was just as overwhelming for John F. W. Herschel, who also saw God's direct action in the fact that so many milliards of atoms are both permanently and precisely identical. As he put it, strict similarity between several things points to a common principle independent of them, and since the number of strictly identical atoms was immense, the argument seemed to acquire in his view an "irresistible force."[88]

Herschel's reasoning was fully endorsed by Maxwell. The fact that the properties of atoms or molecules are always the same, that they are never changed even slightly by the processes of nature, proved for Maxwell that their existence and the identity of their properties through time could not be ascribed to any natural cause. For Maxwell, both the premises and the conclusion had all the appearances of solid finality. The exact equality of each molecule to all others of the same kind displays, according to Maxwell, "the essential character of a manufactured article and precludes the idea of its being eternal and self-existent."[89] Yet, the trouble with all this was that Maxwell based his denial of the eternity and self-existent character of atoms on an appraisal of a particular stage of science and the appraisal was patently false. He was wholly wrong in stating that science has assured herself that the atom "has not been made by any of the processes we call natural."[90] As a scientist Maxwell had no right to state, in connection with the question of the origin of atoms that science by 1873 had reached the end of a "strictly scientific path" beyond which there was the gap of the unknown, and an absolute unknown at that. Placing God directly and immediately beyond the world of atoms was not, as later discoveries showed, an arrangement to the liking of nature.

Such speculations on atoms therefore provided the same essentially unreliable support for proving God's existence as had the unexplained phenomena of the solar system in Newton's time. Also doomed to failure were the efforts of those who throughout the nineteenth century postulated God beyond the nebula that supposedly preceded the all-inclusive system of stars, the Milky Way. In general, proofs based on gaps were discredited with unfailing regularity as science marched on and the partisans of dogmatic materialism could hardly have rejoiced more over such state of affairs. One of their most revered prophets, Engels, wrote with sarcastic satisfaction:

God is nowhere treated worse than by natural scientists who believe in Him. . . . In the history of modern natural science God is treated by his defenders as Frederick William III was treated by his generals and officials in the campaign of Jena. One division of the army after another lowers its weapons, one fortress after another capitulates before the march of science, until at last the whole infinite realm of nature is conquered by science, and there is no place left in it for the Creator. Newton still allowed Him the "first impulse" but forbade Him any further interference in his solar system. Father Secchi bows Him out of the solar system altogether, with all canonical honours it is true, but none the less categorically for all that, and he only allows Him a creative act as regards the primordial nebula."[91]

Engels evidently badly misread Newton. This should not, however, be a cause of surprise. Engels' familiarity with science, as shall be seen in the next chapter, had never passed beyond the rudiments. Yet, he spoke with hardly any second thought about almost everything in science. He could therefore see nothing scientifically wrong about his investing matter with the attributes of God. For all that his scathing words should have served as a sobering warning to unwary physicists and eager theologians alike about the "effectiveness" of proofs based on the provisional incompleteness of scientific knowledge.

The old temptation to locate God in the no-man's-land of science is far from being extinct. It was not long ago that E. N. da C. Andrade took the view that "the electron leads us to the doorway of religion."[92] Being an atomic scientist, he should have been more aware of the rapid changes that characterize scientific research of the subatomic realm of matter. There might be many gaps in the present-day scientific information about the electron, but none of these gaps can serve as doorways to God. "Gaps in knowledge," as Weizsäcker aptly put it, "have a habit of closing — and God is no stopgap."[93] What are meant to be doorways invariably become trapdoors that take as their victims those to whom the lessons provided by the history of science are pretty much a closed book. Instead of indicating clear-cut frontiers of what can be known in science, the advances made by science suggest strongly that there is rather a continuity between what we know and what we do not know.

More reliable than the "gaps," as regards the proofs of God's existence, has been the order in nature evidenced by the laws of physics. While scientific progress has proved consistently detrimental to resting God's case on a gap that sooner or later came to be filled, new discoveries have only enhanced the range and universality of scientific laws. No law, however imperfect or limited, has ever disappeared completely as research progressed. They have rather proved to be particular cases of other laws much wider and deeper in application. That order and lawfulness are the Creator's fingerprints in nature was a conviction shared by all major figures of classical physics. Seeing the world "established in the best order," stated Copernicus at the very outset of his work, was a sure way to lead one to "wonder at the Artificer of all things."[94] For Kepler, discovering laws in nature was nothing short of reading the mind of God himself. It was the desire "to obtain a sample test of the delight of the divine Creator in his work and to partake of his joy"[95] that provided him with the ultimate motivation for his nearly superhuman scientific labors. To Galileo, the inexorable and immutable nature that "never transgresses the laws imposed upon her" brought clear testimony of a Lawgiver.[96] What is more, knowledge of nature's laws expressed mathematically was in Galileo's eyes a sharing in the truthfulness of divine Wisdom.[97] Newton was no less emphatic on this point: "This most beautiful system of the sun, planets, and comets could only proceed from the counsel and dominion of an intelligent and powerful Being."[98]

Impressive as the scientific documentation of the orderliness in nature might have been, it was not without pitfalls. These became evident when certain laws of physics were presented as showing literally the way in which God ran his universe. Euler, for instance, was emphatic in claiming teleological and ultimately theological roots for the maximal and minimal principles.[99] The same direct connection between the Creator's mind and the principle of least action was made by Maupertuis, who wrote: "Our principle, being more in conformity with the ideas we ought to form of things, leaves the world in the continuous care of the power of the Creator, and is a necessary consequence of the most wise use of this power."[100]

Before long, however, these and similar efforts to identify physical axioms and laws with the wisdom of a Supreme Intellect came in for their share of criticism. For as D'Alembert warned, the nature of the Supreme Being transcends man's intellectual grasp too much to permit man to decide whether a particular law of physics is or is not the exact equivalent of God's planning. Not that D'Alembert was unwilling to admit that scientific laws, and particularly their simplicity, manifest in a sense God's wisdom. But he considered as erroneous all endeavors to establish one-to-one correspondence between certain laws of physics and the actual laws of nature as established by God. For D'Alembert, such attempts were merely stuttering speculations about what God's mind ought to be, and at the same time they tended to induce in the physicist a state of mind in which critical awareness was conspicuously absent. Speaking of Descartes explicitly and also having an eye on such contemporaries of his as Maupertuis, he wrote:

It is because he [Descartes] followed this method and because he believed that it was the Creator's wisdom to conserve the same quantity of motion in the Universe always, that Descartes has been misled about the laws of impact. Those who imitate him run the risk of being similarly deceived; or of giving as a principle something that is only true in certain circumstances; or finally of regarding something which is only a mathematical consequence of certain formulae as a fundamental law of nature.[101]

In the mid-1700's, when D'Alembert wrote these lines, the range of physics was pretty much restricted to mechanics. Other branches of physics were in a rudimentary form, and their eventual unification could at that time be only a matter of hopeful belief. Ready to see the reflection of God's wisdom in the laws of mechanics in general, D'Alembert would probably have concurred with those who in the next century ushered in the era of unification of the various branches of physics. The identification of heat processes with the transformation of mechanical energy indicated to Joule that "order is maintained in the universe — nothing is deranged, nothing ever lost . . . and though . . . everything may appear complicated . . . yet the most perfect regularity is preserved — the whole being governed by the sovereign will of God."[102] Mayer and Helmholtz, who shared with him the glory of enunciating the law of the conservation of energy, had similar views.

When we turn to those who elucidated the laws of interaction between magnetism and electricity, their unanimity in seeing behind such unity and order in nature a glimpse of the Supreme Intelligence is no less striking. For Oersted, "every fundamental investigation into nature must issue in a recognition of the existence of God."[103] Ampère[104] and Henry[105] spoke in a similar vein. Faraday was particularly explicit in viewing in such light the manner in which the various forces of nature are correlated. In fact, he liked to speak of the eventual unification of gravitational, electrical, and mechanical forces as a glorious discovery to be made by physics of the wisdom and power of God manifested in the created universe.[106] As for Maxwell, who put the capstone on classical electricity and magnetism, one should read either his "prayer of a scientist"[107] or his letter of November 22, 1876, to Bishop Ellicott, in which he stated that in his opinion "each individual man should do all he can to impress his own mind with the extent, the order, and the unity of the universe, and should carry these ideas with him as he reads such passages as the 1st Chap. of the Ep. to Colossians."[108] The same conviction was expressed in the introductory lecture that Kelvin read with almost no changes at the beginning of each academic year from 1846 on.[109] The eloquent conclusion of J. J. Thomson's address delivered in Winnipeg before the British Association in 1909 struck a similar note in speaking about how the laws of physics keep yielding to ever deeper and more inclusive ones: "In the distance tower still higher peaks, which will yield to those who ascend them still wider prospects, and deepen the feeling whose truth is emphasized by every advance in science, that 'great are the works of the Lord.'"[110]

But Thomson's phrases reflected the philosophic style of an older generation of physicists. The straightforward appeal from nature's order to nature's God was already disappearing from the writings and addresses of the younger generation. Rutherford, Bohr, Einstein, De Broglie, Heisenberg, and the majority of the great names of modern physics no longer made the same direct connection between nature and God. While Stokes, whose acquaintance with scientists was extensive, could state in a letter after the turn of the century that skeptics in religious matters form a very small minority among men of science,[111] this was not so some thirty years later. One could rather notice an increasing diffidence in scientific circles against making inferences touching on natural theology. As a result, the explicit testimonies voiced by physicists on behalf of belief in a Creator began to thin out considerably. Various reasons could be cited for this change, among which the most important are perhaps the changing educational background of physicists as contrasted to that of their predecessors in the nineteenth century and the idealistic leanings evident in the formulations of the basic assumptions and methods of modern physics.

Did this change undermine the legitimate contribution of physics to arguments proving God's existence from the order evidenced by nature? Essentially it did not, though the change brought out forcefully that such arguments depended far more on one's philosophy of physics than on the data of physics itself. On the other hand, the paradoxical status of many basic principles and findings of modern physics illustrates vividly the view that order in nature is not simply the creation of the inquiring mind. More forcefully than ever, physics had to recognize that its laws describing this order were not a priori constructions but had to be tailored meticulously to the stubborn, brute facts of nature. These facts are the actual setup, distribution, quantization of forces, and the sharply defined characteristics of the "fundamental" particles of matter, which simply state that not everything imaginable occurs in nature. Nature is a supreme paragon of drastic limitations of physical possibilities, and the order of the universe is just another aspect of this primordial fact. Democritus once postulated atoms of all imaginable magnitudes, from the invisibly minute ones to giant but equally hard units of matter, for he could not see how matter itself could provide a principle or cause that would impose on it a restriction permitting only vanishingly small units of matter. His followers recoiled from such a deep question and simply accepted as a postulate the proposition that the size of atoms was limited to the infinitesimally small. Yet, the fact of limitation remains inextricably present in the order and correlation of things as we see and interpret them, and of this limitation which can in principle take on so many various forms, nature itself gives no explanation.

Theologians are fond of calling this brute fact of limitation the "contingency of nature," and of this contingency even modern physicists who do not share the traditional Christian views on the Creator bear an indirect testimony. Einstein, for one, believed, "in Spinoza's God, who reveals himself in the harmony of all being, not in a God who concerns himself with the fate and actions of men."[112] How distinct is Spinoza's God from nature either in Spinoza's or in Einstein's interpretation is difficult to tell. Yet, even if in Einstein's belief nature's God was not "playing dice" and was not "malicious" or acting as a person, nature's order and intelligibility as viewed by Einstein the physicist was a fact, an "inexplicable, hard fact," a contingency that demanded faith if science was to exist. The comprehensibility of the world, which for him was the "most incomprehensible thing about the world,"[113] could be grasped according to him by faith alone, and he could not "conceive of a genuine scientist without that profound faith."[114] This faith in nature's intelligibility included, however, no less absolute respect for the facts of nature. Yet, facts like the magnitude and constancy of the speed of light in any reference system, the magnitude of the elementary electric charge, and a host of others are not self-explanatory. One has to accept them, in a sense unconditionally, although recognition of the existence of such facts is not the last stage in human inquiry. It is always legitimate to search for underlying reasons, for logically more exhaustive answers or concepts, even if by doing so one passes into an area of human reflection that is clearly beyond the range of physics.

Proofs of the existence of God based either on the gaps in physical science or on the order in nature evidenced by the laws of physics had been for a long time the customary way of engaging the services of physics on behalf of theology. An entirely new and enticing possibility seemed to offer itself, however, with the formulation of the law of entropy. For the first time in the history of physics, there was a law that could be taken as a direct, positive indication of the temporal limitedness of the universe. It took seventeen years for physics to see Clausius coin the word entropy, after Kelvin hit upon Carnot's almost forgotten paper in 1848. But years before the exact formulation of the entropy function by Clausius, the far-reaching consequences of the law had been clearly recognized. They were hard to accept, for Carnot's theorem indicated that something was always lost in the transformation of heat into mechanical energy or vice versa. Clearly, in an age that witnessed the triumph of the principle of the conservation of energy, such ideas necessarily appeared highly repugnant at first. Joule for one strongly urged the abandonment of Carnot's fundamental theorem, but Kelvin saw that this could not be done without incurring "innumerable other difficulties."[115] As it later turned out, a part of the utilizable energy was lost in every process, leaving the conservation principle intact.

Just as intact, however, remained some ponderous consequences of Carnot's law. In his paper "On a Universal Tendency in Nature to the Dissipation of Mechanical Energy" (1852), Kelvin summarized their consequences in three points.[116] The first stated the presence in the material world of a universal tendency to dissipate mechanical energy. The second declared the impossibility of the restoration of any mechanical energy without causing an even greater dissipation of it. The third enunciated the limited period during which the earth would be fit for human habitation. This was indeed a gloomy state of affairs, and Kelvin took pains to point out that the inexorable course of events could be reversed only by operations "which are impossible under the laws to which the known operations going on at present in the material world are subject."[117]

Two years later, Helmholtz praised the sagacity of Kelvin, who "in the letters of a long known little mathematical formula which only speaks of the heat, volume and pressure of bodies, was able to discern consequences which threatened the universe, though certainly after an infinite period of time, with eternal death."[118] In Helmholtz's view, Carnot's law had to be recognized as a "universal law of nature," which radiated light "into the distant nights of the beginning and of the end of the history of the universe."[119] The end, as Helmholtz phrased the conclusion of physics, had to be a "state of eternal rest," and the only comfort our race could take, thus threatened with a "day of judgment," was that for physics the dawn of that day was "still happily obscured."[120] So spoke not a clergyman but a physicist, and no mean one at that.

Before long some physicists drew from the law of entropy conclusions that were even more theological in character. *The Unseen Universe* of P. G. Tait and B. Stewart, which truly could pass as the "physico-theology" of the late nineteenth century and which in eight years went through nine editions, flatly declared that the law of entropy proved it "absolutely certain" that life had a beginning and will have an end.[121] The universe was inexorably running down and its lifespan, so the authors stated, was staked out between two fixed endpoints determined by the minimum and maximum entropy. "We have then thus reached the beginning as well as the end of the present visible universe, and have come to the conclusion that it began in time and will in time come to an end." [122]

Tait, a colleague of Kelvin, could at least have indicated that Kelvin himself had come in the meantime to have second thoughts about the cosmological implications of the law of entropy. The idea of the possibility of an infinite universe as a reservoir of an infinite amount of energy, which Tait dismissed as meaningless, was upheld by Kelvin as of great importance. As he put it, unlimited extrapolation of the law of entropy would be valid only if "the universe were finite and left to obey existing laws." But, as he added at once, "it is impossible to conceive a limit to the extent of matter in the universe; and therefore science points rather to an endless progress, through endless space, of action involving the transformation of potential energy into palpable motion and thence into heat, than to a single finite mechanism, running down like a clock, and stopping forever." In general Kelvin tried to strike an encouraging tone to offset somewhat the "dispiriting views" inspired by the early interpretation of the law of entropy "about the destiny of the race of intelligent beings by which it [the earth] is at present inhabited." [123]

A similar cautious tack was taken on the question by Maxwell. Not that he tried to minimize at any cost the cosmological implications of the law of entropy. The irreversible character of the dissipation of energy, he told the Liverpool meeting of the British Association in 1870, leads to a moment in which the mathematical formula of entropy reaches a critical or minimum value, and beyond this point the "formula becomes absurd." In other words, there seemed to be no way of inquiring into the state of things that existed during the instant immediately preceding that moment. In the climate of classical physics the idea of an absolute beginning was of course more palatable than the vision of an icy end. Yet, entropy brought both ideas home for the late nineteenth-century physicist with greater surprise, as Maxwell put it, than "any observer of the course of scientific thought in former times would have had reason to expect." [124] Still, Maxwell warned against hasty and rigid conclusions. Although a heated body, he remarked, will eventually settle down into an ultimate state of quiet uniformity, the human mind and speculation follow a course that is highly unpredictable.

Full of scientific wisdom as this remark was, it could hardly bring mental comfort to many of his colleagues. Some were already busy devising theorems aimed at saving the idea of an ever-operating universe from the prospect of slow decay. Some of these efforts were motivated by purely scientific preoccupations, such as the theory proposed by one of the founders of thermodynamics, M. Rankine. In his paper, "On the Reconcentration of the Mechanical Energy of the Universe" (1852), he took the view that "there must exist between the atmospheres of the heavenly bodies a material medium capable of transmitting light and heat; and it may be regarded as almost certain that this interstellar medium is perfectly transparent and diathermaneous."[126] This medium he then imagined to end abruptly at very great distances in space and form parabolic walls that would refocus the radiant heat at various points in space. Thus the recapture of the energy dissipated in interstellar space would be secured. To restart the cosmic process, all that was needed was to assume that at one time or another a cold, dead star may wander into one of those very hot foci of the ether wall and regain almost instantaneously its original high temperature. The supreme irony of all this, however, was that Carnot's principle came to be negated by Rankine's theory, although Rankine's very aim was to harmonize that principle with the concept of the perpetuity of physical processes. Clearly, it was self-defeating to assume that radiant heat would continuously pass from colder bodies (earth or cold stars) or cold interstellar spaces into the ultrahot foci of the ether wall.

Actually, the law of entropy proved more unyielding with every attempt to modify it. A picturesque recognition of this was given by Maxwell, who conceived of "a being whose faculties are so sharpened that he can follow every molecule in its course."[127] Such a being (later dubbed a demon) could separate low-speed molecules from those with high speed by manipulating a microscopic stopcock and thus reestablish original energy states. To handle the traffic of molecules between two small glass containers was an astounding feat even for a demon, but what about the infinite number of molecules wandering in infinite space with all sorts of speeds? So the demon was in fact a reminder, bearing the stamp of Maxwell's gentle sarcasm, of a direction with hardly any promise for scientific speculation.

This was realized by some, such as Mach, who searched for logical inconsistencies in the extension of the law of entropy to the whole universe. In substance he took the view that science can investigate only a limited number of phenomena, and a strict application of the results to the totality of the phenomena is, as he put it, scientifically meaningless. Dubious as it was from the viewpoint of the scientific method to erect such a gap between the parts and the whole of the universe, Mach took delight in giving the argument in several variations. Thus he wrote: "The universe is like a machine in which the motion of certain parts is determined by that of others, only nothing is determined about the motion of the whole machine."[128] Again, he stated that if one part of the universe serves as a clock for the other part, "we have nothing left over to which we could refer the universe as to a clock. For the universe there is no time."[129] In 1872, of course, Mach could not dream of its possible expansion. Thus he could perhaps feel more at ease in classifying the universal cosmic validity of entropy as a statement "worse than the worst philosophical ones."[130]

Mach as usual sounded very assertive though not necessarily convincing. But at least he faced the issue as honestly as he could. Others preferred to be silent, like Tyndall, whose two widely read works, *Heat, a Mode of Motion* and *Fragments of Science*, contained only one short reference to the second law of thermodynamics, in connection with the republication of Carnot's paper in 1878. In truth, Tyndall's silence on this point was indicative of the "metaphysical" uneasiness some felt about the cosmic implications of entropy, debatable as these were on several counts. Others, however, for whom science had to bolster monistic or materialistic tenets, were loudly indignant. For them entropy was a bogey that science had to get rid of as soon as possible. Haeckel, for instance, simply rejected the second law on the ground that it contradicted the first law, which for him meant nothing less than the eternity of substance and of "cosmogenetic" processes. "If this theory of entropy were true," he wrote, "we should have a 'beginning' corresponding to this assumed 'end' of the world."[131] Obviously the thought of a beginning or creation of the universe disturbed him deeply. Others gloated over the fact that the heat death of the universe at absolute zero temperature was in clear contradiction to the biblical image of a world meeting its end in flames. But such witticism could hardly conceal some deep uneasiness. Witness the agitated reaction of Engels, who saw no escape from admitting a creation of the universe if the law of entropy were unrestrictedly true: "Clausius — if correct — proves that the universe has been created, *ergo* that matter is creatable, *ergo* that it is destructible, *ergo* that also force, or motion, is creatable and destructible, *ergo* that the whole theory of the conservation of force is nonsense, *ergo* that all its consequences are also nonsense."[132]

Although such tirades had no scientific merit, the further developments in theoretical physics uncovered aspects of the law of entropy that made it even more difficult to consider the "running down" of the universe an unquestionable conclusion. The developments were due mainly to Boltzmann, the chief architect of the statistical interpretation of the law of entropy. The apparently absolute irreversibility of mechanical processes was shown by him to be only a statistical result, and one had to admit in principle at least that their direction might take at times an opposite sense as well. Much as Boltzmann's monist views could color his phrases, the very strength of his reasoning lay in the concept of probability laws, which stated that both "probable" and "improbable" could occur, and that the macroscopic realm, being the sum of microscopic units, could not be an exception to this rule. As Boltzmann put it: "The laws of probability calculus imply that, if only we imagine the world to be large enough, there will always occur here and there regions of dimensions of the visible sky with a highly improbable state of distribution." [133] There was no denying that statistical theory found a flaw in an argument that simply equated the concept of entropy with the unavoidable running down of the universe as a whole. Poincaré therefore had a point when he referred to his uneasiness about reasonings where "reversibility is found in the premises and irreversibility in the conclusions." [134]

Around the turn of the century, when Boltzmann was seeking acceptance for his views, the utilization of the law of entropy in theology had already enjoyed a short but agitated history. Curiously enough, the first to take up the study of the bearing of entropy on the problem of creation, A. Fick, reached a negative conclusion.[135] But when Father A. Secchi, of solar physics fame, concluded to the merit of the law of entropy in apologetics,[136] a host of Roman Catholic theologians voiced their approval. Only a few, like H. Pesch, A. Rademacher, and F. Diekamp, struck a dissenting note. The reaction of conservative Protestant theologians was divided in much the same proportion. The liberal wing of Protestant theology had by then grown totally indifferent to such topics.

Spiritually oriented philosophers like Bergson greeted the law of entropy as "the most metaphysical of the laws of physics," but before long P. Duhem, an outstanding physicist and a devout Catholic, poured cold water on an enthusiasm mixed with so much wishful thinking. For Duhem, the point where the arguments of the apologists went wrong was the meaning given to a law of physics. Physical law, as Duhem noted, is a mathematical proposition representing with a degree of accuracy the data of observation; therefore, it is subject to revision and can hardly form the basis for unconditionally definitive answers, especially answers concerning the very remote past or the very distant future, about which scientific information is particularly meager. "By its very essence," Duhem warned some unadvised apologists, "experimental science is incapable of predicting the end of the world as well as of asserting its perpetual activity. Only a gross misconception of its scope could have claimed for it the proof of a dogma affirmed by our faith." [137]

Partly under the impact of the caution counseled by eminent physicists, partly because of the demands set by any rigorous proof of

the existence of God, a more critical appraisal began to emerge slowly in theological quarters. C. Isenkrahe[138] and J. Schnippenkötter,[139]
who thoroughly reviewed the apologetic literature on the subject
prior to 1920 and analyzed the arguments, concluded to the fundamental weakness inherent in a proof that tried to establish the creation of the universe and consequently the existence of God on the
basis of the law of entropy. This did not mean, however, that the
cosmological potentialities contained in the concept of entropy no
longer lurked in the back of the mind of many physicists and theologians alike.

Among the former were no lesser names than Eddington and Jeans,
who in their immensely successful popular expositions of modern
physics and astronomy gave a new luster to the cosmological attractiveness of the concept of entropy. Eddington in fact described the
law of entropy as holding "the supreme position among the laws of
Nature."[140] To show that such "exaltation of the second law is not
unreasonable,"[141] Eddington could refer to the incontrovertible fact
that it has always been verified in all measurements relating to the
most widely differing fields of physics. From it an almost illimitable
chain of deductions has been drawn both in theoretical and applied
physics, and with unfailing success at that. The law of entropy gave a
good account of itself even in such fields of research as quantum mechanics, where the mechanism of the individual quantum process
was and still is unknown and is even unimaginable. The law demonstrated to Eddington that at a remote time the energy of the universe
was wholly organized or concentrated and beyond this state it was
impossible "to go back any further under the *present* system of
natural law"[142] (italics mine). But characteristically enough, he
wanted no part in drawing from it conclusions of metaphysical
nature. In fact he urged his readers to regard the law of entropy "as
the working hypothesis of thermodynamics, rather than its declaration of faith."[143] Jeans spoke much in the same vein. According to
him, every physical process pointed "with overwhelming force to a
definite event, or a series of events of creation at some time or times,
not infinitely remote." The universe, he noted, cannot always have
been the same as now, for in this case the whole universe would consist of a uniform cool glow of radiation. "This is indeed, so far as
present day science can see, the final end towards which all creation
moves, and at which it must at long last arrive"[144] (italics mine). Yet,
although Jeans was not reluctant to compare God with a mathematician, he was far from willing to conclude to a Creator from the law
of entropy.

Distinctly more metaphysical was the attitude of Whittaker, who in his Riddell Lectures of 1942 asserted that one can trace "by purely scientific methods" the development of the universe backward in time to a "critical state of affairs beyond which the laws of nature, *as we know them*, cannot have operated: a Creation in fact." (italics mine). Though Whittaker emphasized that science can give no account of Creation itself, he was firmly convinced that "physics and astronomy can lead us through the past to the beginning of things and show that there must have been a Creation."[145] The critical state of affairs to which Whittaker referred was the perfectly ordered state of the universe, with no random motion in any of its parts, and with no energy yet being used up. Prior to this state, the world could exist, if it existed at all, only in an absolutely inert condition, which even if it had existed from past eternity, went Whittaker's reasoning, would be scientifically inconsequential to the determination of the moment that saw the beginning of all physical processes in the universe.[146] A very similar view was adopted by Weizsäcker, who saw no way of escaping the conclusion that each part of the world possesses only a finite reserve of events.[147]

But the picture was far from that simple. Whittaker himself admitted in his Tanner Lectures of 1947 that relativistic thermodynamics permits the case of a universe in which an expansion at a finite rate could take place reversibly.[148] This, he recognized, was a loophole, however narrow, in the argument of forecasting a final heat death for the universe. It is interesting to note that the idea of an unconditional heat death of the universe was strongly disputed by the noted cosmologist, E. A. Milne, who at the same time worked hard on a proof of God's existence based on his theory of kinematic relativity. Milne called attention to the fact that the law of entropy has been validated only for finite portions of the universe, with rather stringent conditions at that. Any of these portions, Milne noted, should always be divisible into two parts, so that the phenomena in one could be regarded as not affecting the phenomena occurring in the other. What is more, in one of the portions such reversible entropy changes should take place as could measure the increase of entropy resulting from the irreversible processes occurring in the other. Now it is by no means demonstrated that there are no processes in some sections of the universe that leave unaffected processes taking place in other sections. Apart from all that, however, if one can conclude to a cosmic clockwinder, Milne remarked, the possibility of repeated rewindings remains also.[149]

The last point made by Milne was of course an argument ad hominem showing that the philosophical amplifications of the entropy law were far from rigorous. Though not claiming a strict demonstrative value to a theological interpretation of the law of entropy, Pius XII in his allocution of November, 1951, to the Pontifical Academy of Sciences gave wholehearted support to Whittaker's views on the subject. This was clear not only from the long passages quoted by the pope from Whittaker's *Space and Spirit* but also from the pope's categorical statement referring to the heat death of the universe: "This unavoidable fate . . . postulates eloquently the existence of a Necessary Being."[150] To be sure, the pope was careful to distinguish between scientific and metaphysical conclusions, but the sweeping phrase seemed to many to have gone too far. The reasons for this will be touched upon in connection with the expansion of the universe, the other major result of modern physical science that affected contemporary religious thought so forcefully.

The discovery of the recession of galaxies by Hubble in 1929 has certainly been one of the major events in the history of science. Just as significant was the fact that it had borne out the daring theoretical suggestion of the concept of an expanding universe made two years earlier by Abbé Lemaître. From the fact of the expansion, one could infer with ease the idea of the universe concentrated within a relatively small space at the beginning of the expansion, and it was only too tempting to view this as the primordial state of the world determined by the act of Creation. To estimate the time elapsed since the expansion began was a matter of straightforward calculation. The results indicated a few billion years and fell in the same order of magnitude as the values calculated for the age of the universe from the analysis of radioactive decay and of the stability of certain stellar systems. On the basis of this truly remarkable convergence of the data, Whittaker felt justified in concluding in 1942 that regardless of whether a time determination of such order would ultimately prove correct or not, the laws underlying these calculations lead us to a moment when the cosmic development begins.[151] Four years later Whittaker sounded even more definitive and referred to this moment of some nine or ten billion years ago as the one which "constitutes the very last limit of science. We refer to it perhaps not improperly as creation."[152]

The whole passage seemed to obtain a special emphasis when Pius XII quoted it with obvious satisfaction in his address of November, 1951. The pope himself concluded that physics with its concreteness confirmed "the well-founded deduction as to the epoch when the cosmos came forth from the hands of the Creator."[153] Present-day science seems to have succeeded, to quote the pope again, "in bearing witness to that primordial *Fiat Lux* uttered at the moment when along with matter, there burst from nothing a sea of light and radiation, while the particles of chemical elements split and formed into millions of galaxies."[154] True, the pope's speech stressed that the data reviewed still needed further research, that they were in need of further development before they could provide a sure foundation for philosophical arguments, that the scientific answer in question was neither explicit nor complete. He also contented himself with assigning to the universe an age of "billions of years," "of some five billion years." Yet, on the whole the pope's phrases clearly displayed an enthusiasm hardly in keeping with the very serious scientific uncertainties inherent in the argument.

When the pope spoke, the determination of the age of the earth and that of the universe stood in serious disagreement. In fact, the data showed the earth to be twice as old as the universe, and Baade's revised distance scale of the galaxies that resolved this discrepancy was still a year away. Furthermore there is nothing to prove that the rate of expansion remained sensibly the same throughout the very remote past. Nor is it any better demonstrated that a universe of an oscillatory type is impossible or that matter existed, before the expansion started, in a state that is inaccessible to scientific investigation. It is even more dangerous to consider the actual state of scientific research as the ultimate word, which needs improvements in some particular aspects only. Scientific explanations are by no means exhaustive, nor can one take it for granted that all the physical forces are known to science today. In the past thirty years alone science has had to stumble upon such unexpected newcomers as nuclear force and the force of weak interactions, and there is no telling what is yet in store. Matter still may have a great number of properties that not only have not yet been observed, but that may remain undiscovered for a long time or forever. Apart from speculations on what is not known or might not be known, it is well to remember that some seventy years ago the consensus of physicists was solid in estimating the age of the sun and the earth at about forty million years. Viewing the situation in retrospect, how unwise would it have been on the part of a body such as the church, claiming supreme authority in matters touching on the ultimate explanation of the universe, to have endorsed the consensus of the physicists of that time. The jolt such consensus received with the discovery of radioactivity has been vividly recalled by J. J. Thomson from the distance of three decades, and his words should be constantly kept in mind by both physicists and theologians when they are tempted to figure out such recondite matters as the date of Creation: "I think this is a warning against taking too seriously speculations about either the remote past or the remote future of the Universe, founded as they must be on the physics of the moment."[155]

Permeated as the pope's address is with a genuine appreciation of the achievements of science, it may very well turn out one day that the same pope, who a year later praised Copernicus and Galileo,[156] may have linked the Creation of the universe to "the physics of the moment" a little too tightly. More detached was the pope's attitude in his address of 1952 to astronomers. There he directly raised the question whether the riddle of the universe will one day yield to man's inquiry. Yet, instead of trying to propose an answer, he merely referred to the view prevailing in the scientific community: "The reply of those hardy spirits who have most studied the physical universe is modest and cautious. We are, they say, only at the beginning. There is still a long way to go, and it seems that the quest will be endless. It is quite unlikely that even the most gifted enquirer will succeed in recognizing (and much less solving) all the mysteries locked up in the cosmos."[157]

If there was any profit in the apologetics the pope tried to draw from the progress of physical research, it was a very general one: "The spirit of man belongs to a higher order than matter, even though the latter is immeasurable, immense, unbounded."[158] Instead of emphasizing some particular recent result in astronomy as perhaps possessing a special apologetic value, he rather pointed to the fact that "astronomical science is still far from having arrived at the end of its thrilling adventure."[159] He noted with no sign of anxiety that the new world picture of physics and astronomy "has reduced man to an atom on this speck of cosmic sludge, relegating both to a corner of the universe."[160] His comment on such an almost frightening vision was rather short and had an almost Pascalian ring: None of the changes in the scientific knowledge of the universe, he warned, can alter the nature of human love and divine compassion, so distinct from and superior to all physical phenomena.

The gist of this speech clearly went beyond the vistas of his address of the previous year when he termed the alleged determination by modern science of the fact and time of the Creation as "the reply we were awaiting from science and which the present human generation is awaiting from it."[161] It may be noted also that this change helped to bring out a more vigorous assertion of that true catholicity or universality in thought that alone can assure a reliable use of the results of science on behalf of theistic convictions. For once God is believed to be truly the Creator of the universality of nature, all reliable results of the scientific inquiry should be equally welcomed regardless of the accidental interpretative garment in which they are first clothed, either by the discoverers themselves or by the cultural ambience in which the discoveries actually took place. None of the findings can be given a preferred "theological" treatment for some specious and ephemeral apologetic considerations. After all, a discovery or new theory that could play havoc with some non-essential theological inferences can in a different context often be turned into an apologetical weapon, although almost invariably into a very doubtful one only. A case in point is the Copernican system, accepted at first, condemned later, and finally molded by Newton into a deceptively strong argument in favor of the Creator's existence. Those who cherished the Newtonian natural theology as the final word in cosmology could not but be shocked and scandalized when Laplace concluded to the stability of the solar system. What Laplace proposed once about the reduction of the beautiful order of planets to another and more universal physical law can and should be done anew with regard to the expansion of the universe: could it not be the partial manifestation of an even more universal phenomenon? For if the history of science is to have any instruction for the future, such a possibility is always to be kept in mind.

The theologian should not feel disappointed because it is not within the range and competence of physics to gaze directly at the moment of Creation or at the end of the universe. He should rather rejoice in the fact that physics so far has found nothing — and, owing to the limitations of its method, can find nothing — that might weaken in any way the fact of the contingency of the material universe and its basic difference from anything spiritual. Above all, the theologian should be aware of the danger inherent in any quasi-scientific attempt aimed at casting physics in the role of a theological timekeeper. It will, more probably than not, once again set physics on a collision course with theology. Yet, this is what one should expect as long as physicists and theologians continue to overstep their respective ranges of competence. To have stated that the earth is immobile was a clear abuse of sound theology even by seventeenth-century standards. But it was no less an abuse of physics when Planck wrote: "The knowledge of nature, continually advancing on incontestably safe tracks, has made it utterly impossible for a person possessing some training in natural science to recognize as founded on truth the many reports of extraordinary occurrences contradicting the laws of nature, of miracles, which are still regarded as essential supports and confirmations of religious doctrines."[162]

Planck was not alone among the physicists in ascribing such consequences to scientific training. Einstein, too, saw an irreconcilable contradiction between science and miracles, and undoubtedly many physicists today have similar convictions. Miracles are of course like mushrooms: they too have their "toadstools" and these probably outnumber the savory kind. To point this out is the right of science, and to accept it is the duty of theology. The scientist, however, is no less obliged to keep in mind that it is one thing not to believe in miracles and another to forget that Galileo, Kepler, Newton, Ampère, Faraday, Maxwell, and a host of others, who had "some training in natural science," did believe in miracles. With Planck's personal beliefs no one should have any quarrel, but his reference to "scientific training" is a distinctly different matter. What Planck had overlooked was that in such questions it is not his or anyone else's training in physics that makes the difference, but rather the individual physicist's philosophy and his information in religious matters, or not infrequently the sheer lack of it. The fallacy is of course widespread that belief in miracles, or in a revealed religion based on them, is a rare phenomenon among men of physical science. The fact, however, is that as regards their religious preferences, physicists as a group do not differ from other groups of comparable professional standing. As H. Lamb of hydrodynamics fame observed some fifty years ago: "Among scientific men, as amongst other people, you may find all shades of belief or of agnosticism." [163] Thirty or so years later, B. Russell, no friend of revealed religion, admitted that "at all times and places . . . the majority of scientific men have supported the orthodoxy of their age." [164] This pattern was clearly evidenced in the late 1940's by the survey conducted by *Fortune* magazine among several thousand randomly selected American men of science. Seventy per cent of them claimed to be theists, the majority of whom maintained active church affiliation. [165] Or if an even more recent appraisal of this question is desired, the conclusion of C. P. Snow, who had a first-hand knowledge of a large segment of physicists on the other side of the Atlantic, might very well serve this purpose: "Statistically, I suppose, slightly more scientists are in religious terms unbelievers, compared with the rest of the intellectual world — though there are plenty who are religious, and that seems to be increasingly so among the young." [166]

Clearly, the fact that all shades of attitude toward religious tenets can be found among physicists should strongly indicate that taking a stand for or against a belief in revealed religion is not an issue that can be decided on the basis of fluxions, or renormalization techniques. The concept of the miracle will prove repulsive to the physicist only if his philosophy of physics confines the rational to the measurable. In such a case miracles have to appear as irrational suspensions of the laws of nature. But miracles are something wholly different, and there were and are now physicists fully aware of the fact that miracles have sense only if they are believed to have been produced by the source of all rationality. Physicists can perhaps take pride in the fact that one of the most lucid formulations of what miracles mean with respect to the laws of physics came from the pen of one of them, Sir George Stokes, of whom that debunker of miracles, T. H. Huxley, wrote: "There is no one of whom I have a higher opinion as a man of science, no one whom I should be more glad to serve under, and support year after year in the chair of the [Royal] Society."[167] What Stokes emphasized with respect to miracles was their complete harmony with the working of nature and with the genuine concept of the Creator of nature: "Admit the existence of God, of a personal God, and the possibility of miracles follows at once; if the laws of nature are carried on in accordance with His will He Who willed them may will their suspension. And if any difficulty should be felt as to their suspension, we are not even obliged to suppose that they have been suspended."[168] In this last sentence Stokes refers to God's knowledge of the modes of interaction among nature's forces as infinitely superior to that of man. To say that miracles result from a set of initial conditions hidden from us closely parallels Stokes's thought, which aimed at preserving the unimpeded course of physical laws even in the production of miracles by the Creator of those very same laws. Independently of the particular way in which miracles are explained, however, the reasonability of miracles is sufficiently evidenced by their acceptance on the part of so many outstanding workers in the field of science. As was acknowledged recently by a prominent physicist of our times, R. P. Feynman, not a believer by his own admission, "many scientists do believe in both science and God, — the God of revelation — in a perfectly consistent way."[169]

Yet, consistency should not issue in an effort aimed at uniting physics and theology in the sense of exploiting specific results of physics in support of theological tenets. Such union is a "misalliance" that can produce in the long run only a troubled atmosphere between the two. It ends invariably in what T. H. Huxley once described so graphically: "Extinguished theologians lie about the cradle of every science as the strangled snakes beside that of Hercules."[170] When a theologian of note, such as K. Heim of the University of Tübingen, can suggest with emphasis that the universe is ten billion years old,[171] both theologians and physicists should feel embarrassment and anxiety. Unfortunately, the temptation to provide fresh and "exact" proofs on behalf of God's existence or his Creation is all too human and will persist stubbornly. What Bishop Ellicott noted in his highly instructive exchange of letters[172] with Maxwell is still the true description of the prevailing attitude: "Theologians are a great deal too

fond of using up the last scientific hypothesis they can get hold of."
They would do well to reflect on the reply Maxwell penned to the
bishop's inquiry whether Maxwell agreed with the statement, often
made on the theological side, that in view of the possibilities offered
by the ether, "the creation of the sun posterior to light involves no
serious difficulty."

Convinced as he was of the existence of the ether and of its "marvelous" properties, Maxwell was highly indifferent about invoking
the ether in support of a "scientific" explanation of the biblical account of the creation. For, as he put it: "The rate of change of scientific hypotheses is naturally much more rapid than that of biblical
interpretations, so that if an interpretation is founded on such a
hypothesis, it may help to keep the hypothesis above ground long
after it ought to be buried and forgotten." For the same reason, Maxwell declined an invitation to join the Victoria Institute, which tried
to exemplify this "harmony" between science and religion by having
on its roster distinguished men from both fields. Such an institutionali-
zation of the harmony between science and religion implied in Maxwell's view the sanctioning of current scientific explanations, which
he knew were changing only too rapidly. This is why he noted in his
draft of the reply to the invitation "that the results which each man
arrives at in his attempts to harmonize his science with his Christianity ought not to be regarded as having any significance except to
the man himself, and to him only for a time."[173]

Today this rapid change is faster than ever and brings out sharply
the lack of finality in many a conclusion and speculation of physical
science. Yet, until this feature of the scientific quest has become
deeply ingrained in the minds of physicists and theologians alike, the
mutual trespassing between theology and physics will go on. Theologians for one will continue building specious superstructures on
quickly shifting grounds and at their own risk. Physicists like Whittaker and others, busy exploiting physics for theology, acknowledged
this unwittingly when they kept adding to their speculations qualifying subclauses of the sort: "as we know the laws of physics" or "as
present day science knows them." For there is no guarantee that those
laws will stay forever in their present form. And this means, as Eddington put it, that a God whose existence is staked on quantum
theory is "liable to be swept away in the next scientific revolution."[174]
Such an outcome is unavoidable, for whatever there is in the statements of theologians that can be measured, or subjected to observations and experiments (and there will be plenty if they keep borrowing from physics), its truth must eventually face the acid test of

further measurements, observations, and experiments, if truth, which is indivisible and universal, is to be served.

If theology is expected to talk about man's destiny without pontificating in matters that belong to physics, so should physical science assert, to quote Eddington once more, "its conclusions as to the geometry of space-time continuum without trespassing on the realm of theology."[175] This is not a particularly taxing injunction, though even Eddington failed to comply with it. For he was the one who asserted in the same breath in speaking about the enunciation of the principle of indeterminacy "that religion first became possible for a reasonable scientific man about the year 1927."[176] This unbelievable, truly half-baked statement, belongs of course in the class of Planck's dictum on miracles and deserves comment only in one respect. Is not such an assertion, putting as it does the gown of theology on the principle of indeterminacy, a very sadly unscientific flouting of the rules of the game Eddington himself laid down?

The picture of theologians hanging on to the coattails of popularizing scientists is hardly a heartening sight. Just as little encouraging, however, is the spectacle of physicists making scientifically coated statements on matters theological. For most of the time, as the collection of essays entitled *Science Ponders Religion* shows,[177] such utterances and amplifications display, to put it bluntly, a shocking lack of proper information about theology. How many physicists today, who talk and write on questions of theological implications, could claim to do what Maxwell did all his life, devote the Sunday of each week to theological studies? But something of this sort is absolutely necessary if a physicist, no matter how eminent in his own field, wishes to pay more than lip service to intellectual honesty when embarking on theological waters. Acquaintance with the nuances and developments of theological reflection is no less a hard-earned achievement than is a thorough familiarity with the intricacies of modern physics. Of this physicists should be keenly aware if they wish to do their part in making the warfare between science and theology a thing of the past.

On the other hand, theologians should not be disturbed by the relentless advances in physical sciences that may bring them face to face in the future with problems far surpassing those of the past. Science cannot renounce its dictum that any meaningful question or any possible area of research ought to be explored. The findings may play havoc for a moment even with physics itself. It was in fact at such a juncture that Kelvin exclaimed: "Do not be afraid of being free-thinkers. If you think strongly enough, you will be forced by science to the belief in God, which is the foundation of all religion.

You will find science not antagonistic but helpful to religion." [178] Trying to ignore issues, scientific or otherwise, is a retreat, but as a prominent physicist, Weizsäcker, warned: "There is no honest retreat from rational thought into naive belief. It is an old saying that the first sip from the cup of knowledge cuts off from God, but in the bottom of the cup God waits for those who seek Him." [179]

Nor should the theologian be fretting over the so-called atheism of science. Incomplete and faulty as Laplace's conclusions about the solar system may have been, it remains true that "every scientist must certainly set himself the goal of making the hypothesis 'God' superfluous in his field." [180] Lodge's contention that "materialistic monism" should be the working hypothesis of scientific explorers in every department of science is undoubtedly sound, provided, as Lodge himself warned, that such hypothesis is not put forward as basic philosophy of the wholeness of existence. [181] The theologian should always be aware that the world picture of physics, which is constructed with the explicit intent of removing from it all realities and values, esthetic, ethical, personal, or spiritual, cannot by definition have any room for a personal God, or for human personality for that matter. It is in this sense that Schrödinger could speak of the atheism of science, [182] without probably knowing that Cardinal Newman had put forward very similar views a hundred years earlier. [183]

For the same reason that there is no place for God in science, or physics to be specific, there is no place in physics for "deified" matter either. Trite slogans to the contrary, classical physics as such was no more materialistic than modern physics is theistic. The theologian should therefore be careful not to make much of the views of several leading figures of modern physics who claim, as, for instance, W. Heisenberg does, "that atomic physics has turned science away from the materialistic trend it had during the nineteenth century." [184] Such claims attest not only a very narrow definition of materialism but also reveal a serious misreading of nineteenth-century physics and of the views of nineteenth-century physicists as well. To see a form of theism bubble forth from atomic physics is no less naïve than it was a hundred years ago to see an antidote to materialism in a revived form of the dynamism of Boscovich. [185] In all such cases, the direction of the reasoning depends not so much on the data or concepts of physics as on the assumptions adopted by those who clearly have an ax to grind. One should not forget that ever since Benedetto Castelli found that some of Galileo's opponents argued against the concept of inertial motion on the ground that it led to atheism, [186] almost all basic notions and laws of physics have been stretched beyond their meaning by people who thought that by doing so they could promote the cause of theism or atheism. Of the innumerable examples, let us recall that the principle of the conservation of energy bespoke the Creator for Joule, while Tyndall argued from it to the eternity of matter. In our times, the meaning of the principle of mass-energy equivalence is being stretched and distorted in the same ambivalent manner. There is indeed as little substance in the contention that natural science is "on the path to religion," as there is in the word-twisting done by Marxism when "interpreting" modern physics. One should not be overimpressed either by Whittaker's view that modern physics provided the thomistic argument of the existence of God with its first unassailable scientific foundation.[187] Apart from the weaknesses in Whittaker's reasoning, which were well shown by E. L. Mascall,[188] one should rather keep in mind that hardly anybody ever finds God in physics, classical or modern, if he had not already found Him on more unchangeable grounds.

The expression "atheism of science," far from indicating a kind of antagonism to religious inspiration, is rather a wholesome reminder of the distinct roles that theology and science play in the quest for understanding. It was in connection with their distinctness that W. F. Libby, a Nobel laureate physicist, stated that "science and religion are not in conflict, nor are they in full cooperation. They are fulfilling very different needs."[189] In fact, only if their respective distinctness is kept intact, can both physics and theology become reliable partners for a fruitful cooperation, of which no one uttered more memorable words than another Nobel laureate, Sir William Bragg. Speaking in the aftermath of the first world conflagration of material and spiritual values, he sketched the use of the physics of sound in warfare and the potential of science in creating a better life. Self-defense and self-improvement, he said, "that is what science stands for." And he added:

It is only half the battle, I know. There is also the great driving force which we know under the name of religion. From religion comes Man's purpose; from science his power to achieve it. Sometimes people ask if religion and science are not opposed to one another. They are: in the sense that the thumb and fingers of my hand are opposed to one another. It is an opposition by means of which anything can be grasped."[190]

Part Four

Physics:
Master or Servant?

CHAPTER ELEVEN

The Fate of Physics in Scientism

"VERY FEW PEOPLE read Newton because it is necessary to be learned to understand him. Yet, everybody talks about him."[1] Thus did Voltaire poke fun at his contemporaries of whom the fashion of the day demanded some familiarity with the *Principia* or at least the pretense of it.[2] The society in which Newton's name had become a household word had already given its assent to a mechanistic or rationalistic conception of the world when Newton's main conclusions began to filter down to a wider segment of the educated public. The ground for the Newtonian triumph over society had been well prepared. The grand vision of what the "new method" was to have in store for mankind was painted in minute detail in Bacon's *New Atlantis*. The Island of Bensalem, hidden in the faraway seas, harbored a people whose leaders were neither kings nor tyrants, but the Society of Solomon's House, a college of men eminently versed in the sciences. Their sacred task consisted in learning as completely as possible the causes of all phenomena and all secret motions hidden in things. Supporting them in their arduous work was the firm belief that scientific expertise not only enlarges the bounds of man's dominion over nature but also provides the secure basis for an enlightened social living. Bacon's description of the scientifically organized society was both alluring and deceiving. As Bacon intimated, such a program was within easy reach of any society or nation that had set its heart upon achieving it. Though fully convinced that the human mind could never unlock the ultimate mysteries of matter, Bacon felt confident that there was a type of scientific progress tailored to man's limited abilities. Thus he took the view that his method would "level all wits," and with the collection of all scientific data done, "the investigation of Nature and of all sciences will be the work of a few years."[3]

461

Bacon's dream undoubtedly helped create an atmosphere of eager expectation, and discoveries did indeed come swiftly as the century of genius ran its course. It was as if a whole age were being "converted" to science, and as regards the individual scientists, the conversion in many cases was literally true. Astronomy lured Brahe from the study of the *humaniora,* and Kepler registered his conversion to astronomy in a frankly religious vein: "I wanted to become a theologian; for a long time I was restless. Now, however, observe, how through my effort, God is being celebrated in astronomy."[4] To the same change of orientation did Huygens refer, though in a rather different tone, when he contrasted the lucidity of geometrical propositions to the imprecision of theological doctrines.[5] Intended for the ministry, both Jacob Bernoulli and Leonard Euler instead went on to distinguish themselves in mathematics. Of Jacques Ozanam, Fontenelle said in his *Eloges* that after four years spent in studying for the priesthood, he left "out of piety and love for mathematics."[6] Not that such a piety was necessarily opposed to piety in its traditional acceptance. Fontenelle for one was careful to recall Ozanam's favorite saying: "It is the business of the doctors of the Sorbonne to argue, of the pope to make pronouncements and of the mathematician to go to heaven along the vertical."[7] This had to be so if one was to accept Fontenelle's view of J. P. Maraldi, an aide of Cassini, the astronomer: "His character was like the one which science gives to those who make its study their exclusive occupation: serious, simple, upright."[8]

If geometry could inspire so much piety and form such splendid characters, how much more would it benefit less lofty endeavors? Indeed, in Fontenelle's *Eloges* one comes face to face with the new outlook in which geometry, as the embodiment of the scientific spirit, was presented as the touchstone and source of every cultural value and accomplishment. That the art of politics could not succeed without geometry was exemplified for Fontenelle by Viviani, whose engineering skill alone could put an end to a dispute between the pope and the grand duke of Tuscany.[9] If denominational prejudices were to be overcome or international diplomacy effectively served, the spirit of geometry would again be the magic device. Witness the case of Leibniz who made great efforts to persuade the Lutheran princes of Germany to accept the leadership of the emperor and the pope. What enabled Leibniz, a devout Lutheran himself, to take a course demanding so high a degree of detachment? Fontenelle gives his answer without any hesitation: It was "the systematic spirit which he possessed in supreme measure that proved so effective in prevailing over religion and partisan spirit."[10] In effect Fontenelle envisioned that one day mathematics would also master the complicated realm of moral questions and conduct. Such was the benefit, he thought, that mankind might ultimately derive from Bernoulli's work on mathematical probability.[11]

Whether Fontenelle was right or wrong is beside the point. He gave his contemporaries what they wanted to hear: An intimation of a new age in which politics, economics, ethics, and all other facets of human life would be governed by the art of calculation. To the always present, practically minded critics, Fontenelle was eager to point out that theoretical science could not fail to produce useful results. Observations of the moons of Jupiter, he recalled, helped navigation, study of the parabola improved artillery, the cycloid was the heart of good pendulum clocks, and even animal anatomy gained new insights from mechanics. But that was not all that Fontenelle said about physics and mathematics in his essay "A Preface on the Utility of Mathematics and Physics . . ." He boldly went on to state the heart of his message in an almost shockingly direct way: "The geometrical spirit is not so tied to geometry that it cannot be detached from it and transported to other branches of knowledge. A work of morals and politics or criticism, perhaps even of eloquence, would be better (other things being equal) if it were done in the style of a geometer."[12] Such had to be the case if, as Fontenelle insisted, eternal laws, not human caprice, were revealed in geometry and physics. After this it was almost anticlimactic to hear him declare: "The true physics is rising to a level where it becomes a kind of theology."[13]

That Fontenelle spoke now of geometry, now of physics, in the same context can easily be understood. In Fontenelle's day — and he lived a long life — nobody doubted the close identity of geometry and physics. In a sense, the *Principia* of Newton was an exercise in geometry, though not an easy one. But whatever its mathematical difficulties, once the contents of Newton's work had become known in a general way, who could set a limit to the attainments of geometry? Could not the marvelous simplicity of the laws governing the heavenly bodies be discerned in the interactions of individuals and of societies also? This was only a dream at first, to be sure, but it carried unlimited confidence as only dreams can do. Naturally enough, it had to appear first in the form of poetry. The title of the poem was "The Newtonian System of the World, the Best Model of Government" (1728), and its author, the Rev. J. T. Desaguliers, a disciple and friend of Newton.

The dream's vision, however, became in no time the central idea of books written on political and social organization. In his essay entitled *That Politics May Be Reduced to a Science*, Hume remarked that the data of political forces, forms, and developments showed so little dependence on human caprice that "consequences almost as general and certain may sometimes be deduced from them as any which mathematical sciences afford us."[14] To implement such hopes one needed, it seemed, only favorable circumstances. Nowhere else had these offered themselves more naturally than in the freshly opened lands of Pennsylvania, where some settlers resentful of Continental and Colonial narrowness began to build their society along the dictates of "purely" scientific reasoning. References to Newton of course were not lacking. Benjamin Franklin, for instance, instructed his fellow frontiersmen on the implications of the Newtonian system in an essay whose title speaks for itself, "On Liberty and Necessity; Man in the Newtonian Universe." In the same vein the Declaration of Independence reflected a political theory guided by the axiomatic (or mechanistic) method adopted by Locke, who was, as Desaguliers informed us, "the first who became a Newtonian philosopher."[15] The chief architect of the American Constitution, Jefferson, liked to call himself a scientist and was regarded as such by friends and foes alike. Some of his opponents declared him unfit for the presidency precisely because of his preoccupation with science. As an anti-Jeffersonian pamphlet put it, "Science and government are two different paths. He that walks in one, becomes at every step less qualified to walk with steadfastness or vigor in the other. The most lamentable prelude, the worst preparation possible for a ruler of men, was a life passed like that of Newton."[16] Unlike Newton, Jefferson felt completely at home in politics. What is more, to forgo politics would have been for Jefferson a betrayal of the Newtonian spirit, which was for him synonymous with progress based on rational planning. Promoters of material and spiritual welfare widely shared this outlook. New cities like Washington were built around a central hub to resemble the most perfect of all constructions, the solar system. As regards the spiritual realm, Cotton Mather, the New England clergyman and author of the first popular exposition of the Newtonian system in America, declared Newton to be "the Perpetual Dictator of the learned World."[17]

On the Continent, Voltaire, in making Newton known to the public, aimed at far more than simply assuring the victory of Newtonian physics over Cartesian. He believed that the finding of regular patterns resembling those of physics should be possible even in the study of history, if historiography was ever to become a trusted branch of learning. Only the concept of regular, observable patterns was acceptable to his philosophical skepticism, which regarded only what could be measured and counted as above any doubt. Everything else, as his letter to Gravesande reveals, he viewed as sheer metaphysical vanity: "We are made to compute, to measure and to weigh; this is what Newton did; this is what you are doing with Musschenbroeck."[18] In this he not only gave a distorted account of Newton, but he also set a trap for himself. For in a short collection of essays entitled *Des singularités de la nature* (1768), he was led to a wholesale rejection of theories "as irresponsible speculations of naturalists,"[19] because they were not based on measurements and calculations. Among these "worthless" theories he included the animal origin of coral and the organic origin of fossils, and he suggested that marine fossils found in the Alps had probably been left there by pilgrims. To make his scientific bungling complete, he summarily rejected Maupertuis' law of least action, which got the Voltairian ax because it seemed to smack of metaphysics.

The day was still far away in which the shortcomings of the quantitative method would be laid bare. For the time being, physics was praised without restraint by philosophers who aimed at nothing less than laying the foundations of a new "heavenly city," which would spread over the globe. "It is to physics and to experience," declared Holbach, "that man must have recourse in all his investigations: he must consult them in matters of religion, ethics, legislation, political government, the sciences and the arts, even in his pleasures and sufferings. Nature acts by simple, uniform and invariable laws which experience enables us to know."[20] It was indeed characteristic of Voltaire's century that the ideas of its great social and political theorists showed a distinct affinity with fundamental concepts of the physical sciences. Condillac's *Treatise on Systems* viewed society as an "artificial body" composed of mutually reacting parts that ought to be kept in equilibrium. Montesquieu's first writings dealt with physical problems, and in his *Spirit of the Laws* the problem of the greatest possible freedom is presented as a problem in statics, where individual forces are kept in balance by counterforces. It was in such scientific considerations that he found the ultimate justification of his doctrine of the "division of powers." The author of *Emile* and of the *Social Contract* also tried his hand at science by penning a treatise on the fundamental laws of chemistry.

Among the eighteenth-century prophets of the universal application of the quantitative method, Condorcet probably displayed the greatest boldness. According to him, human and physical events were "equally susceptible to being calculated and all that is necessary to reduce the whole of nature to laws similar to those which Newton discovered with the aid of the calculus, is to have a sufficient number of observations and mathematics that is complex enough."[21] It was from the radical mathematization of social science that he hoped to derive those "unattackable proofs" that would bring to fruition the efforts of the Revolution: "to repair promptly the dislocations inseparable from every great movement, . . . to ensure the adoption of needed reforms in the face of selfish interests and base faith."[22] Bold indeed if not foolish was the role Condorcet assigned to science. He expected science "to tame the future"[23] as if "taming the elements" were not enough, and the key to progress he saw in the rigorous application of the law of least action. In Condorcet's fancy the calculus of probabilities was supposed to transform ethics into a system of exact laws and to permit the forecasting of majority decisions in politics. The politicians themselves were expected to yield their functions to the scientists, because the state was to be run, as Condorcet put it, by "social mathematics."

Life in such a state would be dull at best, if not outright cruel and inhuman. One of Condorcet's last writings, a commentary on Bacon's *New Atlantis*, lists all the mental and emotional mortifications required of scientists to prevent rivalries and secure total coordination.[24] To keep such a vision alive, Condorcet, like anyone else, needed the conviction that the age of fulfillment in science was close at hand. This conviction he had in full measure. "The Kingdom of truth is near; the duty to say it never was more pressing, as it was never more useful."[25] In his fifth *Mémoire* on public education, Condorcet described the final assault in the great program of conquering nature with a tone of sure, imminent triumph: "Interrogated everywhere, observed in all its aspects, attacked simultaneously by a variety of methods and instruments capable of tearing away its secrets, nature will at last be forced to let them escape."[26] He spoke of the not-so-distant moment when "animals, plants, minerals," or every detail of our planet, will be known with such completeness that new findings "will not present really new phenomena nor will they offer embarrassing results."[27] A science that had no more perplexing discoveries to expect was Condorcet's dream. It was a dream that fully anticipated what is known today as scientism, which acknowledges no assurance, no directives, and no truths, but those offered by the science of the day and takes those truths as forever established. As the prophets of scientism are notorious for their blindness to possible future overhaulings of the edifice of science, Condorcet too could contemplate only such "rejuvenations of science" as would bear on minor details, never on the major propositions of physical science.[28]

When Condorcet submitted these views, nature already appeared to have yielded one of her secrets, which in Condorcet's eyes, and those of his contemporaries, was perhaps her most important one. It concerned the stability of the solar system, shown convincingly by

Lagrange and Laplace. For Condorcet this stability provided the ultimate basis for the limitless perfectibility of human nature and society, as can be seen in his *Sketch of a Historic View of the Progress of Human Mind.* The progress that lay ahead for the human race, Condorcet assured his readers, had "no other limit than the duration of the globe upon which nature has placed us." Nor could this progress take a backward direction, he continued, "as long as the earth occupies the same place in the system of the universe." [29]

The limitless improvement of the human race secured by the stability of the solar system was for Condorcet a "universal law of nature," and from it he extracted with truly evangelical zeal all the good tidings which the human race would henceforth comfort itself with. Man no longer had to consider himself "as a being limited to an isolated and fleeting existence destined to vanish"; man was rather "an active portion of the grand total and a co-worker in an eternal project." [30] What alone was perennial in all this was the shortsightedness unable to suspect at that moment that science itself would a few generations later conjure up the vision of an icy heat death for civilization everywhere in the universe. Strengthened by his illusions, Condorcet could confidently point to the tremendous progress of astronomy, where, as he put it, "the march of the truth is the most secure and where it can be measured with the greatest precision." [31]

Possessed of such confidence in the results and methods of physical science, Condorcet could hardly realize that to squeeze a long array of value judgments out of technical papers on the stability of the solar system was logically a rather questionable procedure. In all fairness to him, his way of thinking was not the peculiar foible of a single student of science. At one time Laplace himself thought that celestial and social engineering were but separate applications of the same system of principles. With his life's work in science largely accomplished, Laplace turned to politics and obtained from Napoleon the post of minister of the interior. He took only six weeks to prove himself a supreme failure in handling human affairs that, after all, seemed to follow rules other than those obeyed by the planets. This truth Napoleon expressed in a biting remark after shifting Laplace to the Senate: "He brought into the administration the spirit of infinitesimals."

Cases like Laplace's failure as a social engineer were not expected to discredit the alleged connection between the supreme perfection of celestial mechanics and the limitless perfectibility of mankind. To turn the magic trick in history and human affairs one only needed, so it was believed, to apply Newton's wit to the cultural programs that planners had been busily drafting for several decades. The cultural crusade of the Enlightenment reached its peak, and in this new

form of religious campaign, Newton's name was the revered shib-
boleth. Saint-Simon's little tract, *Lettres d'un inhabitant de Genève
à ses contemporains* (1803), is a classic example of the cult which
such religious veneration given to science could build around Newton's memory. At the tomb of Newton, Saint-Simon began, a subscrip-
tion should be opened to finance the establishment of a Great Council
of Newton. Three mathematicians, three physicists, three chemists,
three physiologists, three novelists, three painters, and three musicians
would constitute the Council headed of course by a mathematician.
Being elected by all the subscribers, the members of the Council
would represent in their collectivity the supreme authority on earth,
and would act as emissaries of God. The pope, bishops, and priests
would be deposed for their reluctance to embrace the dictates of
science, through which God intended to transform the earth into a
"Garden of Eden." To carry out the orders of the Great Council of
Newton, temples of Newton would be erected everywhere, which
would serve as centers of rational worship, research, and instruction
on the local level.

So far, fortunately for science at least, all this was a one-way proposition. Science was taken as the perfect embodiment of knowledge,
and everything else was to be shaped in accordance with it. But with
the coming of Comte the trend came full circle, and the once-deified
science was told to pattern its program and methods according to the
dictates of what came to be known as scientism. The start was innocent enough, though very ambitious. Convinced at an early age of
his future greatness, Comte lost no time in outlining the program of his
life's work. The title of one of his earliest pamphlets could not have
better epitomized his great dream. The *Prospectus des travaux scientifiques pour réorganiser la société* dates from May, 1822, and already
contained the law of three states and the core of the positivist classification of the sciences. So far all this was nothing more than an echo
of Saint-Simon's aspirations in a more disciplined format. A tone
smacking of intellectual rigor, of course, befitted the former student
of the famed Polytechnique, where Comte later served as instructor
and examiner in mathematics. Some of the greatest figures of French
science served on the same faculty — the creators of mathematical
physics, Laplace, Fourier, Ampère, and Fresnel, to mention only a
few. To grasp the dynamic spirit of physics was therefore within
Comte's reach, but his ambitions did not let him reflect deeply enough
on the genesis of the formulation of physical laws and theories that
were put forward in no small number in the 1820's.

Comte's ambitions could hardly have been surpassed. In an essay
entitled *Considérations sur le pouvoir spirituel* (1826), Comte en-

visioned a "new spiritual power" that was destined to exercise an even greater influence over temporal affairs than did the Church at the height of her influence in the Middle Ages. The new spiritual power was supposed to take full control of education, to determine public opinion, and to place all nations of the West under a "souveraineté spirituelle." Engrossed in such sanguine visions, Comte could hardly have had the necessary time both to achieve a tempered analysis of the results of physics and to keep a finger on the pulse of the physical research of the day.

Comte's was a gamble that stood little chance of succeeding, for he wanted to be both a supreme authority in science and the greatest of social prophets as well. The limitations of human capabilities hardly ever allow for success in two so disparate directions. In Comte's case competence in science — or in physics, to be more specific — fell far short of the ambitious goal. In fact, as Comte's English translator and admirer, H. Martineau, ruefully admitted in 1853, physics was "the weakest part of the *Positive Philosophy* "in regard both to the organization and the details of the subject." [32] Trying to exculpate her hero, she referred to the great advances made in every department of physics between 1835 and 1853, but Comte's account of physics and astronomy was very deficient and erroneous even by 1835 standards. For all the insistence of some past and present admirers of Comte on his allegedly great insights in physical science, the conclusion of the non-physicist Huxley was closer to the mark. Urged by a friend to discover for himself a "mine of wisdom," that is, Comte's *Positive Philosophy*, Huxley found that the accent should be put on "mine" rather than on "wisdom." Huxley's summary of Comte's attitude toward science, however, was full of wisdom: "What struck me was his want of apprehension of the great features of science; his strange mistakes as to the merits of his scientific contemporaries; and his ludicrously erroneous notions about the part which some of the scientific doctrines current in his time were destined to play in the future." [33] More recently, G. Sarton, who urged his readers to respect Comte's genius, called him "crazy," admitting that Comte was neither an accomplished scientist nor a competent historian of science. [34]

Science — astronomy and physics in particular — was for Comte but a vehicle to support his dreams about the final, absolutely valid rules that were to usher in the age of positivism in every facet of human life, individual and social alike. To prove that there were such rules Comte had to make the most of the apparent perfection of physical or astronomical laws. But physical sciences could yield the sorely needed proofs only at the price of being straightjacketed by scientism. For scientism first exploits the results and methods of the

science of the day in support of "social engineering" and then forbids science to outgrow its current phase, lest the prospects of the "scientifically based," final reorganization of society be proved illusory.
Comte's speculations constituted a classic example of the exploitation
of science, this time not by the divines but by one who was later to
invite the pope to join the positivist religion, defined happily by
Huxley as "Catholicism minus Christianity."[35] It was Comte, the
latter-day Bentley, misusing the conclusions of Laplace, the latter-
day Newton. In all fairness to Comte, he was at least utterly frank in
laying bare his deepest motivation as soon as he reached the section
on astronomy in the *Positive Philosophy*. "It is no exaggeration to say
that Social Physics would be an impossible science, if geometers had
not shown us that the perturbations of our solar system can never be
more than gradual and restricted oscillations round a mean condition
which is invariable. If astronomical conditions were liable to indefinite
variations, the human existence which depends upon them could
never be reduced to laws."[36]

Such candid explicitness about the sociological motivation of
scientism and its pseudoscientific basis could hardly be improved
upon, even in the *System of Positive Polity*, where Comte insisted at
length on the total subordination of science to the dictates of positivist sociology. There Comte declared of science in general that only its
social usefulness could give moral justification to its cultivation.[37] No
less revealing was what he said of the sociological roots of the study
of astronomy in particular. From astronomy, he stated, "always will
date our systematic study of the natural order which governs Humanity."[38] The signal service astronomy and physics were to render
on behalf of positivism consisted in the display of an overwhelming
exactitude that was to support the contention that exact prevision is
possible, and in due time it can be achieved in every domain of life.
This was the great aim of positivist studies destined to procure at the
same time freedom from theological and metaphysical slavery. This
all-or-nothing proposition was to stand or fall depending on whether
perfect precision in nature was or was not a fact. This is why Comte,
in speaking of Kepler's laws, does not tire of repeating "our precision
is perfect" and it is "absolute precision."[39]

To Comte it appeared so absolute that he forbade science to improve on it. For deep in his heart Comte was beset by fears. He was
horrified at the prospect that further research and more precise measurements might one day play havoc with what he considered to be
the final word in astronomy. He could not make a truce with that ever
restive drive in science, the quest for greater precision. Haunted by
such fears, he could not restrain himself from making truly desperate

utterances wholly alien to the spirit of scientific investigation. "Natural laws," he warned frantically, "could not remain rigorously compatible in any case with a too detailed investigation."[40] He called overprecise measurements "incoherent or sterile," displaying only "childish curiosity stimulated by vain ambition,"[41] and he equated concern for greater precision "with an active disorganisation" of science.[42]

Comte asserted such hostility toward precision in science even more in the *System of Positive Polity*. There he deplored the "retrograde" trend "exemplified in physics by the unfortunate disposition to abandon previous discoveries in the vain attempt to arrive at absolute precision."[43] Of course Comte wanted to spare the solar system any further discovery that could upset it as a paragon of orderly perfection. Consequently, his research program for planetary and stellar astronomy shrank almost to the vanishing point. As to the solar system, Comte denounced what he termed the "insane enthusiasm" that filled both lay and astronomical circles following the discovery of Uranus.[44] The existence of Uranus, having no influence on the motion of the Earth, can be of any interest only to the inhabitants of Uranus itself, Comte observed. Ignorance that does not result in any practical inconvenience, Comte insisted, should leave us wholly unconcerned as regards scientific pursuits. For the same reason he dismissed as meaningless any attempt to clear up several obscure points about the motion of Saturn and Neptune.[45] To Comte the "normal condition of astronomy" was defined by the "degree of accuracy which was really necessary for practical purposes." Therefore everything in the solar system that was not "visible to the naked eye," so Comte decreed, had to be treated with "the same indifference as the fixed stars."[46]

The stars indeed fared badly in Comte's legislation upon scientific matters. Comte's primary concern in this regard was to preserve the apparently complete lack of any connection between the stars and the solar system. For him the solar system's independence of the fixed stars was "perfectly certain."[47] As supreme proof, he referred to the inability of astronomy to measure the parallax of any star[48] and even spoke of a method "by which it is most certainly established that the parallax of the stars is absolutely insensible."[49] The stars, however, were only pawns in the fantastic chess game that Comtean positivism played with the physical world. To secure the absolute finality of the main tenets of positive philosophy, the whole universe had to fade into insignificance. And this because of Comte's deep apprehension about possible future discoveries that could sweep away his world view, restricted cozily to the solar system.

To allay misgivings Comte could only state in a distinctly authoritarian manner: "There is no difficulty about this to persons who, like myself, admit that our researches are limited by the boundaries of our own system and that positive knowledge cannot go beyond it. The study of the universe forms no part of natural philosophy."[50] For Comte, positive philosophy and the solar system were coextensive so far as meaningful thinking was concerned, and consequently, he could forbid the application of the concept "system" to anything beyond the outermost planet. Blatantly ignoring Herschel's work on our galaxy, Comte stated: "We do not know, more or less, and men will probably never know, whether the innumerable suns that we see compose a general system, or any number, large or small, of partial systems entirely independent of each other."[51] Contrary to the aspirations of all astronomers of his day, Comte declared that the universe was inaccessible to study,[52] that a "clear distinction is forever established between our system and the universe at large."[53] And while from Galileo on it was the recognition of the earth's motion that broke open a closed universe, for Comte the earth's motion meant a rigid enclosure, a gap between the earth and the stars that was never to be bridged. Once in the chains of scientism, science not only had to confine itself to a small corner of the universe but was also made to negate the meaning of its most basic achievements. To crown the comedy, science also had to praise its dictator who decreed that "this circumscription is, as elsewhere, to be regarded as real progress."[54]

To mention what Comte said about stellar physics after all this would be almost anticlimactic. According to him, man can never know with accuracy the orbits of the stars, or their periodic times,[55] chemical compositions, and temperatures.[56] The cataloguing of stars he believed to be unimportant;[57] the separation of the binary stars and their dynamics he declared to remain forever unknowable.[58] He dismissed cosmological theories for the reason that "we find our system to be the only subject of knowledge."[59] In such a framework, could the study of stars be anything but a "gratification of our curiosity"? In Comte's eyes it was far worse than that. The study of fixed stars, he warned — in a sweeping indictment of astronomers like Fraunhofer, Bessel, and Herschel — would lead "to a series of unconnected speculations as irrational as they were useless. These have now been continued for nearly a century, and small as their result has evidently been, astronomers still persevere in the old and useless routine, although the public is now beginning to suspect its frivolity."[60] To claim the alleged consent of the educated public in support of such a verdict was at best a poor exercise in rhetoric on Comte's part. Before long, men of science began to disappear from among those attending his lectures.

Ineffective as were Comte's strictures on the scientific research of his day, they nevertheless remained instructive. In these misguided strictures, the whole inner logic of scientism laid itself bare. Once the results of physical science were taken as the norms to which every principle bearing on human existence should conform, there was necessarily a feedback that threatened to undermine the scientific pursuit itself. If Comte's dictatorial rulings in astronomical matters were highly revealing in this regard, his statements about physics were no less so. Dismally unfortunate in his evaluation of the results of physics, Comte was highly successful in expressing with shocking directness the true nature of scientism: the exploitation of a particular stage of science on behalf of dreams far surpassing the competence or the range of scientific conclusions. Such an attitude toward science had to divide the concepts and results of science into two groups, a "useful" one and a "useless" one — such distinction being made essentially on the ground of what is demanded by a particular form of scientism, which is usually a sociological system. Furthermore, to protect the claim of scientism as the ultimate word in the intellectual quest, science was urged to "freeze" its advance and to reject any dangerous curiosity that might upset the supposedly definitive status quo. It was in this "freeze-in" effect that scientism's feedback on science came to a head.

A definition of scientism like the one given by M. N. Rothbard, as "the profoundly unscientific attempt to transfer uncritically the methodology of the physical sciences to the study of human action,"[61] considers only the social or humanistic side of the symptom of scientism. The scientific side of the problem is perhaps less obvious, but it certainly repays a detailed analysis. To achieve this one need only pose the question: How does science fare in the hands of scientism? Very badly, as Comte's precepts in physics show. In Comte's system, physics is totally subjugated to a sociological theory, if not a dream, and he was utterly frank to admit it: "The question of reorganizing the principles of physical science depends ultimately upon the entire renovation of thought, and it is therefore intimately connected with the great problem of social regeneration."[62]

With Comte this was a lifelong conviction, as witnessed by his last major work, the *System of Positive Polity*. There the nature, concepts, boundaries, and methods of physical science came to be rigidly defined by what he called subjective principles. The supreme subjective principle for Comte was man's nature, individual and social alike. And just as man's contact with his environment is based on his senses, so "the divisions of science are determined more by the multiplicity of our senses than by the corresponding distinctions in the properties of matter. They result from the constitution of Man, not from an objective source."[63]

Consequently, acoustics and optics were decreed by Comte to remain forever distinct, based as they were on two different senses. And since electrology, thermology, and barology, to use Comte's terminology, revealed three different aspects of the sense of touch, in all likelihood they too would preserve their distinctness. Comte of course could have no patience with the aspirations of physicists aiming at the reduction of the various branches of physics to a common ground. This he termed a "visionary notion" and contended that a sounder philosophy indicated "six, and perhaps seven irreducible branches in place of the five recognised at present."[64] Fortunately nineteenth-century physicists were not obligated to make a total turnabout and subscribe to what in effect was a death sentence passed on the very core of physics.

A theory so attached to the way in which man's senses function of course could not come to terms with the atomic theory. While admitting its great logical value, Comte considered it a notion that could not be applied in biology, might have only a limited success in chemistry, and might be useful in physics only in the corpuscular hypothesis.[65] Science, which had been liberated from philosophical dicta only a little over two centuries earlier by an appeal to the senses, was now subjected by Comte to what was nothing short of a tyranny of the senses. Properties of matter other than those discernible by the everyday use of the senses were declared unscientific and attempts to discover them irrational. Chemical research therefore was to be conducted so as not to upset the finality of scientific concepts formed on the basis of macroscopic observations. According to Comte, common sense proved irrefutably the totally different natures of chemical and physical phenomena, and therefore he drew an absolute distinction between chemistry and physics. By the same logic he banned both mathematics and physics from chemical research as being too troublesome to the traditional macroscopic view of matter. Similarly, he decried the tendency of "irrationally assimilating chemical to electrical properties."[66] "Nothing is gained," he warned, "towards explaining the molecular connection by electrical or magnetic theories. . . . Such inventions give no idea whatever of molecular cohesion. Nor is affinity, or the tendency to combination, any better explained by the electro-chemical theory. . . . every attempt to make chemistry as a whole enter into any branch of physics is thoroughly anti-scientific."[67] That electrochemical phenomena had already played an important role in chemical research was of little significance to him. His tenets told him that he "must, once for all, reject the conception through which it has been attempted to transform all chemical into electrical phenomena." [68]

It is not only in retrospect that we know how badly Comte's scientism made him misread the meaning of the chemical research of his day. The very year Comte sent the foregoing passages to press, Faraday wrote in his notebook: "chemical and electrical forces are identical" and "electricity is chemical affinity, chemical affinity is electricity." [69] Comte did not fare any better with organic chemistry. All through his life he rejected it as a legitimate field of inquiry undoubtedly because it pointed toward the atomic structure of matter, which was anathema in the Comtean creed. Seven years after Wöhler's synthesis of urea (1829), Comte called organic chemistry a "heterogeneous and factitious assemblage" that must be destroyed, [70] and two years after the discovery of the compound ammonia by Würtz (1849), he attacked organic chemistry as having no value at all. He deplored bitterly the "degenerate condition" of the chemistry of his day, because it aspired "to nothing less than the complete explanation of the phenomena of nutrition," and he took the biologists to task for not protesting against such encroachments upon their domain. He was even more angered by the attempts to explain animal motion and sensation on the basis of physics. The fact that such topics were given up "to the irrational handling of physicists" constituted in his eyes a fatal step away from "the true progress of vital science." [71] But could rhetorical injunctions stop physics from unfolding the submacroscopic world — that world which Comtean positivism wanted science to ignore forever?

In physics anything that reached into the submacroscopic realm of matter was branded by Comte as metaphysical, this adjective standing for the most reactionary thing in his vocabulary. Any attempt to discover specific, basic properties of matter was in his words "unscientific." [72] To raise questions such as why one metal was a better conductor than another was for him "irrationelle." Yet, contrary to his precepts, problems like the conductivity of metals were destined to mark the road that led to entirely novel and rapidly expanding branches of physics known today as solid-state and semiconductor physics.

Comte could hardly entertain such possibilities, for he was beset with frantic fears that future results would upset the absolute, definite order physical science was to provide his brand of scientism. To search for the cause of gravity, or for its way of operation, was for Comte utterly nonsensical and "metaphysical," whereas no less a physicist than Faraday devoted the last two decades of his career to this problem. True, Faraday did not get anywhere, but at least he had the

courage to state, when pondering over the causes of gravity and related problems: "To leave them untouched, hanging as dead weights
upon our thoughts, or to respect or preserve their existence whilst they
interfere with the truth of physical action, is to rest content with darkness and to worship an idol."[73]

Idolatry was precisely the word Comte liked to hang on people who
were in the stubborn habit of always asking for further causes concerning physical phenomena. If, however, there was any idolatry, it
was scientism that demanded it and science had to bow. How deeply
is seen in the way Comte treats optics. He takes the development of
optics in the seventeenth and eighteenth centuries as an illustration
of how science discarded theology and metaphysics. Yet, most of what
Newton stated in the queries of his *Opticks*, so heavily laden with
"metaphysical" curiosity, is passed over silently by Comte. Worse still,
the truly seminal research carried out in optics in the 1820's and
1830's, the work of Laplace, Poisson, Biot, Arago, Fresnel, and
Cauchy, was of no concern to Comte, who betrayed time and again a
chronic insensitivity to what was best in contemporary research.[74]

Unfortunately for Comte, these distinguished scientists talked a
great deal about the luminiferous ether for the simple reason that
they were extremely curious about the manner light waves were propagated. What is more, they saw in this question the very heart of
physics and only one approach appeared meaningful to them — to
reduce light to the vibrations of a subtle fluid, the ether. For Comte
nothing else could have gone more against the grain. He denounced
bitterly the fact that optics and electricity were still pervaded by
concepts referring to various fluids. The first to advocate the establishment of a chair for the history of science, Comte was not unaware
of the important role such fluids played in the progress of physics. But
for the founder of positivism, history was merely the captive servant
of the present, not a pattern for the future. Indeed the future possessed no genuine historical meaning within the framework of positivism, which claimed the completion of history on the arrival of the
positivist age. Thus fluids, a remnant of the metaphysical past, could
no longer serve a useful purpose after physics had supposedly entered
its positivist and ultimate phase. This arbitrary evaluation of the
role of fluids as specific hypotheses in physics offered itself for easy
generalization, and a systematic mind like Comte's could hardly resist
it. True to his bent of mind Comte codified this generalization in a
most one-sided pronouncement on the nature of the scientific hypothesis. "All scientific hypotheses, to be capable of being judged, must
bear exclusively on the laws of the phenomena and never on their
modes of production."[75] In saying this, Comte could hardly have

made matters more exasperating for those members of the Viennese school of neopositivism who insisted that Comte truly recognized the full meaning of the physical hypothesis.[76] Such an interpretation can be made plausible, however, if one is willing to minimize the importance of a long list of absurd limitations forced on physics by the Comtean definition of hypothesis. When Comte spoke of hypothesis, he meant something altogether different from what physicists by and large used to denote by this word both in their speculations and in their practice.

It was hardly lessons drawn from the history of science that inspired Comte's definition of the scientific hypothesis. It was rather the instinctive fear and awareness in Comte that had told him nothing could be so effective in melting away his rigidly frozen scientific status quo than just asking innocently how this or that happened. Consequently, he sifted, shuffled, or just simply ignored the facts to create the impression that the physical sciences had run their course. Such a fallacy, however, was obvious even to some fellow advocates of scientism, as can be seen in Renan's indictment of Comte's dreams: "Comte believes with us that one day science will endow humanity with a creed; but the science in his mind's eye is that of Galileo, of Newton, of Descartes, remaining as it is."[77] Blindfolded by scientism, Comte was able to bring himself to overlook the obvious fact that the science of his day had already moved far beyond that chimerical final stage which he tried to "protect" from further changes. What is more, the means he advocated to achieve this flouted the spirit of science itself. Thus he pleaded for the creation of an areopagus of scientists charged with the duty of listing scientific topics to which research should be limited. In a similar vein he declared an end to freedom of conscience in astronomy, physics, and chemistry on the ground that it would be absurd not to believe with complete and unconditional confidence in the principles established by competent people.[78] This sadly antiscientific dictum could not but touch off a crushing rebuttal on Huxley's part: "All the great steps in the advancement of science have been made by just those men who have not hesitated to doubt the principles established in the sciences by competent persons."[79]

With his mind so thoroughly dominated by the dogmas of scientism, Comte had to misread history, misunderstand the research of his contemporaries, and miscalculate the course science was to follow in defiance of the Comtean precepts. To forecast the future of science with some accuracy is of course a very risky task, even from a vantage point far better than positivism may provide. Signal achievements in research can also go for some time unrecognized or unappreciated by contemporaries. Yet, even on these counts, Comte's shortcomings were more than ordinary. His reading of the history of science, especially, is beyond any excuse. True, there are always processes at work in history that reveal only much later their true nature and direction. Thus, a relatively recent turn of events might often be needed to make possible deciphering the meaning of a long evolution. It is also possible that scientists might at a particular stage of the past misinterpret what they are really doing. Thus, positivism might have proved in itself a valuable principle for the understanding of the past had Galileo, Newton, and their successors really been positivists without being aware of it. One can even argue that the words of scientists of any epoch are so conditioned by the context of the times that their actual scientific procedures rather than their comments should be given the principal attention. However this may be, there can be no justification whatsoever for the kind of historiography that deliberately chooses to ignore any evidence not in conformity with the preset pattern of interpretation. Nor is the thesis acceptable that equates the contemporary stage of science with truth and regards as meaningless those conclusions of the scientific past that were modified or discarded by later findings. From such faulty premises, only a badly deformed picture can emerge, one presenting only a single, narrow aspect of the past. It is typical of scientism, however, to produce in its devotees a mental attitude that is incapacitated by its own nature to appreciate both sides of a coin. It was only natural that Comte's admirers could not avoid the same pitfall. Thus J. S. Mill found nothing wrong with Comte's claim that anybody who ever contributed anything to science did it in the genuine spirit of positive philosophy. According to J. S. Mill, Bacon, Descartes, and Galileo were the forerunners of positivism, which was first formulated and practiced in all clarity by Newton himself.[80] Such a claim, however, is faulty on two counts. First, it ignores all that is simply fantastic in Comte's comments on science; second, it blatantly falsifies the historical record. Newton for one never once came close in his statements to the absurd one-sidedness of Comtean positivism as regards the search for causes. True, Newton argued against Leibniz that one can make progress in studying gravitation without knowing its cause. In so doing, however, he was far from emulating unknowingly the Comtean program. For Newton's was a mind searching for causes, as the General Scholium or the queries of the *Opticks* could have shown to either Comte or Mill. But where Newton merely distinguished two parts of the scientific procedure, Comte negated the one not to his liking. Unlike Comte, who dismissed hypotheses bearing on the way phenomena are produced, Newton clearly enjoyed constructing them and rejected only those that by definition could not be subjected to experimental tests.[81]

Actually the interplay of observations, mathematical formalism, and hypotheses was the lifeblood of science for over a hundred years before Comte tried to narrow down scientific method and research to a single aspect. Speaking of the respective merits of the one- and two-fluid theories of electricity, Coulomb, for instance, found it perfectly legitimate to express in the very same sentence both his appreciation of the mathematical formalism and the recognition of true physical causes: "On the supposition of two electric fluids, I have no intention other than to present in the simplest manner possible the results of my experiments and calculations, not to indicate the true cause of electricity."[82] Fourier, once an ardent supporter of the scientism of the revolution's early phase, had not anticipated either the language of Comtean positivism when writing as a scientist. In the preliminary discourse introducing his *Analytical Theory of Heat*, Fourier emphasized the overriding importance of continued observations and more refined measurements, stressed the search for physical causes, and spoke of the properties of things. He even entertained ideas about calculating the temperature of the "heavens," or interplanetary space, an endeavor hardly in line with Comte's precepts.[83] Ampère, whom Maxwell called the Newton of electricity, did not confine either his research to the shallows of Comtean scientism but constructed bold hypotheses about atomic mechanisms responsible for magnetism. Nor could Maxwell have ever formulated his abstract equations without the aid of models made up of gears, rods, cylinders, idler wheels, and the like. Or did any physicist, astronomer, or chemist need Comte to convince himself that phenomena regardless of their nature are conformable to law? Did any of them follow him in frowning on microscopes and on the most potent of all scientific instruments of the nineteenth century, the spectroscope? Was Rowland at the close of the century speaking according to Comtean tenets when he asked: "What is matter; what is gravitation; what is ether and the radiation through it; what is electricity and magnetism; how are these connected together and what is their relation to heat? These are the greater problems of the universe. But many infinitely smaller problems we must attack and solve before we can even guess at the solution of the greater ones."[84] Nor could any justification be given on a Comtean basis for Einstein's studies in statistical mechanics, the sole aim of which was, in Einstein's words, "to find facts which would guarantee as much as possible the existence of atoms of definite finite size."[85]

While admiring Mach's "incorruptible skepticism and independence," Einstein had to realize that epistemological positivism is "essentially untenable."[86] Its sociological dreams apart, positivism was as much a blind alley in science in Comte's day as it was later in Mach's time. Witness Mach's wholly erroneous attitude toward atomic theory that was, as Einstein himself pointed out, a direct consequence of positivist principles.[87] The facts of the history of physics hardly bear out the contention of certain representatives of the Viennese school of neopositivism that Comte exerted a decisive influence when classical physics had to yield to modern conceptions. To give up ideas, however cherished, when facts stubbornly demand a change was a guiding light in the thinking of physicists long before Comte and continued to be so regardless of him.

In the parlance of modern physics, this procedure is referred to as the elimination of empirically meaningless statements, and of this no lesser figure of modern physics than M. Born said that it "has nothing to do with positivism, although an association between the two is being asserted both by the adherents and by the enemies of this philosophy. It is a heuristic idea, which has proved its worth in all parts of modern physics."[88] Born's resolute doubts on the allegedly major role of positivism in modern physics carry all the more weight, since they come from one who has been so closely associated with every phase of modern physics in its first five decades and who had a better appreciation of the historical roots of recent developments in physics than did most of his colleagues. This might give second thoughts to those who have accepted the well-known contention that positivism is the most adequate formulation of scientific methodology. And if they find Born's demurrer thought provoking they will perhaps be better disposed to ponder the words of J. J. Thomson, who for all his daring in experimental physics never parted wholly with the frame of mind of the classical physicist.

Speaking of neopositivism toward the end of a truly exceptional scientific career, Thomson rejected most resolutely the contention of those who object "to the introduction of ideas which do not relate to things which can actually be observed and measured." Such a view was in his words "bad Physics as well as bad Metaphysics." "I hold," he added," that if the introduction of a quantity promotes clearness of thought, then even if at the moment we have no means of determining it with precision, its introduction is not only legitimate but desirable. The immeasurable of today may be the measurable of tomorrow. . . . It is dangerous to base philosophy on the assumption that what I know not can never be a knowledge."[89] Clearly Thomson defended the legitimacy of theories and concepts that may not lead immediately to directly observable conclusions. This was, however, a position diametrically opposite to a cardinal tenet in Comte's philosophy of science. Thomson spoke in full agreement with the past and present practice of physicists. History could but support a man who lived the tenets of physics through a long life of extraordinary research, but history sadly deserted the one who read its pages through the distorting lenses of scientism.

Comte's efforts to implement his scientism never got off the ground of inept academic exercise. It was otherwise with the cumbersome argumentation of another visionary of the nineteenth century, Karl Marx, a man no less steeped in scientism than Comte himself. Marxism has grown into a giant political force, and its history amply shows what physical science can expect when scientism has the political power to impose its intellectual tyranny. That dialectical materialism was to rule science and scientific research was already intimated on the pages of *Das Kapital*. Marx not only compared his method to that used by physicists,[90] but he also claimed to have discovered a basic set of rules unconditionally valid in both natural and social sciences. Of this method that was to provide the golden key to any puzzle under the sun, Marx stated that it "includes in its comprehension and affirmative recognition of the existing state of things, at the same time also the recognition of the negation of that state, of its inevitable breaking up."[91]

Characteristically enough, such a sweeping statement about the possession of a universally valid scientific solution to everything came, as is usually the case, from the pen of one who was conspicuously short on scientific training. But it is such barren ground that provides the proper soil for inordinate ambitions that aim at final pronouncements in the realm of science. To get a suddenly needed piece of information in scientific matters, Marx counted on a quick recourse to Engels, who had at least mastered the rudiments. While an almost total lack of competence is usually too obvious to become harmful, just skimming the surface of knowledge over a wide area can be deceivingly dangerous. Engels fits this case perfectly. Even worse, his ramblings in science in time were elevated into a canonical text whereby party philosophers tried to decipher what course science ought to follow.

According to his own admission, Engels made a survey of mathematics and natural sciences to convince himself in detail that "amid the welter of innumerable changes taking place in nature, the same dialectical laws of motion are in operation as those which in history govern the apparent fortuitousness of events."[92] The law in question, the Hegelian dialectic in materialistic gown, was believed by Engels to be supreme with no room left for any exception. Even the processes leading to the emergence of consciousness in man were supposed to have been fashioned in accordance to it. As a true addict of scientism, Engels, as the quotation shows, had already been convinced, exactly as Comte was, of what science had to say before interrogating it. Science's answer was therefore tailored meticulously to Engel's specifications. Under his guidance, physics produced an apparently overwhelming documentation of the validity of the law of dialectic. Once more Comte was echoed when Engels stated the formulation of this law to be of historic significance.[93] And in a dictatorial phrase that intimated all the struggles science had to face under Marxist rule, Engels warned natural scientists: "Whatever pose natural scientists adopt, philosophy rules over them. The question is only whether they want to be ruled by some vile fashionable philosophy, or whether they want to be guided by a form of theoretical thought that is grounded on acquaintance with the history of thought and its achievements."[94]

Marxist dialectic, like Comtean positivism, was not hesitant about laying down a long array of regulations physics was supposed to comply with. That these rules were invariably erroneous was due in no small part to Engels' unbounded admiration for Hegel, a typical pitfall for thousands of amateur scientists in the nineteenth century. It was Hegel whom Engels echoed in claiming that the Newtonian theory of gravitation rested on taking attraction as the essence of matter. It should be no surprise that such mishandling of the facts of scientific history was but a step toward a wholly arbitrary account of the facts of nature. Beside attraction, Engels claimed with Hegel, one should assume repulsion as a basic law in nature. "Hegel is quite right in saying," Engels insisted, "that the essence of matter is attraction and repulsion."[95] This was the a priori sway of scientism that came forth in brute force in Engels' dictum: "Attraction and repulsion are as inseparable as positive and negative, and hence from dialectics itself it can already be predicted that the theory of matter must assign as important a place to repulsion as to attraction, and that a theory of matter based on mere attraction is false, inadequate, and one-sided."[96] If issues in physics are settled by dialectic, experiments will have, of course, to take second place. Scientism, however, is all too eager to pay at least lip service to the importance of observational evidence in physics. Yet, "the evidence" is usually chosen in a most arbitrary manner. In the case of repulsion, the alleged observational evidence came from optics. "Repulsion resides in light," Engels declared, and he referred to the tail of comets that invariably bend away from the sun. Not content with this "experimental feat," Engels claimed a proof from theoretical physics as well. To achieve this he turned inside out some statements in Helmholtz' epoch-making paper of 1847, "On the Conservation of Force." For his "service" to Marxist scientism Helmholtz, however, was not to be given any credit. On the contrary,

Engels emphasized that "ancient philosophy" (the forerunner of dialectical materialism) recognized long before Helmholtz that "the sum of all attractions in the universe is equal to the sum of all repulsions."[97]

This, however, was hardly what Helmholtz meant by formulating the law of conservation of energy. What is more, he was rebuked by Engels for admitting only attractive forces.[98] After all, Engels' idol in physics was no other than Hegel, on whose shoulders Helmholtz rightly placed the blame for the indifference of nineteenth-century physicists toward philosophy.[99] Having given his allegiance to Hegel, Engels could hardly find any fault with Hegel's dicta in physics. Of the over fifty references to Hegel in the *Dialectics of Nature*, only a few strike a note of criticism, while Helmholtz is castigated time and again. One of Helmholtz' opinions is called "childishness"[100] by Engels, and so are dismissed on occasion by Engels, Maxwell, Clausius, and Carnot. Thomson's and Tait's *Treatise on Natural Philosophy* was for him a book in which "the ability of thinking came to a standstill," in which "thinking is forbidden."[101] If Faraday escapes such strictures it is because, in Engels' view, his ideas on electricity closely duplicate those of Hegel.[102] But Newton was called an "inductive ass,"[103] who could only picture the universe but not explain it.[104] Newton, of course, could not learn from Hegel what Engels did, namely, that "if nature itself proceeds exactly like old Hegel, it is surely time to examine the matter more closely."[105]

That nature did in fact follow Hegel was for Engels an article of creed. He went to unbelievable lengths to whitewash even the most stupendous scientific bunglings of his hero. He pleaded stoutly for the Hegelian geocentricism in science, as did Comte for the geocentricism of positive philosophy and science. Engels defended the merits of Hegel's legislation about what the relative distances of the planets ought to be.[106] He even found something basically good with Hegel's definition of electricity as "the angry self of matter."[107] Only a mind totally dedicated to his own brand of scientism could go so far. To read his statements about which conclusions of physics should be regarded as absolutely valid could hardly make matters any worse. Engels had to make such statements because philosophy had in scientism a direct bearing on the conclusions of science. The way Engels presented the respective roles of true philosophy (dialectical materialism) and science, anticipated with frightening concreteness the parlance of future party theoreticians in charge of laying down the line for scientists. Scientists, according to Engels, were of necessity "in bondage to philosophy."[108] That philosophy (dialectical philosophy, of course) could lead of itself to strictly scientific results was for Engels an unquestionable truth. "Scientists," he warned, "could have seen even from the successes in natural science achieved by philosophy that the latter possessed something that was superior to them even in their own special sphere." [109]

It was only natural that Engels, like Comte, was oversensitive about any scientific theory or discovery that seemed to go contrary to his wishes. Thus Clausius became nothing short of a bogeyman in Engels' eyes. He could hardly pen Clausius' name without giving vent to his fears, under the disguise of ridicule, of course. The concept of entropy seemed to pull the rug from under the absolute validity of one of Engels' scientific dogmas: the eternal cycle in which matter moves. His prose nowhere else rises to such solemn heights as in depicting the never-ending cycles in which matter supposedly reproduces "with the same iron necessity . . . its highest creation, the thinking mind . . . somewhere else . . . and at another time again." [110] This was the unmistakable voice of what is deepest in scientism, a blind sectarian drive that would accept science only as a slave and not as a source of objective information. In Engels' brand of scientism, the conclusions of science were no longer arrived at, they were rather dictated: "Physics, like astronomy before it, had arrived at a result that necessarily pointed to the eternal cycle of matter in motion as the ultimate reality." [111] Of this "truth" of physics, the leading physicists of Engels' time, however, were simply unaware.

The pattern remained essentially the same when it became Lenin's turn to wear Marx's mantle. The unexpected changes that began to agitate physics after the turn of the century were carefully watched by Lenin, who felt called to protect the purity of Marxist scientism. In the little breaks that came in his feverish activity as editor, organizer, and agitator, he tried to keep his finger on the pulse of physics that was beating more rapidly with every year. Lenin, however, could get no farther than any other spare-time devotee of science could. He was able to discern in the curious winds blowing across early twentieth-century physics only what he already firmly believed physics ought to bear out:

Modern physics is in travail; it is giving birth to dialectical materialism. The process of childbirth is painful. In addition to a living healthy being, there are bound to be produced certain dead products, refuse fit only for the garbage-heap. And the entire school of "physical idealism," the entire empirio-critical philosophy, together with empirio-symbolism, empirio-monism, and so on, and so forth, must be regarded as such refuse. [112]

Though short on physics, Lenin was long on semantics and could forge, if not convincing arguments, at least phrases bubbling with confidence about the decisive contribution of modern physics to dialectical materialism. In such a vein did he greet the radioactive transmutation of elements and the extension of the electromagnetic spectrum into the range of X-rays and radiowaves. That modern physics had refuted mechanical materialism, he readily admitted, but he was quick to insist with the limitless adaptability required of any good Marxist that "the destructibility of the atom, its inexhaustibility, the mutability of all forms of matter and of its motion, have always been the stronghold of dialectical materialism. All boundaries in nature are conditional, relative, movable, and express the gradual approximation of our reason towards the knowledge of matter." [113] Even within the framework of Marxist aspirations this was a strange inconsistency to say the least. The physical universe, mind, history, and society were, according to Marxist tenets, governed by the same definite laws. Society was believed to be rapidly approaching the well-defined, precisely engineerable, final classless state. But to save physics for Marxist purposes, physics — this symbol of precision — was told to sink its roots into the bottomless haze of indefiniteness.

This was the inconsistency of expediency that scientism, Marxist or Comtean, could not avoid if its borrowed time was to be lengthened even a little. Scientists therefore could have expected some sort of indifference by the party about their scientific views, since according to the utterances of orthodox Marxism quoted above, the conclusions of science were not in any case to be considered definitive. Could this haziness, which supposedly covered the ultimate layer of the universe, not have served as ground for freedom of science in the Marxist realms? Did Lenin not even go so far as to state that the *diamat* (dialectical materialism) defines only one characteristic of matter, namely, that it exists independently of the mind? [114] Did Lenin not repeat the lip service Engels paid occasionally to the priority of observation and experiments over theories? Did he not approve of Engels' dictum that the interpretation of materialism should be revised with every major physical discovery? [115] Could one not discern an apparently sincere flexibility in the authoritative Marxist pronouncements on science, dictated not by "dialectic" but by the recognition of the perplexing paradoxes encountered by science itself?

A definitive answer to this last question will probably never be forthcoming. For it is the nature of Marxist dialectic to bank heavily on the highly adaptable formulation of its theses. The testimony of Marxist action documented in its history, however, is far less ambiguous. For even in the years when the Communist state came close to collapsing, Lenin found it necessary to keep vigil over science in the light of Marxist ideology. Writing in the December 3, 1922, issue of *Pod znamenem marksizma (Under the Banner of Marxism)*, Lenin called attention to the revolution going on in modern science and warned that reactionary philosophical schools and tendencies would arise as a result. What Lenin deplored most of all was that most of the leading scientists appeared to be reluctant to champion materialism. But in a *Marxist* society science and scientists could not remain uncommitted on this score. Lenin therefore gave the journal the task of assuring that modern science and its representatives would prove themselves spirited allies of materialism. Otherwise, as he put it, "fighting materialism would be neither fighting nor materialism."

That this call to arms remained merely words for a while was due only to the fact that the chief battles Marxism had to wage for the time being concerned its crumbling economy and the question of who would succeed Lenin. Thus in the 1920's Soviet physicists were generally left undisturbed. They traveled, studied, and lectured abroad a great deal. Fok worked in Göttingen, Frenkel at the University of Minnesota, and several others received and used Rockefeller grants. Kapitza became Rutherford's right-hand man at the Cavendish Laboratories. In general Russian physicists took an active part in developing modern physics, and their Western colleagues frequently showed up in Russia either to lecture or to attend congresses. As to the philosophical implications of modern physics, the reactions of Russian physicists paralleled the various shades of opinion represented by their Western counterparts. Most of the Russian physicists were in fact wholly indifferent or simply hostile to the Marxist interpretation of science. The only physicist of note to join the party in the early 1920's, A. K. Timiryazev, denounced relativity as Machism, but the great majority of physicists active in the field of relativity, or translating and writing textbooks, such as S. L. Vavilov, gave ideological discussions a wide berth. A. A. Maksimov, the party ideologue who pleaded for a recasting of Einstein's physics by proletarian scientists, characteristically enough had no more than undergraduate training in physics. In general all the Marxist appraisal of relativity in the 1920's had been done by publicists. Regardless of Lenin's prerevolutionary invective against Mach, Machist Marxists — as those with positivist preferences were called — enjoyed some measure of freedom to argue their views up until the fateful year of 1929.

As soon as the party felt slightly less pressed by internal squabbles or by external adversities, however, the drive was on with all the furor of scientism to assert and enforce science's "bondage to philosophy."[116] Stalin declared 1929 "the year of the great break," "the year of shattering transformations . . . on all fronts of socialist construction." The year was heralded as the beginning of the "scientific changeover from bourgeois to red specialists." Apolitism and neutralism were condemned; the Academy of Sciences witnessed the first drastic purges within its ranks; higher education was thoroughly reshaped. The next year saw the union of scientific technicians demonstrating in the streets of Moscow against scientists in particular and the intelligentsia in general. In April, 1931, Bukharin threatened scientists with a "physical and moral guillotine," which meant the merciless imposition of Marxist doctrine in university courses and periodicals. The wild furor of scientism even went to the extreme of trying to create a statistical theory in accordance with Marxist principles. The *Theory of Mathematical Statistics* — the work of an insurance specialist — received sanction from the party for a while, though no teacher of mathematics was forced to adopt it.

The Communist Academy, or the party division of the Academy of Sciences, also began to exercise an ever-growing control over the debates relating to the philosophical implications of physics. The 1930 debate at the Polytechnic Institute on Frenkel's operational representation of the electric and magnetic fields was perhaps the last meeting of this type where specialists could dominate the discussions. The 1930's saw natural scientists subscribing to Marxism en masse and intellectuals shouting Marxist slogans in fervent unison. All this was the supreme parody of Marxist contentions that scientists would spontaneously come around to the Marxist creed. By forcing scientists to revamp their own subjects, Marxism openly belied its claim of being the sole possessor of methods that alone could assure the success of scientific work. The end result of this welter of contradictions was best summed up by D. Joravsky:

Worshipping science, the Bolsheviks had to raise cries of crisis in science. To make dialectical materialism an effective fighting creed in a war against ideologically alien scientists, they had to renounce faith in it as an objective description of the way that scientists discover the natural order. The union of revolutionism and scientism (*nauchnost'*), which Lenin had described in 1894 as the chief power of attraction of Marxism, could hardly be maintained in the face of these contradictions. To believe in one part of their doctrine the Bolsheviks had increasingly to disbelieve another. At the maddening climax of most intense belief and disbelief they shut off further discussion, "disarmed" their intellects (the phrase was a catchword of the great break), made their minds wax in the hands of the General Committee and the chief."[117]

Only the outbreak of World War II could temporarily silence the weirdest cacophony scientism has ever produced. The "authorities" in science, who were never clearly identifiable, had to yield the floor to patriotic rallies, and in the din of arms, declarations of allegiance to Marxist science were reduced to the repetition of Stalin's vague short formulas on scientific orthodoxy. In the life-and-death struggle, there was little if any energy left to implement in scientific matters the injunction of article 126 of the Stalinist constitution, which states that the party is "the leading core of all organizations" and should investigate everything. There was no time left to repeat investigations like the one in 1937 that found the Soviet Astronomical Institution "teeming with the enemies of the people."[118] At any rate too many persons disappeared daily to remember the half dozen astronomers who also were purged or vanished.

The lull, however, was only temporary. Scientism endowed with political power could no longer thrive without feeding on extorted fresh cries of allegiance. The new onslaught of scientism was launched by Zhdanov's speech of June 24, 1947.[119] True to the spirit and style of scientism, he called for a fight against the "countless philosophical weeds," against the "quasi-objectivism of the professors," against neo-Kantianism and agnosticism, against the "whole arsenal of the philosophical lackeys of the imperialism," and of all things, against "smuggling God into science." First hit were the geneticists whom the party forced to subscribe to the fantasies of Lysenko. By 1950, however, the physicists were also under concentrated attack. Relativity was once more diagnosed as the source of evil. The word itself disappeared from the textbooks over which the Ministry of Education had effective control. Physicists were called upon to replace relativity with a so-called high-speed physics. Physics conferences, like the ones held in Kharkov in 1948 and 1952, showed clearly the renewed ideological campaign that reached its high-water mark when A. A. Maksimov, member of the Academy of Sciences, published in *Krasny Flot* (*Red Fleet*, June 13, 1952) a paper with the brazenly revealing title, "Reactionary Einsteinism in Physics." The same year also saw the publication of an important symposium by party ideologues on scientific matters, which became known as the "green book." There, one of the editors, I. V. Kuznetsov, discussed at length the relation of Soviet physics and dialectical materialism. In genuine Marxist style, he claimed of course that "Soviet physics is the standard-bearer for the most modern and progressive ideas of contemporary natural science." The Marxist definition of the progressive, however, entailed some strange consequences. Thus, according to Kuznetsov, the development of science could only be secured by "the total renunciation of Einstein's conception, without compromise or half-measure."[120]

The patience of the physicists was at its end. First-ranking figures of Soviet physics — Landau, Lifshitz, Frenkel, and Fok, who until then conspicuously avoided being involved in ideological disputes — now joined the battle. As Relativity and quantum mechanics had already become indispensable tools of basic military and industrial research, the physicists were able to argue from a position of strength. They held the most important keys to productive effectiveness, which, in theory at least, is so highly valued by Marxist dialectic. Ideologues could talk but only physicists could deliver the absolutely needed atomic warheads, rocket motors, and electronic systems, without which the party could have hardly kept alive its hopes of ever subduing the West.

Maksimov's paper — full of errors — betrayed a thorough lack of expertise in the Theory of Relativity. Picking it to pieces presented no difficulty for any physicist. Fok could speak of its "antiscientific trend," of its "earth-shaking errors." He could easily give the lie to Maksimov's assertion that in the eyes of most physicists Relativity is but a blind alley in modern physics. He could even equate Maksimov's rejection of relativity with the denial of the earth's sphericity. In the 1930's such a harsh rebuttal of a prominent party ideologue would have been highly dangerous, but in the 1950's physicists were protected by the crucial role which was theirs in a modern technological society, and scientism had to retreat before this. The fact was that scientism, which professed itself the sole infallible interpreter of science, took its worst beating yet at the hands of science.

In the conference of physicists and philosophers held in Kiev in 1954, the physicists made another devastating attack on the party philosopher, I. V. Kuznetsov, whose serious shortcomings in modern physics were mercilessly laid bare. The debacle of party scientism was in a sense complete. It was impossible to keep up an all-out effort to enforce ideological claims. The tactics therefore were to be changed, although the basic aim was not given up by any means. To save face as much as possible, the party ideologues beat a speedy retreat. The January, 1955, issue of *Voprosy Filosofii*, (Problems of Philosophy) the leading official organ in philosophical matters, acknowledged the truth of Relativity. Maksimov, the editor, had to resign, only to be replaced by the just as much humiliated Kuznetsov.

Such an evidently meaningless change in the ranks of party ideologues could presage only the reopening of the campaign at the very first opportunity. At the Congress of Philosophers in June, 1956, the leading philosopher of science, M. E. Omelyanovsky, demanded that all philosophers should concentrate on the all-important role played by Lenin's ideas in coping with the basic problems of modern science. With an obstinacy peculiar to a devotee of scientism, Omelyanovsky charged that the ignorance of these principles led many investigators of nature "to gross errors both in the methods of scientific research and in the field of philosophical conclusions."[121] Two years later, the "powerful positive role" of the Marxist-Leninist philosophy in the development of natural sciences was again stressed by Omelyanovsky at the Congress on the Philosophical Questions of Natural Sciences held in Moscow. Even the president of the Academy of Sciences, A. N. Nesmeyanov, joined the chorus praising Lenin's scientific importance. Others, like Fok, more critical of such an attitude, at that time had to disguise their dissent under specious calls for more "creative" interpretations of the scientific tenets of the diamat.

Token lip service to the party line is not infrequent, even in writings of highly technical character. The concluding remarks in *Quantum Electrodynamics*, by A. I. Akhiezer and V. B. Beretetsky, are very characteristic in this regard. For the authors, the swiftly changing state of particle and field physics on the subatomic level corroborates Lenin's theses on the inexhaustibility of the properties of the electron. Thus the very advanced textbook duly closes with two quotations from Lenin, which condemn Western idealism and relativistic agnosticism. Faith in the basic Marxist truths is obediently upheld: Nature is infinite, it exists forever, and the approximate character of our knowledge "is the only categorical, the only unconditional acknowledgment of its [nature's] existence outside the consciousness and perception of man."[122]

Some physicists went much further in paying homage to Marxist scientism. At the height of the campaign of the diamat to force physics into the party line, D. I. Blokhintsev stated that denying the existence of an absolute reference system on the basis of the Theory of Relativity was "mixing good physics with bad philosophy."[123] S. I. Vavilov, for his part, contributed an article to the "green book" with the highly revealing title, "Philosophical Problems of Modern Physics and the Tasks of Soviet Physicists in the Struggle for a Progressive Science."[124] Blokhintsev and Vavilov also took a prominent part in attacking the Copenhagen interpretation of quantum mechanics. This criticism was mainly an exercise in evasive phraseology and contained the usual platitudes, such as, only the Marxist law of the unity and struggle of opposites can account for the wave-particle dualism.

The volunteering of some Soviet physicists to bring the interpretation of physics into line with party doctrine were not the only successes Marxist scientism could chalk up. The party was particularly successful in imposing restrictions on astronomical and cosmological research. From a review of the articles contributed between 1930 and 1956 to the leading astronomical periodical, *Astronomichisky Zhurnal*, the following facts emerge.[125] Not a single author dared to deny the infinity of space. None of them dared to advocate the idea of a creation of matter, as demanded, for instance, by the steady-state theory. Nor was the "heat-death" of the universe ever discussed in an affirmative sense. Known physical laws were never extrapolated or applied by any of the authors to the totality of the universe, nor was the existence of laws unrestrictedly valid for the whole cosmos ever admitted. Relativistic cosmologies were attacked as early as 1931, as idealist, reactionary, bourgeois, and fideist theories. The conventions of Soviet astronomers in 1939 and 1949 rejected the concept of a finite, expanding universe and made compulsory the acceptance of the idea of a universe infinite in space and time.

When professing the infinity of the universe, scientism in fact defends only its own narrowness and draws stiflingly tight horizons for astronomy. This is so because "infinity" and "finitude" are only slogans in the vocabulary of scientism. What is of concern for scientism is its own narrow range, which it wants to foist upon science at any price. This narrow range is mapped out by concepts quite alien to science, as shown by the statement of the Soviet astronomer, M. S. Eigenson: cosmology is either bourgeois or capitalist.[126] Believing as it does in infinite and eternal space and time, Marxist scientism could only be hostile to any form of cosmology that implied ever so slightly a beginning and an end. Such a position was not without pronounced effects on Soviet astronomical research and theories. Until very recently, little or no work was done by Soviet astronomers on such problems of primary importance as the amount and distribution of hydrogen or the origin and age of the elements. They did not propose any large-scale model of the universe. The relativistic cosmological equations they accept, but they are reluctant to apply them to the universe as a whole. The expansion of the universe is admitted by them but only with reference to an infinite space-time frame. They have consistently sought a non-recessional explanation of the red shift, although the best they could do was to hail attempts made in this direction by their Western counterparts.

It is only understandable that Soviet astronomers display extraordinary caution when touching upon the philosophically sensitive problems of cosmology. For while in a scientism that lacks the power of enforcement the scientific problems are either "positive" or "idolatrous," in a scientism with a sword to brandish, the problems of science are "proletarian" or "bourgeois," "progressive" or "reactionary," or in all honesty, "safe" or "unsafe." The sad fact is that as soon as scientism acquires social or political power, it sets up an uninterrupted inquisition beside which Galileo's trial is child's play. Characteristically enough some advocates of scientism are openly proud of this. They are even willing to praise the pope, as the British Marxist J. G. Crowther did, for trying to destroy Galileo. A most influential person in handling British scientific manpower, Crowther deplored the fact that most scientists of today do not realize "that in crises the possession of power is more important than the cultivation of intellectual freedom. . . . The danger and value of an Inquisition depend on whether it is used in behalf of a reactionary or a progressive governing class."[127]

Fate and fortunes of human lives apart, one may well ask whether facts would bear out Crowther's contention that "inquisition is beneficial to science when it protects a rising class."[128] Soviet scientism acted and still acts on this conviction and tries to impose the view that science greatly benefits by being protected from "dangerous" ideological deviations. To see any harm in this of course is almost impossible for any believer in Marxist scientism, for he is bound to interpret facts through glasses ground to the specifications of the diamat. To him even the words of the great old man of Soviet physics, the erstwhile wonder of the Cavendish Laboratories, P. Kapitza, would be of no relevance. But unlike most of the armchair advocates of Marxist scientism, Kapitza had first-hand knowledge of both modern physics and the cruel reprisals practiced by institutionalized scientism. Kept under house arrest for some time, he was later given important assignments in view of his extraordinary talents. Yet, no honor, or position, however high, could make him oblivious of the real situation. In 1962 he took to task the party ideologues, who without adequate knowledge had declared the basic theories of physics invalid because they appeared to conflict with "dogmatically applied dialectical methods."[129] Kapitza of course had to create the impression that he was placing the blame not on the dogmas of Marxism but on some of its dogmatists. Old party pundits, however, were not to be taken in by such obvious tactics. Much less could they be expected to take in stride the question that Kapitza raised about the rejection of the relativistic mass-energy equivalence by the party philosophers: "Would it have been good for physicists if they had followed the conclusions of some philosophers and stopped working on the problem of understanding the Theory of Relativity in nuclear physics? In what position would the physicists have put the country if they had not been prepared for the practical use of the achievements of nuclear physics?"

This was clearly the position of the physicist arguing from a position of strength against institutionalized scientism. How strong this position was can be seen from the sarcastic remarks that Kapitza felt free to make about the necessity of a profound knowledge of modern physics if one were to philosophize on modern physics. The diamat, added Kapitza condescendingly, is without such knowledge nothing else but "the squeaking of a Stradivarius, absolute squeaking." Clearly Kapitza put the emphasis not on the diamat being a Stradivarius, but on its squeaking. And he rounded off his remark with a backhanded compliment to its users: "But in order to play it, one has to be musician and know music. Without this it will be just as false as ordinary squeaking." Kapitza's position was too obvious. It mattered little that he made repeated references to the dialectical balance between theory and practice as demanded by the diamat. Nor was his obsequious reference to Lenin of much help. Only a simpleton could miss seeing that it was the diamat and not the shortcomings of some of its representatives that were indicted by Kapitza.

That Kapitza, his prominent position in Soviet physics notwithstanding, could not avoid an official rebuke shows that institutionalized scientism can give no quarter and must, of its nature, demand unconditional allegiance. Kapitza's rebuke by the party was in strict conformity with the pattern that scientific life has to share in the cage of scientism, even if this cage is at times as comfortable as the English-style home built for Kapitza after he was prevented from returning to Cambridge. From his confinement Kapitza in 1936 wrote to Rutherford words which conveyed with naked strength the mental atmosphere prevailing in scientism. "After all, we are only small particles of floating matter in a stream which we call fate. All that we can manage is to deflect our tracks slightly and keep afloat — the stream governs us."[130] Kapitza's words are even more revealing when set beside the lines of the greatest of all Russian scientists, Mendeleev, who wrote with ebullient enthusiasm: "Knowing how contented, joyous and free life is in the realm of science, one fervently wishes that many enter its portals."[131] But Mendeleev, for all the "reactionarism" of the czarist regime that supported his research, was still spared the dubious "blessings" of institutionalized scientism.

That institutionalized scientism is losing its former grip on Soviet scientists merely shows that intellectual tyranny has to yield sooner or later to man's inborn search for truth or at least to the facts of productivity. Lysenkoism was recently abandoned in favor of Abbot Mendel's genetics, long anathema in Marxist realms, and in 1965 Kapitza again felt free to chastize Marxist scientism, this time for its promotional system.[132] What Kapitza deplored was a government policy reluctant to dismiss non-productive scientists provided they hewed faithfully to the party line. Evidently the blessings of scientism devolve not so much on science as on scientists willing or forced to compromise the principle of scientific integrity. Though this process of compromise does not imply so much violence nowadays, it is well to remember that the true blessings of scientism were in its heyday all too often but simple purge and terror. Western scientists visiting there for only a short time were not permitted to notice this,[133] but it could not remain hidden from those scientists from the West who sought refuge there for reasons other than spying. One prominent physicist, who was able to learn through personal contact of their frightening experiences, had this to say:

In 1938 I met three physicists who had actually lived in Russia in the thirties. All were eminent scientists, Placzek, Weisskopf and Schein; and the first two have become close friends. What they reported seemed to me so solid, so unfanatical, so true, that it made a great impression; and it presented Russia, even when seen from their limited experience, as a land of purge and terror, of ludicrously bad management and of long-suffering people.[134]

Fervent advocates of Marxist scientism still drawing on the huge benefits of free science in the West, however, will be shaken little if at all by any evidence to the contrary. They will keep going to unbelievable lengths to argue the sacred cause. To enhance Engels' credibility, Haldane, the geneticist, was even ready to wade deep into modern physics. He presented the Pauli exclusion principle as the verification of Engels' views on the existence of a repulsive force as universal in nature as is gravitational attraction.[135] The zero-point energy of quantum statistics proved for him the truth of the Marxist dictum about matter eternally in motion. Another Marxist dogma, the qualitative changes resulting from quantitative changes, was proved for him by the various radiations of the electromagnetic spectrum. In Haldane's hands even Dirac's delta function served Marxism by illustrating the Marxist axiom that science has to rely on devices that work though they are not fully understood.[136] One need not be surprised either to hear Haldane state that as early as 1873 Engels had expressed ideas adumbrating Dirac's work on matter and antimatter.[137] And so it goes ad infinitum in the Marxist mills of science writing.

It no doubt requires a carefully conditioned and receptive mind to accept this without great mental discomfort. But this is perhaps the supreme irony of scientism: Its devotees lose the greatest benefit science can provide, a critical, cautious mind, ever willing to pay heed to facts from whatever corner they come and regardless of how sweeping a revision they may demand of well-established concepts.

That physics and its range is drastically curtailed in ever so many ways when scientism gets the upper hand is only the result of that change of intellectual atmosphere that scientism is bound to produce. For the intellectual atmosphere required by scientific growth is freedom but scientism cannot afford it. Consequently, the growth of physics is necessarily stunted whenever it becomes controlled by scientism that can thrive only on a physics frozen into immobility.

That the efforts of scientism have to end in glaring failures and contradictions is only natural. Such is scientism's reward for thwarting the natural course of scientific research. Marxist scientism had to witness the setback of theoretical chemistry in the Soviet Union following the denunciation by party ideologues of the theory of chemical bond proposed by L. Pauling. Recently the party has also had to take a hard look at Lysenkoism. In keeping with the methods of institutionalized scientism, the blame for the historic fiasco of course was laid on the hapless Lysenko. The same party that set him up as a model of the socialist researcher now charged him with having falsified his experimental data.[138]

When scientism is not supported by political power, its most telling effect is disillusionment with science on the part of those whom scientism predisposed to look upon science as the exclusive source of meaning and purpose. Literary history abounds in impressive cases illustrating how disenchantment follows any unbridled love affair with science. Shattered hopes have invariably been the rewards of those who revered science as the true religion. "What is wanted," wrote the young Renan in 1848, "is to look for the perfect beyond, to push science to its furthermost limits. Science and science alone is capable of restoring to humanity that without which it cannot live: a creed and a law."[139] The day of fulfillment — the day "when humanity will no longer believe" — he described as a condition in which the laws of government of human affairs will be known with the same exactitude as physics knows its laws.[140] Clearly, if science were to deliver these goods, all this had to be a most serious matter. If the destinies of mankind and the perfection of the individual were bound up with it, then science was to be served as religion used to be.[141] An allegiance to science nothing less than absolute was the only reasonable attitude if one admitted as Renan did that science is "the solution of the enigma, the final explanation to mankind of the meaning of things, the explanation of man to himself."[142]

Renan's qualifications in science were too superficial to give him second thoughts at that point of the game. But the four decades spent in waiting for the great day to come cooled his fervor toward science-religion. As is known, the manuscript of *The Future of Science* did not see publication until 1890. A friend of the young Renan, whose advice he had taken, thought his ideas to be far ahead of the times. Four decades of a lifetime, however, were enough to make them obsolete on some crucial points. True, the preface of 1891 still professes Renan's religious belief in science: "My religion is now as ever the progress of reason, in other words the progress of science. But in looking over these pages of my youth, I often found a certain confusion which distorted certain deductions."[143] The old Renan's brilliant phrases still upholding the correctness of scientism should not mislead us. The errors he admitted in his early reasoning show that they concerned the very roots of an attitude that took science for religion. For not only did science come to be mutilated in scientism, but human nature as well. And of this Renan, the full-blooded humanist, could have no part. He came to realize that Comte failed "utterly to understand the sciences of humanity,"[144] that Comte wanted man to live "exclusively upon science, nay upon little scraps of phrases, like the theorems of geometry, barren formulas."[145] Comte's concept of man was a parody, and Renan could not repress an impassioned protest: "If human nature were such as it is conceived by Mr. Comte, every noble soul would hasten to commit suicide."[146]

The minor difficulties of scientism proved to be no less significant. As Renan had to recognize, the "science-religion" was as hard to spread as was scientific information, and it could not be expected ever to become more than the possession of a relatively small segment of society. Nor did it afford automatic protection against evil, the reality of which loomed for him far more menacingly than forty years before. But most disappointing for him must have been the recognition that science, which, as he put it, "alone is capable of improving the unhappy lot of man below," was far from ready to give the solution. "I am afraid," he wrote, "that the main contribution of science will be to deliver us from superstitions rather than to reveal the ultimate truth."[147] The final, all-inclusive scientific answer to all the ills of human existence, which he once so eagerly expected, was no longer hoped for.

Anatole France, who went through much the same experience, did not hesitate to lay bare his disgust of scientism. As he recalled in 1910, his days of youth were animated by a belief that science can provide "a system of ethics, social laws, a political constitution and everything else."[148] Soon, however, the heroes of his novels began to strike a different note. So it dawned on Sylvestre Bonnard, in *Le livre de mon ami* (1885), that science might often just be chasing appearances, and in *Les opinions de M. Jérôme Coignard*, France voices his utter disillusion in science: "I hate science for having loved it too much, after the manner of voluptuaries who reproach women with not having come up to the dream they formed of them. I wanted to know everything and I suffer today for my culpable folly."[149]

True, at times, in the period 1898–1906, France recovered some of his erstwhile confidence in science. In *Vers les temps meilleurs,* a collection of anticlerical, socialist, and pacifist pamphlets, he prophesied that "universal peace will be realized, not because man will become better, but because a new order of things, a new science, new economic necessities will impose peace."[150] But France's was an incorrigibly searching mind, and he could not be held captive for long by illusory promises. In a short essay entitled "Mysticism and Science," he admitted that the younger generation openly rejected science as a supreme guide. They found that science could not provide a single moral truth, and its inspiration was powerless in face of the major questions of life. Nay, it was inhuman, and outright cruel, as it reduced man to a cosmic insignificance, to a mere dust speck, and thus robbed him even of the sense of identity.[151] This was a devastating indictment, denying as it did to this "cure-all" concept of science the right to rule humanity. With no less force did it evidence France's own feelings: "The confidence in science which we used to feel so strongly is more than half lost."[152]

Even more dramatic was the demise of science worship in the case of H. G. Wells, the most widely read literary apostle scientism had in the first half of the twentieth century. Instead of bringing emancipation and happiness to man, science did not sustain the rosy visions of *A Modern Utopia* (1905) or the optimistic expectations of the coauthor of *The Science of Life* (1929). The tone of *The Fate of Homo Sapiens* (1939) was already despondent. It voiced the view that science had deprived man of exactly what its one-track-mind devotees expected from it: an indestructible zest for life. Not that science had not delivered an immense amount of information and a superb mastery over nature. On the contrary, Wells firmly believed that "there exist already scattered about the world all knowledge and imaginative material required to turn . . . the whole world into one incessantly progressive and happily interested world community."[153] What then was missing? To hear Wells admit it, science has no means of securing that man would listen to its results. It was beyond the capabilities of science to do what Wells termed the essential theme of his book: to *change* the "mental superstructure, the knowledge, idea and habit system of mankind."[154] Clearly, it began to dawn on Wells that information, however immense, would not of itself breed knowledge or wisdom, that is, the embracing of truth. It was not within the power of science to make man apprehend the real nature of his troubles and make him, collectively at least, less of a fool.

It must have been no less disconcerting for Wells that apart from this existential impotence of science, the scientific world picture contained elements strongly discouraging man from hoping for "a magic change when all the forces, within him as without, are plainly set against it."[155] The universe afforded no justification for hope in a secure, permanent progress. Science failed to establish more meaning to man's existence than to the long extinct icthyosaur or pterodactyl. "In spite of all my disposition to a brave-looking optimism, I perceive that now the universe is bored with him, is turning a hard face to him, and I see him being carried less and less intelligently and more and more rapidly, suffering as every ill-adapted creature must suffer in gross and detail, along the stream of fate to degradation, suffering and death."[156] Thus the science that was to provide the perfect and happy pattern for shaping man's life failed to come through. And no one else could realize the depths of this disappointment more keenly than the one who turned to science in his youth with the hope of a Christian convert. For just as Christianity is meaningless if its hope in eternity is merely wishful thinking, so is scientism if its preconceived pattern of the universe is just an illusion. Capitalism without profit, Communism without classless society, legal code without morality — that was scientism without the presence of finality in science. It was not science's fault. Furthermore, the pattern science finds in the universe is only a part of the great pattern necessary for making meaningful the totality of human existence. When, however, the part is taken for the whole, and the transient for the final, as is always the procedure in scientism, no pattern and no answer is left at all. "Our universe is not merely bankrupt," wrote the author of the *Mind at the End of Its Tether*, "there remains no dividend at all; it has not simply liquidated, it is going clean out of existence, leaving not a wrack behind. The attempt to trace a pattern of any sort is absolutely futile."[157]

It took the holocaust of the two world wars, ending in the glow of atomic explosion, for Wells to realize the disastrous fallacy of taking the partial pattern for the whole. But the logic of this mistake had already been pointed out a half a century before in the Romanes Lecture of 1894 delivered by T. H. Huxley. There it was recognized that no "scientifically determined course of conduct," no "absolute political justice" could usher in a trouble-free life. The best that man could learn was not happiness but the practice of self-restraint and renunciation. Susceptible as the human race was to improvement, it could never achieve perfect harmony with nature. The ends of cosmic forces were not identical with man's deepest aspirations [158] — a far cry indeed from the optimistic vistas described in his lay sermon, "On the Advisableness of Improving Natural Knowledge," of 1866. There scientism was still in full vigor, unshaken, wholly confident in itself. The same Huxley was speaking, who six years before, following the death of his son, wrote in utter defiance: "I know what I mean when I say I believe in the law of the inverse squares, and I will not rest my life and my hopes upon weaker convictions." [159] Huxley at that time stood with absolute firmness for the full equivalence of ethical perfection and scientific knowledge, this standing or falling proposition of scientism. He made no restriction whatsoever on the very first article in the creed of scientism: "There is but one kind of knowledge and but one method of acquiring it." [160] Thirty years later, however, it became clear to him that all this can lead at best to facing bravely but hopelessly the final extinction, when "the State of Nature prevails over the surface of our Planet." [161]

The historical failures of scientism to do justice both to physics as a science and to man's own nature are lessons of which this generation should constantly remind itself. To those who got admission into the inner sanctuary of physics and are rightly enraptured by the unique sweep and beauty of its laws, history should be a constant warning that a false adulation of physics invariably leads to its strangulation. Their purely scientific bearing apart those failures have an important role to play when it comes to the evaluation of the place of physics in the broad context of all the values and factors needed to shape human life both individual and social. The failures of scientism show more palpably than any abstract reasoning that the principles and applicability of physics have a limited range. Most of all physical science — that most wondrous achievement of human reflection — is unable to tell how to fit the pattern it finds into the great pattern of human existence. It is not within its domain to stake out the road leading toward a more secure, more fully human life.

True, there are still people around of high scientific caliber who time and again fall back on the shallows of scientism. One of them is Bertrand Russell, who admitted that for decades one of the principles that governed his wide-ranging reflections was a "prejudice in favor of explanations in terms of physics wherever possible." [162] This prejudice, the hallmark of scientism, has not yet completely left him. To hear Russell say it, science is a liberator not only of the bondage of physical forces but also "from the weight of destructive passions. We are on the threshold of utter disaster or unprecedentedly glorious achievement. No previous age has been fraught with problems so momentous; and it is to science that we must look for a happy issue."[163]

It is rather reassuring that a genuine familiarity with the spirit and method of physics had strengthened in most cases a conviction diametrically opposite to that of Russell. Witness the often eloquent warnings on the part of prominent physicists who particularly in our age diagnosed so accurately the proportion of danger resulting from the deification of physics. For what appeared to Russell as the road to a happy issue, looks to P. Kusch, a Nobel laureate in physics, as something very different: "Science cannot do a very large number of things and to assume that science may find a technical solution to all problems is the road to disaster."[164]

At this point, one cannot help wondering if it had not been possible to come to the realization of this long before wholesale cultural debacles had to strike their blows. Perhaps yes. If only scientists, and especially all those who ever spoke about science, had kept asserting with greater sincerity and insistence what Fontenelle said of physics! Praising the universal utility of geometry and physics, he voiced the view that physics was still in its cradle and that because of the inexhaustible richness of nature, physics will in all probability never become a completed science.[165] Wonderful as these words are, they were hardly more, judging by the rest of Fontenelle's essay, than a stylish lip service to intellectual humility. To rest satisfied with the veneer of intellectual humility might become too costly in our times. Two and one-half centuries after Fontenelle, physics still appears to be in its cradle, and one knows just as little now as then whether it can ever bring its task to completion. Still, just as in Voltaire's time, everybody talks about physics and, what is of far greater importance, everybody depends, for better or for worse, on its marvelous feats in a greater measure than ever.

CHAPTER TWELVE

The Place of Physics in Human Culture

SCIENTISM HAS SEEN in Comtean positivism its most narrow-minded, in Marxism its most militant manifestation. These two systems or ideologies, however, are far from exhausting its full possibilities. Among the less overt forms of scientism one is a major danger symptom of the twentieth-century Western world: the almost complete cleavage between the humanities and the sciences, the phenomenon that C. P. Snow has called the two cultures.[1] This epithet is probably the most concise expression of the cultural diagnosis. Some longer paraphrases of it, however, are also worth remembering. Of these is P. R. Calder's striking simile: scientists leave their discoveries, like foundlings, on the doorstep of society, while the stepparents do not know how to bring them up.[2]

Undoubtedly the rift between science and society, or between science and the humanities, is largely the result of a lack of concern on one hand and a lack of competence on the other. It is only natural that such factors should create a cultural climate in which superstitious beliefs about science crop up in abundance. Foremost is the widely shared conviction that science is an all-powerful, almost magical device. Unlimited effectiveness is often ascribed to science, or to its "infallible" method, which, if properly applied or fed into computers, is expected to solve all conceivable problems, select war strategy, predict business trends, settle the authenticity of the Pauline letters, and even compose symphonies. Science is supposed to have found the way to absolute truth, expressed of course in ever more esoteric mathematical formalism. In due allegiance to the "absolute truth," certain tensor equations claiming to have solved the ultimate riddle of the universe can make front-page headlines in leading newspapers. It matters little if these formulas are Arabic to the ordinary folk. People nowadays are dutifully resigned to the fact that only the limited circle of scientists can ever have access to the fountain-head of knowledge. Although monopoly of knowledge is in most cases highly dangerous, people are reassured that the "scientific spirit," which motivates men of science, would prevent them from abusing their power. Because of the rift between the two cultures, science appears to many a mysterious sacred cow symbolizing the promises of a not-too-distant golden age. A Nobel laureate physicist, P. Kusch, diagnoses this mental atmosphere in these words: "I am quite certain that the mass of men believe that the better world of tomorrow will come through science."[3] Like all other superstitions and utopistic hopes, this too is fraught with grave dangers. Kusch himself felt impelled to warn that such a superstitious faith in science "ought to be publicly combated," and in this connection, he assigned a prominent role to reporters and science writers: "The point that science alone will not create the good life should be endlessly explored by the press."[4]

In paying heed to this call, the press would atone to some extent for the immense harm done by that sort of science news reporting that consistently exaggerates the content and import of scientific discoveries. For obvious reasons science news editors too often yield to a sensationalism that inevitably distorts their accounts of the latest findings of science. Indeed, is it not puzzling that the public can be told at regular intervals that the "latest" discoveries have led science to the very threshold of solving the ultimate riddle of life, of finding the ultimate constituents of matter, and of sighting the true outlines of the stellar universe? Is it not strange that the public's attention is seldom called to the peculiar phenomenon that, in spite of its magical advances, science never quite manages to cross that mythical threshold separating us from the "ultimate" solutions? Is the very concept of scientific progress not made meaningless by constant references to the "ultimate barrier" that in fact never proves to be ultimate? Scrupulous intellectual honesty and profit-seeking sensationalism represent, however, widely differing motivations. As a result, the public is being bombarded with news about "final breakthroughs" in science and is induced to attribute to science a miraculous wisdom that science in fact does not possess. Worse still, sensational science reporting results in the very opposite of what it is supposed to promote — a genuine understanding of science. First, it impairs the ability of the non-scientist to grasp the true meaning of science. Second, it conceals the fact that ignorance of the limitations of science constitutes a dangerous pitfall.

Science news reporters and authors of popular books on science are not alone responsible for the almost schizophrenic split in our culture. C. P. Snow put the major blame on the non-scientific sector of the educated public. In his view, scientists are far better versed in humanistic lore than men of letters, politics, art, and education are in scientific matters. The great majority of the representatives of humanistic learning, according to Snow, are scientific illiterates. The result is that "the great edifice of modern physics goes up, and the majority of the cleverest people in the Western world have about as much insight into it as their neolithic ancestors would have had."[5]

The indifference of many humanists to science and its more recent developments is no doubt deplorable. Rutherford was rightly indignant when he once remarked to Bohr that some humanists were going too far "when expressing pride in their complete ignorance of what happened in between the pressing of a button at their front door and the sounding of a bell in the kitchen."[6] Yet, those who know best the mysteries of the flow of electrons may also display shocking unfamiliarity about non-scientific matters and are at times, like some of their humanist counterparts, apt to brag about it. As a perceptive physicist described this side of the coin, scientists "are often ignorant of the philosophic and indeed social implications of their work, and . . . they sometimes overstate their competence with blatant self-delusion."[7] Disconcerting as is the lackadaisical attitude of many a humanist toward science, no less pitiful is it, as another physicist put it, "to hear a physicist talking one moment with great authority and competence about nuclear physics, and then to find a moment later that he is talking with an air of equal authority about problems of international government."[8] Thus it is questionable whether the humanists should bear the major part of the blame for the cultural schizophrenia. Nor is it probable that the inclusion of a half dozen courses in the physical sciences in the humanistic curriculum, as advocated by Snow, would make a non-scientist college graduate scientifically literate. To understand adequately and properly the language and message of modern physics, years of hard and advanced work are needed, and this will always remain beyond the talents, time, and energies of most humanists. The fact is that familiarity with electromagnetic theory, Special and General Relativity, and quantum mechanics, to name only a few principal topics, are indispensable for gaining an understanding of modern physics. It is therefore simply unrealistic to think that such studies could be accommodated in the program of a candidate for an advanced degree in classical philology or modern history or sociology. Furthermore, could the overburdened humanist be sure that if he covered all these topics, his physics teachers would instruct him clearly and properly, say, about the various differences which the concept of mass has in the principal branches of modern physics? More likely the humanist will find out that many teachers of physics are not even aware of this fact. And if they are, will they tell him that this is only one of the many question marks blossoming all over the fascinating field of modern physics? Snow's plan for the exposure of humanists to an extensive program in scientific education thus seems a very questionable means to a very desirable end.

There remains therefore the only other expedient: to let the humanists skim the surface by subjecting them to an introductory survey course in modern physics. Such a course is always useful, but it can also be very dangerous, for reasons that are not difficult to find. First, a little knowledge is always dangerous unless its littleness is genuinely realized, and such a degree of intellectual self-appraisal, however, is infrequent. The authors of introductory books are not much inclined to keep in focus the meagerness and inadequacy of the information their work is able to give of a very vast and complicated subject. For one thing, a primary means of introductory education — the principle of visualization — cannot be applied to the atomic and subatomic levels of matter. In an introductory survey of modern physics, however, a reliance on pictorial analogies is indispensable, and this procedure leads oftener than not to a basic distortion of the meaning of the methods and results of modern physics. The second reason lies in the very motives behind such survey books on modern physics. Are they not too often written with a subconscious effort to praise physics rather than appraise it, even when their style does not yield to impulses of rhetoric? Is the presentation in such books of the truly spectacular triumphs of modern physics ever tempered with a sincere avowal of the highly transitional character of so many of the allegedly final results, concepts, and conclusions?

In addition to these questions one may raise even more ponderous ones. Do physicists exert a concerted effort to reveal the tentativeness of many of their theories? Do they emphasize enough the provisional aspects of their explanations of the universe on both the cosmic and subatomic levels? The answer would have to be in the negative. Such a conclusion may sound harsh, putting as it does the larger share of the blame for the cultural split on the shoulders of scientists. Had such judgment been reached by a humanist, it might be suspected of arising from bias and incompetence. But it was a former president of the American Association for the Advancement of Sciences, W. Weaver, who formulated the verdict:

I am myself more inclined to place the greater blame upon the scientists. Although some scientists seem almost childishly eager to leave their laboratories to talk about things which they do *not* understand, they have been pretty reluctant to leave their laboratories to talk and write intelligently about what they *do* superbly understand. Far too little have they been concerned with general interpretation of their methods and their results.[9]

Clearly, just as the prime responsibility for a child rests with the parents, so does the basic responsibility for healthy cultural and social assimilation of modern physics remain with the physicist, who is the most competent to interpret his art. This is why the remedy prescribed by Snow to heal the split in our culture is not the cure-all it is purported to be. As we have seen, Snow pleads for more science courses in all phases of Western education and envisions an age where politicians, administrators, and the entire community know enough science to have a sense of what scientists are talking about. But who or what would guarantee that humanists would absorb the proper scientific sense? For, to raise the crucial question: Who or what would guarantee that physicists whose field of study moves away at an ever-increasing rate from the commonsense or visualizable world would also exert effort in the same proportion to hand down a fairly balanced and cautious interpretation of their more recent findings? The past record is unfortunately not encouraging on this score. Thus, because of the popular fashion, which always emphasizes the triumphal note in the fortunes of the scientific quest, the traditional popular images of science are as firmly entrenched as ever. This is all the more deplorable from the cultural point of view, since the surprises that were in store for physics during the past forty to fifty years could have been highly effective, if interpreted and communicated properly, in doing away with those false images. These G. Holton found to number seven.[10] More significant than the number seven in his list is the fact that all these images present a distorted view of science for the very same reason: the unbalanced, exclusively triumphal note which in one way or another they reflect. These images picture science as the supreme tool and power for gaining knowledge; as the irresistible, always righteous iconoclast; as the most effective force in life; as the most noble human instrument perverted only by the evil in man; as the unsurpassed debunker of "absolute" truths and values; as the pattern that should be imitated in all areas of human endeavor; and finally as the great magic that transforms the scientist into a wizard and an oracle.

All these images reflect superhuman features and can thus but deepen the gap between man and science. What is really needed is the determination to restore the truly human features in the portrait of science, and this means that both the successes and the failures of the scientific quest should be given proper emphasis. For, as the historical record shows, failures, limitations, and persistent frustration are not an irrelevant side aspect of the march of science. As the triumphs of physics unfold in connection with solving basic problems of the life sciences, so with each advance of physics are new aspects of the immense complexity of living matter revealed. In the same manner, man's conquest of one layer of matter reveals at the same time the outlines of a hitherto unsuspected one. Man's penetration into the realm of stars and galaxies shows the same characteristics, those of an apparently never-ending process. Again, all claims to the contrary, physical science has relatively little to say about such crucial aspects of human endeavor as those manifested in ethics, social organization, religion, and the arts. Obviously, the metric tools of science are not the instruments for probing the depths of the non-metric dimension of human existence. Science, as Vannevar Bush recently noted, is not a "complete set of facts and relations about the universe, all neatly proved," nor can science provide that firm basis on which men should "securely establish their personal philosophy, their personal religion, free from doubt and error."[11] Far from being the complete truth based on incontrovertible facts, science even in its own field is not wholly objective or infinitely precise or unconditionally permanent or philosophically inescapable as regards its findings and laws. It does not always deal with certainties, but often with probabilities, and it cannot pretend so much as to know even the "true" length of a rod. Science can only inch itself laboriously toward this true length without ever determining it with perfect accuracy.

It is in this laborious process of advance that science demonstrates most convincingly a basic feature of its proper image: its highly provisional character. To concentrate on a single, short phase of science, in particular on the most recent phase of science, might easily lead to dangerous illusions about the finality of the scientific views and conclusions. It is rather the study of the whole life-span of science that will effectively unveil both its human and humanistic aspects. It is there that the conviction will force itself on the student that successful as science is in dealing with a great many phenomena, it is unable to furnish all-inclusive explanations even within its own range. It is through the pages of history that the incontrovertible fact will become evident that science and scientists so far have not succeeded in reaching a consensus about the relation of scientific observations to the underlying reality. While several theorems of mathematical physics are particularly instructive in this regard, for most people their abstract nature can never equal the immediate appeal of a historical illustration of such aspects of the scientific quest. Effective as

Gödel's proof is in showing that no sufficiently wide set of mathematical assumptions can have inner consistency, it will remain beyond the grasp of the average educated man. The same is also the case with a rigorous quantum mechanical proof of the essentially statistical character of many a basic law of physics. Just as difficult to understand for the layman is the mathematically exact formulation of the statement that there is no such thing in physics as a strictly exclusive explanation, because an infinite number of theories can be constructed to correlate a given set of observations and quantitative data.

A careful presentation of the history of physics retains, however, a singular effectiveness in shedding light on the human aspect of the true image of science, which, as anything genuinely human, is a mixture of achievements and failures characterized by incessant changes. As a distinguished physicist of our times, R. E. Peierls, noted: "We must realize that a subject like physics changes all the time. It is not like what it has ever been and it is never going to be exactly like it is. Even what we mean by physics, our views of what is the proper domain of the physicist, also keeps changing."[12] This flux of change is known as history. For the great majority of those who try to discover for themselves a physics that is truly human, the study of its history will remain the royal road. In particular the recent upheavals in the history of physics should be effective in conveying the proper human perspective for the appreciation of physics. A little reflection, say, over a classic passage of Whitehead's own reminiscences might serve this purpose far better than could long abstract reasonings:

When I went up to Cambridge early in the 1880's my mathematical training was continued under good teachers. Now nearly everything was supposed to be known about physics that could be known — except a few spots, such as electromagnetic phenomena, which remained (or so it was thought) to be co-ordinated with the Newtonian principles. But, for the rest, physics was supposed to be nearly a closed subject. Those investigations to co-ordinate went on through the next dozen years. By the middle of the 1890's there were a few tremors, a slight shiver as of all not being quite secure, but no one sensed what was coming. By 1900 the Newtonian physics were demolished, done for! Still speaking personally, it had a profound effect on me; I have been fooled once, and I'll be damned if I'll be fooled again. Einstein is supposed to have made an epochal discovery. I am respectful and interested, but also skeptical. There is no more reason to suppose that Einstein's relativity is anything final, than Newton's *Principia*. The danger is dogmatic thought . . . and science is not immune from it.[13]

To live through a major crisis in physics and to have both the scientific training and the intellectual sensitivity to feel its sobering lessons will never be everyone's share, even among scientists. That physicists for the most part feel at least a mild shock when a scientific revolution sweeps away apparently rock-solid conclusions is most of the time duly mentioned even by the "triumphalistic" accounts of the development of physics. Such accounts, however, remain almost invariably silent about the often slow and reluctant acceptance of new ideas on the part of physicists,[14] a phenomenon which bespeaks so eloquently the truly human aspect of science.

No less effective in showing the human aspect of the image of science is the analysis of the ways scientific theories and discoveries are arrived at. Such an approach to the question represents, as was cogently argued by J. B. Conant, the most illuminating method for integrating the facts and reasoning of science into the whole texture of human culture and thereby bridging the gap between the sciences and the humanities. As he pointed out, technical information about science does not necessarily lead to the understanding of it as a type of intellectual quest. "What is needed," he noted, "are methods of imparting some knowledge of the Tactics and Strategy of Science to those who are not scientists."[15] It is the real complexity of such tactics and strategy that is going to be absent in the so-called single-line accounts about fundamental particles, non-conservation of parity, relativistic time dilation, and kindred items. Actually, such accounts only strengthen the mistaken belief in the straightforward wizardry of physics, instead of honestly revealing the often tortuous conceptual complexities that preceded and followed such discoveries.

It is well to remember that an account of the development of science that tries scrupulously to do justice to the historical truth is more than a matter of academic excellence. It has important cultural merits as well. To neglect the historical approach in science is to lose sight for all practical purposes of the human aspects in the image of science, and such a loss can hardly be condoned in an age engulfed in cultural schizophrenia. It cannot be emphasized enough that if the steps leading to a discovery — to the formulation and acceptance of a scientific law — are carefully retraced, a truly human picture will emerge that shows not only the triumphs, the flashes of intuition, but also the faltering steps in the dark, the mistaken beliefs, the erroneous observations, and faulty generalizations. It was in fact a close look at the history of optics that prompted Mach to remark how little the development of science takes place in a logical and systematic way. This holds true not only of classical physics but also of twentieth-century physics if the record is analyzed in detail. In short, the progress of physics has little in common with the victorious march of a phalanx of soldiers conquering at a steady pace the redoubts of ignorance. If comparisons have any value at all, a most telling one in this respect was drawn in 1888 by the American physicist, S. P. Langley, who did valuable research in the analysis of the infrared spectrum. For him the features of the scientific progress evoked the onrush of a pack of hounds "which in the long run, perhaps, catches its game, but where, nevertheless, when at fault, each individual goes his own way, by scent, not by sight, some running back and some forward; where the louder-voiced bring many to follow them, nearly as often in the wrong path as in the right one; where the entire pack even has been known to move off bodily on a false scent."[16]

The simile is anything but flattering, yet it is certainly more objective than the perennial panegyrics on science. For it is objectivity that counts in the end, and it is only a sober, historical objectivity that can build a reliable bridge between science and the humanities and connect the disrupted channels of communication between the two cultures. To achieve this goal is not the responsibility of the historian of science alone. The main duty to talk about physics and about the way physicists work will always rest with the physicist. Such activity on the part of the physicist, however, will become a truly humanistic contribution only if in his reflections the historical point of view is more strongly emphasized than it has been in the past. Technical study alone of the latest results and theories of physics will not give even the most successful physicist a sufficiently wide, culturally effective perspective. His great achievements, if viewed in isolation from the chain of concepts and investigations leading to them, might indeed give him a false sense of the true proportions and permanence of his discovery. The fact that this present age of physics is so fertile in startling discoveries and at the same time so impotent about conveying a balanced and enlightened account of it is worth pondering. Yet, this is only the reverse of the question: Why is it that the educated general public is unaware of the basically human aspects of physical research? The problem was reduced by the British astronomer and philosopher of science, H. Dingle, to what he called the "unselfconscious automatism" of present-day science. According to him, it is due "to the lack throughout its [science's] history of a critical school working within the scientific movement itself and performing the function, or at least one of the functions, which criticism has performed for the literature from the earliest times."[17]

The type of criticism of which Dingle spoke is probably the most potent means of keeping a historical perspective in scientific reflection, keeping alive an awareness of all the aspects of the scientific endeavor, and developing a balanced image of physics. It was a physicist of no mean stature, J. Larmor, who said in referring to the humanistic reintegration of physics that "the study of the historical evolution of physical theories is essential to the complete understanding of their import" and to their coming to life, so to speak. And as he added, "it is the business of criticism to polish them gradually to the common measure of human understanding."[18]

The call for a standing school of criticism is more than a desire for catering to an antiquarian curiosity, or to the pedantry of epistemologists, or to the praiseworthy efforts of educators aiming at a healthy integration of physics into a humanistic framework of life and thought. A historical criticism has fundamental relevance to the progress of physics itself. It is encouraging that there are not lacking prominent theoretical physicists fully aware of this. Thus F. J. Dyson urged present-day physicists to try putting themselves into the position of the physicists of the 1920's, for reasons that are worth being quoted in full:

We must do this, because we must expect changes in the theory in the future. If a theory is taught and learned dogmatically as it stands, without regard to its origins, then it is in danger of becoming fossilized and of being finally an obstacle to further progress. Science, and even quantum mechanics, is not a body of revealed truth to be piously preserved. We must understand what is essential in the theory and what is not, and *the best way to reach such understanding is by studying its history*"[19] (italics mine).

Actually the development of physics has reached a stage where the very problem of concept formation is beginning to force present-day physicists to recognize the relevance of thorough historical information. The physicist who gained from textbooks and research papers his image of physics as an intellectual pursuit in forming new ideas will be greatly amazed at the actual situation that will be unveiled to him by the history of physics. And it is there alone that he will find the answers to a great number of questions vital to the understanding of physics, questions that are unexplainable in a strictly technical exposition. Why, for instance, is one scientific alternative given preference to another, while decisive proofs are still lacking? Or what makes men of science talk in the most realistic style of a concept, or assumption, such as the existence of the ether, not yet proved by any direct evidence? How is it that the strongest advocates of the inductive method are often still looking for facts in support of a theory which is already firmly adhered to? Such questions, strictly connected with various conclusions of physics, could easily be multiplied. They show that the prominence given to a quantitative statement does not rest exclusively on quantitative evidence and on the exclusive use of the quantitative method. And there are questions in even greater number to be answered if an adequate image of the role of physics in human reality is to be obtained. The answers will show that not only during the emergence of classical physics but up to the present, physics implied far more than pointer readings and "pure logic."

Physics, like any other area of human culture, both created and absorbed the temper of a particular age, and the way in which physics concentrated on certain problems and neglected others shows to what degree it can become, as Schrödinger put it, a "fashion of the times." [20] Our own age is no exception to this pattern. In fact, one can point out several traits that dominate both modern physics and the intellectual preferences of our century.[21] Yet, modern physics, in cultivating these traits with great awareness, is both helping them gain general cultural prominence and at the same time borrowing them from the non-scientific wellsprings of cultural orientation. Physics has always needed and used a generous amount of that not always "purely rational" creative imagination and in doing so showed itself strikingly similar to the humanistic branches of learning. Again, physicists, like other intellectuals, are apt to develop stubborn resistance to fresh viewpoints, often entertain preconceived ideas and arbitrary postulates, and may even make their selection of a hypothesis on the basis of rather subjective aesthetic considerations.[22]

Had a more consistent criticism and awareness of its past course accompanied the development of physics, the true history of physics would not have become buried under the heap of half-truths, over-simplifications, clichés, and outright distortions. For what is of relevance is the actual path, the path full of pitfalls, mistakes, and detours, which the discoverers did in fact follow, and not the path they should have followed according to the ever marvelous hindsight of the epigones. The eighteenth century especially teemed with the latter. It was precisely then and there that the historiography of physics became sidetracked soon after its start. Not that the encyclopedists failed to lay down a set of well-balanced principles on how to read scientific history. Indeed one could hardly improve on D'Alembert's precept which listed knowledge, opinions, disputes, and errors as the four great subjects of the general and classified history of the arts and sciences. What is more, D'Alembert emphasized that the study of these four basic topics should issue in a high degree of awareness of the shortcomings inherent in the scientific undertaking. The history of *knowledge*, by which he meant the history of the well-established results of science, should both elate and humble man, he said, "by showing him how little he knows." The history of *errors* on the other hand should teach us "to mistrust both ourselves and others; it shows us moreover, the ways which have led away from the truth and it helps us to find the right pathway."[23]

Yet, an age that so proudly contrasted its own enlightenment with the darkness of the past could hardly bring itself to admitting that grave errors and absurd views are not the exclusive features of allegedly dark ages. Having treated the scientific ideas of earlier centuries as myths, the men of the Enlightenment had to fall victim to the inner logic of such a stance, and as a result, they treated their own ideas on science as dogmas. Thus, while they painted in black and white the relative merits of the pre-Galilean and post-Galilean phases of science, they failed to notice the often gross imperfections of the early phase of classical physics. Complaisance, however, is not the key for unlocking the meaning of history. Nor is the task of achieving a scientifically enlightened and truly human culture so simple a proposition as the Age of Enlightenment would have had us believe. In our own times the expectation of an age of greater awareness, the hope for a unification of the two cultures will remain just a dream as long as those speaking of physics do not acknowledge at least in principle what R. Dugas wrote as a final remark in his *History of Mechanics*:

Nothing is futile in scientific matters, not even the contemplation of the past. For this embodies the lesson of our vagaries, our scruples, our illusions, and our errors. Science did not progress by that harmonious path, the illusion of which is easily created after the event. The direct knowledge of the old works, however they may be outstripped today, can only enrich the perspective of the future which opens before us.[24]

That a historian of physics should reach such a conclusion is natural, but there is no lack, either, of prominent physicists who have voiced similar appraisals of historical studies in physics. The reconstitution of the knowledge and theoretical concepts of the past was in De Broglie's view the means that makes us "privy to the very basis of science and enables us to see it in *perspective*"[25] (italics mine). Again it was the theme of perspective that was stressed by Oppenheimer at the dedication of the Niels Bohr Library of the History of Physics. A historical study of the road taken by physics in recent decades, he stated, will help students of the human predicament to understand better what has befallen mankind through the twentieth-century revolution in physics. They will see, as Oppenheimer noted, "good and bad things both, and they will see them in a wiser and deeper *perspective* than we who act in it"[26] (italics mine).

Perspective is indeed the decisive factor that also insures for physics the emergence of its proper image, through which physics is expected to function as a servant of man in a genuinely human sense. Perspective, a fruit of criticism and awareness, is truly indispensable if physical science as a way of thinking is to shape in a wholesome manner the general cultural attitude. Physics, as a way of thinking, however, is primarily the physicists' way of thinking about their own craft, in which triumphs, failures, and grave questions stand side by side. To see the just proportion of all these, we need historical perspective that will also show in a reliable way how genuinely the mind of physicists was reflected in a particular age by the views the educated public entertained about the meaning of science. For the assimilation of the results of science can take on different forms. For instance, were not fantastic ideas formed in the eighteenth century about the power of Newtonian mechanics — ideas that would have been repudiated by Newton himself? Did not many of the mistakes in social and political experimentation have their origin in "scientific" convictions that were never shared by the most prominent scientists? Were not cultural upheavals, social confusion, and disorders in part motivated by the gross unawareness of the shortcomings of both the results and methods of science?

These are crucial points to keep in mind if we are to benefit from the study of the first scientific revolution with an eye on the second of which we are both spectators and participants. The impact of science on society has many shades. It does not consist merely in destroying traditional views and values, nor is it the exclusive harbinger of the goods man stands in need of. Among its effects are both the beneficial and the harmful, and the latter usually stem from a misunderstanding and misuse of science. The misunderstanding, however, can be based as much on worshiping science as on ignoring it. This is one of those truths about science that should be assimilated as widely as possible if we are to merge the two cultures again and prepare individuals and society alike to meet in a balanced manner the challenge of the second scientific revolution. Such assimilation does not so much require technical information about science as it does a better insight into its history. It can be done without knowing calculus or matrix mechanics. But it cannot and will not be done if physicists remain as unaware of the varied history of their field as they are today. Physicists are the chief spokesmen of the results and implications of physical science and a great deal depends on whether they will hand down their findings with a genuine awareness of the ever-changing flux of their research.

More than goodwill is needed in this respect. It is rather an outlook thoroughly familiar with the recent and remote scientific past that is necessary. What is more, physicists would do well to remember that the study of the history of science is just emerging in Western culture as a subject whose importance for us will be as great as the study of Greco-Roman culture was for the centuries that saw Europe come of age. Thus the physicist might even be forced by changes in the cultural self-reflection to adopt a perspective that constantly keeps in view the historical background of the discussions and conjectures of the present. Yet, it cannot be emphasized enough that no stereotyped history of science will help achieve the goal: the balanced presentation of the current findings and phases of science. For what has been said about science itself is true of its history as well: either worshiping it or belittling it vitiates the purpose of its study. To present the development of science as Whewell did a century ago as the essential process that gives rise to a set of definitive, unalterable truths is unhistorical, to say the least. Such a view hardly does justice to the manifold abilities of the human intellect; nor does it help the intended humanistic integration of science. "That every great advance in intellectual education has been the effect of some considerable scientific discovery, or a group of discoveries"[27] is again a statement that is simply not borne out by any judicious account of history. Nor is what G. Sarton believed to have found in science any closer to truth: the only source of statements that can develop in a truly progressive fashion and outside which progress can have no definite meaning.[28] Such shallow views can hardly ever suffice in coping with deep cultural problems.

One should not think, however, that the need for a study in depth of the history of physical science is simply a by-product of present-day anguishes over the gap between two cultures. Some fifty years ago, one of the very few who excelled both as a physicist and as a historian of physics outlined the contribution of a historical approach in physics to the intellectual development of a physicist along much the same lines, though with no direct reference to a cultural crisis. According to Duhem, a historical perspective will keep the physicist both from the "mad ambitions of dogmatism" and from the "despair of Pyrrhonian skepticism."[29] The first of these — scientific dogmatism — in both its milder and its cruder forms, was a major factor in fostering the false images of science which in turn have been so effective in transforming the difference between science and the humanities into a cultural crisis. It was dogmatism that prevented so many physicists from developing a keen awareness of the long series of errors, hasty conclusions, and deflated theories so evident in the history of physics. As a result, the unwary physicist could naïvely adopt that cocksure style which, let us remember, is not exclusively a feature of the writings of nineteenth-century physicists. To overcome dogmatism, the lessons of history are needed to keep before the physicist's eyes the sudden demise of universally accepted, highly praised theories and to make him somewhat reluctant to endorse current theories, however successful, as absolutely final. Clearly, if a physicist feels called to discharge his cultural responsibility, that is, to talk of the meaning and significance of physics, he cannot be content, to use Duhem's words, "with knowing physics through the gossip of the moment."[30]

On the other hand, the physicist's study of the history of his field will guard against his considering theories as mere artificial creations of the mind, suitable today and useless tomorrow. History will show him that although theories may quickly replace one another, all that was valuable in one will be incorporated into the next one, and as new theories are adopted, the observational data will be classified in an increasingly more comprehensive manner. Thus the conviction of the physicist will be strengthened that physics is making progress in learning more about the physical world and that physical theory is not just a construct having no real relation to the physical reality. For strict operationalism is just as much an extreme as mechanistic realism, and the smugness and despair they generate respectively can create only an intellectual atmosphere that will widen further the gap between science and the humanities. Whatever other means are available for the physicist to restrain him from extremist statements and attitudes, the study of the history of his subject is no doubt one he can hardly dispense with. As Duhem wrote: "Every time the mind of the physicist is on the point of going to some extreme, the study of history rectifies him by means of an appropriate correction. . . . History thus maintains him in that state of perfect equilibrium in which he can soundly judge the aim and structure of physical theory."[31]

The equilibrium of which Duhem spoke can also be defined as an awareness of the simultaneous existence of two main lines in the progress of physics, one convergent and one divergent. To search for convergence is the very soul of any scientific quest, and faith in the basic unity or correlation of all phenomena and laws of nature is an attitude no physicist can live without. During its history, physics witnessed the merger of apparently wholly disparate fields of research, and there were even times when hopes were soaring about a not-too-distant convergence of all partial results. One of these times was the closing decades of the nineteenth century. In drawing the picture of the ultimate physical theory, Kelvin spoke of the "many different roads converging towards it from all sides," though he added that "this consummation may never be reached by man."[32] Michelson, for one, sounded far more optimistic. To him this consummation, the elucidation of the nature of the ether, was close at hand. "The day," he said, "seems not far distant when the converging lines from many apparently remote regions of thought will meet on this common ground," and all branches of physics "will be marshaled into a single compact and consistent body of scientific knowledge."[33]

In physics, however, experiments and theory do not always point in the same direction, nor do they always converge. At times their divergence is nothing short of violent. As J. J. Thomson pointed out in 1930, the additions physics makes to man's knowledge do not become "smaller or less fundamental or less revolutionary, as one generation succeeds another." The progress of physics is not like a convergent series, where only the first terms count. Physics, he warned, "corresponds rather to the other type of series, called divergent, where the terms which are added one after another do not get smaller and smaller, and where the conclusions we draw from the few terms we know, cannot be trusted to be those we should draw if further knowledge were at our disposal."[34] Thomson probably would have been reluctant or perhaps even incapable of forming such a view forty years earlier. He, like all his colleagues, needed the unique experience of seeing the converging lines of physics suddenly veer away from one another and new avenues open up, the direction of which no one could guess with any certainty.

Living through a major turn in physics, however, is not the only means of recognizing this basic double aspect of the growth of physical science. Duhem, the physicist and the penetrating historian of science, wrote clearly of this tension at a time when the emergence of sharply divergent lines was far from being fully evident. In 1906 he referred to the often conflicting results of the experimenter and the theorist and characterized the development of physics as a "continual struggle between nature that does not tire of providing and reason that does not wish to tire of conceiving."[35] It is in this unfinished struggle that one might find the most balanced aspect of physics as an intellectual quest. This aspect is worlds apart from that spurious atmosphere that knows only of triumphs and "final" results in physics and that even today befogs the vision of many scientists and of the general public about science.

In all likelihood, without a deeper wading into the study of the history of physics, the fostering of this false atmosphere will go on. It is highly improbable that the epigones, the successors of any major figure in physics, will change heart. They will, by transferring scientific theories into absolute certainties, continue the process of "dehumanizing" science by making it superhuman. They will have for the most part a much more rigid point of view than had the originators of the major physical theories. They were the ones who, as Newton bitterly remarked, spoke of his views on the corporeity of light "with absolute positiveness," although he had used the word perhaps.[36] Again they were the ones who spoke of the action-at-a-distance in a tone that had little in common with Newton's cautious attitude on this point.[37] They offered the greatest resistance to the Theory of Relativity and today they cry shame when the Copenhagen interpretation of quantum mechanics is questioned. In their eagerness to possess something definitive, they are unable to perceive that a conclusion, however brilliant, might be but a link in the long chain of successive approximations toward the complete truth. But this is the type of intellectual detachment that epigones usually fail to learn from the great masters. "They are," as De Broglie once noted, "rash disciples . . . blinded by uncritical enthusiasm."[38]

Unfortunately these gentlemen will always remain vociferous and will often succeed in creating the impression that their conclusions represent the true status in science. The history of physics offers many classic cases where the non-scientific attitude of "disciples" is quite unmistakable, and the study of such cases might very well give the physicist a "feel" for recognizing similar patterns occurring in our days. It was reflecting upon one such case — the difference of attitude between Newton and his successors — that made Einstein remark: "Newton himself was better aware of the weaknesses inherent in his intellectual edifice than the generation of learned scientists which followed him. This fact has always aroused my deep admiration."[39]

The distortion of the image of physics by the far-fetched statements of some physicists and the cultural damage caused thereby is only compounded by the intellectual mischief of those whom Tait once called "the quasi-scientific writers." As regards their unscientific scripts, Tait felt impelled to warn in a most serious vein: "Eschew popular science, whose dicta are pernicious just in proportion as they are the outcome of presumptuous ignorance."[40] Stokes also felt it important to deplore the fact that "there may be wild scientific conjectures put forward by some, chiefly those whose science is only at second hand, as if they were well established scientific conclusions."[41] Such dubious, self-appointed spokesmen of science, however, were not restricted to the Victorian age of physics. The principate of Newton was teeming with them, and they seemed to thrive best when startling discoveries in physics came, as in our age, thick and fast. Their success is largely due to the high respect that modern culture pays in increasing measure to science. As a result, as Maxwell noted, "the most absurd opinions may become current provided they are expressed in language, the sound of which recalls some well-known scientific phrase." The spectacle of a society flooded with "all kinds of scientific doctrines" prompted him to remind his fellow physicists of their cultural duty "to provide for the diffusion and cultivation not only of true scientific principles but of a spirit of sound criticism, founded on an examination of the evidences on which statements apparently scientific depend."[42]

A resolute stand in this regard on the part of physicists would certainly make its beneficial effect felt in non-popular writings on physics, such as textbooks and research papers. It may sound strange but these too frequently betray the marks of scientism either by omission or in some direct way. Committed to convey technical information, physics textbooks often treat with shocking superficiality the historical aspects of crucial discoveries in physics. One would also look in vain in most textbooks for even a hint about the very serious problems implicit in so many of the basic concepts used in physics. Again, it was in reference to textbooks that P. Frank found it necessary to remark that one cannot prevent the misinterpretation of scientific principles unless "in every statement found in books on physics and chemistry, one is careful to distinguish an experimentally testable assertion about observable facts from a proposal to represent the facts in a certain way by word or diagram."[43] Experimental proofs of new theories are presented time and again without any reference to the uncertainties that make certain "classic" experiments far from conclusive. In this respect the handling by most textbooks of two of the three most important experimental evidences of the General Theory of Relativity should provide a thought-provoking case history.[44] It seems that many authors of physics textbooks view certain experimental proofs not with scientific detachment but with a generous dose of wishful thinking all too eager to speed the "triumph" of a particular theory. At times this attitude comes to the surface in really startling form. Speaking of the necessity of adopting the Theory of Relativity, a widely used textbook on optics declared: "Relativity is . . . now accepted as a faith. It is inadvisable to devote attention to its paradoxical aspects."[45] Relativity, however, like any other physical theory, neither needs nor deserves such faith. The faith presupposed by scientific research is of a different kind. Particular theories can rest only on observations evaluated without the confusing influence of "faith" and "loyalty" given all too often to one or another theoretical construct in physics.

Maxwell's and others' pleas for a spirit of sound criticism should appear all the more meaningful in view of the fact that even scientific papers and monographs seldom contain a detailed assessment of the uncertainties implied in the inferences drawn from observational evidence. As H. Bondi pointed out in connection with astronomical research, such papers often do not make clear enough in each case the connection of a statement to observations, assumptions, and subsidiary hypotheses. To those who think that only "demonstrated facts" appear in scientific papers, it may come as a shock that Bondi pleaded for making general the habit "of clearly stating uncertainties and assumptions."[46] It may sound strange, but even in the twentieth century, the progress of astronomy or physics still demands an advance not only in the refinement of instrumental technique but also in the rigor and precision of scientific reasoning and presentation. Prejudices even about the scientific method are still with us. Observations, as Bondi remarked, are too often regarded as "facts," while theories are taken as "bubbles," and it is not well enough realized that both are liable to error. The separation of facts from hypotheses, of high probabilities from mere plausibilities and conjectures, is not merely an obligation to scientific objectivity. For as Faraday warned more than a hundred years before the emergence of the "two cultures," it is not for his own safe progress but also for that of others that the scientist should be most attentive to this aspect of his work.[47] Clearly, without a more cautious evaluation pervading everything written or said about scientific results — about so-called truths and final conclusions of science — the cultural atmosphere will never be clear enough to let the genuine image of the scientific quest emerge before the public's eyes.

Culture is the art of finding the true proportion in things, situations, and human affairs. Consequently, any ingredient in culture must take its place in the whole according to its own proportion of truth, uncertainty, and error. By ignoring history, it is easy to forget that errors, blind alleys, wrong assumptions, and illusions in physics far outnumbered the successful efforts. Faraday, for one, found that even in the most successful instances not a tenth of his preliminary ideas and conclusions could be carried to satisfactory completion. In his diaries failures were recorded as faithfully as successes, in the conviction that an awareness of failures was indispensable for progress. No one upheld this view more resolutely than Maxwell, whose electromagnetic theory was deeply rooted in the study of Faraday's notes. Comparing the methods of Ampère and Faraday, Maxwell warned his students that it was necessary to study both in order to get a view in depth of a scientific theory. Ampère, said Maxwell, does not show the steps by which he arrived at his perfect demonstration: "He removed all traces of the scaffolding by which he had raised it."[48] Faraday, on the other hand, made known both his successful and his unsuccessful experiments, both his crude and his developed ideas. Therefore, if Ampère's research should be read, to hear Maxwell state it, as a "splendid example of scientific research," Faraday's writings should be studied "for the cultivation of a scientific spirit."[49]

Maxwell made this point even more explicit soon after the opening of the Cavendish Laboratories. "The history of science," he said, "is not restricted to the enumeration of successful investigations. It has to tell of unsuccessful inquiries, and to explain why some of the ablest men have failed to find the key of knowledge, and how the reputation of others has only given a firmer footing to the errors into which they fell."[50] That the study of the story of its errors is an indispensable means of progress for physics itself regardless of more general cultural considerations should of course be highly reassuring to anyone pondering over the finality and limits of the message of physics. Yet, it must be admitted that the story of their errors is rarely told by physicists. Scientific diaries like those of Faraday are few and far between. One of the few exceptions is Kepler's work, where one truly gets a close view of the tortuous paths that led him to the formulation of his famous laws. Those paths and the errors marking them were for him "almost as worthy of wonder" as the great insights of astronomy into celestial matters.[51] Most physicists, however, went no further in talking about their trials and errors than Helmholtz went. Referring to his solution in 1891 of some long-standing problems in mathematical physics, he compared himself to "a wanderer on the mountains who not knowing the path . . . often has to retrace his steps" before finding the access to the summit. "In my works," he added, "I naturally said nothing about my mistake to the reader, but only described the made track by which he may now reach the same heights without difficulty."[52]

Such reticence about one's errors in scientific matters is somewhat strange, since it is usually errors that make possible a start, as Bacon noted long ago: "Truth will sooner come out from error than from confusion."[53] Great discoveries are preceded almost invariably by many tentative steps, the knowledge of which might give the best insight into the intricate process of the formulation of a revolutionary scientific concept. Unfortunately not much is known about the details of "the long and many times curving path"[54] to which Planck referred in his Nobel Prize acceptance speech, where he recalled some of the details of his discovery of the quantum of radiation. Einstein too lifted only a little the veil over the "years of searching in the dark" that according to his own admission preceded the formulation of the General Theory of Relativity.[55] Curiously enough, such short allusions are all that physicists disclose of the "negative" aspects of their research, though most of them would readily agree with Davy's opinion: "The most important of my discoveries have been suggested to me by my failures."[56]

Errors have evidently had a more basic role in science than to justify secretiveness about them. A greater awareness of this on the part of both the general public and the scientific community would perhaps obviate much of the disbelief, ridicule, and even contempt that at first often greets many a bold step in science. The same awareness might also be very effective in laying to rest legends that still decorate certain accounts of the history of physics.[57] For an objective image of physics to prevail, awareness both of the errors made by physicists and of the fallacies of the historiography of physics is indispensable. Its absence can only obstruct clear vision and sound perspective.

Illusory and misleading images of the past can be created with little effort and with equally meager information. Hard work and a detached attitude is needed to restore the true features of a distorted past. Furthermore, constant attention is required to keep the hard-earned lessons and startling experiences from falling into oblivion. In fact, major turns and upheavals in physics that once so forcefully reminded complacent generations of the highly revisable character of physical theories can lose their impressiveness with amazing rapidity. The notion of the quantum of energy was not yet thirty years old when, as Rutherford noted in 1929, it was already difficult to realize how "strange and almost fantastic" this new conception of radiation appeared to scientists in the first decade of the century.[58]

Rutherford spoke of his generation, for whom the shock of a lifetime experienced only two decades or so earlier already began to turn into a fading memory. If the passing of a relatively short time could so effectively weaken the impact of great events in those who had lived through them, what then about the younger generation of physicists born in the years when De Broglie, Schrödinger, Heisenberg, and Dirac startled the world of science with their papers? Clearly, a stereotyped account of the major aspects of the birth of modern physics could hardly impart to them the vividness of a great personal experience. For those who have grown up during or after World War II and inherited quantum mechanics as *the* physical theory, it contained no element of surprise. As one of them, F. J. Dyson, disclosed in 1954: "It is extremely difficult to imagine the state of mind of the men who were creating the theory before 1926."[59]

It is of course always difficult to recreate the state of mind of past generations and to recapture vividly their hopes, their beliefs, their brilliance, and also their often stunning blindness. Only a few can do this — those familiar with the writings of the physicists of the remote and recent past and also gifted with the creative talent to conjure up effectively past events and conditions. In a sense an artistic gift is needed, for science is an artifact produced not only by sharp intellect and keen observation but also by emotions, preferences, and all the beauty and limitations of human existence. As an artifact science reflects a state of mind no less than do other branches of creative work, and every stage of science is also a monument to a state of mind represented by past generations. The practical, technical wizardry secured by its discoveries is not the whole story of physics. All too often it is not even its most telling aspect. Speaking of the closing phase of classical physics, P. R. Heyl, an outstanding experimental physicist, said in a memorable phrase that "the most remarkable thing about the late nineteenth-century physics was the state of mind of the physicists."[60] His words, however, apply equally well to any phase of the history of physics, whether it is connected with the names of Aristotle, Galileo, Laplace, and Kelvin, or with those of Einstein, Heisenberg, and Yang. Once this state of mind is laid bare, physics will appear as human as do works of art or political systems. Thus even for the physics of today, a historical approach will be highly illuminating, for the state of mind of the living generation of physicists cannot be properly diagnosed unless set off against the mentality of their predecessors. In such a setting, it might begin to dawn on us what will be perfectly clear fifty or a hundred years from now, that Kelvin and his colleagues were not the last to raise methods and views to the status of sacred cows. From sacred cows, however, only false images of science can be milked, and these will lack the perspective of a truly objective understanding of the achievements of physics.

This is not to be construed as a plea for antiquarianism in science teaching, or as a suggestion that physics textbooks be turned into historical narrations. But parallel to the systematic and technical explanation of physics, there should be available in every department of physics a set of courses geared to shed light on the development of physics, on the changes that took place in the state of mind of physicists, and in general on all the factors that are necessary for the formation of a proper image of physics as a branch of the human quest for understanding. Today our great universities are vying with one another to enhance the stature of their departments of history of science and are steadily increasing the number of science courses aimed at the humanists. At the same time, however, hardly an effective step has been made toward securing a place for the history of physics within the departments of physics. Although it has already become a truism in most branches of learning that their study is inconceivable without historical perspective, the programs of the departments of physics still reflect a concept of physics that has no historical dimension. The spectacle of physicists teaching humanists a "physics with human perspective" is puzzling to say the least. A few physicists are perhaps equipped to come to grips with such a task. But by and large, physicists, before engaging in such a venture, should become familiar through historical studies with that full image of physics that alone can cure humanists of their three main defects in relation to the physical sciences: indifference, hostility, and inordinate admiration. Thus if any change is absolutely necessary in the undergraduate or graduate curriculum to help heal the splitting of one culture into two, this change should be made in the way science, or physics to be specific, is being taught to students of physics. While lack of familiarity of many a physicist with important humanistic topics can be cured only by exposing them to courses in the humanities, such a remedy is neither the most fundamental nor is it without pitfalls. The contrast between the historically oriented presentation of humanistic subjects and the alleged "ahistoricity" of physics might further strengthen the illusion of the physics student that in the humanities everything is "relative" and highly debatable, whereas in physics the truth is independent of the fashion of the times.

Exposure to an objective and full history of what went on in physics and in the minds of physicists is an indispensable condition for securing the emergence of a new generation of physicists possessed of a greater awareness of both the grandeur and limitations of their craft. This awareness physicists must have in a high degree if it is proper to care about human culture as a whole and of physics' organic role in it. The role played by physics and physicists in today's world is taking on ever-increasing importance. Ever more frequently one hears physicists being called a select class in the making of modern culture. As a matter of fact, some classical physicists had actually felt this way. In 1899 Rowland spoke of physicists as a "small and unique body, a new variety of human race . . . an aristocracy . . . of intellect and of ideals." [61] Today, the wondrous effectiveness and esoteric aloofness of the physicists' work evokes frequently among ordinary folk the image of a magical priesthood. One scientist, W. Weaver, describes this popular belief in the following words:

There is a special, small priesthood of scientific practitioners; they know the secrets and they hold the power. The scientific priests themselves are wonderful but strange creatures. They admittedly possess mysterious mental abilities; they are motivated by a strange and powerful code known as the "spirit of science," one feature of which seems to be that scientists consider that they deserve very special treatment by society.[62]

Whether it is appropriate to speak of the body of physicists as a new priesthood is not the point of importance. Physicists by and large would probably find little taste for it, though some of them clearly enjoy being admired by the gullible public as "initiates" in matters both scientific and non-scientific. It was in fact a physicist, J. Franck, who described theoretical physicists as a "kind of international brotherhood, comparable in many respects to a religious order." For as he said, "Their common goal is truth."[63] At any rate, the comparison is not without some plausibility and can be carried without exaggeration as far as finding the counterparts of theologians, monks, and secular clergy in theoretical physicists, research scientists, and engineers, respectively. What is crucial in this comparison was well emphasized by its originator, R. E. Fitch:[64] even a "priesthood," supposedly devoted to facts, observations, and cold logic alone, can at the same time be particularly blind to the true character of some of its cherished tenets. It is well to remember that to Rowland the sacred rule of this "new variety of human race" was that "the scientific mind should never recognize the perfect truth or the perfect falsehood of any supposed theory or observation. It should carefully weigh the chances of truth and error and grade each in its proper position along the line joining absolute truth and absolute error."[65] Yet Rowland and his colleagues held as absolute truths many conclusions of classical physics that have since been overthrown.

Worse, however, was that they could be very dogmatic about it and that they were not at all reluctant to draw from their "absolute" truths peremptory inferences concerning non-scientific matters. Indeed, only a few years after Rowland made the foregoing statement, another distinguished physicist, O. Lodge, felt called upon to warn his fellow scientists that it would be folly "to imitate the old ecclesiastical attitude." "Let us not fall into the mistake of thinking," he told the British Association meeting in Birmingham in 1913, "that ours is the only way of exploring the multifarious depths of the universe, and that all others are worthless and mistaken."[66] That such an admonition was not amiss can perhaps best be gathered from the persistence with which A. N. Whitehead, himself a scientist of no mean stature, kept referring to the dogmatism noticeable among men of science. Scientists, he noted in 1929, are what large segments of the clergy were a few generations earlier, "the standing examples of obscurantism."[67] Ten years later, he again singled out the scientists as the chief representatives of that "self-satisfied dogmatism with which mankind at each period of its history cherishes the delusion of the finality of its existing modes of knowledge."[68] His main reasons for such remarks were based on firsthand experience. He had to find "after escaping the certainty and dogma of ecclesiastics" that "the scientists, from whom he expected an elastic and liberal outlook, were the same people in a different setting."[69]

The reason why "absolute truths," dogmatic attitudes, and even oversensitive reactions to criticism can develop among scientists was spelled out long ago by Bishop Berkeley, that keen observer of the patterns of the "scientific" state of mind feeding on the still fresh synthesis of Newton. Berkeley had already seen how prejudices in favor of a particular system could forestall the sober recognition of paradoxes and perplexities burdening it. He could observe at close range how easily in the younger generation the repetition of inherited views can breed quick familiarity and how "this familiarity at length passeth for evidence." As he noted, for a mind dominated by this sort of "evidence," the "arts and shifts" of a physical or mathematical theory remain largely insignificant and such a mind can hardly ever appreciate words advising a fair measure of caution or doubt about an idolized theory.[70]

This healthy dose of doubt is one of the most valuable benefits that one may derive from the study of the historical record. Yet, to secure it one should learn that record with an unprejudiced mind and with an eye open for the less flattering details too. One of these should be the realization of the fact that our times are no exception to the pattern in which S. Newcomb described around 1900 the relation of the physics of his day to that of the next century. Discussing the recent discoveries of radioactivity, he conjured up the likelihood that there were probably even more startling results in store for science and warned: "Perhaps, before future developments, all the boasted achievements of the nineteenth century may take the modest place which we now assign to the science of the eighteenth century — that of the infant which is to grow into a man."[71] No age of course likes to think of itself as a child. From the seventeenth century on, that is, since modern science has been with us, each century has looked upon itself as the grown-up phase of history. Each century, blind to its blindness, was willing to be astonished only at the blindness of times past. But on each of them Seneca's remarks came to be fulfilled: "The day will come when our children will wonder at our ignorance."[72] The age has yet to come that is possessed of enough maturity not to take itself too seriously at times. We are still captives of a cultural myopia and often forget that several centuries of biased, exaggerated presentation of the results and ideas of science have been a chief factor in perpetuating this self-deception. The modern version of Seneca's statement may indeed claim an astounding degree of validity: "We are indeed a blind race and the next generation, blind to its own blindness, will be amazed at ours."[73] Moderate voices and wise warnings issued by the more responsible representatives of science are still being drowned out by the noise of the peddlers and worshipers of science. The general public is still under the barrage of views that picture the role, power, and wisdom of scientists in a manner thoroughly at variance with the real situation. Hardly anyone recalls how resolutely a prominent physicist, like Lord Rayleigh, protested against the fallacy implied in such contentions: "I do not think that he [the scientist] has a claim superior to that of other educated men, to assume the attitude of a prophet. In his heart he knows that underneath the theories he constructs there lie contradictions he cannot reconcile. The higher mysteries of being, if penetrable at all by human intellect, require other weapons than those of calculation and experiment."[74]

These are humble words indeed that by echoing the sentiments of those who excelled most in natural philosophy might very well serve as a motto for any exposition of the results of physical science to the general audience. In Lord Rayleigh's words one can read those principles of which no scientific discourse can ever be forgetful. The first of these concerns the incompetence of the quantitative method in patently non-quantitative problems and considerations. It is little remembered that a very impressive recognition of this elementary yet often ignored truth can be found in the introduction of the book that first succeeded in putting the theory of limits — that cornerstone of classical physics — on a logically unobjectionable ground. For it was in the opening pages of his *Cours d'analyse* that Cauchy found it highly appropriate to emphasize the fact that whatever the merits of mathematical analysis, it could not "suffice in all branches of reasoning." As he put it, "It would be a grave error to think that one can find certainty in geometrical demonstrations alone or in the testimony of the senses," and he reminded his reader that historical facts, such as the existence of Augustus or Louis XIV, are unquestioningly accepted though never proved by calculus. And he added:

What I have said here of historical events can be applied equally well to a great number of questions in religion, in ethics, in politics. Therefore let us remain convinced that there are truths other than those of geometry and realities other than those of sensible objects. Let us therefore cultivate with fervor the mathematical sciences without trying to apply them beyond their range. And let us not imagine that one could attack the problems of history with mathematical formulas, or that one could try to sanction the principles of morality by the theorems of algebra and calculus.[75]

Berkeley no doubt would have immensely relished these words of wisdom, as well as Cauchy's candid admission that while everybody agreed on who governed France in the seventeenth century, scientists themselves were not in agreement upon, say, the extent of the validity of Maclaurin's theorem. For was it not precisely the awareness of the uncertainties in so many corners of scientific reasoning that prompted Berkeley's apprehensive question: "Whether it might not become men who are puzzled and perplexed about their own principles, to judge warily, candidly and modestly concerning other matters"?[76] A truly timeless question. For has not our own age seen many a physicist claim superior wisdom about major national, cultural, geopolitical, and moral issues clearly unmanageable by the methods of nuclear physics? The dangerous illusions that were indeed created in the public at large by such claims can be measured by the words of E. P. Wigner, who found it appropriate to warn a press conference convened after he was announced a Nobel Prize recipient. Winning the prize, he noted, "does not make me a person of wisdom. . . . It is a great danger if statements of scientists outside their field are taken too seriously."[77]

Yet, even within the field of physics proper the wisdom of physicists falls far short of solving many problems implied in the complexities of the physical universe. Physics as a science is as much a set of precise answers as it is a labyrinth of puzzles, and this is what Lord Rayleigh's reference to the contradictions that continually arise in physical speculations wanted to emphasize. Physics is basically an unfinished business, and the most obvious feature of its theories consists, to quote J. J. Thomson, in their power of suggesting new fields of investigation. All the history of physics, but especially that of the last sixty years, confirms that a physical theory is, in Thomson's words, a tool or policy but not a creed, not "something which it is a heresy to doubt," but rather an instrument and compass "which will lead the observer further and further into previously unexplored regions."[78]

But for how far and how long, no one can tell. Hopes for a quick completion of this journey have always failed to see their fulfillment. Results that were once considered the utmost that man can achieve in a given area of science have fared no better, than Bacon's *Instances of Ultimity or Limit*. Among them he listed gold as the heaviest possible substance, iron as the hardest, and the dog's nose as the best possible detector of odors.[79] The march of science left all these behind and many other far better argued "final" words in science regardless of how unobjectionable they might have seemed in their own day. As time passes on, so increases the range of both the attainment and the tasks of science, and the relentless march of history could justly inspire the comment made by De Broglie:

For the scientist to believe that science is complete, is always as much of an illusion as it would be for the historian to believe that history was ended. The more our knowledge has developed the more nature has appeared to us as possessing an almost boundless wealth in its different manifestations; even in the realm of a science already far advanced, such as physics, we have no reason to think that we have exhausted the treasures of nature or that we are near to having completed an inventory of it.[80]

Whether the intricacies of nature are strictly inexhaustible is not the crucial point in this connection. The never-ending character of the physicist's search derives just as much from the nature of the questions he poses about the riddles of the physical universe. These questions must confine themselves to a narrow sector of what is knowable in the realm of existence, but at the same time they have no way of defining the limits of what is investigable in such a way in the domain of reality. Thus the certitude of science cannot exceed the narrow limits of the scientific answer or truth, which, being partial, has to be at the same time always subject to modification. Humiliating as this consideration might appear to the worshiper of science, it is just as much an integral aspect of the scientific quest as the fact that new discoveries on the whole help science toward better defined and more valid conclusions. But like it or not, a sincere feeling of humility must needs arise from the awareness of the ever-partial character of the answers science can offer. The unwary scientist, the shallow popularizer, and the overawed layman might do well to reflect on what Franklin admitted after making some unexpected discoveries: I have "become a little diffident of my hypothesis, and asham'd that I have express'd myself in so positive a manner. In going on with these experiments, how many pretty systems do we build, which we soon find ourselves oblig'd to destroy! If there is no other use discover'd of electricity, this, however, is something considerable, that it may help to make a vain man humble."[81]

In the past two centuries the pattern of discoveries has not changed much. Still it seems that a mood of humility is more likely to assert itself within scientific circles when, as in our times, unexpected puzzles turn up at an increasing rate, bringing home forcefully the unfinished character of physical research. Thus soon after the announcement of the indeterminacy principle, Bridgman felt impelled to speak of a vision never before envisaged by the physicist, the possibility that the world may ultimately elude him and this, he warned, "must forever keep him [the physicist] humble."[82] In much the same vein De Broglie suggested that the ever-limited nature of scientific knowledge has to inspire the scientist "with some humility in the face of the immense task which always remains to be accomplished."[83] Perhaps scientific history is entering a phase that will be marked, to use R. P. Feynman's words, with the scientist's "intellectual humility" in the face of the unanswerable secrets of the universe, which ought to remain unanswered.[84] Perhaps we shall see reasserted that "instinctive horror of the folly of prediction" that, according to Bridgman, seized the physicists in the late 1920's.[85] Perhaps there will be a decrease in the number of those who time and again permit themselves complacent statements that science is in sight of its final all-inclusive triumph.

Would that "the turn from arrogance to humility," as P. Jordan described the change of mood in the scientific community of our times,[86] may before long become a widely shared cultural asset. Would that genuine humility may soon supersede that veneer of it that consists in cliché references to the littleness of man in the vast universe. The type of humility scientists as such should cultivate ought rather to be anchored in the constant awareness of the limitations of the methods and results of the scientific research. Little is gained with the dubious type of "scientific humility" that rambles far beyond the competence of science by stating, for instance, that the great architect of the vast universe can hardly have any concern for puny man.[87] For is it within the competence of the quantitative method to state, as Jeans did, that the universe is terrifying because of the vast, meaningless distances, because of inconceivably long stretches of time?[88] Is it right for physics or astronomy to declare one distance more "meaningless" than another and thereby intimate a distinction of values? Is it physics' business to conclude that the universe is terrifying "because it appears to be indifferent to life like our own"?[89] Can it follow logically from the physical method that "emotion, ambition and achievement, art and religion all seem equally foreign" to the plan of the universe?[90] Should one forget suddenly that science itself excluded problems of this sort from its domain over three hundred years ago? Does not the gap between science and the humanities arise precisely from confusing respective areas of competence in this way? If there is a valley of humiliation into which man has to go still lower, to recall a remark of Eddington,[91] it is not because of man's discovery of the immense number of island universes, as he thought, but rather because of the high degree of revisability of scientific theories and systematizations. If there is a humility science ought to foster, it should be the realization of the fact, to quote Oppenheimer, "that we know only a tiny part of what we need to know"[92] and that this is one of the basic truths our discoveries teach us every day.

Still that tiny bit of knowledge is extremely precious and truly genuine. None of that strong faith in the reality of the external world, a faith so indispensable for science, can ever derive from a view that pokes fun at the ordering of observations by the rules of logic, as merely being "an artificial procedure" of which man is said to be so inordinately proud as to insist that "its results should be considered as 'laws of nature.'"[93] No sustained progress will ever come forth from the mental atmosphere generated by strict operationalism that should believe, if it is to be consistent, that "logic, mathematics, physical theory, are all only our inventions."[94] The history of physics does not support the contention that "our only justification for hoping to penetrate at all into the unknown with these inventions is our past experience that sometimes we have been fortunate enough to be able to push on a short distance by acquired momentum."[95] On the contrary, history shows that to let such momentum evidence itself, a prior and firm conviction was needed in the objective value and meaning of scientific investigation.

True, it can be argued, as Poincaré did in 1890, that the number of possible scientific explanations of a physical phenomenon is limitless.[96] Yet, as Boltzmann pointed out a few years later, possibly with an eye on Poincaré, these explanations are not equally simple and do not represent the appearances equally well.[97] Had Poincaré succeeded in overcoming his skepticism about the existence of a theory simpler than the rest, he could have, as De Broglie remarked, easily discovered the principle of relativity.[98] It is indeed not without deep significance that this discovery was reserved for one whose reading of the history of science was not hobbled by a debilitating type of relativism. Einstein's was a firm belief "that the history of scientific development has shown that of all thinkable theoretical structures a single one has at each stage of advance proved superior to all the others."[99] Such a system can always be recognized by its high degree of simplicity, and this is, as Einstein remarked, what physics looks for: "the simplest possible system of thought which will bind together the observed facts."[100]

Whether a theory of ultimate simplicity — a theory answering all questions in physics — will ever be achieved, no one can say. Certainly no responsible observer of the recent phase of physics would entertain hopes of any impending final success. Cocksure optimism no longer seems to square with the facts of physics. On the contrary, the course of scientific history has brought out forcefully to what a surprising degree finality is absent in physical research. Physics in all of its main types — organismic, mechanistic, mathematical — has failed to find the definitive key to the complete intelligibility of the physical world. Again, physics appears to be caught up in an endless process in its search to find the final pattern of the material universe either on the cosmic or on the atomic scale. And neither is finality the exclusive benefit that physics can derive from using its most characteristic tool, the relentless quest for greater precision that is no less effective in creating new problems than it is in settling old ones. Physics, as has been amply illustrated, is not the source of final answers for other branches of learning and nowhere does it find itself more drastically hampered than in a scientism that takes a specific phase and type of physics as absolutely final and definitive. All this seems to bear out the view that physics will be properly understood and appraised by those alone who remain aware of the fact that physics is a paradoxical mixture of carefully established results, which work marvelously, and of a chain of never-ending puzzles.

To respond properly to this state of affairs one has to cultivate a state of mind that on the one hand has a firm faith in physics as one of the human intellect's finest lines of advance, but which on the other hand refuses to believe in any type or result of physics as the last word in scientific investigation. To have faith and yet not to believe blindly is paradoxical enough but no more so than to cultivate intellectual humility and boldness at the same time. In fact the lessons provided by the changes at work in physical science have rarely been more conducive than in our times to make one espouse wholeheartedly the attitude that Conant described with these words: "With humility we recognize the vast oceans of our ignorance . . . yet we can see no limits to the future expansion of the 'empire of the mind'."[101] Such indeed is the state of mind that physicists and those speaking of physics should try to develop in themselves.

In imposing such a demand physics is not unique among the various areas of learning. Yet, there is some difference. Physical research leads to observations and laws that are characterized by a special measure of well-defined precision. Consequently, their impact on general cultural reflection and behavior is also marked by a special definiteness. This definiteness accompanies both the true and permanent results of physics and those which, as research forges ahead, will be shown to be patently inadequate. Yet, in the meantime these unsatisfactory results, concepts, and half-truths will exert their hampering impact on both physics and the general framework of human reflections and attitudes. The physicist can take some dubious comfort in the fact that physics has a built-in self-correcting factor that sooner or later will eliminate from its realm spurious ideas, fallacies, and conceptual idols. Yet, does this attitude not betray a distinctively non-scientific smugness? Would physics itself not profit if fashionable ideas could be prevented by a greater sense of criticism from becoming unchallengeable idols in physics? What is more, would not a greater degree of awareness of the essential revisability of physics also blunt the harmful edge of the decisive impact that so many not really definitive results of physics have on human culture and thinking in general? All this will not be achieved, however, in any measurable degree unless many of those who speak about physics undergo a change of mind and develop a more balanced and more universal outlook about the nature of physical science and the relative proportions of its triumphs, failures, and unsolved puzzles, or in short, about what is the *true* image or story of physics.

Only an adequate knowledge of history can afford hopes that future physicists by and large will not be as much the victims of fallacies and fashions as were their predecessors. A proper account of history will show for physics that not only do historians repeat one another but that in large measure history also repeats itself. This means that since history is enacted by men and the history of physics by physicists, all too often the same mental attitudes, biases, follies, presumptions, and illusions also repeat themselves as the history of physics unfolds its new phases. Awareness of the pattern of the past course alone can provide that genuinely humanistic renaissance in the state of mind of the spokesmen of physics if it is deemed meaningful to diminish as much as possible the cultural misuses of physics and to secure its benefits for man. Modern culture, in which the role of physics is paramount, is shaped at least as much by the state of mind of the physicist as by the products of his research. Einstein's remarks on the theoretical physicists — "don't listen to their words, fix your attention on their deeds"[102] — are very much to the point as regards the field of physics proper. In the broad cultural context, however, it is their words, their state of mind, that act as the chief catalyzing factor when it comes to inserting the results of science into the cultural whole.

Physics, like the countless other projects man pursues, is not cultivated for its own sake, but for man's sake. The cultivation of physics is a human act, and as any human act, physics too will retain its full beauty and meaning only insofar as its cultivation is properly coordinated to the wholeness of human reality. For physics does not lack intellectual vistas of arresting beauty, and Boltzmann's comparison of Maxwell's equations to a symphony was far more than an exercise in rhetoric. Physics has meaning, the depth of which was never so strikingly indicated as in Whitehead's remark that only the babe born in the manger created a stir greater than the one caused by the emergence of physical science in the seventeenth century. Yet, great as the beauty of physics is and deep as its meaning ought to be, beauty and meaning can exist only in the coordination of the part to the whole, of particular experiences to the totality and universality of human reality. Thus ultimately both the beauty and the meaning of physics will be secured if the wisdom of Pascal's words are fully appreciated: "People should not be able to say of anyone that he is a mathematician, or a preacher, or an eloquent man, but that he is a man. Only this universal quality pleases me . . . for this universality is the finest thing."[103] Undoubtedly this universality is also the only key to one culture, because culture is harmony, or the coordination of parts in an all-pervading unity.

Abbreviations

Bacon, F., *Works*:	*The Works of Francis Bacon*, edited by J. Spedding, R. L. Ellis, and D. D. Heath (new ed.; London, 1870).
Berkeley, G., *Works*:	*The Works of George Berkeley, Bishop of Cloyne*, edited by A. A. Luce and T. E. Jessop (London: T. Nelson, 1948–57).
Boyle, R., *Works*:	*The Works of the Honorable Robert Boyle*, edited by T. Birch (London, 1772).
Brit. Assn. Rep.:	*Report of the –th Meeting of the British Association for the Advancement of Science* (London, 1833–1938). The year of the respective meeting is given in parenthesis.
Copernicus, N., *Revolutions*:	*On the Revolutions of the Heavenly Spheres* ("Great Books of the Western World"), translated by C. G. Wallis, XVI (Chicago: Encyclopaedia Britannica, 1952).
Descartes, R., *Oeuvres*:	*Oeuvres de Descartes*, edited by C. Adam and P. Tannery (Paris: L. Cerf, 1897–1913).
Diels, H.:	*Die Fragmente der Vorsokratiker*, edited by W. Kranz (6th ed.; Berlin: Weidmann, 1951–52).
Galileo, G., *Dialogue*:	*Dialogue Concerning the Two Chief World Systems*, translated by Stillman Drake (Berkeley: University of California Press, 1953).
Galileo, G., *Discoveries*:	*Discoveries and Opinions of Galileo*, translated with an introduction and notes by Stillman Drake (Garden City, N.Y.: Doubleday, 1957).

Helmholtz, H. von, *PLSS*:   *Popular Lectures on Scientific Subjects,* translated by E. Atkinson (New York, 1873).

Huygens, C., *Oeuvres*:   *Oeuvres complètes de Christian Huygens* (The Hague: M. Nijhoff, 1888–1950).

Kant, I., *Werke*:   *Kant's Gesammelte Schriften* (Berlin: G. Reimer, 1902–55).

Kelvin, *PLA*:   *Popular Lectures and Addresses* (London, 1891–94).

Kepler, J., *Werke*:   *Johannes Kepler, Gesammelte Werke,* edited by W. von Dyck and M. Caspar (Munich: C. H. Beck, 1938–).

Laplace, P. S., *Oeuvres*:   *Oeuvres complètes de Laplace* (Paris: Gauthier-Villars, 1878–1912).

LCL: The Loeb Classical Library.

Leibniz Selections: *Leibniz Selections,* edited by P. P. Wiener (New York: Charles Scribner's Sons, 1951).

Maxwell, J. C., *Sci. Papers*:   *The Scientific Papers of James Clerk Maxwell,* edited by W. D. Niven (Cambridge, 1890).

Newton, I., *Principia*:   *Sir Isaac Newton's Mathematical Principles of Natural Philosophy and His System of the World,* translated by A. Motte; revised and edited with notes by F. Cajori (Berkeley: University of California Press, 1934).

Newton, I., *Opticks*:   Dover reprint of the 4th edition (1730) (New York, 1952).

Voltaire, *Oeuvres*:   *Oeuvres complètes de Voltaire,* edited by L. Moland (Paris, 1877–85).

Notes

CHAPTER I. The World as an Organism

1. *Early Greek Philosophy* (4th ed.; London: A & C Black, 1930), p. v.
2. See Galileo, *Dialogue*, pp. 131, 320–21.
3. W. Heisenberg, *Physics and Philosophy* (New York: Harper, 1962), p. 160.
4. *Ibid.*, p. 166.
5. E. Schrödinger, *What Is Life and Other Scientific Essays* (Garden City, N.Y.: Doubleday, 1956), p. 91.
6. *Ibid.*, p. 101.
7. *Histoire des sciences exactes et naturelles dans l'antiquité gréco-romaine* (Paris: A. Blanchard, 1924), p. 211.
8. In a letter of April 23, 1953, to Mr. J. E. Switzer of San Mateo, California; see D. J. de Solla Price, *Science since Babylon* (New Haven, Conn.: Yale University Press, 1961), p. 15.
9. H. Frankfort and H. A. Frankfort, *The Intellectual Adventure of Ancient Man* (Chicago: University of Chicago Press, 1946), p. 20.
10. *The Cessation of Oracles*, in W. Lloyd Bevan (ed.), *Plutarch's Complete Works*, I (New York: T. Y. Crowell, 1909), 476.
11. Diels 68, B164.
12. *Metaphysics*, 983b.
13. Diels 13, B2.
14. *Metaphysics*, 985b.
15. Diels 44, B11.
16. Diels 47, B1.
17. Diels 67, A14.
18. Diels 68, A49.
19. Diels 68, B6.
20. *Phaedo*, translated by H. N. Fowler, LCL (1914), see especially sections 96–101 Subsequent references will be to this translation.
21. An impressive reconstruction of Anaxagoras' thought, if not of his book, is available in D. E. Gershenson and D. A. Greenberg, *Anaxagoras and the Birth of Physics* (New York: Blaisdell, 1964), pp. 1–51.
22. *Phaedo*, 99C.
23. *Ibid.*, 99D-E.
24. *Laws*, 889.
25. *Timaeus*, 30B.
26. *Ibid.*, 46B-E.
27. See F. M. Cornford, *From Religion to Philosophy* (London: Longmans, Green, & Co., 1912), p. 53.

28. *Laws*, 967.
29. *Physics*, 252b–253a; *Aristotle's Physics*, translated by R. Hope (Lincoln: University of Nebraska Press, 1961). Subsequent references will be to this translation.
30. *Ibid.*, 199a.
31. *On the Parts of Animals*, translated by A. L. Peck, LCL (1937), 641a. Subsequent references will be to this translation.
32. *Ibid.*, 640b.
33. *Ibid.*, 645a.
34. *On the Heavens*, translated by W. K. C. Guthrie, LCL (1960), 294b. Subsequent references will be to this translation.
35. *Physics*, 192b.
36. *On the Heavens*, 268b.
37. *On the Parts of Animals*, 641b.
38. *On the Heavens*, 300b.
39. *Ibid.*, 271a.
40. *Ibid.*, 276b.
41. *Ibid.*, 279a.
42. *Ibid.*, 284a.
43. *Ibid.*, 284b.
44. *Ibid.*, 285a-b.
45. *Ibid.*, 285b.
46. *Ibid.*, 286a-b.
47. *Ibid.*, 287b.
48. *Ibid.*, 288a.
49. *Ibid.*, 290b.
50. *Ibid.*, 292a.
51. *Ibid.*, 293b.
52. *Ibid.*, 295b.
53. *Ibid.*, 297a.
54. *Ibid.*, 301b.
55. *Ibid.*, 302b.
56. *Ibid.*, 303b.
57. *Meteorologica*, 338b, in *The Works of Aristotle Translated into English*, translated by E. W. Webster (Oxford: Clarendon Press, 1923). Subsequent references will be to this translation.
58. *Aristotle's System of the Physical World* (Ithaca, N.Y.: Cornell University Press, 1960), pp. 400–405.
59. *Meteorologica*, 346b.
60. *Ibid.*, 341b.
61. *Ibid.*, 344b.
62. *Ibid.*, 348b.
63. *Ibid.*, 360a.
64. *Ibid.*, 367b.
65. *Ibid.*, 349a.
66. *Ibid.*, 360a.
67. *Ibid.*, 361a.
68. *Ibid.*, 360a.
69. *Ibid.*, 353a.
70. *Ibid.*, 355b.
71. *Ibid.*, 366b.
72. *Ibid.*
73. *Ibid.*, 368a.
74. *Ibid.*, 366b.
75. *Ibid.*, 351a-b.
76. *On Fire*, in *Theophrasti Eresii opera*, edited by F. Wimmer, III (Leipzig, 1862), 50–73.
77. *Ibid.*, p. 51.

78. *Meteorologica*, 345b.
79. *Ibid.*, 340b.
80. *On the Heavens*, 294b.
81. *Ibid.*, 271b.
82. *Ibid.*, 287b.
83. *De generatione et corruptione* (*On Coming-to-be and Passing Away*), translated by E. S. Forster, LCL (1955), 316a.
84. *Meteorologica*, 344a.
85. *Ibid.*, 361b.
86. *Ibid.*, 339a.
87. *Ibid.*, 366a.
88. *On the Heavens*, 306a.
89. *Ibid.*
90. Darwin's comment on reading William Ogle's translation of *The Parts of Animals*. See F. Darwin, *The Life and Letters of Charles Darwin*, III (London, 1888), 252.
91. *From Euclid to Eddington: A Study of Conceptions of the External World* (Cambridge: Cambridge University Press, 1949), p. 46.
92. *On the Parts of Animals*, 641b.
93. *On the Heavens*, 307b.
94. *Ibid.*, 293b.
95. *Physics*, 248b.
96. *On the Heavens*, 304b.
97. *Ibid.*, 294a.
98. *Ibid.*, 301a.
99. *Ibid.*, 311a.
100. "Corollarium de Inani," in *Philoponi in physicorum octo libros commentaria*, edited by H. Vitelli (Berlin, 1888), p. 683, 10–17.
101. *Theophrastus, Metaphysics*, translated by W. D. Ross and F. H. Fobes (Oxford: Clarendon Press, 1929), p. 31.
102. *Ibid.*, p. 39.
103. *On the Heavens*, 294a.
104. *Ibid.*, 310b.
105. *Ibid.*, 271b.
106. Lib. 1, cap. 1, in *Opera astronomica minora*, edited by J. L. Heiberg (Leipzig: B. G. Teubner, 1907), p. 71.
107. *Ibid.*, lib. 2, cap. 7, p. 120.
108. *Ibid.*
109. *De natura deorum*, lib. 2, cap. 22, edited by A. S. Pease, II (Cambridge, Mass.: Harvard University Press, 1958), 601–2.
110. Lib. 3, cap. 15; lib. 6, cap. 14; lib. 6, cap. 24; in *L. Annaei Senecae opera*, edited by C. R. Fickert, III (Leipzig, 1845), 536–38, 654, 668–69.
111. *Ad Lucilium epistularum moralium libri XX*, lib. 7, epist. 3; in *Opera*, edition cited, I, 260.
112. *Confessionum libri XIII*, lib. 4, cap. 16, edited by J. Capello (Turin: Marietti, 1948), p. 125.
113. See especially chap. 72 of part I, "A Parallel between Universe and Man," translated by M. Friedländer (2d ed.; London: G. Routledge, 1919), pp. 113–19.
114. "De disciplinis mathematicis" (1574), in *Tychonis Brahei Danis opera omnia*, edited by J. L. E. Dreyer, I (Copenhagen: In Libraria Gyldendaliana, 1913–29), 156.
115. *On the Loadstone*, translated by P. Fleury Mottelay (New York, 1893), pp. 64–71, 105–15, 308–12.
116. "The ancient opinion that man was Microcosmus, an abstract or model of the world, hath been fantastically strained by Paracelsus and the alchemists, as if there were to be found in man's body certain correspondences and parallels, which should have respect to all varieties of things, as stars, planets, minerals, which are extant in the great world." *Of the Advancement of Learning*, book 2, in *Works*, III, 370.

117. *Dialogue*, p. 462.
118. See, for instance, "On the Principal Parts of the World," in the *Epitome of Copernican Astronomy*, Book 4, part 1, chap. 1, or the "Epilogue concerning the Sun, by Way of Conjecture" at the end of the *Harmonies of the World*.
119. *Epitome of Copernican Astronomy*, Book 1, part 5, chap. 3, "What Are All the Other Indications of a Soul Operating in the Body of the Earth."
120. *The Monadology* (1714), in *Leibniz Selections*, p. 547.
121. *Ibid.*, p. 546.
122. *Ibid.*
123. *Sämmtliche Werke*, edited by B. Suphan, XIII (Berlin, 1887), 68.
124. *Conversations of Goethe with Eckermann and Soret*, translated by J. Oxenford (London, 1850), 145.
125. *Truth and Fiction Relating to My Life*, translated by J. Oxenford, II (London, 1903), 108–10.
126. *Italienische Reise*, in *Sämmtliche Werke*, XXX (Munich: Müller, 1909–32), 309. Subsequent references to this edition of Goethe's works will be referred to as *Werke*.
127. *Tag- und Jahreshefte*, in *Werke*, XXXVIII, 203.
128. *Zur Farbenlehre. Historischer Teil*, in *Werke*, XXII, 145.
129. *Italienische Reise*, in *Werke*, XXX, 358.
130. *Zur Farbenlehre. Historischer Teil*, in *Werke*, XXII, 382.
131. *Zur Farbenlehre. Polemischer Teil*, in *Werke*, XXI, 243.
132. *Schriften zur Naturwissenschaft*, in *Werke*, XXXIX, 91.
133. *Werke*, XX, 21.
134. *Zur Farbenlehre. Historischer Teil*, in *Werke*, XXII, 233.
135. *Zur Farbenlehre. Polemischer Teil*, in *Werke*, XXI, 243.
136. *Schriften zur Naturwissenschaft*, in *Werke*, XXXIX, 92.
137. *Ibid.*
138. *Ibid.*, p. 93.
139. *Conversations of Goethe with Eckermann and Soret*, II, 373.
140. *Bruno, ein Gespräch*, in *Sämmtliche Werke*, IV (Stuttgart, 1859), 271.
141. *Fernere Darstellungen aus dem System der Philosophie*, in *Werke*, IV, 501.
142. *System der gesammten Philosophie und der Naturphilosophie insbesondere*, in *Werke*, VI, 492.
143. In *Sämmtliche Werke*, I (Stuttgart, 1927), 1–12.
144. See R. Wolf, *Geschichte der Astronomie* (Munich, 1877), p. 685.
145. *System der Philosophie. Zweiter Teil, Die Naturphilosophie*, in *Werke*, IX, 150.
146. *Ibid.*, p. 123.
147. *Ibid.*, p. 139.
148. *Ibid.*, p. 375.
149. *Hegel: A Re-examination* (London: Allen, 1958), p. 282.
150. *Ibid.*, p. 281.
151. Fechner, *Zend-Avesta* (Leipzig, 1851–54); see especially chap. 3, "Vergleichende physische Erd- und Himmelskunde," and chap. 5, "Die Erde, unsre Mutter."
152. While Oersted's various addresses show time and again the stamp of Naturphilosophie, his historic paper announcing the deflection of a magnetic needle by a current-carrying wire could not be more factual. Nor did he take kindly to the speculations of some Naturphilosophers who belittled the role of experiments. See, for instance, his "Critical Notice of Steffens' Polemical Journal for the Furtherance of Speculative Physics," reprinted in H. C. Oersted, *The Soul in Nature*, translated by L. Horner and J. B. Horner (London, 1852), pp. 257–88.
153. As amply documented in the monumental study by J. Needham, *Science and Civilization in China*, II: *History of Scientific Thought* (Cambridge: Cambridge University Press, 1956). See especially pp. 496–505 and 572–82. Yet, the author of this most impressive illustration of the impasse of early Chinese science could go as far in his engrossment with organismic concepts as to claim that the future of Western physics depends on its rejuvenation by organismic views emphasized in Chinese intellectual tradition.

CHAPTER II. The World as an Mechanism

1. *Le livre du ciel et du monde*, Book 2, chap. 2, edited by A. D. Menut and A. J. Denomy, *Medieval Studies* 4 (1942):170.
2. *Narratio prima*, translated by E. Rosen, *Three Copernican Treatises* (New York: Dover, 1959), p. 137.
3. Letter to Herwart, Feb. 10, 1605, in *Werke*, XV, 146.
4. M. Caspar, *Kepler*, translated by C. Doris Hellman (London: Abelard-Schuman, 1959), p. 135.
5. W. von Dyck and M. Caspar, "Nova Kepleriana," *Abhandl. Bayer. Akad. Wis.* 31, No. 1 (1927):107.
6. *Discoveries*, p. 63.
7. *History of the Inductive Sciences*, I (New York, 1858), 338.
8. *Le monde ou traité de la lumière*, in *Oeuvres*, XI, 10.
9. *Ibid.*, p. 11.
10. Letter to Mersenne, in *Oeuvres*, II, 497.
11. *Les principes de la philosophie*, I. 28, in *Oeuvres*, IX, 37.
12. *Ibid.*, II, 53, in *Oeuvres*, IX, 93.
13. *La dioptrique*, discours II, in *Oeuvres*, VI, 98.
14. *Les météores*, discours VIII, in *Oeuvres*, VI, 340.
15. Letter to Mersenne, Oct. 11, 1638, in *Oeuvres*, II, 380.
16. *Harmonie universelle*, II: *Traité de méchanique* (Paris, 1635), 112.
17. *Ibid.*
18. *Traité de physique*, I (4th ed.; Paris, 1682), 166.
19. "Remarques de Huygens sur *La vie de Descartes* par Baillet," in M. V. Cousin, *Fragments philosophiques*, II (5th ed.; Paris, 1866), 118.
20. *Traité de la lumière* (1690), in *Oeuvres*, XIX, 461.
21. In *Oeuvres*, VII, 298.
22. *About the Excellency and Grounds of the Mechanical Hypothesis*, in *Works*, IV, 68–69.
23. *Ibid.*, p. 73.
24. Letter to Huygens, Dec. 29, 1691; in *Leibniz Selections*, p. xxv.
25. *Nouveaux essais*, Book IV, chap. xii, in J. E. Erdmann (ed.), *God. Guil. Leibnitii opera philosophica* (Berlin, 1840), p. 383.
26. *New Experiments Physico-Mechanical, touching the Spring of the Air*, in *Works*, I, 12.
27. *Origin of Forms and Qualities according to the Corpuscular Philosophy*, in *Works*, III, 1–113.
28. *Experiments, Notes, etc., about the Mechanical Origin or Production of Divers Particular Qualities*, in *Works*, IV, 230–354.
29. *Micrographia* (1665), preface; reprinted in R. T. Gunther (ed.), *Early Science in Oxford*, XIII (Oxford: Clarendon Press, 1938), p. g.
30. Quoted in J. C. Gregory, *A Short History of Atomism from Democritus to Bohr* (London: Black, 1931), p. 35.
31. *Principia*, p. 21.
32. See L. T. More, *Isaac Newton: A Biography* (New York: C. Scribner's Sons, 1934), pp. 294–95.
33. *Principia*, p. xviii.
34. See Leibniz's letter of Feb. 10, 1711, to Hartsoeker in *Memoirs of Literature*, IV (2d ed.; London, 1722), 453–60.
35. "Four Letters from Sir Isaac Newton to Doctor Bentley, containing some Arguments in Proof of a Deity," in *The Works of Richard Bentley*, III (London, 1838), 212.
36. This letter was first printed in Boyle's *Works* edited by Thomas Birch, I (London, 1744), 70.
37. Newton's letter to the editor of *Memoirs of Literature*, in D. Brewster, *Memoirs of the Life, Writings and Discoveries of Sir Isaac Newton*, II (Edinburgh, 1855), 283.

38. *Memoirs of Literature,* pp. 457–58.
39. *Principia,* p. 400.
40. *The Science of Mechanics,* translated by T. J. McCormack (La Salle, Ill.: Open Court, 1960), p. 226.
41. M. Delambre, "Notice sur la vie et les ouvrages de M. le comte J. L. Lagrange," in *Oeuvres de Lagrange,* I (Paris, 1867), xx.
42. "Über ein neues allgemeines Grundgesetz der Mechanik" (1829), in *Werke,* V (Göttingen, 1877), 25.
43. *Essai de cosmologie,* in *Oeuvres de Maupertuis,* I (Lyon, 1756), 45.
44. "Account of the Changes that have happened during the last Twenty-five Years in the relative Situation of Double-stars," *Phil. Trans.* (London) 93(1803):340.
45. *Brit. Assn. Rep.* (1850), p. xxxii.
46. *A Philosophical Essay on Probabilities,* translated by F. W. Truscott and F. L. Emory (New York: Dover, 1951), p. 4.
47. *The New Organon,* Book 2, chap. 20, in *Works,* IV, 149.
48. *A History of the Theories of Aether and Electricity,* I (London: Nelson, 1951), p. 40.
49. "An Inquiry concerning the Source of Heat which is excited by Friction," *Phil. Trans.* (London) 88 (1798):99.
50. *Ibid.,* p. 80.
51. *Ibid.,* p. 100.
52. "An Essay on Heat, Light and the Combinations of Light" (1799), in *The Collected Works of Sir Humphry Davy,* II (London, 1839), 14.
53. *Encyclopaedia Britannica,* XI (8th ed.; Boston: Little, 1860), 260.
54. "On the Conservation of Force" (1862), in *Popular Lectures on Scientific Subjects,* translated by E. Atkinson (New York, 1873), p. 342.
55. "The Place of Modern Physics in the Mathematical View of Nature" (1910), in *A Survey of Physical Theory,* translated by R. Jones and D. H. Williams (New York: Dover, 1960), p. 28.
56. "Grundzüge einer Theorie der Gase," *Pogg. Ann.* 99 (1856):316.
57. *Sci. Papers,* I, 377.
58. "On Physical Lines of Force" (1861–62), in *Sci. Papers,* I, 486.
59. *Electric Waves,* translated by D. E. Jones (London, 1893), p. 20.
60. "A Dynamical Theory of the Electromagnetic Field," in *Sci. Papers,* I, 564.
61. "On the Dynamical Evidence of the Molecular Constitution of Bodies" (1875), in *Sci. Papers,* II, 419.
62. *Ibid.*
63. *Ibid.*
64. *Ibid.,* p. 418.
65. "On the Conservation of Force. A Physical Memoir" (1847), translated by J. Tyndall, in *Taylor's Scientific Memoirs* (London, 1853), p. 117.
66. *Ibid.,* pp. 117–18.
67. "On the Aim and Progress of Physical Science" (1869), in *Popular Lectures on Scientific Subjects,* p. 375.
68. *Principles of Mechanics,* translated by D. E. Jones and J. T. Walley (London, 1899), p. xxi.
69. *Applications of Thermodynamics to Physics and Chemistry* (New York, 1888), p. 1.
70. *Treatise on Thermodynamics,* translated by A. Ogg (New York, 1897), p. ix.
71. See L. Poincaré, *The New Physics and Its Evolution* (New York; Appleton, 1908), p. 10.
72. *Ibid.,* p. 9.
73. "On the Aim and Progress of Physical Science," p. 375.
74. "On the Relations between Light and Electricity" (1889), in *The Miscellaneous Papers of Heinrich Hertz,* translated by D. E. Jones (London, 1896), p. 314.
75. *Vorlesungen über mathematische Physik* (Leipzig, 1876), p. 1.
76. *Über das Ziel der Naturwissenschaften* (Heidelberg, 1865), p. 24.
77. See "On Faraday's Lines of Force," in *Sci. Papers,* I, 156.

78. *A Treatise on Electricity and Magnetism* (3rd ed.; Oxford, 1892), II, 175.
79. *Physik des Aethers auf elektromagnetische Grundlage* (Stuttgart, 1894), p. vii.
80. "The Highest Aim of the Physicist" (1899), in *The Physical Papers of Henry Augustus Rowland* (Baltimore: Johns Hopkins Press, 1902), p. 673.
81. Sir Rayleigh [R. J. Strutt], *The Life of Sir J. J. Thomson* (Cambridge: Cambridge University Press, 1942), p. 203.
82. Letter to Weber, March 19, 1845, in *Werke*, V, 629.
83. *Scientific Writings*, I (Washington, D. C., 1886), 305.
84. "Notes of Lectures on Molecular Dynamics and the Wave Theory of Light," stenographic report by A. S. Hathaway (Baltimore, 1884), p. 131.
85. *Baltimore Lectures* (Baltimore: Publication Agency of Johns Hopkins University, 1904), p. 187.
86. *Ibid.,* p. 14.
87. "Notes of Lectures on Molecular Dynamics," p. 132.
88. *Baltimore Lectures,* p. 270.
89. *Ibid.,* p. vii.
90. "On Atmospheric Electricity," in *Reprint of Papers on Electrostatics and Magnetism* (2d ed.; London, 1884), p. 224.
91. *Baltimore Lectures,* p. 159.
92. In *Oeuvres*, XXII, 451.
93. *Lettres d'Euler à une princesse d'Allemagne*, edited by E. Saisset, I (Paris, 1866), 232.
94. *Die Geschichte und die Wurzel des Satzes von der Erhaltung der Arbeit* (Prague, 1872), p. 32.
95. *Exposition du système du monde*, Book IV, chap. 17, in *Oeuvres*, VI, 343.
96. "On the Origin of Force," *The Fortnightly Review* 1 (1865):438.
97. "Atom," in *Sci. Papers*, II, 474.
98. *Ibid.,* p. 476.
99. See Maxwell's review of *An Essay on the Mathematical Principles of Physics* by J. Challis, in *Sci. Papers*, II, 341.
100. Introduction by H. von Helmholtz to H. Hertz, *Principles of Mechanics*, translated by D. E. Jones and J. T. Walley (London, 1899), p. xvi.
101. "On Atmospheric Electricity," p. 224.
102. B. Stewart and P. G. Tait, *The Unseen Universe* (9th ed.; London, 1880), p. 146.
103. "A Dynamical Theory of the Electromagnetic Field" (1864), in *Sci. Papers*, I, 571.
104. "A Discourse of the Nature of Comets" (1682), in *Posthumous Works* (London, 1705), pp. 184–85.
105. *Principia*, p. 547.
106. *Lettres d'Euler à une princesse d'Allemagne*, I, 232.
107. "On the Theories of the Internal Friction of Fluids in Motion and of the Equilibrium and Motion of Elastic Solids," in *Mathematical and Physical Papers of George Gabriel Stokes*, I (Cambridge, 1880), 121, 126.
108. *Baltimore Lectures*, p. 11.
109. "On the Laws of Crystalline Reflexion and Refraction," *Trans. Roy. Irish Acad.*, 18 (1837):68.
110. For their exchange of letters, see L. Campbell and W. Garnett, *The Life of James Clerk Maxwell* (London, 1892), pp. 392–96.
111. *Sci. Papers*, II, 775.
112. *Heat a Mode of Motion* (4th ed.; New York, 1873), p. 258.
113. *Ibid.,* p. 351.
114. *Baltimore Lectures*, p. 533.
115. "Elasticity Viewed as Possibly a Mode of Motion" (1881), PLA, I, 146.
116. "On the Relation between Light and Electricity," p. 314.
117. *Ibid.*
118. *Ibid.*
119. "Presidential Address to the Mathematical and Physical Section of the British Association," *Nature* 38(1888):446.

120. *Ibid.*, p. 448.
121. *Ibid.*
122. *Principles of Mechanics*, p. xvi.
123. "Presidential Address to the Mathematical and Physical Section," p. 448.
124. "The Highest Aim of the Physicist," p. 672.
125. *Light Waves and Their Uses* (Chicago: University of Chicago Press, 1902), p. 162.
126. "The Ether and Its Functions," in O. Lodge, *Modern Views of Electricity* (London, 1889), p. 358.
127. W. F. G. Swann, "What Has Become of Reality in Modern Physics," *Sigma Xi Quarterly* 27 (1939):28.
128. *La science et l'hypothèse* (Paris: Flammarion, 1902), p. 200.
129. "The Discharge of a Leyden Jar," in O. Lodge, *Modern Views of Electricity*, pp. 382–83.
130. "On the Relations between Light and Electricity," p. 327.
131. *The Progress of Physics during 33 Years (1875–1908)* (Cambridge: Cambridge University Press, 1911), pp. 11–12.
132. Letter to Heaviside, Feb. 4, 1889; see *A History of the Theories of Aether and Electricity*, I, 148.
133. S. P. Thompson, *The Life of William Thomson*, II (London: Macmillan, 1910), 984.
134. *Ibid.*, p. 1013.
135. *Baltimore Lectures*, p. 486.
136. *Ibid.*, p. 492.
137. "Autobiographical Notes," in P. A. Schilpp (ed.), *Albert Einstein: Philosopher-Scientist* (Evanston, Ill.: Library of Living Philosophers, 1949), p. 21.
138. "Presidential Address to the British Association" (1871), *PLA* II, 163.
139. "Elasticity Viewed as Possibly a Mode of Motion" (1881), *PLA* I, 146.
140. "Presidential Address to the Institution of Electrical Engineers" (1889), in *Mathematical and Physical Papers*, III (London, 1890), 484.
141. *Ibid.*, p. 511.
142. "On the Dynamical Evidence of the Molecular Constitution of Bodies," (1875), in *Sci. Papers*, II, 433.
143. *Phil. Mag.* (London) 49 (1900):118.
144. *Baltimore Lectures*, p. 527.
145. *The Theory of Sound* (2d ed.; London, 1894–96).
146. See G. Hennemann, *Naturphilosophie im 19. Jahrhundert* (Freiburg: K. Alber, 1959), p. 113.
147. *Where Is Science Going?*, translated by J. Murphy (New York: Norton, 1932), p. 214.
148. Thompson, *The Life of William Thomson*, II, 1014.
149. "The Unity of the Physical Universe" (1908), in Planck, *A Survey of Physical Theory*, pp. 23–25.
150. "On the Scientific Use of the Imagination" (1870), in J. Tyndall, *Fragments of Science* (2d ed.; London, 1871), p. 136.
151. *The Science of Mechanics*, p. 559.
152. *Travaux du Congrès International de Physique, Paris 1900*, edited by C.-E. Guillaume and L. Poincaré, IV (Paris: Gauthier-Villars, 1901), 7.
153. "Über die Prinzipien der Mechanik" (1900), in L. Boltzmann, *Populäre Schriften* (Leipzig: Barth, 1905), p. 317.
154. *La théorie de la physique chez les physiciens contemporains* (Paris: Alcan, 1907), pp. 16–17, 39.
155. *Opticks*, p. 398.
156. Thompson, *The Life of William Thomson*, II, 1125.
157. *Ibid.*, 1077.
158. "Autobiographical Notes," in *Albert Einstein: Philosopher-Scientist*, p. 19.
159. As stated by Henry Power, an English physician and member of the nascent Royal Society, in *Experimental Philosophy* (London, 1664), p. 192.

CHAPTER III. The World as a Pattern of Numbers

1. A. S. Eve, *Rutherford* (New York: Macmillan, 1939), p. 223.
2. *Ibid.*, p. 226.
3. *Ibid.*, p. 221.
4. See the Obituary Notice of N. R. Campbell, in *Proc. Phys. Soc.* (London) 62B (1949):857.
5. *Atomic Theory and the Description of Nature* (Cambridge: Cambridge University Press, 1934), pp. 103–4.
6. *Nobel Lectures: Physics 1942–1962* (Amsterdam: Elsevier, 1964), p. 27.
7. P. A. Schilpp (ed.), *Albert Einstein: Philosopher-Scientist* (Evanston, Ill.; Library of Living Philosophers, 1949), pp. 45–47.
8. *Timaeus* 36D.
9. *De communi mathematica scientia*, chap. 23, edited by N. Festa (Leipzig, 1891), p. 73.
10. *Ibid.*, chap. 32, p. 94.
11. *The New Organon*, Book 2, chap. 8, in *Works*, IV, 126.
12. *Ibid.*, Book 1, chap. 96, in *Works*, IV, 105.
13. Letter to Herwart von Hohenburg, April 9, 1599, in *Werke*, XIII, 309.
14. *Harmonice mundi*, Book 4, chap. 1, in *Werke*, VI, 223.
15. Letter to Mersenne, July 27, 1638, in *Oeuvres*, II, 268.
16. *Les principes de la philosophie*, in *Oeuvres*, IX, 101.
17. Letter to Mersenne, March 11, 1640, in *Oeuvres*, III, 39.
18. *Discours sur la méthode*, Part 6, in *Oeuvres*, VI, 69.
19. *Dialogue*, p. 207.
20. *Ibid.*
21. *The Assayer* (1623), in *Discoveries*, p. 238.
22. *Dialogue*, p. 203.
23. *Ibid.*, p. 406.
24. Letter to Herzog Johann Friedrich, Oct. 1671, in *Leibniz Selections*, p. xx.
25. *Works*, IV, 102.
26. *Works*, IV, 115.
27. *A Fair, Candid and Impartial State of the Case between Sir Isaac Newton and Mr. Hutchinson* (Oxford, 1753), p. 72.
28. *Defence of Free Thinking in Mathematics*, in *Works*, IV, 116.
29. *The Analyst*, in *Works*, IV, 68.
30. See his *Cours d'analyse* (Paris, 1821), pp. 26–33.
31. *Werke*, IV, 470.
32. Translated by A. Freeman (Cambridge, 1878), p. 7.
33. Letter to Legendre, July 2, 1830, in *C. G. J. Jacobi's Gesammelte Werke*, I (Berlin, 1881), 454.
34. For Bode's account of this law, see H. Shapley and H. E. Howarth (eds.), *A Source Book in Astronomy* (New York: McGraw-Hill, 1929), p. 180.
35. For the conceptual background of Balmer's formula, see the reminiscences by E. Hagenbach's son, A. Hagenbach, "J. J. Balmer und W. Ritz," *Naturwissenschaften* 9 (1921):451–55.
36. Kronecker made this statement in 1866 at the Naturforscher Versammlung in Berlin; see H. Weber, "Leopold von Kronecker," *Jahresbericht der Deutschen Mathematiker Vereinigung* 2 (1891–92):19.
37. Letter to Sonja Kowalewsky, March 24, 1855; see G. Mittag-Leffler, "Weierstrass et Sonja Kowalewsky," *Acta Mathematica* 39 (1923):194.
38. "On Series in Spectra," *Sci. Proc. Roy. Soc.* (Dublin) 11 (1907):181–83.
39. "The Constitution of the Solar Corona," *Monthly Notices Roy. Astr. Soc.* 72 (1912):677–93.
40. *Die Bedeutung der Röntgenstrahlen für die heutige Physik* (Munich: R. Oldenbourg, 1925), p. 11. One might add that some perceiving humanists quickly noticed this aspect of the new orientations of physics. In 1926 the Austrian playwright, Hugo von Hoffmanstahl, made the following comment in an address

delivered at a meeting devoted to the promotion of classical studies in secondary education: "In the central region of the natural sciences . . . where only that exists for us that can be measured, precisely there do we find arising out of the fog of theories, like the light of the ancient and eternal day, Plato's vision of a numerical theory of nature and within it the wisdom of Pythagoras." "Vermächtnis der Antike," in *Gesammelte Werke*, IV (Frankfurt am Main: Fischer, 1955), 318.

41. On these details, see the Nobel Prize acceptance speech by J. Hans D. Jensen, "The History of the Theory of Structure of the Atomic Nucleus," *Science* 147 (1965):1419–23.

42. Translated by H. L. Brose (London: Methuen, 1922), p. 312.

43. *From Euclid to Eddington: A Study of Conceptions of the External World* (Cambridge: Cambridge University Press, 1949), p. 186. The *Fundamental Theory* was published by the same press in 1946.

44. On these details, see A. Vibert Douglas, *The Life of Arthur Stanley Eddington* (London: Nelson, 1957), pp. 170–71.

45. Maestlin's letter of May, 1596, to the Tübingen Senate, in Kepler's *Werke*, XIII, 84.

46. *Space, Time and Gravitation* (Cambridge: Cambridge University Press, 1929), p. 198.

47. *Traité de la lumière*, in *Oeuvres*, XI, 47.

48. *Relativity Theory of Protons and Electrons* (New York: Macmillan, 1936), p. 327.

49. *Philosophy of Mathematics and Natural Science* (Princeton, N.J.: Princeton University Press, 1949), p. 289. The earliest statement of Weyl on this subject is in his article, "Eine neue Erweiterung der Relativitätstheorie," *Annalen der Physik* 59 (1919), p. 129.

50. *Open Vistas: Philosophical Perspectives of Modern Science* (New Haven, Conn.: Yale University Press, 1961), p. 216. The first detailed consideration of the role of pure numbers in cosmological theory is in the paper of P. A. M. Dirac, "A New Basis for Cosmology," *Proc. Roy. Soc.* (London) 165A (1938):199–208.

51. "An Empirical Mass Spectrum of Elementary Particles," *Progr. Theor. Physics* 7 (1952):595–96.

52. "A Periodic Table for Fundamental Particles," *Ann. N.Y. Acad. Sci.* 76 (1958):1–16.

53. *The Beginning and End of the World* (London: Oxford University Press, 1942), p. 25.

54. *A History of the Theories of Aether and Electricity*, II (London: Nelson, 1952), 192.

55. "Space and Time," in A. Einstein *et al.*, *The Principle of Relativity*, translated by W. Perrett and G. B. Jeffrey (London: Methuen, 1923), p. 91.

56. *The Mysterious Universe* (New York: Macmillan, 1930), p. 106.

57. *Ibid.*, p. 138.

58. *Ibid.*, p. 144.

59. *Traité de calcul différentiel et de calcul intégral*, I (Paris, 1864–70), i.

60. *Leçons sur les coordonnées curvilignes et leurs diverses applications* (Paris, 1859), pp. 367–68.

61. Translated by W. K. Clifford, in *Mathematical Papers by William Kingdon Clifford*, edited by R. Tucker (London, 1882), p. 69.

62. "Akademische Antrittsrede," in *Mathematische Werke*, I (Berlin, 1894), 225.

63. *The Theory of Groups and Quantum Mechanics*, translated by H. P. Robertson (London: Methuen, 1931), p. xxi.

64. See D'Arcy W. Thompson, *On Growth and Form* (new ed.: Cambridge: Cambridge University Press, 1948), pp. 10–11.

65. "The Six Gateways of Knowledge" (1883), in *PLA*, I, 273.

66. "The Highest Aim of the Physicist" (1899), in *The Physical Papers of Henry Augustus Rowland* (Baltimore: John Hopkins Press, 1902), pp. 674–75.

67. See B. Jaffe, *Michelson and the Speed of Light* (Garden City, N.Y.: Doubleday, 1960), p. 93.

68. *The Aims of Mathematical Physics* (Oxford: Clarendon Press, 1929), p. 9.
69. From a paper of Maxwell on determinism and contingency published posthumously in L. Campbell and W. Garnett, *The Life of James Clerk Maxwell* (London, 1882), p. 444.
70. "On the Relations between Light and Electricity" (1889), in *The Miscellaneous Papers of Heinrich Hertz*, translated by D. E. Jones (London, 1896), p. 318.
71. *Électricité et optique* (2d ed.; Paris: Gauthier-Villars, 1901), p. iii.
72. *The Aim and Structure of Physical Theory*, translated by P. P. Wiener (Princeton, N.J.: Princeton University Press, 1954), p. 91.
73. See C. S. Hastings, "Josiah Willard Gibbs," *Biographical Memoirs*, VI (Washington, D. C.: National Academy of Sciences, 1909), 392.
74. "Atomic Theory and Mechanics," *Atomic Theory and the Description of Nature* (Cambridge: Cambridge University Press, 1934), pp. 34–35.
75. *Nuclear Physics* (London: Methuen, 1953), p. 30.
76. *The Principles of Quantum Mechanics* (2d ed.; Oxford: Clarendon Press, 1935), p. 12.
77. *Elementary Wave Mechanics* (Oxford: Clarendon Press, 1945), p. 70.
78. *The Principles of Quantum Mechanics*, p. 10.
79. *The Physical Principles of the Quantum Theory*, translated by C. Eckart and and F. C. Hoyt (Chicago: University of Chicago Press, 1930), p. 11.
80. *Brit. Assn. Rep.* (1878), p. 31.
81. "Linear Associative Algebra" (1870), published in the *American Journal of Mathematics* 4 (1881):97.
82. *An Introduction to the Study of Experimental Medicine*, translated by H. C. Green (New York: Macmillan, 1927), p. 41.
83. *Critical, Historical and Miscellaneous Essays*, IV (New York, 1871), 302.
84. "Academical Education" (1857), *Orations and Speeches*, III (Boston, 1859), 514.
85. *Adventures of Ideas* (New York: Macmillan, 1933), p. 295.
86. See E. T. Bell, *Mathematics: Queen and Servant of Science* (New York: McGraw-Hill, 1951), p. 21.
87. "Recent Work on the Principles of Mathematics," *The International Monthly* 4 (1901):84.
88. *The Principles of Mechanics*, translated by D. E. Jones and J. T. Walley (London, 1899), p. 9.
89. See F. Cajori, *A History of Physics* (New York: Macmillan, 1929), p. 301.
90. *Mathematical Foundations of Quantum Mechanics*, translated by R. T. Beyer (Princeton, N.J.: Princeton University Press, 1955), pp. 327–28.
91. "Beyond the Electron," *Classics in Science* (New York: Philosophical Library, 1960), p. 191.
92. *The Aims of Mathematical Physics*, p. 9.
93. *The Theory of Groups and Quantum Mechanics*, p. xx.
94. "Geometry and Experience" (1921), *Sidelights on Relativity* (London: Methuen, 1922), p. 28.
95. "On the Method of Theoretical Physics" (1933), *The World as I See It* (New York: Covici, 1934), p. 33.
96. *Ibid.*
97. "The Problems of Space, Ether, and the Field in Physics" (1930), *Ibid.*, p. 89.
98. *Ibid.*, p. 95.
99. *Ibid.*
100. "The Unreasonable Effectiveness of Mathematics in the Natural Sciences," *Communications on Pure and Applied Mathematics* 13 (1960):12.
101. "The Evolution of the Physicist's Picture of Nature," *Sci. Amer.*, 208 (May, 1963):50.
102. *Ibid.*
103. *Newsweek*, Nov. 1, 1965, p. 56.
104. "Atomic Structure," M. H. Shamos and G. M. Murphy (eds.), *Recent Advances in Science* (New York: New York University Press, 1956), p. 46.

105. "Die logischen Grundlagen der Mathematik," *Gesammelte Abhandlungen,* III (Berlin: Springer, 1935), 178.
106. B. Russell, "Recent Work on the Principles of Mathematics," p. 100.
107. "Neubegründung der Mathematik," *Gesammelte Abhandlungen,* III, 157.
108. "Du role de l'intuition et de la logique en mathématiques," *Compte rendu du deuxième congrès internationale des mathématiciens,* edited by E. Duporcq (Paris: Gauthier-Villars, 1902), p. 122.
109. *Philosophy of Mathematics and Natural Science,* p. 219.
110. "Die Grundlagen der elementaren Zahlenlehre," *Gesammelte Abhandlungen,* III, 193.
111. Gödel's paper is now available in an excellent English translation by B. Meltzer, the usefulness of which is further enhanced by R. B. Braithwaite's introduction. (Kurt Gödel, *On Formally Undecidable Propositions of Principia Mathematica and Related Systems* [Edinburgh: Oliver and Boyd, 1962]).
112. "Mathematics and Logic," *Amer. Math. Monthly* 53 (1946):13.
113. "On the Method of Theoretical Physics," p. 33.
114. *Ibid.*
115. "Clerk Maxwell's Influence on the Evolution of the Idea of Physical Reality" (1931), *The World as I See It,* p. 60.
116. "Quo Vadis," *Daedalus* 87 (Winter, 1958):91.
117. *Mathematical Papers by William Kingdon Clifford,* p. 69.
118. "Space and Time," p. 75.
119. "Geometry and Experience," p. 32.
120. J. A. Wheeler, *Geometrodynamics* (New York: Academic Press, 1962), p. xi.
121. *Ibid.,* p. xiii.
122. *Ibid.*
123. "The Unreasonable Effectiveness of Mathematics in Natural Science," p. 2.
124. "David Hilbert and His Mathematical Work," *Bull. Amer. Math. Soc.,* 50 (1944):653.
125. "Inertia and Energy," *Albert Einstein: Philosopher-Scientist,* p. 533.
126. *Experiment and Theory in Physics* (Cambridge: Cambridge University Press, 1943), p. 35.
127. *The New Age in Physics* (New York: Harper & Brothers, 1960), p. 15.
128. "Quantised Singularities in the Electromagnetic Field," *Proc. Roy. Soc.* (London) 133 (1931):60.
129. "The Evolution of the Physicist's Picture of Nature," p. 50.
130. See R. Kronig, "The Turning Point," *Theoretical Physics in the Twentieth Century: A Memorial Volume to Wolfgang Pauli,* edited by M. Fierz and V. F. Weisskopf (New York: Interscience, 1960), p. 22.
131. *The Mysterious Universe,* pp. 150–51.
132. *Pensées sur l'interprétation de la nature,* in *Oeuvres complètes de Diderot,* II (Paris, 1875), 11.
133. "A Half-century of Mathematics," *Amer. Math. Monthly* 58 (1951):523.
134. *The Principles of Quantum Mechanics* (Oxford: Clarendon Press, 1930), p. 7.
135. *Philosophy* (New York: Norton, 1927), p. 157.
136. "On Faraday's Lines of Force" (1855), in *Sci. Papers,* I, 156–57.
137. *Science and First Principles* (New York: Macmillan, 1931), pp. 19–20.
138. *Adventures of Ideas,* p. 161.
139. See, for instance, P. A. M. Dirac, "The Evolution of the Physicist's Picture of Nature," p. 53; J. Jeans, *The Mysterious Universe,* p. 144.
140. *Traité de métaphysique* (1734), chap. viii, in *Oeuvres,* XXII, 223.

CHAPTER IV. The Layers of Matter

1. *Time,* Sept. 15, 1961, p. 83.
2. *Posterior Analytics,* 79a.
3. Diels 68A 37.

4. *De rerum natura*, Book 1, lines 315–30.

5. Diels 68B 125.

6. See for instance *Physics* 231a, and *On the Heavens* 303a.

7. *Timaeus* 30B.

8. *On the Heavens* 303a.

9. *The New Organon*, Book 1, chap. 64, in *Works*, IV, 65.

10. For some very characteristic details, see his *Sylva sylvarum or a Natural History in Ten Centuries*, pars. 789, 846, and 847, in *Works*, II, 595–96, 618–19.

11. *Les principes de la philosophie*, II, 23, in *Oeuvres*, IX, 75.

12. *Ibid.*, II, 4, in *Oeuvres*, XI, 65.

13. *Ibid.*, IV, 103, in *Oeuvres*, IX, 256.

14. *Traité de physique*, I (4th ed.; Paris, 1682), 166.

15. *Works*, I, 562.

16. See especially par. 14, "Progymnasma meteori," in his *Ortus medicinae, id est initia physicae*, edited by his son, F. M. Van Helmont (Amsterdam, 1652), pp. 54–59.

17. *The Sceptical Chymist*, in *Works*, I, 562.

18. *Origin of Forms and Qualities according to the Corpuscular Philosophy* (1666), in *Works*, III, 5.

19. *Ibid.*, p. 15.

20. *The Sceptical Chymist*, in *Works*, I, 494–95.

21. *Ibid.*, p. 477.

22. *Certain Physiological Essays* (1661), in *Works*, I, 356.

23. *Elements of Chemistry*, translated by R. Kerr in Great Books of the Western World, XLV (Chicago: Encyclopaedia Britannica, 1952), 3.

24. See D. McKie, *Antoine Lavoisier: The Father of Modern Chemistry* (Philadelphia: Lippincott, 1935), pp. 121–22.

25. "Réflexions sur le phlogiston pour servir de suite à la théorie de la combustion et de la calcination," in *Oeuvres de Lavoisier*, II (Paris, 1862), 640.

26. *Ibid.*, p. 623.

27. *Elements of Chemistry*, p. 3.

28. *Ibid.*

29. *Ibid.*, p. 10.

30. *Ibid.*, p. 55.

31. "Mémoire sur la combustion en général" (1777), in *Oeuvres*, II, 225.

32. *Elements of Chemistry*, p. 55.

33. *Ibid.*, p. 2.

34. *The Collected Works of Sir Humphry Davy, Bart.*, edited by John Davy, IV (London, 1840), 358.

35. *Ibid.*, p. 364.

36. *Ibid.*

37. "The Aim and Progress of Physical Science," in *Popular Lectures and Addresses*, translated by E. Atkinson (New York, 1873), p. 373.

38. *Ibid.*, p. 375.

39. "The Periodic Law of the Chemical Elements," *Journal of Chemical Society* 55 (1889):634–56.

40. *Ibid.*, p. 645.

41. *Ibid.*, p. 644.

42. *Brit. Assn. Rep.* (1887), pp. 558–76.

43. *Ibid.*, p. 559.

44. *Ibid.*

45. *Ibid.*, p. 558.

46. *Ibid.*, p. 560.

47. *Ibid.*

48. *Ibid.*, p. 576.

49. For details on their views in this regard, see Chap. 10.

50. *Brit. Assn. Rep.* (1887), p. 568.

51. *Ibid.*, p. 569.

52. *Ibid.*, p. 576.
53. Sec. 1, Book 3, chap. 6, in *Opera omnia*, I (Lyons, 1658), 268–69.
54. *Physiologia Epicuro-Gassendo-Charltoniana, or a Fabrick of Science Natural upon the Hypothesis of Atom* (London, 1654), p. 114.
55. *Ibid.*
56. *Ibid.*, p. 118.
57. *Ibid.*, p. 102.
58. *Ibid.*, p. 101.
59. *Ibid.*, p. 89.
60. *Opticks*, p. 400.
61. *Ibid.*, p. 402.
62. *Ibid.*, p. 376.
63. "On Aristotle's and Descartes' Theories of Matter" (*ca.* 1671), in *Leibniz Selections*, p. 91.
64. Fourth Letter to Clarke (1716), in *Leibniz Selections*, p. 236.
65. *Dictionnaire philosophique*, "Atomes," in *Oeuvres*, XVII, 477.
66. *Ibid.*
67. *A New System of Chemical Philosophy* (Manchester, 1808), p. 141.
68. *Memoirs of the Literary and Philosophical Society of Manchester*, 1 (1805):271.
69. "On the Relation between the Specific Gravities of Bodies in their Gaseous State and the Weights of their Atoms," *Thomson's Annals of Philosophy* 6 (1815):321.
70. *Thomson's Annals of Philosophy* 7 (1816):113.
71. "Researches on the Mutual Relations of Atomic Weights" (1860), translated in *Prout's Hypothesis* (Edinburgh: Alembic Club Reprints, 1932), p. 42.
72. *Ibid.*, p. 44.
73. *Ibid.*, p. 47.
74. *Ibid.*
75. "On the Electro-chemical Decomposition," in *Experimental Researches in Electricity*, in Great Books of the Western World, XLV (Chicago: Encyclopaedia Britannica, 1952), 390.
76. 2d ed.; Paris, 1878. See especially the sixth and seventh lectures.
77. *Ibid.*, p. 314.
78. *History of the Inductive Sciences*, III (3d ed.; London, 1857), 127.
79. *A History of the European Thought in the Nineteenth Century*, I (Edinburgh, 1896), 403.
80. Letter of Jan. 28, 1835, to Berzelius. See A. Findlay, *A Hundred Years of Chemistry* (London: Duckworth, 1937), p. 23.
81. *Lehrbuch der organischen Chemie*, II (Erlangen, 1861), 56.
82. *Les origines de l'alchimie* (Paris, 1885), pp. 290–93.
83. In his Faraday Lecture of April 19, 1904, *Nature* 70 (1904):15. Only extreme positivism could champion a cause that was hopeless even by 1904 standards. Before long Ostwald had to capitulate. In the introduction written to the 1912 edition of his *Lehrbuch der allgemeinen Chemie* he admitted that the agreement between the Brownian movement and the kinetic theory of gases raised the atomic hypothesis "to the position of a scientifically well-founded theory."
84. *The Concepts and Theories of Modern Physics* (2d ed., reprinted with notes by P. W. Bridgman; Cambridge, Mass.: Harvard University Press, 1960), pp. 125–26.
85. *The Science of Mechanics*, translated by T. J. McCormack (La Salle, Ill.: Open Court, 1960), p. 589.
86. *Ibid.*, p. xiv. Charges and countercharges of dogmatism were rather frequent between Mach and his critics. A classic instance of this was Mach's indignant reaction to Planck's charges that positivist physicists tried to impose their rejection of atoms on their colleagues in an authoritative manner. "We can see," wrote Mach putting on record Planck's charge, "that the physicists are on the surest road to becoming a church and are already appropriating all the customary means to this end. To this a simple answer: 'If belief in the reality of atoms is so essential to you I thereby abandon the physicists' manner of thought. . . .

I will be no regular physicist. . . . I will renounce all scientific recognition. . . . In short, the communion of the faithful I will decline with best thanks. For dearer to me is freedom of thought.'" (See P. Carus, "Professor Mach and his Work," *Monist* 21 [1911]:33.) The way Mach reversed his antiatomic views was truly dramatic. To Stefan Meyer who witnessed it, the event remained one of the most striking memories of his life. He brought to the bedside of the partly crippled Mach a scintillation screen and exposed it to radiation by alpha particles. The sight of little stars of light that suddenly appeared on the screen overwhelmed almost instantaneously the man who much his life used to scoff at talks about atoms with the question: "Have you seen one?" This time, however, he was speechless and finally he said, "Now I believe in the existence of atoms." Indeed, as Meyer put it, a whole world vision changed in him in a few minutes. (On Meyer's account, see his paper, "Die Vorgeschichte der Gründung und das erste Jahrzehnt des Institutes für Radiumforschung," in *Sitzungsberichte der Österreichischen Akademie der Wissenschaften* Abteilung II.a, 159 [1950]: Heft 1, 1–26.)

87. "Über die Unentbehrlichkeit der Atomistik in der Naturwissenschaft," *Annalen der Physik und Chemie* 60 (1897):231–47.
88. *Ibid.*, p. 235.
89. *Ibid.*
90. "On Loschmidt's Experiments on Diffusion in Relation to the Kinetic Theory of Gases," *Sci. Papers*, II, 343.
91. *PLA*, I, 148.
92. "Molecules" (1873), *Sci. Papers*, II, 377.
93. *Ibid.*, p. 361.
94. *Ibid.*
95. "On the Size of Atoms" (1870), reprinted in W. Thomson and P. G. Tait, *Treatise on Natural Philosophy*, II (new ed.; Cambridge, 1883), 495.
96. Presidential address to the British Association, *PLA*, II, 167–68.
97. *Baltimore Lectures* (Baltimore: Publication Agency of Johns Hopkins University, 1904), p. 11.
98. *Ibid.*, p. 12.
99. "Ether" (1878), *Sci. Papers*, II, 767–68.
100. *Baltimore Lectures*, p. 533.
101. Presidential address, p. 164.
102. Address to the Mathematical and Physical Section of the British Association (1870), *Sci. Papers*, II, 223.
103. *A History of the European Thought*, II, 66.
104. *The Corpuscular Theory of Matter* (London: A. Constable, 1907), p. 2.
105. *Lectures on Some Recent Advances in Physical Science* (London, 1876), p. 298.
106. "Deep Water Ship-Waves," *Proc. Roy. Soc.* (Edinburgh) 25 (1905):565.
107. "On the Ascertained Absence of Effects of Motion through the Aether" (1904), *Mathematical and Physical Papers by Sir Joseph Larmor*, II (Cambridge: Cambridge University Press, 1929), 277.
108. *The Principles of Chemistry*, translated by G. Kamensky, II (London: Longmans, Green & Co., 1905), 522–28.
109. *Aether and Matter* (Cambridge: Cambridge University Press, 1900), p. vi.
110. *Electricity and Matter* (London: A. Constable, 1907), p. 50.
111. *The Foundations of Science*, translated by G. B. Halsted (Lancaster, Penn.: The Science Press, 1913), p. 7.
112. *Exposition du système du monde*, Book 5, chap. 5, *Oeuvres*, VI, 471–72. That the patterns and structure of matter in the range of the invisibly small would differ only in size from the features of ordinary matter was emphasized through the whole life span of classical physics. According to Descartes "there is nothing more in conformity with reason than to think of things, which due to their smallness cannot be perceived by the senses, according to the example and pattern of things we see." Letter to Plempius, Oct. 3, 1637, *Oeuvres*, I, 421. In the very last phase of classical physics Gibbs expressed the same belief when

he remarked: "It is my firm conviction that every little molecule or atom has a definite atmosphere about it very much like the earth's atmosphere about it (the earth) . . . such a concept would explain so many things." See L. P. Wheeler, *Josiah Willard Gibbs* (rev. ed.; New Haven, Conn.: Yale University Press, 1952), p. 195.

113. "Molecules" (1873), *Sci. Papers*, II, 374.
114. "Cathode Rays," *Phil. Mag.* (London) 44 (1897):310–11.
115. "The Electrical Structure of Matter," *Science* 58 (1923):211.
116. H. Nagaoka, "On a Dynamical System Illustrating the Spectrum Lines and the Phenomena of Radio-activity," *Nature* 69 (1904):392–93.
117. "The Scattering of α and β Particles by Matter," *Phil. Mag.* (London) 21 (1911):688.
118. "Forty Years of Physics" (1936), *Background to Modern Science*, edited by J. Needham and W. Pagel (New York: Macmillan, 1938), pp. 67–68.
119. *Matter and Energy* (New York: H. Holt, 1912), p. 143.
120. *Nature* 81 (1909):257.
121. "Nuclear Constitution of Atoms," *Proc. Roy. Soc.* (London) 97A (1920):374–400.
122. "The Electrical Structure of Matter," p. 219.
123. *The Nature of the Physical World* (Cambridge: Cambridge University Press, 1928), p. 3.
124. Broadcast Lecture (1930), in Sir Rayleigh [R. J. Strutt], *The Life of Sir J. J. Thomson* (Cambridge: Cambridge University Press, 1942), p. 266.
125. *Electrons (+ and −), Protons, Photons, Neutrons, and Cosmic Rays* (rev. ed.; Chicago: University of Chicago Press, 1947), p. 320.
126. See N. R. Hanson, *The Concept of the Positron* (Cambridge: Cambridge University Press, 1963), p. 138.
127. *Time*, June 25, 1965, p. 64.
128. R. B. Lindsay and H. Margenau, *The Foundations of Physics* (New York: John Wiley & Sons, 1936), p. 513.
129. "The Highest Aim of the Physicist" (1889), *The Physical Papers of Henry Augustus Rowland* (Baltimore: John Hopkins Press, 1902), p. 673.
130. "A Survey of the Sciences," *Science* 106 (1947):137.
131. Murray Gell-Mann and A. Rosenfeld, "Hyperons and Heavy Mesons (Systematics and Decay)," *Annual Review of Nuclear Science* 7 (1957):408.
132. *The Constitution of Matter* (Eugene: Oregon State System of Higher Education, 1956), p. 2.
133. *Ibid.*, p. 31.
134. W. Pauli, *Collected Scientific Papers*, edited by R. Kronig and V. F. Weisskopf, I (New York: Interscience, 1964), xvii.
135. C. W. Kilmister, *The Environment in Modern Physics: A Study in Relativistic Mechanics* (New York: American Elsevier Publishing Company, 1965), p. 131.
136. Most authoritatively documented in the lists published in *Reviews of Modern Physics* 35 (1963):314–23, 36 (1964):977–1004, 37 (1965):633–61.
137. *Time*, April 26, 1963, p. 60.
138. *Scientific Uncertainy, and Information* (New York: Academic Press, 1964), p. 41.
139. *Physical Review Letters*, Dec. 1, 1962, p. 472.
140. *The New York Times*, Feb. 20, 1964, p. 12, col. 8.
141. The frustration of such hopes was noted for instance in the *Britannica Book of the Year: 1965* (Chicago: Encyclopaedia Britannica, Inc., 1965), pp. 637–38.
142. "Quantum Theory and Elementary Particles," *Science* 149 (1965):1189.
143. Luke C. L. Yuan (ed.), *Nature of Matter: Purposes of High Energy Physics* (Upton, N.Y.: Brookhaven National Laboratory, 1965).
144. *Ibid.*, pp. 21–22.
145. *Time*, May 5, 1958, p. 53.
146. *Newsweek*, Aug. 24, 1959, p. 82.
147. *Philosophical Problems of Nuclear Science*, translated by F. C. Hayes (London: Faber, 1952), p. 101.
148. *Cosmology* (2d ed.; Cambridge: Cambridge University Press, 1960), p. 144.

149. Then physicists said either that the principle of the conservation of energy and matter presupposed a creation once and for all, or that, as J. Tyndall contended, it excluded "both creation and annihilation." See his *Heat, a Mode of Motion* (4th ed.; New York, 1873), p. 467.
150. *Les origines de l'alchimie*, p. v.
151. *Physics and Philosophy* (New York: Harper, 1958), p. 165.
152. *The Constitution of Matter*, p. 36.
153. *Lectures on Some Recent Advances in Physical Science*, p. 284.
154. *Ibid.*, pp. 284–85.
155. "Our Conception of Matter," *What Is Life and Other Scientific Essays* (Garden City, N.Y.: Doubleday, 1956), p. 161.
156. *Ibid.*, p. 162.
157. R. Oppenheimer, *The Constitution of Matter*, p. 37.

CHAPTER V. The Frontiers of the Cosmos

1. *The Realm of the Nebulae* (New Haven, Conn.: Yale University Press, 1936) p. 21
2. Diels 12, A10.
3. Diels 12, A18.
4. Diels 22, B3.
5. Diels 31, A61. Anaxagoras, who dared to defy hallowed traditions by speaking of the sun as a red-hot stone, was less bold concerning his views on the dimensions of the sun. "The sun," he reportedly stated, "exceeds the Peloponnesus in size" (Diels 59, A42). There are no indications that he had ever compared this magnitude to the size and distance of other parts and objects of the skies.
6. *On the Heavens*, 298b.
7. *Meteorologica*, 362b.
8. *On the Heavens*, 298b.
9. *Meteorologica*, 345b.
10. *Ibid.*, 340a.
11. *Ibid.*, 345b.
12. *Simplicii in Aristotelis de caelo commentaria*, edited by I. L. Heiberg (Berlin, 1893), p. 488.
13. For further details and references, see J. L. E. Dreyer, *A History of Astronomy from Thales to Kepler*, revised edition by W. H. Stahl (New York: Dover, 1953), chaps. 6 and 8.
14. *Ibid.*, p. 175.
15. *Natural History*, Book 2, chap. 21, translated by H. Rackham (LCL, 1938), p. 288. The passage, as Pliny gives it, is rather obscure and seems to indicate that Posidonius satisfied himself with round numbers. In the same context, Posidonius puts the width of the atmosphere or cloudy region at 40 stadia, or less than 4 miles, and the moon's distance at 2 million stadia, or less than 200,000 miles.
16. *Hypothèses et époques des planètes de C. Ptolémée et hypotyposes de Proclus Diadochus*, edited by M. l'Abbé Halma (Paris, 1820), pp. 145–46.
17. Dreyer, *A History of Astronomy*, p. 257.
18. T. L. Heath, *The Works of Archimedes* (Cambridge, 1897), p. 222.
19. *The Almagest*, Book 1, chap. 6, in Great Books of the Western World, XVI (Chicago: Encyclopaedia Britannica, 1952), 10.
20. *Ibid.*, Book 1, chap 7, pp. 10–12.
21. See Plutarch, *Concerning the Face which Appears in the Orb of the Moon*, chap. 6, 922F-923A, in *Plutarch's Moralia*, edited and translated by H. Cherniss and W. C. Helmbold, XII (LCL, 1957), 55.
22. *Ad Lucilium naturalium quaestionum libri vii*, Book 1, prologue, in *Annaei Senecae opera*, edited by C. R. Fickert, III (Leipzig, 1845), 401.
23. *Natural History*, Book 2, chap. 8.
24. The misguided efforts of Theodore and his followers exploiting scriptural texts to

prove the flatness of the earth were exposed by John Philoponus in his *De mundi creatione*, Book 3, chap. 10. In the preceding chapter Philoponus lists various proofs of the sphericity of the earth. See *Bibliotheca veterum patrum*, edited by A. Gallandius, XII (Venice, 1778), 528–32.

25. *Revolutions*, Book 1, chap. 6, pp. 516–17.
26. *Ibid.*, p. 517.
27. *Ibid.*, p. 529.
28. *De stella nova in pede serpentarii*, chap. 21, in *Werke*, I, 253.
29. Letter of Dec. 16, 1598, to Herwart von Hohenburg, in *Werke*, XIII, 268.
30. *Dissertatio cum nuncio sidereo* (1610), in *Werke*, IV, 307.
31. In the "Dissertatio" added to his edition of Galileo's *Starry Messenger*, Kepler came face-to-face with the most burning of all questions posed by the likely existence of an immensely, if not infinitely, large number of stars: "In what way then are all things for man's good? How may we be said to be the masters of God's works?" See *Werke*, IV, 307.
32. "An Anatomie of the World. The First Anniversary," in *Complete Poetry and Selected Prose of John Donne*, edited by J. Hayward (London: Nonesuch Press, 1949), p. 202.
33. *De stella nova*, in *Werke*, I, 253.
34. *Epitomes astronomiae copernicanae*, Book 4, Part 1, chap. 4, in *Werke*, VII, 288.
35. *Dissertatio cum nuncio sidereo*, in *Werke*, IV, 305.
36. *Dialogue*, p. 56.
37. *The Starry Messenger*, in *Discoveries*, p. 31.
38. *Ibid.*, p. 36.
39. Feb. 28, 1615, *ibid.*, p. 158.
40. *Discours prouvant la pluralité des mondes* (Geneva, 1657).
41. *Histoire comique, contenant les estates et empires de la lune* (Paris, 1659).
42. *Systema saturnium*, in *Oeuvres*, XV, 299.
43. *Il newtonianismo per la dame: ovvero dialoghi sopra la luce e i colori* (Naples, 1737).
44. *Entretiens sur la pluralité des mondes* (1686), edited by R. Shackleton (Oxford: Clarendon Press, 1955), p. 113.
45. In *Oeuvres*, XXI, 680.
46. *Ibid.*, p. 704.
47. *Alciphron*, in *Works*, III, 172.
48. *Cosmologische Briefe über die Einrichtung des Weltbaues* (Augsburg, 1761), pp. 63, 166.
49. *Dialogen über die Mehrheit der Welten* (Berlin, 1780), p. 190.
50. *Allgemeine Naturgeschichte und Theorie des Himmels*, in *Werke*, I, 360.
51. *Revolutions*, Book 1, chap. 10, p. 528. Of the absence of such a unitary explanation of the planetary motions in a geocentric system, Ptolemy himself said: "I do not profess to be able thus to account for all the motions at the same time; but I shall show that each by itself is well explained by its proper hypothesis." See *Hypothèses et époques des planètes*, p. 151.
52. *Epitomes astronomiae copernicanae*, Book 4, Part 1, in *Werke*, VII, 266.
53. Letter of Dec. 16, 1598, to Herwart, in *Werke*, XIII, 267.
54. Book 4, Part 1, chap. 4, in *Werke*, VII, 286.
55. *Paradise Lost*, viii, 78–80.
56. *Kosmothéoros*, in *Oeuvres*, XXI, 816.
57. *Principia*, pp. 596–97.
58. J. B. Morin, *Astrologia gallica principiis et rationibus propriis stabilita*, Book 9, Sec. 3, chap. 6 (The Hague, 1661), p. 189.
59. *An Original Theory or New Hypothesis of the Universe* (London, 1750).
60. Such seminal views, however, accounted for only a small part of Kant's work, which was also marked in places by gross ignorance of some basic laws of mechanics and by fantastic speculations.
61. *Cosmologische Briefe*; see especially the twelfth and thirteenth letters.
62. It was the absence of these in Lambert's work that later prompted Herschel to

refer to it as a book "full of the most fantastic imaginations." *Phil. Trans.* (London) 95 (1805):255.

63. "Account of Some Observations Tending to Investigate the Construction of the Heavens," *Phil. Trans.* (London) 74 (1784):438.

64. "Astronomical Observations Relating to the Construction of the Heavens," *Phil. Trans.* (London) 101 (1811):269.

65. *Ibid.*, p. 270.

66. "Astronomical Observations and Experiments Selected for the Purpose of Ascertaining the Relative Distances of Clusters of Stars," *Phil. Trans.* (London) 108 (1818):463.

67. *Allgemeine Naturgeschichte und Theorie des Himmels*, in *Werke*, I, 255.

68. *Cosmologische Briefe*, p. 263.

69. *History of Physical Astronomy* (London, 1852), p. 568.

70. *The Plurality of Worlds* (new ed.; Boston, 1854), p. 161.

71. "The Nebular Hypothesis" (1858), in *Essays Scientific, Political and Speculative*, I (New York, 1899), 113.

72. In a note added in 1898 to the Essay quoted above, *ibid.*, I, 181.

73. Huggins was, however, all too aware of such a possible bearing of his findings, as can be seen in his communication "On the Spectrum of the Great Nebula in the Sword-handle of Orion," *Proc. Roy. Soc.* (London) 14 (1865):39–42.

74. "The New Astronomy. A Personal Retrospect," *The Nineteenth Century* 41 (1897):916–17.

75. I. S. Bowen, "The Origin of the Nebular Lines and the Structure of the Planetary Nebulae," *Astrophysical Journal* 67 (1928):1–15.

76. "Über den Fraunhoferschen Linien," *Monatsberichte der Akademie der Wissenschaften* (Berlin), October, 1859, pp. 662–65. The report of Bunsen and Kirchoff had an immediate dramatic impact. As Huggins recalled: "This news was to me like the coming upon a spring of water in a dry and thirsty land. A feeling of inspiration seized me. I felt as if I had it now in my power to lift a veil which had never before been lifted; as if a key had been put into my hands which would unlock a door which had been regarded as forever closed to man — the veil and door behind which lay the unknown mystery of the true nature of the heavenly bodies." In "The New Astronomy. A Personal Retrospect," p. 911.

77. Such a suggestion was unpalatable to many of Herschel's colleagues as it implied the concept that the actual shape of the skies was not the work of a split-second Creation. As Constance A. Lubbock, a descendant of Herschel remarked, this might have been one of the reasons why Herschel's papers bearing on this subject were received with distinct coolness by the Royal Society. See C. A. Lubbock, *The Herschel Chronicle* (Cambridge: Cambridge University Press, 1933), p. 197.

78. "On the Sun's Heat" (1887), in *PLA*, I, 415.

79. *The System of the Stars* (2d ed.; London: A & C Black, 1905), p. 349.

80. *History of Astronomy during the Nineteenth Century* (3d ed.; London, 1893), p. 511.

81. "Spectrum Analysis," *The Contemporary Review* 11 (1869):489.

82. *The System of the Stars*, p. 16.

83. These conclusions were submitted by A. Clerke, *A History of Astronomy during the Nineteenth Century*, p. 505.

84. Olbers' paper first published in Bode's *Astronomisches Jahrbuch für das Jahr 1826* (Berlin, 1825), pp. 110–21, was reprinted in C. Schilling (ed.), *Wilhelm Olbers: Sein Leben und seine Werke*, I (Berlin, 1894), 133–41. The English translation appeared in *The Edinburgh New Philosophical Journal* 1 (1826):141–50; for the French translation, see *Bibliothèque universelle des sciences, belles-lettres, et arts* (Geneva) 31 (1826):102–15. Olbers mentions in his paper that the problem was already discussed by E. Halley. Although Olbers fails to give any reference to Halley's papers, he was obviously familiar with Halley's solution to the problem and also right in declaring it erroneous. Halley called the paradox a "metaphysical paradox" and hoped to solve it on the assumption that "as the Light of the Fix'd Stars diminishes, the intervals between them decrease in a less

proportion, the one being as the Distances, and the other as the Squares thereof, reciprocally." (Halley's two short papers on the subject appeared in the *Philosophical Transactions of the Royal Society* 31 [1720]:22–24, 24–26; quotation from p. 24.) Modern cosmologists who rediscovered Olbers' paradox discuss it with apparently no direct knowledge of Olbers' paper. Thus they are not only unaware of Halley's part in the paradox, but they also attribute to Olbers ideas he never entertained, or ideas that are in direct opposition to what he actually stated. The result is a most disconcerting growth in modern cosmological literature of half-truths, and simple untruths about Olbers' thought. It seems that the scientist's habit of carefully rechecking the facts of the laboratory is no guarantee that he will do the same about the facts of scientific history, let alone about facts not related to science. This might perhaps be called the paradox of the scientist's unscientific habits. (In this connection, see my paper "Olbers', Halley's, or Whose Paradox?" [to be published in the *American Journal of Physics*].)

85. "The Wave Theory of Light" (1884), in *PLA*, I, 314–15.
86. "Untersuchungen über Gegenstände der Höheren Geodaesie" (1844), in *Werke*, IV (Göttingen, 1877), 259–300.
87. See H. Poincaré, *La science et l'hypothèse* (Paris: Flammarion, 1902). "Experiments," concluded Poincaré, "do not refer to space but to bodies" (p. 105).
88. "Über das Newton'sche Gravitationsgesetz," *Sitzungsberichte der königlichen bayerischen Akademie der Wissenschaften*, Math.-phys. Classe 26 (1896):373–400.
89. "Cosmological Considerations on the General Theory of Relativity" (1917), in A. Einstein *et al.*, *The Principle of Relativity*, translated by W. Perrett and G. B. Jeffrey (London: Methuen, 1923), p. 178.
90. *Principia*, p. 422.
91. "Considerations on the Change of the Latitudes of Some of the Principal Fixed Stars," *Phil. Trans.* (London) 30 (1717):736.
92. "On the Proper Motion of the Sun and the Solar System," *Phil. Trans.* (London) 73 (1783):269.
93. On the younger Herschel's account of the general disbelief that greeted his father's conclusion, see C. A. Lubbock, *The Herschel Chronicle*, p. 188.
94. "A Relation between Distance and Radial Velocity among Extra-galactic Nebulae," *Proc. Nat. Acad. Sci.* (Washington, D.C.) 15 (1929):173. A number of theories were offered to explain the nebular red-shift on a basis other than recessional velocity. Some proposed the Einstein effect as an explanation but such a solution implies that the gravitational field of galaxies would increase with distance. Others invoked the Compton effect, and some assumed that light-rays become "fatigued" with age. Again, others assigned the effect to the absorption of light by intergalactic dust and to its reradiation at lower wavelengths. On the other hand, not only is the nebular red-shift fully and simply explained by the Doppler effect but also, as Einstein remarked, recessional red-shift as a physical process and relativity theory form an interlocking unity. Again, if the red-shift is due to a cause other than recession, one would have to explain why such a cause is not operating within individual galaxies. Strong as these considerations are in favor of recessional velocity as the cause of the nebular red-shift, they clearly do not render unnecessary alertness to new theoretical possibilities.
95. Four years later an English translation of Lemaître's paper was published under the title, "A Homogeneous Universe of Constant Mass and Increasing Radius Accounting for the Radial Velocity of Extra-galactic Nebulae," *Monthly Notices Roy. Astr. Soc.* 91 (1931):483–90.
96. Lemaître's priority was acknowledged emphatically by Eddington himself. See A. S. Eddington, "On the Instability of Einstein's Spherical World," *Monthly Notices Roy. Astr. Soc.* 90 (1930):668.
97. *The History of Nature*, translated by F. D. Wieck (Chicago: University of Chicago Press, 1949), p. 71.
98. *An Essay on the Foundations of Geometry* (Cambridge, 1897), p. 49.

99. "The Milky Way and the Theory of Gases," in *Science and Method*, translated by F. Maitland (London: T. Nelson, n.d.), p. 255.
100. See H. Wright, *Palomar, the World's Largest Telescope* (New York: Macmillan, 1952), p. 185.
101. *The Individual and the Universe* (New York: Harper & Brothers, 1959), pp. 108–9.
102. "The Supergalaxy," *Scientific American* 191 (July, 1954):30–35.
103. H. Bondi, "Astronocy and Cosmology," in J. R. Newman (ed.), *What Is Science?* (New York: Simon & Schuster, 1955), p. 91.
104. *Opticks*, p. 111.
105. B. K. Harrison, K. S. Thorne, M. Wakano, and J. A. Wheeler, *Gravitation Theory and Gravitational Collapse* (Chicago: University of Chicago Press, 1965), p. 147.
106. *Time*, June 18, 1965, p. 66.
107. In his Lowell Lectures of 1931, published under the title, *Kosmos* (Cambridge, Mass.: Harvard University Press, 1932), p. 3.
108. *Stellar Movements and the Structure of the Universe* (London: Macmillan, 1914), p. 261.
109. *Ibid.*, p. 31.
110. *Kosmos*, p. 134.
111. "Electrical Disturbance Apparently of Extraterrestrial Origin," *Proceedings of the Institute of Radio Engineers* 21 (1933):1387. This conclusion forced itself on Jansky only almost a year after he first discussed his findings in the December, 1932, issue of the same journal.
112. Quoted in J. Pfeiffer, *The Changing Universe: The Story of the New Astronomy* (New York: Random House, 1956), p. 14.
113. *The Nature of the Physical World* (Cambridge: Cambridge University Press, 1928), p. 167.
114. *Ibid.*, p. 166.
115. "A Homogeneous Universe of Constant Mass," p. 489.
116. *Problems of Cosmogony and Stellar Dynamics* (Cambridge: Cambridge University Press, 1919), p. 290.
117. *Monthly Notices Roy. Astr. Soc.* 82 (1922):279.
118. *Ibid.*
119. *Astronomy and Cosmogony* (Cambridge: Cambridge University Press, 1928), p. 401.
120. *Ibid.*, p. 402.
121. *The Universe around Us* (4th ed.; Cambridge: Cambridge University Press, 1943), p. 242.
122. *The Science of Life*, I (Garden City, N.Y.: Doubleday, 1931), 13.
123. *Annals of Science* 1 (1936):430.
124. "Über die Entstehung des Planetensystems," *Zeitschrift für Astrophysik* 22 (1944):319–55.
125. "Anwendungen der Hydrodynamik auf Probleme der Kosmogonie," in *Festschrift zur Feier des zweihundertjährigen Bestehens der Akademie der Wissenschaften in Göttingen* (Berlin: Springer, 1951), pp. 86–122.
126. "Stars, Ethics and Survival," in H. Shapley (ed.), *Science Ponders Religion* (New York: Appleton-Century-Crofts, 1960), p. 8.
127. "Searching for Interstellar Communication," *Nature* 184 (1959):844–46.
128. *Ibid.*, p. 846.
129. "Astronomy," *Scientific American* 183 (Sept., 1950):24–26.
130. *Ibid.*, p. 24.
131. *Problems of Cosmogony and Stellar Dynamics*, p. vi.
132. *Eos; or, the Wider Aspects of Cosmogony* (London: K. Paul, Trench, Trubner, 1928), pp. 17–18.
133. G. H. de Vaucouleurs, *Discovery of the Universe* (New York: Macmillan, 1957), p. 230.
134. See H. Wright, *Palomar, the World's Largest Telescope*, p. 183.
135. *Dialogue*, p. 455.

136. *The Theoretical Significance of Experimental Relativity*, (New York: Gordon & Breach, 1964), p. 122.
137. *Galaxies, Nuclei, and Quasars* (New York: Harper & Row, 1965), p. 10.
138. See H. Alfvén, "Antimatter and the Development of the Metagalaxy," *Rev. Mod. Phys.* 37 (1965):652–65.
139. See R. H. Dicke, "The Earth and Cosmology," *Science* 138 (1962):653–64.
140. "Physics in the Last Twenty Years," *Science* 151 (1966:)1055.
141. *Dialogue*, p. 67.
142. *Relativity, Thermodynamics and Cosmology* (Oxford: Clarendon Press, 1934), p. 488.

CHAPTER VI. The Edge of Precision

1. "Electrical Units of Measurements" (1883), in *PLA*, I, 72–73.
2. *A Preliminary Discourse on the Study of Natural Philosophy* (London, 1831), p. 122.
3. As recorded by E. Warburg, "Friedrich Kohlrausch: Gedächtnisrede," *Verhandlungen der Deutschen Physikalischen Gesellschaft* 12 (1910):920.
4. Presidential Address to the British Association (1871), in *PLA*, II, 157.
5. *The Nichomachean Ethics of Aristotle*, 1094b, translated by D. Ross (Oxford: Oxford University Press, 1954), pp. 2–3.
6. *De communi mathematica scientia*, chap. 27, edited by N. Festa (Leipzig, 1891), p. 86.
7. "Precision Instruments to 1500," in C. J. Singer (ed.), *A History of Technology*, III (Oxford: Clarendon Press, 1954–58), 582–619.
8. *Narratio prima*, translated by E. Rosen, *Three Copernican Treatises* (2d ed.; New York: Dover, 1959), p. 111.
9. *Ibid.*, p. 116.
10. See Tychius Brahe, *De nova stella* (Copenhagen: Regia Societas Scientiarum Danica, 1901), pp. 8–9.
11. See M. Caspar, *Kepler*, translated by C. Doris Hellman (London: Abelard-Schuman, 1959), p. 102.
12. *Mysterium cosmographicum*, chap. 19, in *Werke*, I, 65–68.
13. *Astronomia nova*, Part 2, chap. 19, in *Werke*, III, 178.
14. In his letter of July 21, 1612, to Galileo, Frederico Cesi mentions as common knowledge the elliptical orbits of planets established by Kepler. See *Opere di Galileo Galilei*, edited by A. Favaro, XI (Firenze, 1890–1909), 170. There were several factors for Galileo's showing no proper appreciation of Kepler's achievement. First, his proverbial jealousy that made him so reluctant to recognize anybody as his peer in science; second, the morbid mysticism of Kepler, so alien to Galileo's outlook; and finally Galileo's fidelity to the Aristotelian principle of the "naturalness" of circular motion as contrasted to the "artificiality" of an ellipse.
15. *Dialogue*, p. 296.
16. *Ibid.*, p. 223.
17. The apparatus of inclined planes provided Galileo with a reliable means of establishing experimentally the law of free fall and of arriving at a fairly good value of g. This was convincingly shown by T. B. Settle, "An Experiment in the History of Science," *Science* 133 (1961):19–23. That Galileo did not make a sustained effort to obtain a better value for g was in all probability owing to Galileo's greater confidence in geometry than experiments, a point emphasized by A. Koyré, "An Experiment in Measurement," *Proc. Amer. Phil. Soc.* 97 (1953):222–37.
18. *Principia*, p. 433.
19. *Ibid.*, p. 304.
20. R. H. Dicke, "The Eötvös Experiment," *Sci. Amer.* 205 (Dec., 1961):84–92.
21. *Principia*, p. 422.
22. *Ibid.*

23. See P. A. Schilpp (ed.), *Albert Einstein: Philosopher-Scientist* (Evanston, Ill.: Library of Living Philosophers, 1949), p. 100.
24. "What Is the Theory of Relativity?" in *The World as I See It* (New York: Covici, 1934), p. 80.
25. See L. T. More, *Isaac Newton: A Biography* (New York: C. Scribner's Sons, 1934), p. 330.
26. J. F. W. Herschel, "President's Report, 1840," *Monthly Notices Roy. Astr. Soc.* 5 (1840):97.
27. *A Preliminary Discourse on the Study of Natural Philosophy*, p. 122.
28. See especially the chapter, "Experiments Touching the Weight of Bodies Frozen and Unfrozen," in *Works*, II, 621–28.
29. "An Account of Some Experiments on the Loss of Weight in Bodies on Being Melted or Heated," *Phil. Trans.* (London) 75 (1785):361–65.
30. "An Inquiry concerning the Weight Ascribed to Heat," *Phil. Trans.* (London) 89 (1799):179.
31. *Ibid.*, p. 194.
32. *Ibid.*, p. 192.
33. *Ibid.*
34. *Ibid.*, p. 194.
35. *Ibid.*, p. 192.
36. "Experiments to Determine the Density of Earth," *Phil. Trans.* (London) 88 (1798):470.
37. "Construction et usage d'une balance électrique" (1785), in *Collection de mémoires relatifs à la physique*, I, *Mémoires de Coulomb* (Paris, 1884), p. 107.
38. *The History and Present State of Electricity with Original Experiments* (London, 1767), p. 732.
39. *A Treatise on Electricity and Magnetism*, I (3d ed.; Oxford, 1892), 81–83.
40. "A Very Accurate Test of Coulomb's Law of Force between Charges," *Phys. Rev.* 50 (1936):1071.
41. See Sir Rayleigh [R. J. Strutt], *The Life of Sir J. J. Thomson* (Cambridge: Cambridge University Press, 1942), p. 203.
42. J. Larmor (ed.), *Memoir and Scientific Correspondence of the Late Sir George Gabriel Stokes*, II (Cambridge: Cambridge University Press, 1907), 83.
43. "On the Calorific Effects of Magneto-Electricity and on the Mechanical Value of Heat," *Phil. Mag.* (London) 23 (1843):441.
44. *Ibid.*, p. 442.
45. "On the Changes of Temperature Produced by the Rarefaction and Condensation of Air," *Phil. Mag.* (London) 26 (1845):369.
46. "On the Existence of an Equivalent Relation between Heat and the Ordinary Forms of Mechanical Power," *Phil. Mag.* (London) 27 (1845):206.
47. *Ibid.*
48. "On the Mechanical Equivalent of Heat as Determined by the Heat Evolved by the Friction of Fluids," *Phil. Mag.* (London) 31 (1847):176.
49. In a note of 1885 prefixed by Joule to the reprint of his joint paper with W. Thomson "On the Thermal Effects Experienced by Air Rushing through Small Apertures" (1852), in J. P. Joule, *Scientific Papers*, II (London, 1887), 215.
50. "On the Mechanical Equivalent of Heat," *Phil. Mag.* (London) 140 (1850):64.
51. Letter to Griesinger, July 20, 1844, in *Die Mechanik der Wärme in Gesammelten Schriften*, edited by J. J. Weyrauch (Stuttgart, 1893), p. 145.
52. *Essai théorique et expérimental sur le Galvanisme avec une série d'expérience*, Proposition XIII (Paris, 1804), p. 31.
53. "On Physical Lines of Force," *Sci. Papers*, I, 500.
54. "Introductory Lecture in Experimental Physics," *Sci. Papers*, II, 244.
55. *The Progress of Physics during 33 Years, 1875–1908.* (Cambridge: Cambridge University Press, 1911), p. 7.
56. "Introductory Lecture in Experimental Physics," p. 244.
57. *Light Waves and Their Uses* (Chicago: University of Chicago Press, 1902), p. 24.
58. See *Principia*, p. 594.

59. Letter of March 19, 1879, in *Nature* 21 (1880):315.
60. J. D. Bernal, *Science in History* (London: Watts, 1954), p. 525.
61. J. P. Cedarholm and C. H. Townes, "A New Experimental Test of Special Relativity," *Nature* 184 (1959):1350–51.
62. On this, see Einstein's conversation with Dr. N. Balázs, one of his assistants, in Princeton in the summer of 1953. For the letter of Balázs reporting the essence of this conversation, see M. Polanyi, *Personal Knowledge: Towards a Post-Critical Philosophy* (Chicago: University of Chicago Press, 1958), pp. 10–11.
63. See B. Jaffe, *Michelson and the Speed of Light* (Garden City, N.Y.: Doubleday, 1960), p. 168. The verification of another aspect of Special Relativity, the dependence of mass on velocity, tells a similar story about the role of precision. The first experimental evidence grew out from W. Kaufmann's efforts to test Lorentz's electron theory according to which the electromagnetic mass increased with velocity in the same $(1-v^2/c^2)^{\frac{1}{2}}$ measure as later appeared in Einstein's theory. Kaufmann's first experiments using radioactive materials failed, however, to verify the formula. Only after his apparatus was considerably improved could he announce that "the mass of electrons originating in Becquerel-rays depends on the speed" and that the dependence agrees in full with the formula as given above. "Die elektromagnetische Masse des Elektrons," *Physikalische Zeitschrift* 4 (1903):54.
64. Presidential Address to the British Association (1909), *Nature* 81 (1909):257.
65. See V. F. Hess and J. Eugster, *Cosmic Radiation and Its Biological Effects* (2d ed.; New York: Fordham University Press, 1949), p. 173.
66. "Über Beobachtungen der durchdringenden Strahlung bei sieben Freiballonfahrten," *Physikalische Zeitschrift* 13 (1912):1090.
67. J. Linsley, "Evidence for a Primary Cosmic Ray Particle with Energy 10^{20} eV," *Phys. Rev. Letters* 10 (1963):146–48.
68. *Time*, Aug. 19, 1957, p. 58.
69. Quoted in V. F. Hess and J. Eugster, *Cosmic Radiation and Its Biological Effects*, p. 173.
70. "Note on the Fine Structure of Hα and Dα," *Phys. Rev.* 54 (1938):1113.
71. Nobel Prize acceptance speech, 1946, in *Nobel Lectures: Physics 1942–1962* (Amsterdam: Elsevier, 1964), p. 42.
72. *Fundamental Constants of Physics* (New York: Interscience, 1957), p. 183.
73. D. Halliday and R. Resnick, *Physics for Students of Science and Engineering* (New York: Wiley, 1961), p. 907.
74. See J. Weber, "Gravitational Waves," in Hong-Yee Chiu and William F. Hoffmann (eds.), *Gravitation and Relativity* (New York: W. A. Benjamin, 1964), pp. 90–105.
75. See the suggestion of G. Cocconi and E. Salpeter in *Phys. Rev. Letters* 4 (1960):176–77. For further details, see V. W. Hughes, "Mach's Principle and Experiments on Mass Anisotropy," in *Gravitation and Relativity*, pp. 106–20.
76. "Kernresonanzfluoreszenz in Gammastrahlung im Ir^{191}," *Zeitschrift für Physik* 151 (1958):124–43.
77. See the communication by R. V. Pound and G. A. Rebka in *Phys. Rev. Letters* 4 (1960):337.
78. "Untersuchungen über etwaige Änderungen des Gesamtgewichtes chemisch sich umsetzender Körper," *Zeitschrift für physikalische Chemie* 12 (1903):1–11.
79. *Inaugural Lecture on Astronomy and Papers on the Foundations of Mathematics*, translated by G. Waldo Dunnington (Baton Rouge; Louisiana State University Press, 1937), p. 78.
80. "On the Illumination of Lines of Molecular Pressure and the Trajectory of Molecules," *Phil. Trans.* (London) 170 (1879):164.
81. For an authoritative treatment of the developments of high and ultrahigh vacuum techniques since the turn of the century, see S. Dushmann, *Scientific Foundations of Vacuum Technique* (2d ed.; New York: Wiley, 1962), pp. 118–80 and 673–90.
82. On the origin and development of the "personal equation," see E. G. Boring, *A History of Experimental Psychology* (New York: Century, 1929), pp. 135–50.
83. *British Assn. Report* (1904), p. 427.

84. A very clear illustration of this is the elimination of much of the discrepancy that existed since Newton's time between the predicted and observed values of the speed of sound in air. The discrepancy was about 20 per cent and its possible source was first indicated by Laplace in the heating of a gas when suddenly compressed. When two outstanding experimentalists, Delaroche and Berard succeeded in incorporating Laplace's suggestion in their experiments, the discrepancy was reduced to 2.5 per cent. Such and similar achievements could only corroborate the belief that with steady improvements in precision, theory and experiment were bound to meet in the end in perfect convergence. For the details of this case, see T. S. Kuhn, "The Caloric Theory of Adiabatic Compression," *Isis* 49 (1950):132–40.

85. It is on this point that rests M. Born's incisive reasoning about the gratuitousness of absolute determinism in mechanics, a tenet never seriously questioned by classical physics. See M. Born, "Is Classical Mechanics in Fact Deterministic?" in *Physics in My Generation* (London: Pergamon Press, 1956), pp. 164–70.

86. "A Brief Account of Microscopical Observations," *Phil. Mag.* (London) 4 (1828):161–73.

87. "Mouvement brownien et réalité moléculaire" (1909), in *Oeuvres scientifiques de Jean Perrin* (Paris: Centre national de la recherche scientifique, 1950), p. 176.

88. "The New Visions of Science," *Harper's Magazine* 158 (1929):450.

89. *Ibid.*

90. It is possible, however, that there will be some slowdown in the rate by which the discrepancies present in the correlation of fundamental constants are reduced. As J. W. M. DuMond noted recently ("Pilgrim's Progress in Search of the Fundamental Constants," *Physics Today* 18 [October, 1965]:26–43), progress in establishing more precise values for fundamental constants is possible only if experiments are duplicated in as many laboratories and in as many different ways as possible. Needless to say, such experiments increasingly demand more money and time. As a result scientists might be all the more tempted to adopt current standardizations as definitive and "lose sight of the fact that scientific truth in this field is only to be reached by looking for discrepancies and disagreements" (p. 43). Now, the effectiveness of the edge of precision depends essentially on the measure of accuracy by which the values of fundamental constants are known. This knowledge can, however, be perfected if work in this field is viewed, to quote DuMond again, "a free battlefield, no place for worshippers" (p. 43). Only this openness of mind may help one understand that a revised list of fundamental constants implies at times more than taking into account recent improvements. The possibility is always present that systematic errors will be discovered in the available set of data, as happened to the older set of values for the speed of light, a case that stands today as a monument of the "most astonishing systematic errors in the history of physics." On this last point, see E. R. Cohen and J. W. M. DuMond, "Our Knowledge of the Fundamental Constants of Physics and Chemistry in 1965," *Rev. Mod. Physics* 37 (1965):537–94.

91. "The Highest Aim of the Physicist" (1899), in *The Physical Papers of Henry Augustus Rowland* (Baltimore: Johns Hopkins Press, 1902), p. 675.

92. Letter of Aug. 7, 1962, to B. Waldman, director of Midwestern Universities Research Association.

93. "Evidence for High-Energy Cosmic-Ray Neutrino Interactions," *Phys. Rev. Letters* 15 (1965):429–33.

94. One of the best illustrations of this is the several "generations" of detection chambers, ranging from the simple cloud chambers to the highly elaborate bubble chambers and spark chambers.

95. *Kosmos* (Cambridge, Mass.: Harvard University Press, 1932), p. 134.

96. On Priestley's and other eighteenth-century attempts to measure the pressure of light, see J. Priestley, *The History and Present State of Discoveries Relating to Vision, Light and Colours* (London, 1772), pp. 383–90. Priestley discussed the experiments of Homberg, Mairan, and Michell and took the view that Michell's experiments had demonstrated the effects of the impulse of light rays (p. 389).

97. Still with all the advances of modern scientific instruments, "waiting periods" of several decades are not uncommon. The detection of antiprotons and neutrinos took well over two decades and several of the predictions of General Relativity are still in need of experimental verification.
98. *The Pensées*, translated by J. M. Cohen (Baltimore, Md.: Penguin, 1961), p. 64.
99. *Science and the New Civilization* (New York: Scribner, 1930), p. 164.

CHAPTER VII. Physics and Biology

1. I. B. Hart, *The Mechanical Investigations of Leonardo da Vinci* (2d ed.; Berkeley: University of California Press, 1963), p. 148.
2. *Ibid.*, p. 153.
3. *De humani corporis fabrica*, preface. For an English translation of the preface by B. Farrington, see *The Autobiography of Science*, edited by F. R. Moulton and J. J. Schifferes (Garden City, N.Y.: Doubleday, 1946), pp. 94–104.
4. Letter of January, 1630, in *Oeuvres*, I, 105–6.
5. *Traité de l'homme*, in *Oeuvres*, XI, 120, 202.
6. *Discours de la méthode*, in *Oeuvres*, VI, 201.
7. *Traité de l'homme*, in *Oeuvres*, XI, 202, 165.
8. Letter of Nov. 23, 1646, to the Marquis of Newcastle, in *Oeuvres*, IV, 575.
9. *Traité de l'homme*, in *Oeuvres*, XI, 131.
10. *The Assayer* (1623), in *Discoveries*, p. 277.
11. *Ibid.*, p. 276.
12. *Ibid.*
13. *Movement of the Heart and Blood in Animals* (1628), translated by K. J. Franklin (Oxford: Blackwell, 1957), p. 39.
14. *Ibid.*
15. *A Disquisition about the Final Causes of Natural Things*, in *Works*, V, 427.
16. *Movement of the Heart and Blood in Animals*, p. 57.
17. *Micrographia* (1665), reprinted in R. T. Gunther (ed.), *Early Science in Oxford*, XIII (Oxford: Oxford University Press, 1938), 154.
18. *Ibid.*
19. *Ibid.*, p. 133.
20. Proposition 186, (2d ed.; Leyden, 1685), p. 282.
21. *Ibid.*, preface (n.p.).
22. Fifth letter to Clarke (1716), in *Leibniz Selections*, p. 276.
23. "Preface to the General Science" (1677), *ibid.*, p. 13.
24. "Considerations on the Principles of Life" (1705), *ibid.*, p. 197.
25. "New System of Nature" (1695), *ibid.*, p. 107.
26. *Ibid.*, p. 112.
27. See the "General Scholium" of the *Principia* and Queries 24, 28, and 31 of the *Opticks*.
28. *Statical Essays Containing Vegetable Staticks* (2d ed.; London, 1731), p. 2.
29. London, 1694, p. 234.
30. *An Essay on Regimen, together with Five Discourses Medical, Moral and Philosophical* (2d ed.; London, 1740), p. 2.
31. "A Discourse concerning the Modern Theories of Generation," *Philosophical Transactions* (London) 16 (1691):474.
32. *The Desideratum, or Electricity Made Plain and Useful* (London, 1760), p. 12.
33. *Elementa physiologiae corporis humani*, Book 10, sec. 8, par. 15, IV (Lausanne, 1757–66), 378.
34. *Ibid.*, I, v–vi.
35. *Oeuvres de Maupertuis*, II (Lyons, 1756), 3–133.
36. *Ibid.*, pp. 169–84.
37. References are to the English translation, *Natural History*, published by J. S. Barr, I (London, 1792), 2.
38. *Ibid.*, X, 361.

39. *Ibid.*, 366.
40. *Ibid.*, II, 272.
41. *Ibid.*, 256.
42. *Ibid.*, X, 351.
43. *Ibid.*, 375.
44. Translated by E. Griffith and others, I (London, 1827), 11–24.
45. Part I, art. vii, par. 1., "Différences des forces vitales, d'avec les lois physiques" (Paris, An VIII [1799/1800]), pp. 93–98.
46. *Leçons d'anatomie comparée*, I (Paris, 1800), 47.
47. *Microscopical Researches into the Accordance in the Structure and Growth of Animals and Plants*, translated by H. Smith (London, 1847), p. 268.
48. Translated by H. C. Green (New York: Macmillan, 1927), p. 1.
49. See pp. 139, 185, 209–18.
50. *Ibid.*, p. 95.
51. *Ibid.*, p. 73.
52. *Ibid.*, p. 69.
53. *Ibid.*, p. 185.
54. *Ibid.*, p. 75.
55. *Ibid.*, p. 93.
56. *Ibid.*, p. 69.
57. *Ibid.*, p. 93.
58. "Die organische Bewegung in ihrem Zusammenhange mit dem Stoffwechsel" (1845), in *Gesammelte Schriften*, edited by J. J. Weyrauch (Stuttgart, 1893), p. 76.
59. "Über notwendige Konsequenzen und Inkonsequenzen der Wärmemechanik" (1869), *ibid.*, p. 355.
60. "The Aim and Progress of Physical Science" (1869), in *PLSS*, p. 394.
61. *Ibid.*, p. 396.
62. "On the Application of the Law of the Conservation of Force to Organic Nature" (1861), in *Wissenschaftliche Abhandlungen*, III (Leipzig, 1895), 579.
63. "Hermann Ludwig Ferdinand Helmholtz" (1879), in *Sci. Papers*, II, 593.
64. *Experimental Researches in Electricity*, Fifteenth Series, par. 1749–95, "Notice of the Character and Direction of the Electric Force of the Gymnotus." See Great Books of the Western World, XLV (Chicago: Encyclopaedia Britannica, 1952), 540.
65. S. P. Thompson, *The Life of William Thomson*, II (London: Macmillan, 1910), 1103.
66. "Atom" (1876), in *Sci. Papers*, II, 461.
67. "Biogenesis and Abiogenesis" (1870), in *Discourses Biological and Geological Essays* (New York: Appleton, 1925), p. 260.
68. Letter of Dec. 19, 1844; quoted in B. Jones, *The Life and Letters of Faraday*, II (Philadelphia, 1870), 188.
69. *Memoir and Scientific Correspondence of the late Sir George Gabriel Stokes*, edited by J. Larmor, I (Cambridge: Cambridge University Press, 1907), 33.
70. See P. Thompson, *The Life of William Thomson*, II, 1099.
71. *On the Origin of Species* (reprint of the first edition; London: Watts, 1950), pp. 114–16, 144.
72. Edinburgh, 1795, I, 200.
73. *Illustrations of the Huttonian Theory of the Earth* (1802), in *The Works of John Playfair*, I (Edinburgh, 1822), 431–32.
74. *Ibid.*, p. 130.
75. Anonymous, "Lyell's Principles of Geology," *Quarterly Review* 35(1835):410.
76. Letter of Sept. 5, 1857, in Charles Darwin and Alfred Russel Wallace, *Evolution by Natural Selection* (Cambridge: Cambridge University Press, 1958), p. 265.
77. *On the Origin of Species*, p. 241.
78. *Ibid.*, p. 244.
79. "On the Secular Cooling of the Earth" (1862), in *Mathematical and Physical Papers*, III (London, 1890), 295.

80. *PLA*, I, 368.
81. *Ibid.*
82. *Mathematical and Physical Papers*, III, 295–311.
83. *Ibid.*, p. 300.
84. "The Doctrine of the Uniformity in Geology Briefly Refuted," in *PLA*, II, 6–9.
85. *PLA*, II, 10–72.
86. *Ibid.*, p. 44.
87. S. Haughton, *Manual of Geology* (London, 1865), p. 82.
88. See J. Marchant, *Alfred Russell Wallace: Letters and Reminiscences* (New York: Harper and Brothers, 1916), p. 220.
89. Letter of Jan. 31, 1869, in *More Letters of Charles Darwin*, edited by F. Darwin and A. C. Seward, II (New York: D. Appleton, 1903), 164.
90. *Lay Sermons: Addresses and Reviews* (5th ed.; London, 1874), pp. 228–54.
91. *Ibid.*, p. 249.
92. *Ibid.*, p. 244.
93. "Of Geological Dynamics" (1869), in *PLA*, II, 89.
94. "On the Age of the Earth," *Nature* 51 (1895):225.
95. *Ibid.*, p. 226.
96. *Ibid.*, p. 227.
97. *Ibid.*
98. "Twenty-Five Years of Geological Progress in Britain," *Nature* 51 (1895):369.
99. "The Age of the Earth," *American Journal of Science* 45 (1893):1–20.
100. *Nature* 51 (1895):440.
101. "An Estimate of the Geological Age of the Earth." For a summary, see *Nature* 62 (1900):235–37.
102. "Darwin and the Origin of Species" (1867), in F. Jenkin, *Papers, Literary, Scientific etc.*, edited by S. Colvin and J. A. Ewing (London, 1887), pp. 215–63.
103. *Ibid.*, p. 229.
104. "The Theory of Selection from a Mathematical Point of View," *Nature* 3 (1870):30–33.
105. *Ibid.*, p. 32.
106. "Origin of Life on Earth and Elsewhere," in *The Logic of Personal Knowledge: Essays Presented to Michael Polanyi on His Seventieth Birthday* (London: Routledge and Kegan Paul, 1961), p. 209.
107. M. Florkin (ed.), *Aspects of the Origin of Life* (London: Pergamon Press, 1960), p. vii.
108. Letter of March 29, 1863, to J. D. Hooker, in F. Darwin, *The Life and Letters of Charles Darwin*, III (London, 1888), 18.
109. *Ibid.*
110. L. Huxley, *Life and Letters of Thomas Henry Huxley*, I (London: D. Appleton, 1900), 244.
111. "Biogenesis and Abiogenesis," p. 259.
112. "The Factors of Organic Evolution," in *Essays, Scientific, Political and Speculative*, I (New York, 1899), 458–59.
113. "Discussion sur la fermentation," in *Oeuvres de Pasteur*, edited by Pasteur Vallery-Radot, VI (Paris: Masson 1922–39), 57.
114. Quoted by W. S. Lilly, "Darwinism and Democracy," *The Fortnightly Review* 39 (1886):35.
115. "Vitalism," *The Nineteenth Century* 44 (1898):403–4.
116. *Essays upon Heredity* (Oxford, 1889), p. 33.
117. *Mechanisch-physiologische Theorie der Abstammungslehre* (Munich, 1884), p. 83.
118. *Brit. Assn. Rep.* (1874), p. XCII.
119. In his "Twenty-third Letter on Chemistry" (1856). See J. von Liebig, *Chemische Briefe* (6th ed., Leipzig, 1878), p. 182.
120. "The Kinetic Theory of the Dissipation of Energy," *Proc. Roy. Soc.* (Edinburgh) 8 (1874):326.
121. *Ibid.*

122. "On Mechanical Antecedents of Motion, Heat and Light" (1854), in *Mathematical and Physical Papers*, II (Cambridge, 1884), 37–38.
123. "On the Age of the Sun's Heat," p. 350.
124. "On the Sun's Heat" (1887), *PLA*, I, 415.
125. "Lord Kelvin's Address on the Age of the Earth as an Abode Fitted for Life," *Science* 9 (1899):889–901; 10 (1899):11–18. It was in a sense anticlimactic that Chamberlin's paper pointed out in detail all the non sequiturs with which Kelvin's three famous arguments were teeming. As the revolutionary aspects of radio-activity came to unfold, the physical data in Kelvin's arguments began to appear rather irrelevant to the problem of the age of the sun and the earth. All this could only add further weight to the incisive remark of Chamberlin on the shortcomings of mathematical physics "The fascinating impressiveness of rigorous mathematical analysis, with its atmosphere of precision and elegance, should not blind us to the defects of the premises that condition the whole process. There is, perhaps, nc beguilement more insidious and dangerous than an elaborate and elegant mathe, matical process built upon unfortified premises" (p. 890).
126. *Inorganic Evolution as Studied by Spectrum Analysis* (London: Macmillan, 1900) p. 73.
127. "A Suggested Explanation of Radioactivity," *Nature* 70 (1904):101.
128. "Sur la chaleur dégagée spontanément par les sels de radium," *Comptes rendus* 136 (1903):673–75.
129. "Heating Effect of the Radium Emanation," *Nature* 68 (1903):622.
130. A. S. Eve, *Rutherford* (New York: Macmillan, 1939), p. 107.
131. "Radium as the Cause of the Earth's Heat" (1905), in *The Collected Papers of Lord Rutherford of Nelson*, edited by J. Chadwick, I (New York: Interscience, 1962), 776–85.
132. Eve, *Rutherford*, p. 107.
133. B. H. Hoyer, B. J. McCarthy, and E. T. Bolton, "A Molecular Approach in the Systematics of Higher Organisms," *Science* 144 (1964):959–67.
134. In a note from 1875, in *Oeuvres de Pasteur*, I, 364–66. The most detailed presentation of Pasteur's thought on life and dissymmetry is contained in his two great lectures given in early 1860 before the Chemical Society of Paris, in *Oeuvres*, I, 314–44.
135. Collected in the volume, *Physico-chemical Evolution*, translated by J. R. Clarke (London: Methuen, 1925), see especially pp. 49–56.
136. "Function and Design," *Brit. Assn. Rep.* (1926), pp. 208–18.
137. "On the Molecular Structure of Chromosomes," *Protoplasma* 25 (1936):550–69.
138. *The Origin of Life*, translated by S. Morgulis (New York: Macmillan, 1938), pp. 59–60.
139. G. Wald, "The Origin of Life," in *The Physics and Chemistry of Life* (New York: Simon and Schuster, 1955), p. 12.
140. *Ibid.*
141. *Ibid.*
142. *Ibid.*, p. 5.
143. *Time's Arrow and Evolution* (Princeton, N.J.: Princeton University Press, 1955), p. 170.
144. *What Is Life and Other Scientific Essays* (Garden City, N.Y.: Doubleday, 1956), p. 61.
145. *Ibid.*, p. 83.
146. "The Nature of Viruses, Cancer, Genes and Life — A Declaration of Dependence," *Smithsonian Institution Report* (Washington, D.C.: U.S. Government Printing Office, 1958), pp. 357–70.
147. "Chemical Evolution and the Origin of Life," *American Scientist* 44 (1956): 247–63.
148. *Ibid.*, p. 263.
149. R. Beutner, *Life's Beginning on the Earth* (Baltimore: Williams, 1938), p. 78.
150. "Some Assumptions Underlying Discussion on the Origins of Life," *Annals of the New York Academy of Sciences* 69 (1957):376.

151. "Chemical Diversity and the Origins of Life," in M. Florkin (ed.), _Aspects of the Origin of Life_, p. 62.
152. "Zur Frage einer spezifischen Anziehung zwischen Genmolekülen," _Physikalische Zeitschrift_ 39 (1938):711–14.
153. "The Nature of the Intermolecular Forces Operative in Biological Processes," _Science_ 92 (1940):77–79.
154. "The General and Logical Theory of Automata," in _Cerebral Mechanism in Behavior: The Hixon Symposium_, edited by L. A. Jeffress (New York: Wiley, 1951), p. 1.
155. "Molecular Structure of Nucleic Acids. A Structure for Deoxyribose Nucleic Acid," _Nature_ 171 (1953):737–38.
156. _The Physical Foundation of Biology_ (New York: Pergamon Press, 1958), p. 87.
157. _Ibid._, p. 129.
158. _Ibid._, 131.
159. "The Probability of the Existence of a Self-reproducing Unit," in _The Logic of Personal Knowledge_, pp. 231–37.
160. _Ibid._, p. 233.
161. See I. Bernard Weinstein, "Comparative Studies on the Genetic Code," in _Cold Spring Harbor Symposia on Quantitative Biology_, Vol. XXVIII: _Synthesis and Structure of Macromolecules_ (Cold Spring Harbor, N.Y.: 1963), pp. 579–80.
162. "DNA and the Chemistry of Inheritance," _American Scientist_ 52 (1964):387.
163. _Of Men and Galaxies_ (Seattle: University of Washington Press, 1964), p. 36.
164. _Time_, Jan. 11, 1960, p. 32. W. W. Howells, a past president of the American Anthropological Association, challenged no less emphatically the hopes of some physicists on this point. In his book, _Mankind in the Making_ (Garden City, N.Y.: Doubleday, 1959), p. 345, he wrote: "I will lay a small bet that the first men from Outer Space will be neither bipeds nor quadrupeds but bimanous quadrupedal hexapods. (I have just invented the last word, in the hope that it means six limbs.)"
165. _Mathematical Biophysics: Physico-mathematical Foundations of Biology_, II (3d rev. ed.; New York: Dover, 1960), 405.
166. L. Brillouin, "Life, Thermodynamics and Cybernetics," _American Scientist_, 37 (1949):554.
167. "Humanistic Biology," _American Scientist_, 53 (1965):4–19.
168. R. W. Gerard (ed.), _Concepts of Biology_ (Washington, D.C.: National Academy of Sciences, 1958), p. 145. It might be noted that E. Nagel, who voiced his doubt that life-sciences have succeeded in proving that they are concerned also with characteristics wholly different from those present in the subject matter of physics and chemistry, welcomed the protests of organismic biologists "against the dogmatism often associated with the mechanistic standpoint in biology." _The Structure of Science: Problems in the Logic of Scientific Explanation_ (New York: Harcourt, Brace and World, 1961), p. 445.
169. Gerard, _Concepts of Biology_, p. 115.
170. See his essay "Zoology" in A. E. Heath (ed.), _Scientific Thought in the Twentieth Century_ (London: Watts, 1951), pp. 188–89.
171. "Light and Life," _Nature_ 131 (1933):457.
172. _Physics and Philosophy_ (New York: Harper, 1958), p. 155.
173. "Light and Life Revisited," in _Essays 1958–1962 on Atomic Physics and Human Knowledge_ (New York: Interscience, 1963), p. 26.
174. _De anima_ 415b.

CHAPTER VIII. Physics and Metaphysics

1. In a paper of 1873 on determinism and contingency published posthumously in L. Campbell and W. Garnett, _The Life of James Clerk Maxwell_ (London, 1882), p. 436.
2. Letter to Belisario Vinta (1610), in _Discoveries_, p. 64.

3. D. Brewster, *Memoirs of the Life, Writings and Discoveries of Sir Isaac Newton,* I (Edinburgh, 1855), 95. True to the same semantics, Hooke urged Newton in 1679 to communicate to the society "of what shall occur to you that is philosophical." See A. Koyré, "An Unpublished Letter of Robert Hooke to Isaac Newton," *Isis* 43 (1952):312–37.
4. Translated by R. Kerr, in Great Books of the Western World, XLV (Chicago, 1952), 3.
5. S. P. Thompson, *The Life of William Thomson,* II (London: Macmillan, 1910), 1124.
6. "On Atmospheric Electricity" (1860), in *Reprint of Papers on Electrostatics and Magnetism* (2d ed.; London, 1884), p. 222.
7. *The Life of William Thomson,* II, 1183.
8. "On the Modern Development of Faraday's Conception of Electricity" (1881), in *Wissenschaftliche Abhandlungen,* III (Leipzig, 1895), 53.
9. "Ueber den Ursprung der richtigen Deutung unserer Sinneinsdrücke" (1894), in *Wissenschaftliche Abhandlungen,* III, 536.
10. "The Architecture of Theories," *The Monist* 1 (1891):174.
11. Preface, in *Werke,* II, 66.
12. *Die Mechanik in ihrer Entwicklung* (Leipzig, 1883).
13. 2d ed.; London: A & C Black, 1900, p. 15.
14. *Ibid.,* p. 17.
15. "Remarks on Bertrand Russell's Theory of Knowledge," in P. A. Schilpp (ed.), *The Philosophy of Bertrand Russell* (Evanston, Ill.: Library of Living Philosophers, 1946), p. 289.
16. "The Moral Un-neutrality of Science," *Science* 133 (1961):257.
17. 6th ed.; New York: Hafner, 1957, p. 312.
18. "The Evolution of the Physicist's Picture of Nature," *Sci. Amer.* 208 (May, 1963):48.
19. O. J. Lee, *Measuring Our Universe* (New York: Ronald, 1950), pp. 149–50.
20. Such mistaken view about science inspires most of the time those authors who in discussing non-scientific topics try to emulate the exactness of science. One typical case should suffice as an illustration. According to H. Miller, Darwin's demonstration of the evolution of natural types "fulfilled the intention of empirical thought, and closed the portals forever upon traditional philosophy. Metaphysicians may continue to announce their speculations about the everlasting structure of things, and about the universal criteria of knowledge; but their devotions are a wake, administered to a corpse." *History and Science: A Study of the Relation of Historical and Theoretical Knowledge* (Berkeley: University of California Press, 1939), p. 85.
21. See H. Hartmann, *Max Planck als Mensch und Denker* (3d rev. ed.; Basel: Ott, 1953), p. 49.
22. Nov. 1, 1844, in *Werke,* XII (Göttingen, 1877), 62–63.
23. "On the Relation of Natural Science to General Science" (1862), in *PLSS,* pp. 7–8.
24. L. Koenigsberger, *Hermann von Helmholtz,* I (Braunschweig: F. Vieweg, 1902), 292.
25. Address to the Mathematical and Physical Section of the British Association (1870), in *Sci. Papers,* II, 216.
26. In *Works,* IV, 47.
27. In a letter of Nov. 1, 1844, in *Werke,* XII, 63.
28. *The Life of James Clerk Maxwell,* p. 436.
29. "Science, Philosophy and Religious Wisdom" (1952), in A. C. Pegis (ed.), *A Gilson Reader* (Garden City, N.Y.: Doubleday, 1957), p. 216.
30. *The Nature of the Physical World* (Cambridge: Cambridge University Press, 1928), p. 352.
31. *An Essay Concerning Human Understanding,* edited by A. C. Fraser, II (Oxford, 1894), 215.
32. *Ibid.,* p. 218.
33. *A Treatise of Human Nature,* edited by L. A. Selby-Bigge (Oxford: Clarendon Press, 1951), p. 636.

34. *Untersuchung über die Deutlichkeit der Grundsätze der naturlichen Theologie und der Moral* (1764), in *Werke*, II, 283.
35. *Ibid.*, p. 286.
36. *Kant's Prolegomena to Any Future Metaphysics*, edited by P. Carus (Chicago: Open Court, 1949), p. 2.
37. *De l'intelligence* (Paris, 1870), preface.
38. In *Gesammelte Werke*, XIX (Munich: Musarion Verlag, 1920–29), 373.
39. *The Evolution of Matter*, translated by F. Legge (London: Walter Scott Publishing Co., 1907), p. 18.
40. Translated into English under the title, *The Evolution of Forces* (London: Kegan Paul, 1908).
41. *The Evolution of Forces*, p. 112
42. *Ibid.*
43. *Ibid.*, p. 11.
44. *Ibid.*
45. *Ibid.*, pp. 15–16.
46. "Ist die Trägheit eines Körpers von seinem Energieinhalt abhängig?" *Annalen der Physik* 18 (1905):639–41.
47. J. von Olivier, *Monistische Weltanschauung* (Leipzig: C. G. Naumann, 1906), p. 2.
48. *Course of Experimental Philosophy*, I (3d ed.; London, 1763), viii.
49. *The Mysterious Universe* (London: Macmillan, 1930), p. 136.
50. For a detailed evaluation of the measure of Kant's competence in mathematics, physics, and other branches of science, see E. Adickes, *Kant als Naturforscher* (Berlin: De Gruyter, 1924).
51. *Popular Scientific Lectures*, edited by M. Kline (New York: Dover, 1962), pp. 223–49.
52. *Dialogue*, p. 53.
53. In *Oeuvres*, XXI, 120.
54. *Ibid.*
55. "Science, Philosophy and Religious Wisdom," p. 217.
56. Letter to Bossuet (1694), in *Correspondence de Bossuet*, edited by C. Urban and E. Levesque, VI (new ed.; Paris: Hachette, 1909–25), 523.
57. *Siècle de Louis XIV*, in *Oeuvres*, XIV, 534.
58. *Essai sur les élémens de philosophie*, in *Oeuvres philosophiques, historiques et littéraires de d'Alembert*, II (Paris, 1805), 326.
59. *Traité de dynamique* (Paris, 1758), preface.
60. *Essai sur les élémens de philosophie*, p. 404.
61. *Ibid.*, pp. 323–24.
62. *The Metaphysical Foundations of Modern Physical Science* (2d ed.; New York: Harcourt, 1932), p. 208.
63. "Espace" (1755), in *Encyclopédie, ou dictionnaire raisonné des sciences, des arts et des métiers*, V (Paris, 1755), 949.
64. *Essai sur les élémens de philosophie*, p. 321.
65. "On the Hypotheses which lie at the Bases of Geometry," translated by W. K. Clifford, in R. Tucker (ed.), *Mathematical Papers by W. K. Clifford* (London, 1882), p. 56.
66. "On the Generalized Theory of Gravitation" (1950), in *Ideas and Opinions by Albert Einstein* (New York: Crown, 1954), p. 342.
67. *The Metaphysical Foundations of Modern Physical Science*, p. 227.
68. *The Principle of Relativity* (Cambridge: Cambridge University Press, 1922), p. 6.
69. *The Mysterious Universe*, p. vii.
70. *Ibid.*
71. *Physics and Philosophy* (Cambridge: Cambridge University Press, 1943), preface.
72. *The Principle of Relativity*, p. 6.
73. "Physics and Reality" (1936), in *Out of My Later Years* (New York: Philosophical Library, 1950), p. 58.
74. *Physics and Microphysics*, translated by M. Davidson (New York: Pantheon Books, 1955), p. 238.

75. "On the Method of Theoretical Physics" (1933), in *The World as I See It* (New York: Covici, 1934), p. 36.
76. "Physics and Reality," p. 78.
77. "The Problem of Space, Ether, and the Field in Physics" (1930), in *The World as I See It*, p. 92.
78. "Clerk Maxwell's Influence on the Development of the Conception of Physical Reality" (1931), in *The World as I See It*, p. 60.
79. "Address to the Conference on Science, Philosophy and Religion" (1940), in *Out of My Later Years*, p. 26.
80. *Dialogue*, p. 328.
81. *Where Is Science Going?*, translated by J. Murphy (New York: Norton, 1932), p. 214.
82. "A Scientist's Case for the Classics," *Harper's Magazine* 216 (May, 1958):29.
83. *Philosophy of Mathematics and Natural Science* (Princeton, N.J.: Princeton University Press, 1949), p. vi.
84. See E. P. Wigner, "The Unreasonable Effectiveness of Mathematics in the Natural Science," *Communications on Pure and Applied Mathematics* 13 (Feb., 1960):12.
85. *The Evolution of Physics* (New York: Simon, 1938), pp. 312–13.
86. *The Constitution of Matter* (Eugene: Oregon State System of Higher Education, 1956), p. 37.
87. *Open Vistas: Philosophical Perspectives of Modern Science* (New Haven, Conn.: Yale University Press, 1961), p. 76.
88. In the foreword to M. Born, *Physics and Politics* (Edinburgh: Oliver, 1962), p. v.
89. *The Scientific Outlook* (New York: Norton, 1931), pp. 97, 95.
90. *Dialogue*, p. 341.
91. *Exposition du système du monde*, Book IV, chap. 18, in *Oeuvres*, VI, 356.
92. *Brit. Assn. Rep.*, (1876), p. 38.
93. Letter of Jan. 10, 1880; see L. P. Wheeler, *Josiah Willard Gibbs: The History of a Great Mind* (rev. ed.; New Haven, Conn.: Yale University Press, 1952), p. 89.
94. "Mémoire couronné sur la diffraction," in *Oeuvres complètes d'Augustin Fresnel*, I (Paris, 1866), 248.
95. *The Principles of Mechanics*, translated by D. E. Jones and J. T. Walley (London, 1899), p. xxi.
96. *Ibid.*, p. 23.
97. "On the Generalized Theory of Gravitation" (1950), in *Ideas and Opinions by Albert Einstein*, p. 342.
98. See Part I of his *Universal Natural History and Theory of Heavens*, "Sketch of a Systematic Constitution among the Fixed Stars, likewise of the Plurality of such Star Systems," in *Werke*, I, 247–58.
99. *An Original Theory or New Hypothesis of the Universe* (London, 1750), p. 62.
100. Letter to Maskelyne (1782); see J. B. Sidgwick, *William Herschel, Explorer of the Heavens* (London: Faber & Faber, 1953), p. 124.
101. *Experimental Researches in Electricity*, Nineteenth Series, par. 26; in Great Books of the Western World, XLV (Chicago, 1952), 595.
102. Entry of July 28, 1836, *Faraday's Diary*, edited by T. Martin, III (London: Bell, 1932–36), 68.
103. *The Science of Mechanics*, translated by T. J. McCormack (La Salle, Ill.: Open Court, 1960), p. 35.
104. *Philosophy of Mathematics and Natural Science*, p. 160.
105. *Where Is Science Going?*, p. 215.
106. "Intuition and Abstraction in Scientific Thinking," *Annals of the Japan Association for Philosophy of Science* 2 (1962):97.
107. Letter to Maxwell, Nov. 13, 1857; see B. Jones, *The Life and Letters of Faraday*, II (Philadelphia, 1870), 392–93.
108. *Sci. Papers*, II, 789.
109. *A Treatise on Electricity and Magnetism*, II (3d ed.; Oxford, 1892), 176.
110. "On the Modern Development of Faraday's Conception of Electricity," p. 53.
111. See J. Tyndall, *Faraday as a Discoverer* (5th ed.; London, 1894), p. 54.

112. "Über statistische Mechanik," in *Populäre Schriften* (Leipzig: Barth, 1905), p. 363.
113. *Space, Time and Gravitation* (Cambridge: Cambridge University Press, 1920), p. vi.
114. *Relativity Theory of Protons and Electrons* (New York: Macmillan, 1936), p. 8.
115. Sir Cyril Hinshelwood in his presidential address to the British Association, *Nature* 207 (1965):1060.
116. *Scientific Autobiography*, translated by F. Gaynor (New York: Philosophical Library, 1949), pp. 101–2.
117. *Philosophy and Logical Syntax* (London: Kegan Paul, 1935), pp. 15–18.
118. "On the Method of Theoretical Physics," p. 34.
119. *The Nature of the Physical World* (Cambridge: Cambridge University Press, 1928), p. xiv.
120. *Ibid.*
121. *Ibid.*, p. 344.
122. *Ibid.*, p. xiv.
123. *Mysticism and Logic, and Other Essays* (New York: Longmans, Green & Co., 1921), p. 129.
124. *The Concept of Nature* (Cambridge: Cambridge University Press, 1930), p. 44.
125. *Adventures of Ideas* (New York: Macmillan, 1933), p. 186.
126. *Religion and Science* (New York: Holt, 1935), p. 153.
127. "The Quantum Theory of Radiation," *Phil. Mag.* (London) 47 (1924):790.
128. For the details of these experiments, see R. S. Shankland, *Atomic and Nuclear Physics* (2d ed.; New York: Macmillan, 1960), pp. 208–13.
129. See our discussion of the indeterminacy principle in Chapter VI.
130. "Über den anschaulichen Inhalt der quantentheoretischen Kinematik und Mechanik," *Zeitschrift für Physik* 43 (1927):197.
131. *Mathematical Foundations of Quantum Mechanics*, translated by R. T. Beyer (Princeton, N.J.: Princeton University Press, 1955), p. 327.
132. Letter to the Editor, *Nature* 126 (1930):995.
133. *The Nature of Physical Reality* (New York: McGraw-Hill, 1950), p. 418. It only shows the philosophical poverty of our times that the theory of relativity could be turned by so many into the self-contradictory dogma of "everything is relative." Yet, did physicists emphasize enough that basic features of the theory, such as the constancy of the speed of light, the invariance of the basic laws of physics, are anything but "relative" in character?
134. W. Kauzmann, *Quantum Chemistry* (New York: Academic Press, 1957), pp. 239–40.
135. *Natural Philosophy of Cause and Chance* (Oxford: Oxford University Press, 1949), p. 4.
136. *Handbuch der physiologischen Optik*, III (3d ed.; Hamburg: L. Voss, 1910), 30.
137. *Metaphysics* 982b.
138. *Our Knowledge of the External World* (New York: Norton, 1929), p. 223.
139. "On the Notion of Cause," (1912), in *Mysticism and Logic, and Other Essays*, p. 180.
140. "Physics and Reality," p. 61.
141. *Ibid.*
142. *Physics and Microphysics*, pp. 208–9.
143. Quoted as a private communication of Schrödinger to B. Bertotti in the latter's "Quantum Mechanics and the Uniqueness of the World," *Nuovo Cimento* (Supplemento) 17(1960):7.
144. *The Guide for the Perplexed*, translated by M. Friedländer (London: Routledge, 1904), p. 121. The number is large enough to permit comparison with 10^{24}, which is the order of magnitude of the number of time atoms, or chronons, in one second, as estimated by modern theory.
145. Book II, chap. 15, edition cited, I, 169.
146. *De la contingence des lois de la nature* (Paris, 1874).
147. *Physics and Microphysics*, pp. 190–91.

148. *The Physical Foundation of Biology* (New York: Pergamon Press, 1958), p. 142.
149. *Science and Wisdom* (New York: Scribner, 1940), p. 53.
150. Quoted by A. Sommerfeld, "To Albert Einstein's Seventieth Birthday," in P. A. Schilpp (ed.). *Albert Einstein: Philosopher-Scientist* (Evanston, Ill.: Library of Living Philosophers, 1949), p. 99.
151. Quoted by R. Jungk, *Brighter than a Thousand Suns*, translated by J. Cleugh (New York: Harcourt, 1958), p. 53.
152. "Einstein's Conception of Reality," in *Albert Einstein: Philosopher-Scientist*, p. 246.
153. "The Internal and the External Worlds," *Nature* 184 (1959):1835.
154. M. B. Hesse, *Forces and Fields* (London: Nelson, 1961), p. 303.
155. *Ibid.*
156. *Ibid.*
157. *Space, Time, Matter*, translated by H. L. Brose (London: Methuen, 1922), p. 10.
158. *Science and the Modern World* (New York: Macmillan, 1926), p. 27.
159. *Hermann von Helmholtz*, II, 162.
160. "En marge d'un texte," in *Louis de Broglie, physicien et penseur* (Paris: Michel, 1953), p. 158.
161. *Space, Time, Matter*, p. 10.
162. "Remarks on Bertrand Russell's Theory of Knowledge," p. 289.
163. "Physics and Reality," p. 59.
164. "Introductory Lecture on Experimental Physics" (1871), in *Sci. Papers*, II, 255.
165. N. R. Campbell, *Physics, the Elements* (Cambridge: Cambridge University Press, 1920), p. 11.

CHAPTER IX. Physics and Ethics

1. *Brit. Assn. Rep.* (1938), p. 20.
2. *The Laws*, Book 10, in *The Dialogues of Plato*, translated by B. Jowett (New York, 1871), pp. 907–8.
3. *Les principes de la philosophie*, prefatory letter, *Oeuvres*, IX, 14.
4. *On True Method in Philosophy and Theology* (ca. 1686), in *Leibniz Selections*, p. 58.
5. *The Monadology* (1714), par. 89, in *Leibniz Selections*, p. 551.
6. *Ibid.*
7. *An Essay on Regimen together with Five Discourses, Medical, Moral and Philosophical* (2d ed.; London, 1740), pp. 18–19.
8. *Philosophical Commentaries*, in *Works*, I, 85.
9. *Works*, III, 165.
10. See D. W. Singer, *Giordano Bruno: His Life and Thought. With Annotated Translation of His Work On the Infinite Universe and Worlds* (New York: Henry Schumann, 1950), pp. 263–64.
11. *Les principes de la philosophie*, in *Oeuvres*, IX, 27.
12. *The Monadology*, par. 79, 86, 87, in *Leibniz Selections*, pp. 549–51.
13. *Concerning Government and Society*, dedicatory epistle, in *The English Works of Thomas Hobbes*, edited by W. Molesworth, II (London, 1839–45), iv.
14. *Ethica, ordine geometrico demonstrata*, Part II, prop. 48.
15. *Ibid.*, Part II, prop. 49, *ad fin.*
16. *Le philosophe ignorant* (1766), in *Oeuvres*, XXVI, 55.
17. *Ibid.*, p. 56.
18. *Ibid.*, p. 78.
19. *Oeuvres de Maupertuis*, I (Lyon, 1756), 197.
20. Maupertuis' method was strongly disputed by Kant in his pre-critical *Attempt to Introduce the Concept of Negative Quantities into Philosophy* (*Werke*, I, 165–204). Kant insisted that only homogenous quantities can be added. According to him, human feelings were too complicated and variegated to be easily grouped in a few homogeneous classes. Later, in developing his own ethical theory, Kant

rejected eudaemonism as a basis for ethics and stressed the independence of ethical values and of the ethical imperative from any consideration of pleasure and pain.

21. *Man, a Machine,* French text and English translation with notes (La Salle, Ill.; Open Court, 1912), p. 128.
22. *Ibid.*
23. *Ibid.,* p. 127.
24. *Ibid.,* pp. 147–48.
25. *Ibid.,* p. 128.
26. New ed.; London, 1775. Of the thirty-six chapters, only three bear on physics proper; the rest are "mechanistic moralizing" and atheistic propaganda embellished by tireless references to Newtonian physics.
27. *Ibid.,* pp. 18–19.
28. *Ibid.,* p. 19.
29. *Essai sur l'application de l'analyse à la probabilité des décisions rendues à la pluralité des voix* (Paris, 1785), p. I.
30. *Ibid.,* p. CLXXXIX.
31. See F. E. Manuel, *The New World of Henri Saint-Simon* (Cambridge, Mass.: Harvard University Press, 1956), p. 119.
32. See especially his *Essay on the Foundations of Geometry* (Cambridge: Cambridge University Press, 1897), p. 49., and his *Principles of Mathematics* (Cambridge: Cambridge University Press, 1903), p. 465.
33. "A Free Man's Worship" (1903), in *Mysticism and Logic, and Other Essays* (London: Longmans, Green & Co., 1921), pp. 47–48.
34. "The Departure from Classical Thought in Modern Physics," in P. A. Schilpp (ed.), *Albert Einstein: Philosopher-Scientist* (Evanston, Ill.: Library of Living Philosophers, 1949), p. 196.
35. "La morale et la science," in *Dernières pensées* (Paris: Flammarion, 1913), pp. 240–41.
36. *Les origines de l'alchimie* (Paris, 1885), p. v.
37. See J. Lavrin, *Dostoevsky and His Creation* (London: W. Collins, 1920), pp. 170–71.
38. In L. Campbell and W. Garnett, *The Life of James Clerk Maxwell* (London, 1882), pp. 434–44.
39. *Ibid.,* p. 441.
40. *Sci. Papers,* I, 756–62.
41. *Ibid.,* p. 759.
42. *Ibid.,* p. 760.
43. *Ibid.,* p. 759.
44. *Ibid.,* p. 762.
45. *Ibid.*
46. S. P. Thompson, *The Life of William Thomson,* II (London: Macmillan, 1910), 1097.
47. G. Waldo Dunnington, *Carl Friedrich Gauss, Titan of Science* (New York: Exposition Press, 1955), p. 198.
48. L. Koenigsberger, *Hermann von Helmholtz,* I (Braunschweig: F. Vieweg, 1902), 291–92.
49. *A Survey of Physical Theory,* translated by R. Jones and D. H. Williams (New York: Dover, 1960), p. 68.
50. *Ibid.*
51. "Kausalgesetz und Willensfreiheit" (1923), in *Wege zur physikalischen Erkenntnis* (Leipzig: S. Hirzel, 1933), pp. 87–127.
52. "Positivismus und reale Auszenwelt" (1930), *ibid.,* pp. 208–32.
53. *Ibid.,* p. 231.
54. *Ibid.,* p. 232.
55. *Ibid.*
56. *Scientific Autobiography and Other Papers,* translated by F. Gaynor (London: Williams & Norgate, 1950), p. 185.
57. *Out of My Later Years* (New York: Philosophical Library, 1950), p. 25.

58. *Ibid.*
59. In a conversation with Manfred Clynes, during the winter of 1952–53. See P. Michelmore, *Einstein: Profile of the Man* (New York: Dodd, 1962), p. 251.
60. *Out of My Later Years,* p. 113.
61. "La morale et la science," in *Dernières pensées,* pp. 223–47.
62. *Ibid.,* p. 225.
63. English translation by G. B. Halstead in H. Poincaré, *The Foundations of Science* (Lancaster, Penn.: Science Press, 1913), p. 206.
64. Among the dissenters is C. H. Waddington, who insisted in his *Science and Ethics* (London: G. Allen & Unwin, 1941) that he could not see "that there is any real distinction between the problems 'how did I make my ethical choice' and 'how shall I make it now'" (p. 101).
65. *Brit. Assn. Report* (1931), p. 15.
66. E. A. Milne, *Sir James Jeans: A Biography* (Cambridge: Cambridge University Press, 1952), p. 44.
67. "Man and the Universe," in Sir James Jeans and others, *Scientific Progress* (London: Allen, 1936), p. 38.
68. *Physics and Philosophy* (Cambridge: Cambridge University Press, 1943), p. 216.
69. *The Nature of the Physical World* (Cambridge: Cambridge University Press, 1928), p. 295.
70. *Science and the Course of History,* translated by R. Manheim (New Haven, Conn.: Yale University Press, 1955), pp. 110–12. Many other examples could be quoted. N. Bohr, for instance, stated in his lecture, "Unity of Knowledge" (1954): "Above all, the recognition of inherent limitations in the notion of causality has offered a frame in which the idea of universal predestination is replaced by the concept of natural evolution." *Atomic Physics and Human Knowledge* (New York: Wiley, 1958), p. 81.
71. *New Pathways in Science* (Cambridge: Cambridge University Press, 1934), p. 88. In his Tarner Lectures of 1938, published under the title, *The Philosophy of Physical Science* (London: Macmillan, 1939), Eddington recognized that such an approach to the problem of the freedom of will is, as he put it, "nonsense" (p. 182).
72. *New Pathways in Science,* p. 88.
73. *The Nature of the Physical World,* p. 339.
74. *Ibid.,* p. 338.
75. *Ibid.,* p. 317.
76. "A Survey of the Sciences," *Science* 106 (1947):139.
77. *The Freedom of Man* (New Haven, Conn.: Yale University Press, 1935), p. 26.
78. *Science and Humanism: Physics in Our Time* (Cambridge: Cambridge University Press, 1951), p. 62.
79. *On Life and Letters. Third Series,* translated by D. B. Stewart, in *The Works of Anatole France,* XXIX (New York: Wells, 1924), 209.
80. *Astronomy and Cosmogony* (Cambridge: Cambridge University Press, 1928), p. 414.
81. *Of Stars and Men: The Human Response to an Expanding Universe* (Boston: Beacon Press, 1958), pp. 148–50.
82. *Astronomy and Cosmogony,* p. 414.
83. *Ibid.*
84. *Nature and the Greeks* (Cambridge: Cambridge University Press, 1954), p. 18.
85. Address by P. Kusch to Pulitzer Prize jurors, Columbia University, 1961, in *New York Herald Tribune,* April 2, 1961, sec. 2, p. 3, col. 4.
86. J. Tyndall, *Faraday as a Discoverer* (New York: D. Appleton, 1870), p. 32.
87. *An Essay on the First Principles of Government* (2d ed.; London, 1771), pp. 4–5.
88. *Nature, Addresses and Lectures,* in *Emerson's Complete Works,* I (Boston, 1889), 38. When it came to the details, Emerson's sweeping dictum bogged down, of course, in trivialities.
89. Sunday magazine section, p. 7.
90. *The Evolution of Matter,* translated by F. Legge (London: Walter Scott Publishing Co., 1907), p. 51.

91. "Physical Force — Man's Servant or His Master?" (1915), in F. Soddy, *Science and Life* (London: John Murray, 1920), p. 36.
92. *Ibid.*
93. *Science and the New Civilization* (New York: Scribner, 1930), p. 59.
94. *Nature* 132 (1933):432–33.
95. "Constitution of the Stars," in *Smithsonian Report* (1937), p. 141.
96. W. Laurence, *Men and Atoms* (New York: Simon & Schuster, 1959), pp. 5, 8–9.
97. "The Electrical Structure of Matter," *Science* 58 (1923):209.
98. See R. Jungk, *Brighter than a Thousand Suns*, translated by J. Cleugh (New York: Harcourt, 1958), p. 199.
99. E. Curie, *Madame Curie*, translated by V. Sheean (New York: Literary Guild of America, 1937), p. 204.
100. See E. W. Ashcroft, *Faraday* (London: British Electrical and Allied Manufacturers Association, 1931), pp. 24–25.
101. *Atomic Quest: A Personal Narrative* (New York: Oxford University Press, 1956), p. 128.
102. *Brighter than a Thousand Suns*, p. 199.
103. *Ibid.*
104. Quoted in L. Lamont, *Day of Trinity* (New York: Atheneum, 1965), p. 242.
105. In his Arthur Dehon Little Memorial Lecture, delivered at Massachusetts Institute of Technology, November 25, 1947. See J. Robert Oppenheimer, *The Open Mind* (New York: Simon, 1955), p. 88.
106. Einstein's personal communication to A. Vallentin, author of *Le drame d'Albert Einstein* (Paris: Plon, 1954), p. 229.
107. *Science and Life*, p. 36.
108. *Ibid.*
109. *In the Matter of J. Robert Oppenheimer* (Washington, D.C.: U.S. Government Printing Office, 1954), p. 326 (hereafter referred to as *Oppenheimer Hearings*). By "moral implications" Bethe obviously meant those distinct from the all-important original motivation, which was just as deeply moral in nature: to have the bomb before Nazi Germany would have it. Most scientists involved in the production of the bomb would have probably subscribed to Einstein's well-known apology: "If I had known that the Germans would not succeed in constructing the atom bomb, I would never have lifted a finger." R. Jungk, *Brighter than a Thousand Suns*, p. 87.
110. In an interview published in *The New York Times Magazine*, August 1, 1965, p. 9.
111. A particularly convincing analysis of this question was given recently by Masataka Iwata, a former lieutenant colonel in the Japanese army, and one of the five who plotted unsuccessfully to prevent Emperor Hirohito from declaring the surrender of Japan. See *The New York Times*, August 14, 1965, p. 2, col. 1.
112. *Oppenheimer Hearings*, pp. 251, 229.
113. *Ibid.*, p. 81.
114. "The Social Task of the Scientist," *Bulletin of the Atomic Scientists* 3 (1947):70.
115. *Oppenheimer Hearings*, p. 649.
116. *Ibid.*, pp. 79–80.
117. "The Search for Signals from Other Civilizations," *Science* 134 (1961):1839–43.
118. *Science et éducation* (Paris: Société française d'imprimérie et de libraire, 1901), p. 13.
119. *The Grammar of Science* (2d ed.; London: A & C Black, 1900), p. 8.
120. *Ibid.*, p. 12.
121. *Ibid.*, p. 7.
122. "The Prospect for Intelligence," *Yale Review* 34 (1945):450.
123. W. J. Wiswesser, "New Horizons in Science," *Journal of Chemical Education* 24 (Jan., 1947):21.
124. As one may read it in the latest recommendations of the Educational Policies Commission, published under the title, *Education and the Spirit of Science* (Washington, D.C.: National Educational Association, 1966), p. 16.
125. *On Understanding Science* (New Haven, Conn.: Yale University Press, 1947), p. 10.

126. "Science as a Means of International Understanding" (1946), in W. Heisenberg, *Philosophic Problems of Nuclear Science*, translated by F. C. Hayes (London: Faber, 1952), pp. 109–20.
127. *Ibid.*, p. 118.
128. *Science and Human Values* (New York: Messner, 1956), pp. 48–49.
129. *Ibid.*, p. 48.
130. *Religion and Science* (New York: Holt, 1935), p. 255.
131. "Responsibilities of Scientists in the Atomic Age," *Bulletin of the Atomic Scientists* 15 (Jan., 1959):6.
132. *Ibid.*
133. *On Understanding Science*, p. 8.
134. "A New Approach to Moral Philosophy," *EUROS*, No. 2 (Spring 1965):42.
135. *All in One Lifetime* (New York: Harper, 1958), p. 284.
136. Actually the needles did not present any obstacle to radioastronomy. See *The New York Times*, March 28, 1964, p. 7, col. 1.
137. *Science and Government* (New York: New American Library, 1962), p. 119.
138. E. Rabinowitch, "Responsibilities of Scientists in the Atomic Age," p. 6.
139. *The Ethical Dilemma of Science and Other Writings* (New York: Rockefeller Institute Press, 1960), p. 84.
140. Quoted by P. Duhem, *The Aim and Structure of Physical Theory*, translated by P. P. Wiener (Princeton, N.J.: Princeton University Press, 1954), p. 218.
141. *Opticks*, p. 405 (Query 31).
142. "A Discourse Introductory to a Course of Lectures on Chemistry," in *The Collected Works of Sir Humphry Davy*, II (London, 1839), 326.
143. "Introductory Lecture on Experimental Physics" (1870), in *Sci. Papers*, II, 252.
144. *Dernières pensées*, pp. 223–47.
145. *Ibid.*, p. 241.
146. *Ibid.*
147. On her admiration for Quetelet's *Essai de physique sociale*, as a book revealing the plans of God, see E. T. Cook, *The Life of Florence Nightingale*, I (London: Macmillan, 1913), 480–82.
148. *An Essay on the Foundations of Our Knowledge* (1851), translated by M. H. Moore (New York: Liberal Arts Press, 1956), p. 50.
149. These lectures were published under the title, *Der energetische Imperativ* (Leipzig: Akademische Verlagsgesellschaft, 1912).
150. *The Data of Ethics* (New York, 1883), p. 62.
151. *Ibid.*, p. 74.
152. *Ibid.*, p. 72.
153. *Human Nature and Conduct: An Introduction to Social Psychology* (New York: Modern Library, 1930), p. 12.
154. *The Scientific Monthly* 60 (1945):254.
155. *Evolution and Ethics* (London, 1893), p. 32.
156. "Entropy Consumption and Values in Physical Science," *American Scientist* 47 (1959):376–85.
157. As done, for instance, by M. Planck, "The Principle of Least Action," in *A Survey of Physical Theory*, pp. 69–81, or by M. Born, "Cause, Purpose and Economy in Natural Laws," in *Physics in My Generation* (London: Pergamon Press, 1956), pp. 55–79.
158. As done by W. Heisenberg, *Physics and Philosophy* (New York: Harper, 1958), pp. 196–97; see also M. Born, *Physics in My Generation*, pp. 52–53, and N. Bohr, *Atomic Physics and Human Knowledge*, p. 81.
159. *Ethics and Science* (Princeton, N.J.: Van Nostrand, 1964), pp. 128–29.
160. "Symbol and Reality," in *Natural Philosophy of Cause and Chance* (New York: Dover, 1964), p. 234.
161. *Ethics and Science*, p. vii.
162. *Ibid.*
163. For a work on human psychology replete with meaningless references to electron-proton interactions, see A. P. Weiss, *A Theoretical Basis of Human Behavior*

(Columbus, Ohio: R. G. Adams, 1925). On ethics conceived on such basis, see especially pp. 400–403.

164. "Zur Geschichte der physikalischen Naturerklärung" (1932), in *Wandlungen in der Grundlagen der Naturwissenschaft* (2d. ed.; Leipzig: S. Hirzel, 1936), p. 28.

165. Tahsueh, Liki, chap. XLII, in *The Wisdom of Confucius*, edited and translated with notes by Lin Yutang (New York: Modern Library, 1938), p. 140.

166. From an interview with Michael Amrine, *The New York Times Magazine*, June 23, 1946, p. 44.

167. *Ibid.*, p. 42.

168. *The Impact of Science on Society* (New York: Columbia University Press, 1951), p. 59.

CHAPTER X. Physics and Theology

1. *The Clouds*, 380–84; in *Aristophanes*, translated by B. B. Rogers, II (LCL, 1960), 301.

2. *Plutarch's Lives*, translated by B. Perrin, III (LCL, 1916), 289–91.

3. *Ibid.*, p. 291.

4. *Ibid.*

5. *De rerum natura*, Book 1, lines 1–101.

6. Book 10, 886 D; in *The Dialogues of Plato*, translated by B. Jowett, II (New York: Random House, 1937), 628.

7. Book 12, 967 A, *ibid.*, p. 701.

8. *Ibid.*

9. *On the Heavens*, 279a, translated by W. K. C. Guthrie (LCL, 1960).

10. *Ibid.*, 270b.

11. *Ibid.*, 284a.

12. "Letter to Menoeceus" 134, in C. Bailey, *Epicurus: The Extant Remains* (Oxford: Clarendon Press, 1926), p. 91.

13. "Letter to Pythocles" 85–87, *ibid.*, p. 59.

14. *Ibid.*, p. 65.

15. *The Physical World of the Greeks*, translated by M. Dagut (London: Routledge and Kegan Paul, 1956), p. 167.

16. For an informative discussion of the criticism of Aristotelian physics by Philoponus, see S. Sambursky, *The Physical World of Late Antiquity* (New York: Basic Books, 1962), chap. 6.

17. *Simplicii in Aristotelis physicorum libros quattuor posteriores commentaria*, edited by H. Diels (Berlin, 1895), pp. 1151–52.

18. *Le système du monde*, VI (Paris: A. Hermann, 1913–59), 66.

19. As late as 1687 Halley found it important to mention in his ode prefixed to Newton's *Principia* that man henceforth should not "quail beneath appearances of bearded stars."

20. *Quaestiones super octo libros physicorum*, Book 8, question 12. For an English translation of question 12, see M. Clagett, *The Science of Mechanics in the Middle Ages* (Madison: University of Wisconsin Press, 1959), p. 535.

21. *Science and the Modern World* (New York: Macmillan, 1926), pp. 18–19.

22. *Science and Civilization in China*, II, *History of Scientific Thought* (Cambridge: Cambridge University Press, 1956), 581.

23. *Advancement of Learning*, Book 2, in *Works*, III, 350, 358.

24. *The New Organon*, Book 1, aph. 125, in *Works*, IV, 111.

25. *Of the Dignity and Advancement of Learning*, Book 3, chap. 4, in *Works*, IV, 365.

26. *Ibid.*, pp. 363–64.

27. *The New Organon*, Book 1, aph. 10, in *Works*, IV, 48.

28. *The Great Instauration*, preface, in *Works*, IV, 18.

29. *Advancement of Learning*, Book 2, in *Works*, III, 357.

30. *Ibid.*, p. 478.

31. *The New Organon*, Book 1, aph. 124, in *Works*, IV, 110.

32. *Advancement of Learning*, Book 2, in *Works*, III, 356.
33. *The New Organon*, Book 1, aph. 89, in *Works*, IV, 88.
34. *Advancement of Learning*, Book 2, in *Works*, III, 485.
35. See especially *Of the Dignity and Advancement of Learning*, Book 9, chap. 1, in *Works*, V, 111–12.
36. *Dialogue*, pp. 103–4.
37. *The Assayer* (1623), in *Discoveries*, p. 240.
38. *Letter to the Grand Duchess Christina* (1615), in *Discoveries*, pp. 182–83.
39. *Ibid.*, p. 187.
40. *Ibid.*, p. 183.
41. *Robert Bellarmine, Saint and Scholar* (London: Burns, 1961), p. 378.
42. *On the Book of the Heavens and the World of Aristotle*, Book 2, chap. 25, English translation in M. Clagett, *The Science of Mechanics in the Middle Ages*, pp. 600–609.
43. *Astronomiae instauratae progymnasmata* (1582), in *Tychonis Brahe Dani opera omnia*, edited by J. L. E. Dreyer, III (Copenhagen: In Libraria Gyldendaliana, 1916), 175.
44. *Sämtliche Schriften und Briefe*, IV (Darmstadt: O. Reichl, 1923–), 320, 336, 344, 347.
45. *The Pensées*, translated by J. M. Cohen (Baltimore, Md.: Penguin, 1961), p. 37.
46. *Ibid.*, p. 57.
47. *Ibid.*, p. 100.
48. *Ibid.*, p. 284.
49. *Ibid.*, p. 124.
50. *Ibid.*, p. 53.
51. *Ibid.*, p. 51.
52. *Syntagma philosophiae Epicuri* (1649), in *Opera omnia* I (Lyons, 1658), 293.
53. *Physiologia Epicuro-Gassendo-Charltoniana*, Book 2, art. 7 (London, 1654), p. 87.
54. See H. Butterfield, *The Origins of Modern Science* (new ed.; London: Macmillan, 1957), p. 84.
55. *Traité de la lumière*, in *Oeuvres*, XI, 47.
56. Extract of a letter to Mr. Bayle (1687), in J. E. Erdmann (ed.), *God. Guil. Leibnitii opera philosophica* (Berlin, 1840), p. 106.
57. Fourth letter to Clarke, in *Leibniz Selections*, p. 236.
58. Letter to Antoine Arnauld (n.d.), in *Philosophische Schriften*, edited by C. I. Gerhardt, I (Berlin, 1875), 75.
59. *Systema saturnium* (1659), in *Oeuvres*, XV, 213.
60. *Leibniz Selections*, p. 596.
61. *Seraphick Love*, in *Works*, I, 262.
62. In *Oeuvres*, XXI, 688.
63. "Four Letters from Sir Isaac Newton to Doctor Bentley," in *Works of Richard Bentley*, III (London, 1838), 203.
64. *Ibid.*, p. 204.
65. *Ibid.*, p. 206.
66. *Ibid.*
67. *Ibid.*, p. 210.
68. *Ibid.*, p. 215.
69. *Ibid.*
70. *Opticks*, p. 352.
71. *Ibid.*, p. 402.
72. B. Willey, *The Eighteenth Century Background* (London: Chatto, 1940), p. 136.
73. See, for instance, F. C. Lesser, *Lithotheologie* (Hamburg, 1735). His *Insectotheologie* (Frankfurt, 1738) went into several editions and was translated into Italian and French as well.
74. "Some Considerations about the Cause of the Universal Deluge," *Phil. Trans* (London) 33 (1724–25):118–25. The long delay in the publication was because of Halley's apprehension lest he "might incur the censure of the sacred orders.'
75. First letter to Clarke, in *Leibniz Selections*, p. 216.

76. *Ibid.*, p. 60.
77. Part 2, sec. 8, in *Werke*, I, 331–47.
78. *Éléments de la philosophie de Newton*, in *Oeuvres*, XXII, 404.
79. *Traité de métaphysique*, in *Oeuvres*, XXII, 223.
80. *Notice de la vie et des écrits de George-Louis Le Sage de Genève*, edited by P. Prevost (Geneva, 1805), p. 594.
81. *Exposition du système du monde*, in *Oeuvres*, VI, 479.
82. *Indications of the Creator* (2d ed.; London, 1846), pp. 62–63.
83. *Exposition du système du monde*, in *Oeuvres*, VI, 477.
84. Albany, 1803, p. 258.
85. "Von dem neuen, zwischen Mars und Jupiter entdeckten achten Hauptplaneten des Sonnensystems" (1802). For a partial English translation, see *A Source Book in Astronomy*, edited by H. Shapley and H. E. Howarth (New York: McGraw-Hill, 1929), p. 180.
86. S. P. Thompson, *The Life of William Thomson*, II (London: Macmillan, 1910), 1090.
87. Presidential address to the British Association (1871), in *PLA*, II, 199–200.
88. *A Preliminary Discourse on the Study of Natural Philosophy* (London, 1831), pp. 37–38.
89. "Molecules" (1873), in *Sci. Papers*, II, 376.
90. *Ibid.*
91. *Dialectics of Nature*, translated by C. Dutt (New York: International Publishers, 1940), pp. 176–77.
92. *The Listener*, July 10, 1947; quoted by C. A. Coulson, *Science and Christian Belief* (Chapel Hill: University of North Carolina Press, 1955), p. 23.
93. *The History of Nature*, translated by F. D. Wieck (Chicago: University of Chicago Press, 1949), p. 127.
94. *Revolutions*, Book 1, introduction.
95. *Harmonice mundi*, Book 5, chap. 7, in *Werke*, VI, 388.
96. *Letter to the Grand Duchess Christina*, in *Discoveries*, p. 182.
97. *Dialogue*, p. 103.
98. *Principia*, p. 544.
99. *Methodus inveniendi lineas curvas maximi minive proprietate gaudentes*, Additamentum, I, edited by C. Carathéodory (Bern: Orell, Füssli, Turici, 1952), p. 231.
100. *Essai de cosmologie*, in *Oeuvres de Maupertuis* I (Lyons, 1756), 44.
101. *Traité de dynamique*, I (2d ed., 1758; reprinted Paris: Gauthier, 1921), XXXVI.
102. "On Matter, Living Force and Heat" (1847), in *The Scientific Papers of James Prescott Joule*, I (London, 1884), 273.
103. See the collection of his various essays and addresses, *The Soul in Nature*, translated by L. Horner and J. B. Horner (London, 1852), pp. 88, 171, 189, 446.
104. *Essai sur la philosophie des sciences*, II (Paris, 1838–43), 24–25.
105. See his statement of belief drafted in 1878, in T. Coulson, *Joseph Henry: His Life and Work* (Princeton, N.J.: Princeton University Press, 1950), pp. 296–98.
106. "On Magnetic and Diamagnetic Bodies" (1849), in B. Jones, *The Life and Letters of Faraday*, II (Philadelphia, 1870), 385.
107. L. Campbell and W. Garnett, *The Life of James Clerk Maxwell* (London, 1882), p. 323.
108. *Ibid.*, p. 395.
109. S. P. Thompson, *The Life of William Thomson*, I, 192.
110. *Nature* 81(1909):257.
111. See A. H. Tabrum, *Religious Beliefs of Scientists* (new ed.; London: Hunter and Longhurst, 1913), p. 8.
112. In a telegram to Rabbi H. S. Goldstein, in *The New York Times*, April 25, 1929, p. 60, col. 4.
113. "On Physical Reality," *Journal of the Franklin Institute* 221 (1936):351.
114. "Science and Religion" (1941), in *Out of My Later Years* (New York: Philosophical Library, 1950), p. 26.

115. "An Account of Carnot's Theory of the Motive Power of Heat" (1849), in *Mathematical and Physical Papers* I (Cambridge, 1882), 119.
116. *Ibid.*, pp. 511–14.
117. *Ibid.*, p. 514.
118. "On the Interaction of Natural Forces" (1854), in *PLSS*, p. 173.
119. *Ibid.*, p. 193.
120. *Ibid.*
121. 9th ed.; London, 1880, pp. 127–28.
122. *Ibid.*, p. 128.
123. "On the Age of the Sun's Heat" (1862), in *PLA*, I, 350.
124. *Sci. Papers*, II, 226.
125. In *Miscellaneous Scientific Papers*, edited by W. J. Millar (London, 1881), pp. 200–202.
126. *Ibid.*, p. 201.
127. *Theory of Heat* (new ed.; New York, 1875), pp. 328–29.
128. *History and Root of the Principle of the Conservation of Energy*, translated by P. E. B. Jourdain (Chicago: Open Court, 1911), p. 62.
129. *Ibid.*, p. 63.
130. *Ibid.*
131. *The Riddle of the Universe*, translated by J. McCabe (New York: Harper & Brothers, 1900), p. 247.
132. *Dialectics of Nature*, p. 205.
133. "Über statistische Mechanik," in *Populäre Schriften* (Leipzig: Barth, 1905), p. 362.
134. "Le mécanisme et l'expérience," *Revue de métaphysique et de morale* 1 (1893):537.
135. *Die Naturkräfte in ihrer Wechselbeziehung* (1869), in *Gesammelte Schriften* I (Würzburg: Stahel's Verlag, 1903), 296–361.
136. *Die Grösze der Schöpfung*, translated by C. Gürtler (Leipzig, 1885).
137. "Physics of a Believer," in P. Duhem, *The Aim and Structure of Physical Theory*, translated by P. P. Wiener (Princeton, N.J.: Princeton University Press, 1954), p. 290.
138. *Energie, Entropie, Weltanfang, Weltende* (Trier: F. Lintz, 1910).
139. *Der entropologische Gottesbeweis* (Bonn: A. Marcus and E. Weber, 1920).
140. *The Nature of the Physical World* (Cambridge: Cambridge University Press, 1928), p. 74.
141. *Ibid.*
142. *Ibid.*, p. 84.
143. *Ibid.*, p. 85.
144. *Eos, or the Wider Aspects of Cosmogony* (London: K. Paul, 1928), pp. 55–56.
145. *The Beginning and End of the World* (London, Oxford University Press, 1942), p. 4.
146. *Ibid.*, p. 63.
147. *The History of Nature*, p. 127.
148. *From Euclid to Eddington: A Study of Conceptions of the External World* (Cambridge: Cambridge University Press, 1949), p. 46. For an early discussion of the bearing of relativistic thermodynamics on cosmological studies, see R. C. Tolman, *Relativity, Thermodynamics and Cosmology* (Oxford: Clarendon Press, 1934), pp. 319–26, 432–34.
149. E. A. Milne, *Sir James Jeans: A Biography* (Cambridge: Cambridge University Press, 1952), pp. 164–65.
150. *The Proofs of the Existence of God in the Light of Modern Natural Science*, Address of Pope Pius XII to the Pontifical Academy of Sciences (Washington, D.C.: National Catholic Welfare Conference, 1952), p. 11.
151. *The Beginning and End of the World*, p. 63.
152. *Space and Spirit* (Chicago: Henry Regnery, 1947), p. 118.
153. *The Proofs of the Existence of God*, p. 17.
154. *Ibid.*, p. 15.

155. "Tendencies of Recent Investigations in the Field of Physics" (1930), in Sir Rayleigh [R. J. Strutt], *The Life of Sir J. J. Thomson* (Cambridge: Cambridge University Press, 1942), p. 265.
156. Pope Pius XII, "Discourse to Astronomers," Sept. 7, 1952. English text in P. J. McLaughlin, *The Church and Modern Science* (New York: Philosophical Library, 1957), pp. 184–94. Reference to Copernicus and Galileo on p. 194.
157. *Ibid.*, p. 192.
158. *Ibid.*
159. *Ibid.*, p. 190.
160. *Ibid.*, p. 194.
161. *The Proofs of the Existence of God*, p. 17.
162. "Religion and Natural Science" (1937), in *Scientific Autobiography and Other Papers*, translated by F. Gaynor (London: Williams & Norgate, 1950), p. 154.
163. Letter to H. A. Tabrum, in *Religious Beliefs of Scientists*, pp. 118–19.
164. *Religion and Science* (New York: Holt, 1935), pp. 178–80.
165. "The Scientists," *Fortune* 38 (October, 1948), p. 176.
166. *The Two Cultures and the Scientific Revolution* (Cambridge: Cambridge University Press, 1959), p. 10.
167. *Memoir and Scientific Correspondence of the late Sir George Gabriel Stokes*, edited by J. Larmor, I (Cambridge: Cambridge University Press, 1907), 179.
168. *Natural Theology* (London, 1891), pp. 24–25.
169. "The Relation of Science and Religion," *Engineering and Science* 20 (June, 1956), p. 20.
170. "The Origin of Species" (1860), in *Collected Essays*, II (New York, 1894), 52.
171. *Weltschöpfung und Weltende* (2d ed.; Hamburg: Furche Verlag, 1958), p. 38.
172. *The Life of James Clerk Maxwell*, pp. 392–96.
173. *Ibid.*, pp. 404–5.
174. *The Nature of the Physical World*, p. 353.
175. *Ibid.*, p. 351.
176. *Ibid.*, p. 350.
177. Edited by H. Shapley (New York: Appleton-Century-Crofts, 1960).
178. In a short speech given at the University College, May, 1903, reported in *The Nineteenth Century* 53 (1903):1069.
179. *The History of Nature*, p. 177.
180. C. F. Von Weizsäcker, *The World View of Physics*, translated by M. Grene (Chicago: University of Chicago Press, 1952), p. 157.
181. *Life and Matter: A Criticism of Professor Haeckel's "Riddle of the Universe"* (New York: G. P. Putnam's Sons, 1906), p. v.
182. *What Is Life and Other Scientific Essays* (Garden City, N.Y.: Doubleday, 1956), p. 228.
183. "Christianity and Physical Science," in *The Idea of a University* (London, 1888), pp. 428–55.
184. *Physics and Philosophy* (New York: Harper, 1958), p. 59.
185. As Barré de Saint-Venant did in his paper, "De la constitution des atomes," *Annales de la Société Scientifique de Bruxelles* 2 (1878):74.
186. Letter to Galileo, April 1, 1607, in *Opere di Galileo Galilei*, edited by A. Favaro, X (Firenze, 1890–1909), 170.
187. *Space and Spirit*, pp. 118–25.
188. *Christian Theology and Natural Science* (New York: Ronald, 1956), pp. 146–49.
189. *Time*, Jan. 4, 1963, p. 54.
190. *The World of Sound* (London: G. Bell, Sons, 1920), pp. 195–96.

CHAPTER XI. The Fate of Physics in Scientism

1. *Lettres philosophiques* (1734), in *Oeuvres*, XXII, 130.
2. Perhaps the most popular of the books explaining "Newtonianism" in the style of drawing-room chatter was F. Algorotti's, *Il newtonianismo per le dame: ov-*

vero, dialoghi sopra la luce e i colori (Naples, 1737), of which a French and an English translation were published in 1738 and 1739, respectively, with several editions to follow.

3. *Parasceve*, in *Works*, IV, 252. See also *The New Organon*, Book 1, chap. 112, in *Works*, IV, 105.
4. Letter to Maestlin, Oct. 3, 1595, in *Werke*, XIII, 40.
5. Letter to A. Tacquet, Aug., 1660, in *Oeuvres*, III, 105.
6. *Oeuvres de Fontenelle*, VI (Paris, 1790), 135.
7. *Ibid.*, VI, 515.
8. *Ibid.*, VII, 350.
9. *Ibid.*, VI, 109.
10. *Ibid.*, 454.
11. *Ibid.*, 152.
12. "Préface sur l'utilité des mathématiques et de la physique et sur les travaux de l'Académie des sciences," in *Oeuvres*, VI, 67.
13. *Ibid.*, p. 70.
14. *Essays and Treatises on Several Subjects*, I (new ed.; London, 1767), 16.
15. *Course of Experimental Philosophy*, preface (3d ed.; London, 1763), p. viii.
16. [A. Dickens,] *The Claims of Thomas Jefferson to the Presidency Examined at the Bar of Christianity by a Layman* (Philadelphia, 1800). For long excerpts from this pamphlet, see *The Monthly Magazine and American Review* 3 (1800):354–62. For quotation, see p. 362.
17. *The Christian Philosopher: A Collection of the Best Discoveries in Nature with Religious Improvements* (London, 1721), p. 56.
18. June 1, 1741, in *Oeuvres*, XXXVI, 65.
19. In *Oeuvres*, XXVII, 127.
20. *Système de la nature ou des lois du monde physique et du monde moral*, I (London, 1775), 19.
21. Manuscript notes, quoted in F. E. Manuel, *The Prophets of Paris* (Cambridge, Mass.: Harvard University Press, 1962), p. 65.
22. Condorcet summarized his views on this point in his *Tableau général de la science qui a pour objet l'application du calcul aux sciences politiques et morales*, published posthumously in 1795.
23. "Taming the future" was not only Condorcet's ambition. Laplace claimed, for instance, in the lengthy introduction of his *Théorie analytique des probabilités* (2d ed.; Paris, 1814, p. liv) that there was a close parallelism between eternal reasons evident in the calculus of probabilities and those manifested in the course of history. Suppressed nations, he noted in this connection, never failed to form powerful coalitions to shake off the yoke imposed on them by conquerors. Laplace of course meant Napoleon whose favors he had lost shortly before, and to whom the first edition (1812) of the work was dedicated. The incident hardly throws a favorable light on Laplace. To criticize a ruler who happened to be out of power demanded little if any courage. Nor was much to be commended in Laplace's tactics pulling the science of probabilities down to the level of daily politics.
24. *Fragments sur l'Atlantide ou efforts combinés de l'espèce humaine pour le progrès des sciences*, in *Oeuvres de Condorcet*, VI (Paris, 1847), pp. 608–10.
25. *Sur l'instruction publique. Cinquième mémoire*, in *Oeuvres*, VII, 436.
26. *Ibid.*, p. 432.
27. *Ibid.*, pp. 425–6.
28. *Ibid.*, p. 429.
29. *Esquisse d'un tableau historique des progrès de l'esprit humaine*, in *Oeuvres*, VI, 13.
30. *Sur l'instruction publique. Premier mémoire*, in *Oeuvres*, VII, 183.
31. *Sur l'instruction publique. Troisième mémoire*, in *Oeuvres*, VII, 374.
32. *The Positive Philosophy of Auguste Comte*, translated by H. Martineau I (London, 1875), viii. The fact that J. S. Mill found fault as late as 1871 only with Comte's treatment of chemistry makes a rather unfavorable reflection on Mill's

familiarity with physical science. See his *Positive Philosophy of Auguste Comte* (Boston, 1871), p. 56.

33. "The Scientific Aspects of Positivism" (1869), in *Lay Sermons, Addresses and Reviews* (5th ed.; London, 1874), p. 149.
34. "Auguste Comte, Historian of Science," *Osiris* 10 (1952):328–57.
35. "The Scientific Aspects of Positivism," p. 153.
36. *The Positive Philosophy*, I, 118.
37. *System of Positive Polity*, translated by J. H. Bridges and others, I (London, 1875), 340–41.
38. *Ibid.*, 406–7.
39. *The Positive Philosophy*, I, 146.
40. *Cours de philosophie positive*, VI (3d ed.; Paris, 1869), 638.
41. *Ibid.*
42. *Ibid.*
43. *System of Positive Polity*, I, 425.
44. *Ibid.*, 414.
45. *The Positive Philosophy*, I, 166.
46. *System of Positive Polity*, I, 413.
47. *The Positive Philosophy*, I, 170.
48. *Ibid.*, p. 141.
49. *Ibid.*, p. 126.
50. *Ibid.*, p. 171.
51. *Ibid.*, pp. 141–42.
52. *Ibid.*, p. 141.
53. *Ibid.*
54. *Ibid.*
55. *Ibid.*, p. 173.
56. *Ibid.*, p. 115.
57. *Ibid.*, p. 127.
58. *Ibid.*, p. 173.
59. *Ibid.*, p. 174.
60. *System of Positive Polity*, I, 412.
61. "The Mantle of Science," in H. Schoeck and J. W. Wiggins (eds.), *Scientism and Values* (New York: Van Nostrand, 1960), p. 159.
62. *System of Positive Polity*, I, 425.
63. *Ibid.*
64. *Ibid.*, p. 427.
65. *Ibid.*, p. 421.
66. *The Positive Philosophy*, I, 284.
67. *Ibid.*
68. *Ibid.*, p. 285.
69. See B. Jones, *The Life and Letters of Faraday*, II (Philadelphia, 1870), 384.
70. *Cours de philosophie positive*, III, 174.
71. *System of Positive Polity*, I, 460–61.
72. *Cours de philosophie positive*, II, 491.
73. Letter of June 9, 1857, to the Rev. E. Jones; see B. Jones, *The Life and Letters of Faraday*, II, 388.
74. Comte's shockingly defective account of optics was briefly noted even by W. Ostwald in his work, *Auguste Comte: Der Mann und sein Werk* (Leipzig: Verlag Unesma, 1914), p. 129. Yet, in a most curious way, Ostwald remained silent about Comte's mishandling of other parts of physics and especially of chemistry. This was a strange procedure on the part of a Nobel laureate chemist. The reason for it is not difficult to find. In the introduction of his book (pp. vii–viii), Ostwald praised Comte's "unique contribution" to the "religion of scientific monism." Evidently, the type of scientism that prevented Comte from seeing science objectively prevented Ostwald from seeing his hero with at least a minimum of scientific objectivity.
75. *Cours de philosophie positive*, II, 312.
76. See, for instance, P. Frank, "Einstein, Mach and Logical Positivism," in P. A.

Schilpp (ed.), *Albert Einstein: Philosopher-Scientist* (Evanston, Ill.: Library of Living Philosophers, 1949), pp. 271–86.

77. *The Future of Science* (Boston, 1891), p. 139.
78. *Cours de philosophie positive*, IV, 44.
79. "The Scientific Aspects of Positivism," p. 172.
80. *Auguste Comte and Positivism* (2d ed.; London, 1866), p. 6. Mill insists that Comte did not claim originality for his conception of scientific method and knowledge. However this may be, Comte's disciples did claim this originality for their master, as was pointed out by H. Spencer in his "Reasons for Dissenting from the Philosophy of M. Comte" (1864), in *Essays, Scientific, Political and Speculative*, II (New York, 1899), pp. 118–44. It is interesting to note that Spencer was unable to accept precisely those points in Comte's thought which were undisputably Comtean, such as the law of three states and the corresponding genesis of science, the denial of organic evolution, and the like.
81. One of the early critics of Comte's views on what science should and should not do was W. Whewell, who pointed out with direct reference to Newton that inquiring into the causes of things is inavoidable in scientific research. See his *Philosophy of the Inductive Sciences*, II (2d ed.; London, 1847), 320–25.
82. "Sixième mémoire" (1788), in *Collection de mémoires relatifs à la physique*, Vol. I: *Mémoires de Coulomb* (Paris, 1844), p. 252.
83. Translated by A. Freeman (Cambridge, 1878), pp. 1–13.
84. "The Highest Aim of the Physicist" (1899), in *The Physical Papers of Henry Augustus Rowland* (Baltimore: Johns Hopkins Press, 1902), p. 670.
85. "Autobiographical Notes," in P. A. Schilpp, *Albert Einstein: Philosopher-Scientist*, p. 47.
86. *Ibid.*, p. 21.
87. *Ibid.*, p. 49.
88. *Physics and Politics* (Edinburgh: Oliver, 1962), p. 34.
89. "Tendencies of Recent Investigations in the Field of Physics" (1930), in Sir Rayleigh [R. J. Strutt], *The Life of Sir J. J. Thomson* (Cambridge: Cambridge University Press, 1942), p. 265.
90. *Capital: A Critical Analysis of Capitalist Production*, author's preface to the first edition, translated from the third German edition by S. Moore and E. Aveling (New York, 1889), pp. xvi–xvii.
91. *Ibid.*, pp. xxx–xxxi (author's preface to the second edition).
92. *Herr Eugen Dühring's Revolution in Science*, translated by E. Burns (New York: International Publishers, 1939), p. 15.
93. *Dialectics of Nature*, translated by C. Dutt (New York: International Publishers, 1940), p. 34.
94. *Ibid.*, p. 243.
95. *Ibid.*, p. 259.
96. *Ibid.*, pp. 257–58.
97. *Ibid.*, pp. 37–38.
98. *Ibid.*, p. 65.
99. See L. Königsberger, *Hermann von Helmholtz*, I (Braunschweig: F. Vieweg, 1902), 292.
100. *Dialectics of Nature*, p. 195.
101. *Ibid.*, p. 75.
102. *Ibid.*, p. 86.
103. *Ibid.*, p. 155.
104. *Ibid.*, p. 172.
105. *Ibid.*, p. 127.
106. *Ibid.*, p. 246.
107. *Ibid.*, p. 86.
108. *Ibid.*, p. 184.
109. *Ibid.*, p. 154.
110. *Ibid.*, p. 25.
111. *Ibid.*, p. 11.

112. *Materialism and Empirio-criticism*, (New York: International Publishers, 1927), pp. 323–24.
113. *Ibid.*, pp. 289–90.
114. *Ibid.*, pp. 268–69.
115. *Ibid.*, p. 257.
116. For an account of the fortunes of Russian physics and physicists in the 1920's and early 1930's, see D. Joravsky, *Soviet Marxism and Natural Science: 1917–32* (New York: Columbia University Press, 1961), pp. 275–95.
117. *Ibid.*, p. 313.
118. *Izvestia*, Dec. 16, 1937, p. 4, quoted in M. W. Mikulak, "Soviet Philosophic-cosmological Thought," *Philosophy of Science* 25 (1958):39.
119. For this phase of the running feud between Soviet physicists and the party, see S. Müller-Markus, *Einstein und die Sovietphilosophie: Krisis einer Lehre* (Dordrecht: D. Reidel Publishing Co., 1960), pp. 43–63.
120. "Sovietskaia fizika i dialekticheskii materializm," in *Filosofskie voprosy sovremennoi fiziki*, edited by A. A. Maksimov and others (Moscow: Akademii Nauk USSR, 1952), pp. 34, 72.
121. See his report on scientific life in *Voprosy filosofii* No. 6 (1956), p. 55.
122. Translated by Consultants Bureau, Inc., Technical Information Service Extension, Oak Ridge, Tenn., pp. 493–94. An obsequious quotation from Lenin's *Materialism and Empirio-criticism* also forms the conclusion of D. I. Blokhintsev's work, *Grundlagen der Quantenmechanik*, translated by Detlof Lyons (2d ed.; Berlin: Deutscher Verlag der Wissenschaften, 1953), p. 506. Many other examples could be quoted.
123. "Za leninskoe uchenie o dizhenii" ("On Lenin's doctrine of motion"), *Voprosy filosofii* No. 1 (1952), p. 243.
124. "Filosofskie problemi sovremennoi fiziki i zadachi sovietskikh fizikov v borbe za peredovuiu nauk," in *Filosofskie voprosy sovremennoi fiziki*, pp. 5–30.
125. See M. W. Mikulak, "Philosophic-cosmological Thought," p. 47.
126. "Krizis burzhuaznoi kosmologii," *Priroda* (Nature) No. 7 (1950), pp. 12–18.
127. *The Social Relations of Science* (New York: Macmillan, 1941), p. 331.
128. *Ibid.*, p. 333.
129. In a full-page article in the *Ekonomicheskaya Gazeta* (Economic Gazette), March 26, 1962, p. 10.
130. A. S. Eve, *Rutherford* (New York: Macmillan, 1939), p. 400.
131. Quoted in D. Q. Posin, *Mendeleyev* (New York: Whittlesey House, 1948), p. ii.
132. In a speech reported by *The New York Times*, April 25, 1965, p. 18, col. 5. A year later, writing in the January 20, 1966, issue of *Komsomolskaya Pravda*, the Communist youth newspaper, Kapitza called the attention of his countrymen to the fact that as regards both quantity and quality, the Soviet scientific productivity was not catching up with that of scientists in the United States.
133. A typical case is that of J. Cockcroft who stated, following a brief visit in Soviet Russia, that he found no notable difference between scientific life and research surrounded by Marxism or by Western democracy. See C. P. Snow, *The Two Cultures and the Scientific Revolution* (Cambridge: Cambridge University Press, 1959), p. 48.
134. *In the Matter of J. Robert Oppenheimer* (Washington, D.C.: U.S. Government Printing Office, 1954), p. 10. The persecution of Jewish and anti-Nazi physicists in Germany originated in an ideology that stressed biological and political concepts rather than a scientism based on some interpretation of physics. Thus the ideological recasting of the teaching of physical sciences in Germany did not constitute an immediate objective in the plans of the Nazi government. Misguided efforts, however, were not lacking to assert racism in physics too. In fact, two of Germany's Nobel laureate physicists volunteered their services to this deplorable end. P. Lenard penned a prize-winning essay on the working habits of the German scientist and on his struggle to have Nordic research prevail in science. *Der deutsche Naturforscher: Sein Kampf um nordische Forschung* (München: J. F. Lehmanns Verlag, 1937). Less harmful, though just as indicative, was the fact

that Lenard gave the title, *German Physics*, to his four-volume textbook on phys-
ics. Another Nobel laureate, J. Stark, wrote on "The Pragmatic and Dogmatic
Spirit in Physics." For an English translation see *Nature* 141 (1938):770–72.
There he decried "Jewish dogmatism" (Relativity), to which he opposed the
"German experimental approach." Instead of boasting that he had advocated
such views a decade before Nazism came to power, Stark would have done better
to take a look at the saddening condition of physics in his homeland in 1938.
Official German statistics made no secret of the fact that the number of students
majoring in mathematics and physics dropped sharply during the first four years
of the Nazi regime. See E. Y. Hartshorne, "The German Universities," *Nature* 142
(1938):175–76.

135. Note by J. B. S. Haldane in F. Engels, *Dialectics of Nature*, p. 257.
136. J. B. S. Haldane, *The Marxist Philosophy and the Sciences* (New York: Random
House, 1939), p. 56.
137. *Ibid.*, p. 73.
138. The misfortunes of genetics in the Soviet Union were especially closely followed
by the American Nobel laureate geneticist, H. J. Muller. He worked from 1933
to 1937 as senior biologist at the Institute of Genetics of the Russian Academy
of Science of which he was elected a member. In 1948 he tendered, however, his
resignation in a famous open letter protesting the oppression of free research by
the Communist party. For a list of his publications on the fate of genetics under
Marxism, see *Studies in Genetics: The Selected Papers of H. J. Muller* (Blooming-
ton: Indiana University Press, 1962), p. 559.
139. *The Future of Science* (Boston, 1891), p. 24.
140. *Ibid.*, p. 81.
141. *Ibid.*, p. xxi.
142. *Ibid.*, p. 17.
143. *Ibid.*, p. x.
144. *Ibid.*, p. 140.
145. *Ibid.*, p. 139.
146. *Ibid.*, p. 138.
147. *Ibid.*, p. xi.
148. See L. P. Shanks, *Anatole France* (Chicago: Open Court, 1919), p. 52.
149. *The Opinions of Jerome Coignard*, translated by W. Jackson, in *The Works of
Anatole France*, IV (New York: Wells, 1924), 114.
150. Quoted in L. P. Shanks, *Anatole France*, p. 167.
151. In the essay, "Mysticism and Science," translated by B. Miall, in *The Works of
Anatole France*, XXX, 40–46.
152. *Ibid.*, p. 43.
153. *The Fate of Man* (New York: Alliance Book Corporation, 1939), p. 187.
154. *Ibid.*, p. 173.
155. *Ibid.*, p. 247.
156. *Ibid.*
157. *Mind at the End of Its Tether and the Happy Turning: A Dream of Life* (New
York: Didier Publishers, 1946), p. 17.
158. *Evolution and Ethics and Other Essays*, in *Selected Works of Thomas H. Huxley*,
IX (New York, 1893–94), 44–46.
159. Letter to Charles Kingsley, in *Life and Letters of Thomas Henry Huxley*, edited
by L. Huxley, I (London: D. Appleton, 1900), 218.
160. "On the Advisableness of Improving Natural Knowledge" (1866), in *Lay Sermons, Addresses and Reviews* (London, 1874), pp. 1–19.
161. Romanes Lecture, 1894, in *Evolution and Ethics and Other Essays*, p. 45.
162. "My Philosophical Development," *Encounter* XII (February, 1959): 24.
163. "Science and the Human Life," in J. R. Newman (ed.), *What Is Science?* (New
York: Simon, 1955), p. 18.
164. Address at the annual luncheon of Pulitzer Prize jurors at Columbia University,
New York Herald Tribune, April 2, 1961, sec. 2, p. 3, col. 4.
165. "Préface sur l'utilité des mathématiques," pp. 71–72.

CHAPTER XII. The Place of Physics in Human Culture

1. *The Two Cultures and the Scientific Revolution* (New York: Cambridge University Press, 1959).
2. "The Fragmentation of Science," *The Advancement of Science* 12 (Dec., 1955): 328.
3. Address to the Pulitzer Prize jurors, Columbia University, 1961, in *New York Herald Tribune*, April 2, 1961, sec. 2, p. 3, col. 5.
4. *Ibid.*
5. *The Two Cultures and the Scientific Revolution*, p. 16.
6. N. Bohr, *Essays 1958–1962 on Atomic Physics and Human Knowledge* (New York: Interscience, 1963), p. 72.
7. H. Margenau, *The Nature of Physical Reality* (New York: McGraw-Hill, 1950), p. 18.
8. L. A. DuBridge, "Science and National Policy," *American Scientist* 34 (1946):227.
9. "The Imperfections of Science," *Proc. Am. Phil. Soc.* 104 (1960):427.
10. "Modern Science and the Intellectual Tradition," *Science* 131 (1960):1187–93.
11. "Science Pauses," *Fortune*, 71 (May, 1965):116.
12. "The State of Physics," in S. K. Runcorn (ed.), *Physics in the Sixties* (Edinburgh: Oliver and Boyd, 1962), p. 97.
13. *Dialogues of Alfred North Whitehead as Recorded by Lucien Price* (Boston: Little, Brown, 1954), p. 345.
14. To what extent facts can differ from cliché accounts of the acceptance of a major discovery in physics is well illustrated by what followed the introduction by Planck in 1900 of the quantum of radiation. For years it was ignored by physicists most interested in the problem of black-body radiation. In 1902 Gibbs wrote in the preface of his famous work on statistical mechanics that no solution had so far been found to the problem. Professor E. B. Wilson of Yale heard nothing, as he later recalled, of Planck's formula during his stay in Paris in 1902–3. In 1904, the first edition of J. Jeans's *Dynamical Theory of Gases* contained no reference to it. In the same year Kelvin's *Baltimore Lectures* still spoke of the unsolved problem of the principle of the equipartition of energy. The next year Lord Rayleigh [R. J. Strutt] stated that he did not understand Planck's reasoning, and as late as 1911 he was worrying how the related problem of specific heats could be solved. After all, one should not forget that in 1915 Planck was still trying to find fault with the concept of quantum, which formed the basis of the only possible derivation of his epoch-making formula.
15. *On Understanding Science* (New Haven, Conn.: Yale University Press, 1947), p. 26.
16. "The History of a Doctrine," *Science* 12 (1888):74.
17. "The Missing Factor in Science" (1947), in H. Dingle, *The Scientific Adventure* (London: Pitman, 1952), p. 8.
18. From his introduction to the English translation of H. Poincaré, *Science and Hypothesis* (London: Walter Scott Publishing Co., 1905), p. xviii.
19. In a review of *A History of the Theories of Aether and Electricity* by E. T. Whittaker; *Scientific American* 190 (March, 1954):92.
20. "Is Science a Fashion of the Times?", in *Science, Theory and Man*, translated by J. Murphy (London: Allen, 1957), pp. 81–105.
21. In his paper, "Physical Science and the Temper of the Age," Schrödinger listed five such traits: emphasis on simplicity, desire for change and freedom, preference of the relative to the absolute, generalization of formulas to provide more effective mass control, and universal application of statistical methods. See *Science, Theory and Man*, pp. 106–32.
22. For an interesting documentation of these points, see B. Barber, "Resistance by Scientists to Scientific Discovery," *Science* 134 (1961):596–602, and T. S. Kuhn, "The Function of Dogma in Scientific Research," in A. C. Crombie (ed.), *Scientific Change* (New York: Basic Books, 1963), pp. 347–69. As to the choice made on aesthetic grounds, a case is Eddington's who argued that the cosmic process should

have a "beginning not too unaesthetically abrupt." *The Expanding Universe* (Cambridge: Cambridge University Press, 1933), p. 80. In doing so, he merely provided one more corroboration for A. N. Whitehead's remark that "Nature is probably quite indifferent to the aesthetic preferences of mathematicians." *Science and the Modern World* (New York: Macmillan, 1925), p. 179.

23. *Essai sur les élémens de philosophie ou sur les principes des connoissances humaines*, in *Oeuvres de d'Alembert*, I (Paris, 1821), 123–25.
24. Translated by J. R. Maddox (Neuchâtel: Du Griffon, 1955), p. 640.
25. *New Perspectives in Physics*, translated by A. J. Pomerans (New York: Basic Books, 1962), p. 228.
26. From a mimeographed text of the address.
27. W. Whewell, *Influence of the History of Science upon Intellectual Education* (Boston, 1854), p. 7.
28. *The History of Science and the New Humanism* (New York: Holt, 1931), p. 24.
29. *The Aim and Structure of Physical Theory*, translated by P. P. Wiener (Princeton, N.J.: Princeton University Press, 1954), p. 270.
30. *Ibid.*, p. 304.
31. *Ibid.*, p. 270.
32. "Steps toward a Kinetic Theory of Matter" (1884), in *PLA*, II, 233.
33. *Light Waves and Their Uses* (Chicago: University of Chicago Press, 1902), pp. 162–63.
34. "Tendencies of Recent Investigations in the Field of Physics" (1930), in Sir Rayleigh [R. J. Strutt], *The Life of Sir J. J. Thomson* (Cambridge: Cambridge University Press, 1942), p. 265.
35. *The Aim and Structure of Physical Theory*, p. 23.
36. Letter to Oldenburg, July 11, 1672, in *Isaaci Newtoni opera quae exstant omnia*, edited by S. Horsley, IV (London, 1770–85), 324.
37. This was already noted by Euler. See *Lettres d'Euler à une princesse d'Allemagne*, edited by E. Saisset, I (Paris, 1866), 230. The letter in question dates from October 18, 1760.
38. "L'invention scientifique. Les sciences expérimentales: la théorie," in *L'invention. IXe Semaine Internationale de Synthèse* (Paris: Alcan, 1938), p. 122.
39. "The Mechanics of Newton and Their Influence on the Development of Theoretical Physics" (1927), in *Ideas and Opinions by Albert Einstein* (New York: Crown, 1954), p. 257.
40. *Lectures on Some Recent Advances in Physical Science* (London, 1876), p. 285.
41. Letter to A. H. Tabrum quoted in the latter's work, *Religious Beliefs of Scientists* (new ed.; London: Hunter and Longhurst, 1913), p. 8.
42. "Introductory Lecture on Experimental Physics" (1871), in *Sci. Papers*, II, 242.
43. *Between Physics and Philosophy* (Cambridge, Mass.: Harvard University Press, 1941), pp. 4–5.
44. Reference was made in chapter 6 to the uncertainties in connection with the observed value of the advance of the perihelion of Mercury as a proof of the General Theory of Relativity. As to the bending of light rays observable during solar eclipses, most textbooks flatly reported the confirmation of the theory by experiments. The fact is, however, that "in spite of the great labor spent on the problem, to this day no definitive confirmation of the theory has been obtained in this way." This evaluation of the experimental data is due to F. Schmiedler, of the Munich University Observatory, who himself made an unsuccessful attempt at observing the effect from the shore of Great Slave Lake in Canada during the eclipse of July 20, 1963. See his article, "The Einstein Shift — An Unsolved Problem," *Sky and Telescope* 27 (1964):217–19. Again, since 1925 the assertion has often been made that the gravitational red shift had been observed in the spectrum of Sirius. Yet, even today the most convincing evidence in this respect is provided not by astronomical observations but by laboratory experiments utilizing the Mössbauer effect. See note 77 of chapter 6.
45. R. A. Houston, *A Treatise on Light* (6th ed.; New York: Longmans, Green & Co., 1930), p. 467.

46. "Fact and Inference in Theory and in Observation," in *Vistas in Astronomy*, edited by A. Beer, I (London: Pergamon Press, 1955), 161.
47. "A Speculation touching Electric Conduction and the Nature of Matter," in *Experimental Researches in Electricity*; Great Books of the Western World, XLV (Chicago: Encyclopaedia Britannica, 1952), 851.
48. *A Treatise on Electricity and Magnetism*, II (3d ed.; Oxford, 1892), 175.
49. *Ibid.*, p. 176.
50. "Introductory Lecture on Experimental Physics" (1871), *Sci. Papers*, II, 251.
51. *Astronomia nova*, Part 4, chap. 45, in *Werke*, III, 47.
52. *Vorträge und Reden*, I (5th ed.; Braunschweig: F. Vieweg, 1903), 14–15.
53. *The New Organon*, Book 2, aph. xx, in *Works*, IV, 149.
54. "Die Entstehung und bisherige Entwicklung der Quantentheorie" (1920), in *Wege zur physikalischen Erkenntnis* (Leipzig: S. Hirzel, 1933), p. 68.
55. "Origins of the General Theory of Relativity" (1933), in *The World as I See It* (New York: Covici, 1934), p. 108. This is all the more strange as Einstein was fully aware of the "immense advantage" a discoverer has over historians when it comes to the detailed recording of the many tentative steps leading to a scientific discovery. He even urged that "one ought not to have the opportunity [of keeping such an account] unused out of modesty." (*Ibid.*, p. 101).
56. Quoted in W. I. B. Beveridge, *The Art of Scientific Investigation* (rev. ed.; New York: Norton, 1957), p. 80.
57. Among such legends is the classification of the pre- and post-Galilean phases of physics as that of pure ignorance and sheer enlightenment, respectively. A favorite theme with the encyclopedists, it obtained its classic formulation in Kant's famous preface to the second edition of his *Critique of Pure Reason*. Its present-day version can be studied in the pages of G. Gamow's *Biography of Physics* (New York: Harper, 1961), a work that amply illustrates what happens when a prominent physicist compiles a "history" of physics from long discredited secondary sources. Gamow was obviously unfamiliar with the works of Duhem, Maier, Crombie, and Clagett, to mention only a few important names, when he dismissed the whole of medieval science with the remark that during those centuries "primitive Lysenkoism was flourishing all over Europe" (p. 25). It is therefore only natural that he clung to the legend of Galileo dropping weights if not from the Leaning Tower, at least from the roof of his house (p. 35.) Apparently not every physicist writing about the history of physics dares to warn his readers with the candidness of M. Born that "an active scientist has little time to spend on the history of science." *Experiment and Theory in Physics* (Cambridge: Cambridge University Press, 1943), p. 2. So the almost naïve handling by some physicists of the history of physics goes on, as can be seen, for instance, in the "information" offered by Sir George Thomson, a Nobel laureate, according to whom the two balls which Galileo dropped from the Leaning Tower, weighed 1 and 100 pounds respectively. *The Inspiration of Science* (London: Oxford University Press, 1961), p. 69. Such spurious details are hardly inspirational, especially more than three decades after L. Cooper had exploded both the substance and the details of this legend in his *Galileo and the Tower of Pisa* (Ithaca, N.Y.: Cornell University Press, 1935). The cliché claim to the contrary, the proverbial respect of scientists for the facts of the laboratory does not of necessity imply a critical sense for the facts of scientific history, let alone for the facts of life with no connection to physical science. The reader may also recall in this connection what was said in chapter 5 about the unprecise references to Olbers' paradox.
58. In a note written on the fiftieth anniversary of Planck's doctorate, *Naturwissenschaften* 17 (1929):483.
59. *Scientific American* 190 (March, 1954):92.
60. *New Frontiers of Physics* (New York: D. Appleton, 1930), p. 1.
61. "The Highest Aim of the Physicist," in *The Physical Papers of Henry Augustus Rowland* (Baltimore: Johns Hopkins Press, 1902), p. 668.
62. "Science and People," *Science* 122 (1955):1256.
63. "The Social Task of the Scientist," *Bulletin of the Atomic Scientists* 3 (1947):70.

64. "The Scientist as Priest and Savior," *The Christian Century* 75 (1958):368–70.
65. "The Highest Aim of the Physicist," p. 676.
66. *Brit. Assn. Rep.* (1913), p. 41.
67. *The Function of Reason* (Princeton, N.J.: Princeton University Press, 1929), pp. 34–35.
68. "John Dewey and His Influence," in P. A. Schilpp (ed.), *The Philosophy of John Dewey* (2d ed.; New York: Tudor Publishing Co., 1951), p. 478.
69. As G. Jefferson, the British neurosurgeon reported, in his Lister Oration of June 9, 1949, Whitehead's words to him, *British Medical Journal*, June 25, 1949, p. 1105.
70. *A Defence of Free-thinking in Mathematics* (1735), in *Works*, IV, 117–18.
71. *Sidelights on Astronomy* (New York: Harper & Brothers, 1906), p. 18.
72. *Ad Lucilium naturalium quaestionum libri vii*, Book vii, chap. 25; in *Opera*, III (Leipzig, 1845), 709.
73. L. L. Whyte, *Accent on Form* (New York: Harper, 1955), p. 33.
74. Presidential address to the British Association, *Brit. Assn. Rep.* (1884), pp. 22–23.
75. *Cours d'analyse* (Paris, 1821), p. vii.
76. *The Analyst*, in *Works*, IV, 102.
77. *The Daily Princetonian*, Nov. 7, 1963, p. 1.
78. *The Corpuscular Theory of Matter* (London: A. Constable, 1907), pp. 1–2.
79. *The New Organon*, Book 2, aph. xxxiv., in *Works*, IV, 175.
80. *Physics and Microphysics*, translated by M. Davidson (New York: Pantheon Books, 1955), p. 235.
81. Letter of Aug. 14, 1747, to P. Collinson, in *The Writings of Benjamin Franklin*, edited by A. H. Smyth, II (New York: Macmillan, 1905), 324–25.
82. "The New Vision of Science," *Harper's Magazine* 158 (1929):450.
83. *Physics and Microphysics*, p. 214.
84. "The Relation of Science and Religion," *Engineering and Science*, 20 (June, 1956):23.
85. "The New Vision of Science," p. 450.
86. *Science and the Course of History*, translated by R. Manheim (New Haven, Conn.: Yale University Press, 1955), p. x.
87. *The Mysterious Universe* (London: Macmillan, 1930), p. 12.
88. *Ibid.*, p. 3.
89. *Ibid.*
90. *Ibid.*
91. *The Nature of the Physical World* (Cambridge: Cambridge University Press, 1928), p. 165.
92. *The Constitution of Matter* (Eugene: Oregon State System of Higher Education, 1956), p. 37.
93. L. Brillouin, *Scientific Uncertainty, and Information* (New York: Academic Press, 1964), p. vii.
94. P. W. Bridgman, *The Nature of Physical Theory* (Princeton, N.J.: Princeton University Press, 1936), p. 136.
95. *Ibid.*
96. *Électricité et optique* (Paris, 1890), pp. ix–xiv.
97. "Die Grundprinzipien und Grundgleichungen der Mechanik," in *Populäre Schriften* (Leipzig: Barth, 1905), p. 255.
98. "Henri Poincaré et les théories de la physique," in *Savants et découvertes* (Paris, 1951), p. 55.
99. Prologue by Einstein to M. Planck, *Where Is Science Going?*, translated by J. Murphy (New York: Norton, 1932), p. 10.
100. Address at Columbia University, in *The World as I See It* (New York: Covici, 1934), p. 138.
101. *Modern Science and Modern Man* (New York, Columbia University Press, 1952), p. 111.
102. "On the Method of Theoretical Physics" (1933), in *The World as I See It*, p. 30.
103. *The Pensées*, translated by J. M. Cohen (Baltimore: Penguin, 1961), pp. 39–40.

Index

NAME INDEX

Abel, N. H., 104
Abraham, M., 340
Ackermann, W., 127
Adams, J. C., 304
Addison, J., 290
Akhiezer, A. I., 490
Aldini, J., 253
Alexander, S., 332
Al-Farghani, 192, 194, 201
Alfvén, H., 230
Algorotti, F., 198, 580n2
Alvarez, L. W., 141
Ambartsumian, V. A., 222
Ampère, A. M., 73–74, 298, 438, 452, 468, 479, 519–20
Anaxagoras, 6–9 *passim*, 10–11, 15, 39, 412, 553n5
Anaximander, 6, 10, 142, 188–89, 190
Anaximenes, 6, 7, 15, 142
Anderson, C. D., 175, 177
Andrade, E. N. da C., 436
Andronicus, 357
Aquinas, Saint Thomas, 36, 417
Arago, F., 476
Archimedes, 3, 34, 99, 110, 112, 158, 192
Archytas, 7, 8, 190
Aristophanes, 283, 412
Aristarchus, 190–91, 192
Aristotle, 3–35 *passim*, 39–43, 48–49, 53, 58, 137, 143, 144–45, 189, 190, 196, 238, 329, 335, 356, 357, 365, 412, 413, 414, 415, 417, 420, 427
Arrhenius, S. A., 229
Aston, F. W., 173, 266
Augustine, Saint, 36, 423
Avogadro, A., 155, 161, 164, 349

Baade, W., 449
Bacon, F., 37, 67, 99, 146, 148, 373–74, 420–22, 461, 465, 478, 527
Bacon, R., 53
Bainbridge, K., 395
Balard, A. J., 154
Balfour, A. J., 331
Balmer, J. J., 96, 106, 107, 180
Barghoorn, E. S., 314
Becher, J., 150
Becquerel, A. H., 173, 338
Bellarmine, Saint Robert, 423
Bennett, A. W., 308–9
Bentham, J., 407
Bentley, R., 340, 429–30
Bérard, J. E., 561n84
Beretetsky, V. B., 490
Bergson, H., 366, 445
Bergstrand, E., 260
Berkeley, G. (Bishop), 102–3, 199, 200, 235, 335, 373, 525, 527
Bernard, C., 121, 295–96, 404
Bernoulli, D., 115, 133, 250
Bernoulli, J., 462
Berthelot, M., 163, 380, 399
Bertrand, J., 115
Berzelius, J., 154, 161
Bessel, F. W., 115, 248, 271, 472
Bethe, H., 397
Bichat, X., 293–94
Biot, J. B., 476
Birge, R. T., 266
Bjerknes, C. A., 78
Blackett, P. M. S., 175
Blokhintsev, D. I., 490
Blum, H., 319
Bode, J. E., 47, 105, 199, 434

591

Boehme, J., 38
Boerhaave, H., 151
Bohr, N., 89–90, 95–98, 107–8, 120, 132, 180, 328–29, 361, 393, 409, 439, 503, 573n70
Bok, B., 226
Boltzmann, L., 91, 164, 212, 316, 355, 444–45, 530
Bolyai, J., 134, 212
Bondi, H., 185, 519
Borel, P., 198
Borelli, G. A., 76, 288, 290, 293
Born, M., 132, 333, 364, 409, 480, 561n85, 588n57
Bothe, W., 361
Boscovich, R. J., 92, 456
Boutroux, E., 366
Bowen, I. S., 217
Boyle, R., 58–60, 62, 67, 148–50, 249, 287, 348, 374, 428–29
Bradley, J., 203, 214, 247–48
Bragg, W., 457
Brahe, Tycho, 37, 89, 194, 195, 200, 240–41, 265, 424, 462
Bridgman, P. W., 129, 275, 400, 528–29
Brillouin, L., 182
Brinkley, J., 248
Brodie, B., 155
Brodrick, J., 423
Bronowski, J., 423
Brown, R., 272
Bruno, G., 37, 195, 196, 217, 229, 374
Bucherer, A. H., 265
Buffon, G., 291–92
Bukharin, N. I., 487
Bunsen, R. W., 208
Buridan, J., 418
Burnet, J., 3
Burtt, E. A., 342, 344
Bush, V., 233–34, 506
Byrnes, J. F., 403

Calder, P. R., 501
Calandrelli, I., 248
Calvin, M., 303, 320–21, 326
Campbell, N. R., 96, 370
Cannizzaro, S., 164
Cardan, J., 37
Carnap, R., 356
Carnot, S., 253, 440–41, 483
Cassini, G. D., 462
Castelli, B., 423, 456
Cauchy, A. L., 103, 116, 125, 164, 476, 526–27
Cavendish, H., 250–51
Cayley, A., 117
Cesi, F., 558n14

Chadwick, J., 174
Chamberlin, T. C., 216, 313, 565n125
Chancourtois, A. B., 154
Charleton, W., 158, 427
Cheyne, G., 289, 373
Chiu, H-Y., 221
Ciampoli, G., 197
Cicero, 35
Clairaut, A. C., 66, 115, 133, 245
Clarke, S., 431
Clausius, R., 164, 440, 444, 483, 484
Clerke, A., 209–10
Cocconi, G., 231–32
Cockcroft, J., 584n133
Cohen, E. R., 260
Columbus, 194
Commoner, B., 324
Compton, A. H., 258, 259, 388, 395
Comte, A., 345, 468–80, 482, 483, 484, 496
Conant, J. B., 401, 402, 508, 531
Condillac, E. B., 465
Condorcet, M. J., 378, 379, 406, 427, 465–67
Conduitt, J., 247
Confucius, 410
Convay, A. W., 107
Copernicus, N., 37, 53, 89, 194, 199–202, 239–41, 248, 265, 284, 346, 348, 352, 424, 436, 450
Corneille, P., 342
Cornu, A. M., 72, 90, 260
Coster, D., 97
Coulomb, C. A., 250–51, 479
Couplet, C. A., 243
Courant, R., 117
Cournot, A., 406
Cowan, C. L., 178, 278
Craig, J., 431
Crick, F. H. C., 323
Crookes, W., 155–57, 268–69
Crowe, K. M., 260
Crowther, J. G., 492
Crueger, P., 54
Curie, M. S., 394
Curie, P., 313, 394
Cusa, Nicholas of, 36, 53
Cuvier, G., 30, 293, 294
Cyrano de Bergerac, S., 198

D'Alembert, J., 103, 133, 336, 342–43, 376, 437–38, 511
Dalton, J., 153, 160–62, 279, 349
Dante, 193
Darwin, C., 30, 299–309
Davisson, C. J., 109
Davy, H., 68, 153, 394–95, 404, 521

Day, D. T., 392
De Broglie, L., 109, 176, 344, 365, 366, 370, 439, 512, 517, 528, 529, 530
Dee, J., 37
Delaroche, F., 561n84
Delaunay, C. H., 304
Delbrück, M., 322–23
Democritus, 6–15 *passim*, 143, 158, 420, 439
Derham, W., 430
Desaguliers, J. T., 340, 463, 464
Descartes, R., 55–58, 62, 92, 100–101, 104, 111, 130, 146–48, 285–86, 289, 335, 348, 366, 372, 374, 424, 428, 437, 477, 478, 551n112
De Sitter, W., 224, 227
Dewey, J., 407
Dicke, R. H., 234, 244
Diderot, D., 133–34, 291, 343
Diekamp, F., 445
Digby, Sir Kenelm, 264
Digges, T., 196
Dingle, H., 509
Dirac, P. A. M., 109–10, 120–21, 125, 132–33, 134, 174, 175, 259–60, 333, 351
Döbereiner, J., 154
Donne, J., 195
Dostoevski, F., 380
Drude, P., 74
Du Bois-Reymond, E., 89
Dubos, R., 327
Dugas, R., 512
Duhem, P., 91, 119, 418, 445, 514–15, 516
Dumas, M., 154, 162
Dumbleton, J., 284
DuMond, J. W. M., 260, 561n90
Dyson, F. J., 510, 521

Eckermann, J. P., 39, 43, 45
Eddington, A., 96, 110–12, 127, 174, 215, 224, 227, 228–29, 335, 355, 357–58, 386–87, 393, 446, 454, 455, 529, 573n71, 586–87n14
Edison, T. A., 225
Einstein, A., 4, 70, 86, 93, 95, 98, 112, 114, 123–24, 127–30, 132, 212, 224, 245, 246, 247, 257, 261, 262, 272, 332, 337, 339, 343–47, 350, 351, 352, 357, 365, 367, 370, 384, 385, 393, 396, 410, 411, 439, 440, 451, 479–80, 507, 517, 520, 530, 532, 574n109, 588n55
Eigenson, M. S., 491
Ellicott, C. J., (Bishop), 80, 438, 453
Elsasser, W. M., 109, 323, 366
Emerson, R. W., 391

Empedocles, 6, 144, 189
Engels, F., 49, 435–36, 444, 481–85, 494
Eötvös, L., 244
Epicurus, 158, 160, 372, 415
Eratosthenes, 191
Ernest, Landgrave of Hesse, 424
Eudoxus, 189–90
Euler, L., 64, 76, 79, 102, 104, 115, 133, 437, 462, 587n37
Euripides, 283
Everett, E., 122

Faraday, M., 155–56, 162, 279, 298, 351, 353–55, 438, 452, 475–76, 483, 519–20
Fechner, G. T., 49
Fermat, P., 65
Fermi, E., 278, 393, 395, 397
Feynman, R., 125, 453, 529
Fick, A., 445
Findlay, J. N., 49
Fitch, R. E., 524
Fitzgerald, G. F., 82–83, 85, 305
Fizeau, A. H. L., 254
Flammarion, C., 228, 389
Flamsteed, J., 247
Fok, V. A., 486, 488, 489, 490
Fontenelle, B. de, 198, 229, 462–63, 500
Fordyce, G., 249
Fourier, J., 73, 104–5, 116, 164, 468, 479
France, A., 389, 496–97
Franck, J., 398, 524
Frank, P., 518
Franklin, B., 336, 464, 528
Fraunhofer, J., 248, 472
Frenkel, Ia, I., 486, 488
Fresnel, A. J., 105, 349, 468, 476
Friend, J., 151

Gaede, W., 269
Galen, 8–9, 144, 145
Galileo, 35, 37, 55, 57–58, 60, 93, 101–2, 195–98, 234, 235, 241–42, 264, 286, 330, 336, 341, 346, 348, 374, 422–24, 437, 450, 452, 477, 478, 492, 558n14, 588n57
Galle, J. G., 66
Galvani, L., 253
Gamow, G., 588n57
Garden, G., 289–90
Gassendi, P., 157–58, 427
Gauss, K. F., 65, 71, 74, 116–17, 134, 212, 231, 266, 271, 334, 335, 382
Gay-Lussac, L. J., 161, 349
Geiger, H., 172, 361
Geikie, A., 306
Geissler, H., 268
Gell-Mann, M., 179, 182, 183

Germer, L. H., 109
Gibbs, W., 118–19, 120, 349, 551–52n112, 586n14
Gilbert, W., 37, 101, 146
Gill, Sir David, 232
Gilson, E., 335, 341–42, 369
Gödel, K., 127–30
Gockel, A., 258
Goethe, J. W., 39–45
Graham, G., 247
Grant, R., 206
Grassman, H. G., 134
Gravesande, W. J. van's, 464
Gray, P. L., 263
Grebe, J., 114
Greenstein, J. L., 222
Grey, A., 302
Grosseteste, R., 99
Guye, C. E., 317

Haeckel, E. H., 229, 310, 444
Hagenbach, E., 106
Haldane, J. S., 310, 494
Hales, S., 289, 290
Hall, A., 47, 245
Haller, A., 290
Halley, E., 210, 214, 335, 431, 555–56n84
Hamilton, W., 65, 105, 116, 134
Harnack, A., 366–67
Harrison, J., 264
Hartsoeker, N., 292
Harvey, W., 286–87, 292
Heaviside, O., 305
Hegel, G. W. F., 45, 47–49, 333, 334, 337, 482–83
Heim, K., 453
Heisenberg, W., 4, 117, 120–21, 133, 184, 185, 273, 328, 340, 346, 362, 364, 367, 386, 396, 401, 410, 439, 456
Heitler, W., 120, 320, 380
Helmholtz, H., 68–69, 71–72, 78, 82–83, 88, 130, 154, 167–69, 297–98, 303, 306, 331, 334, 341, 354, 364, 368–69, 382, 438, 441, 482–83, 520
Henderson, T., 248
Henry, J., 74, 438
Heraclitus, 189
Herder, J. G., 38–39
Hero (of Alexandria), 28, 52
Herschel, J. F. W., 77, 156, 236, 248, 249, 435
Herschel, W., 66, 204–6, 208, 213, 214, 235, 350, 472, 554–55n62
Hertz, H., 70, 72, 73, 81–86 *passim*, 119, 122, 225, 349–50
Hesiod, 5, 6

Hess, V. F., 258–59
Hevesy, G., 95, 97
Heyl, P. R., 263, 522
Hilbert, D., 117, 122, 124, 126–28
Hill, A. V., 404
Hinshelwood, Sir Cyril, 367
Hipparchus, 191, 214, 237–38
Hippocrates, 25
Hirohito, Emperor, 574n111
Hitler, A., 384
Hobbes, T., 335–36, 374, 428
Hoerner, S. von, 398–99
Hoffmanstahl, H. von, 545n40
Hofstadter, R., 181, 361
Holbach, Baron d', 39, 40, 292, 377, 379, 465
Holton, G., 505
Homer, 5
Honecourt, V. de, 53
Hooke, R., 59, 60, 67, 76, 79, 158, 247–88, 567n3
Horace, 193
Horne, G., 103
Horrebow, P., 247
Howells, N. W., 566n164
Hoyle, F., 234, 325
Hubble, E., 188, 213, 214–15, 448
Huggins, W., 207, 555n76
Hume, D., 336, 375, 463
Hutton, J., 301
Huxley, J. S., 229
Huxley, T. H., 299, 304–5, 309, 408, 453, 469, 470, 477, 498–99
Huygens, C., 38, 57–58, 60–61, 63, 76, 92, 198, 199, 202–3, 205, 229, 242, 264, 340, 348, 428, 462

Iamblichus, 99, 239
Infeld, L., 347
Isenkrahe, C., 446
Iwata, Masatuka, 574n111

Jacobi, M. H., 104–5
Jacopo, da Forli, 284
Jaki, S. L., 556n84
Jamin, E. V., 72
Jansky, K. G., 225–26
Jeans, J., 114–15, 133, 213, 216, 228–29, 232–33, 313, 340, 344, 385–86, 389–90, 446, 529, 586n14
Jefferson, T., 464
Jenkin, F., 307–8
Jensen, J. H. D., 108
Jerome, Saint, 423
Jolly, P. G. von, 68, 88
Joly, J., 307, 314
Joravsky, D., 487

Jordan, P., 322, 387, 393, 402, 529
Joule, J., 68, 69, 252–53, 438, 441, 457

Kant, I., 104, 155, 199–200, 203–4, 205–6, 216, 331–32, 335, 337, 340–41, 350, 432, 554n60, 571–72n20, 588n57
Kapitza, P., 395, 396, 486, 492–94
Kapteyn, J. C., 213, 232
Kaufmann, W., 265, 340, 560n63
Kayser, H., 106
Keil, J., 293
Kekulé, F. A., 163
Kelvin, William Thomson, Lord, 59, 74–76, 78, 79–89 *passim*, 92, 93, 118, 164–69, 208, 211–12, 217, 236–37, 252, 253, 268, 298, 300–307, 311–14, 331, 382, 434, 438, 440–42, 455–56, 483, 515–16, 522, 586n14
Kepler, J., 37–38, 47, 53–54, 89, 97, 99–100, 102, 105, 109, 111, 117, 194–96, 200–201, 202, 240–42, 436–37, 452, 462, 520, 554n31, 558n14
Kettering, C. F., 83
King, C., 306
Kirchoff, G., 73, 85, 88, 208, 254
Klein, F., 117
Kohlrausch, F., 237, 254, 354
Koyré, A., 558n17
Kramers, H. A., 361
Kronecker, L., 106–7
Krönig, A., 69
Kuiper, G. P., 230
Kundt, A., 237
Kusch, P., 390, 500, 502
Kuznetsov, I. V., 488–89

Laborde, A., 313
Lactantius, 193
Legrange, J. L., 64, 65, 103, 105, 115, 133, 249, 433, 467
Lamb, H., 271, 452
Lamb, W. E., 259
Lambert, J. H., 199, 204, 206, 213, 554–55n62
Lamé, G., 116
La Mettrie, J. O. de, 292, 376–77, 379
Landau, L. D., 488
Landolt, H., 265–66
Langley, S. P., 509
Langmuir, I., 269, 270
Laplace, P. S., 55, 66, 67, 77, 93, 164, 170–71, 207, 216, 249, 271, 275, 296, 304, 348, 379, 433–34, 451, 456, 467, 468, 476, 561n84, 581n23
Larmor, J., 168, 169, 298, 305, 510
Laue, M. von, 132
Laurence, W. L., 393

Lavoisier, A. L., 67, 151–53, 249, 291, 293, 331
Lawrence, E. O., 277
Lawton, W. E., 251
Leake, C. D., 407–8
Leathes, J. B., 318
Le Bon, G., 338–40, 392
Lederman, L. M., 175, 181
Lee, T. D., 180, 181, 340
Legendre, A. M., 104, 271
Leibniz, G. W., 38, 59, 61, 63, 64, 92, 102, 121, 159, 288–89, 342, 348, 372–73, 374, 424, 428, 431–32, 433, 462, 478
Lemaître, G., 215, 227, 448
Lenard, P., 171, 584–85n134
Lenin, V. I., 484–86, 487, 489, 490, 493
Leonardo da Vinci, 53, 99, 284, 391
Le Sage, G. L., 77, 432–33
Lescarbault, E. M., 244
Leucippus, 143–44
Leverrier, U. J., 66, 244–45
Libby, W. F., 457
Lie, S., 117
Liebig, J. von, 300, 301, 311
Lifshitz, E. M., 488
Lindsay, R. B., 408
Linnaeus, C., 30, 40
Lister, Joseph, Sir, 515
Littrow, K. L., 231
Lobachevsky, N., 118, 134, 212
Locke, J., 60, 336, 340, 366, 373, 464
Lockyer, N., 155, 208, 313
Lodge, O., 84, 225, 298, 305, 456, 524
London, F., 320
Lonsdale, A. K., 347–48
Loschmidt, J., 164
Lovell, A. C. B., 218
Lucian, 198
Lucretius, 144, 160, 168, 413
Lyell, C., 301–2
Lysenko, T. D., 488, 495

Macaulay, T. B., 121–22
MacCullagh, J., 80
Mach, E., 64, 77, 90, 91, 163–64, 332, 345, 352, 443–44, 479, 480, 508, 550–51n86
MacNeice, L., 321
Maestlin, M., 111
Maimonides, 36, 366, 417
Maksimov, A. A., 486, 488–89
Maraldi, J. P., 462
Marconi, G., 391
Margenau, H., 113, 347, 363, 367, 409
Maritain, J., 366
Marsden, E., 172

Martineau, H., 469
Marx, K., 481
Mascall, E. L., 457
Maskelyne, N., 271
Massey, H., 132
Mather, C., 464
Mathews, P. T., 183
Maupertuis, P. L. M., 65, 133, 290–91, 375–76, 407, 437, 465
Maxwell, J. C., 69, 70–82 *passim*, 86–87, 105, 112, 119, 135, 136, 156, 164–66, 168, 171, 251, 254, 255–57, 298, 330, 334, 335, 348–49, 353–54, 370, 381–82, 391, 404–5, 435, 438, 442–43, 452, 453–54, 455, 479, 483, 517–18, 519–20
Mayer, C., 214
Mayer, R., 68, 69, 253, 296–97, 438
Mayow, J., 150
Mayr, E., 327
McColley, G., 229
McIntyre, S. A., 361
McVittie, G. C., 215
Medawar, P. B., 327
Meitner, L., 395
Melloni, M., 351
Mendel, G., 308, 493
Mendeleev, D. I., 97, 154–55, 169, 180, 493
Menzel, D., 266
Mersenne, M., 55, 57, 100, 242, 285, 336, 427
Merz, J. T., 162, 168
Meyer, S., 550–51n86
Michell, J., 250
Michelson, A. A., 79, 83–86 *passim*, 256, 257, 266, 267, 280, 516
Mie, G., 127
Mill, J. S., 344, 478, 581–82n32
Miller, H., 567n20
Miller, S. L., 320
Miller, W. A., 252
Millikan, R., 174–75, 280, 393
Milne, E. A., 119, 123, 447
Milton, J., 202
Minkowski, H., 114, 130
Moleschott, J., 382
Molyneux, W., 247
Montesquieu, C. L., 465
Morrison, P., 231–32
Moseley, H., 97, 394
Moulton, F. R., 216
Mössbauer, R., 263–64
Muller, H. J., 325, 585n138
Musschenbroeck, J. van, 465

Nagel, E., 566n168

Nägel, C. von, 310–11
Nambu, Y., 113
Napoleon, I., 433, 467
Navier, C. L., 116
Neddermeyer, S. H., 177
Needham, J., 419, 540n153
Needham, J. Turberville, 292
Nesmeyanov, A. N., 490
Neumann, J. von, 122–23, 127, 323, 362, 398
Newcomb, S., 78–79, 225, 232, 245, 265, 306, 525
Newlands, J. A. R., 154
Newman, J. H., 456
Newton, I., 35, 38, 39, 42, 44, 60–68 *passim*, 73, 76–79, 89, 92, 102–3, 151, 152, 153, 158–60, 174, 200, 203, 205, 210, 213–14, 219–20, 242–44, 247, 248, 256, 261, 265, 289, 330, 340, 342, 343, 348, 373, 404, 429–30, 431, 432, 433, 436, 451, 452, 461, 462, 464, 465, 468, 476, 477, 478, 483, 507, 513, 517, 567n3
Ney, E. P., 259
Nicholson, J. W., 107
Nietzsche, F., 338
Nightingale, F., 406
Northrop, F. S. C., 135
Nouy, Lecomte du, 319

Occhialini, G. P., 175
Oersted, H. C., 50, 438, 540n152
Ohm, G. S., 254, 334
Olbers, W., 210–11, 555–56n84
Oldenburg, H., 61, 330
Omelyanovsky, M. E., 489–90
Oort, J. H. van, 222
Oparin, A. I., 309, 318
Oppenheimer, J. R., 109, 179–80, 186, 187, 188, 222, 347, 395–97 *passim*, 512, 530
Oresme, N., 53, 284, 418, 423–24
Ostwald, W., 163, 406, 550n83, 582n74
Ozanam, J., 462

Pais, A., 179
Paley, W., 434
Paracelsus, 36, 38, 146, 150, 539n116
Parmenides, 6, 359
Pascal, B., 280, 342, 425–27, 533
Pasternack, S., 259
Pasteur, L., 180, 296, 299, 310, 317
Pauli, W., 97, 108, 133, 180, 181, 260
Pauling, L., 322–23, 495
Pearson, K., 332, 399–400
Peierls, R. E., 507
Peirce, B., 121

Peirce, C. S., 331
Perrault, P., 58
Perrin, J., 272
Pesch, H., 445
Peter the Great, 428
Pettenkofer, M., 154
Philolaus, 7
Philoponus, J., 32, 416–17, 553–54n24
Piazzi, G., 248, 271, 434
Picard, J., 201
Pirie, N. W., 321
Pius XII, 448–50
Placzek, G., 494
Planck, M., 33, 69, 72, 88–89, 334, 346,
    353, 356, 367, 382–84, 451, 452, 455,
    520, 550–51n86, 586n14
Plato, 9–13 *passim*, 98–99, 142–43, 145,
    365, 372, 412, 413, 414, 415, 418, 420,
    546n40
Playfair, J., 301
Pliny, 191, 193, 553n15
Plotinus, 36
Plutarch, 5, 6, 76, 198, 412–13
Plimpton, S. J., 251
Poincaré, H., 84, 119, 127, 170, 216, 232,
    338, 380, 384–85, 405–6, 445, 530
Poisson, S. D., 105, 116, 164, 476
Posidonius, 191, 553n15
Powell, C. F., 177
Poynting, J. H., 263, 298
Price, D. J. de Solla, 239
Priestley, J., 251, 391
Pritchard, C., 209
Proclus, 191
Prout, W., 105, 154, 161, 162, 173
Ptolemy, C., 34–35, 192, 240–41, 554n51
Pythagoras, 98, 105, 109, 546n40

Quetelet, L., 406

Rabi, I. I., 126
Rabinowitch, E., 401–2, 403–4
Rademacher, A., 445
Ramsay, W., 208, 255
Ranke, L. von, 121
Rankine, M., 73, 443
Rashevsky, N., 326
Rayleigh, Lord J. W. Strutt, 371
Rayleigh, Lord R. J. Strutt, 88, 95, 255,
    371–72, 526–27, 586n14
Reber, G., 226
Reines, F., 278
Renan, J. E., 477, 495–96
Retherford, R. C., 259
Rey, A., 91
Rey, J., 146
Reymond, A., 4

Reynolds, O., 305
Rheticus, G. J., 53, 240, 284
Riccioli, G. B., 242
Richer, J., 201, 243
Riemann, B., 116, 124, 130, 343
Ritz, W., 106
Robinson, T. R., 66
Roemer, O., 247
Rohault, J., 57, 148
Rolle, M., 103
Rosse, Lord William Parsons, 206
Rothbard, M. N., 473
Rousseau, J. J., 465
Rowland, H. A., 74, 79, 83–84, 118, 170,
    176, 275–76, 479, 523, 524
Rumford, Count B. Thompson, 67–68,
    249–50
Runge, C., 106
Russell, B., 122, 135, 216, 348, 358, 360,
    365, 379, 401, 411, 452, 499–500
Rutherford, E., 95–96, 122, 171, 172, 174,
    273, 313–14, 332–33, 338, 393, 439,
    486, 493, 503, 521
Rydberg, J. R., 106, 168
Ryle, M., 226

Sachs, R. G., 183–84, 346
Saint-Simon, C. H., 379, 468
Sakurai, J. J., 182
Sambursky, S., 415
Sandage, A. R., 223
Sarton, G., 469, 514
Schein, M., 494
Schelling, F. W. J., 45–47, 333–34, 337
Schmiedler, F., 587n44
Schnippenkötter, J., 446
Schramm, M., 322
Schrodinger, E., 4, 109, 117, 174, 187,
    266, 320, 365, 388, 390, 456, 511
Schumacher, H. C., 334
Schuster, A., 85, 255
Schwann, T., 294
Schwarzschild, K., 394
Schweidler, E., 273
Secchi, A., 436, 445
Seeliger, H., 212, 213, 245
Segrè, E., 234
Seneca, 35, 193, 194, 525–26
Settle, T. B., 558n17
Shaftesbury, Lord Anthony Ashley Coo-
    per, 374
Shankland, R. S., 361
Shapley, H., 213, 230–31, 232, 389
Shaw, P., 151
Simplicius, 190, 417, 418
Slater, J. C., 361
Slipher, V. M., 214

Smith, H., 223
Smoluchowski, M., 272
Smuts, J. C., 385
Snow, C. P., 333, 403, 452, 501–5 *passim*
Socrates, 9–14 *passim*, 34, 39, 412, 415
Soddy, F., 122, 173, 273, 392, 396
Solmsen, F., 23
Sommerfeld, A. J. W., 97, 98, 107–8, 112, 245
Spallanzani, L., 292
Spencer, H., 155, 206–7, 309, 407, 583n80
Spinoza, B., 59, 359, 374, 440
Spottiswoode, W., 121
Stahl, G. E., 150, 151
Stalin, J., 486
Stallo, J. B., 163
Stanley, W. M., 320
Stark, J., 584–85n134
Stas, J. S., 154, 161–62
Stewart, B., 441
Stokes, G. G., 80, 252, 298, 300, 439, 453, 517
Strutt. *See* Rayleigh
Struve, F. G. W. von, 248
Struve, O., 229–30
Struve, O. W. von, 232
Swammerdam, J., 292
Swift, J., 50
Szilard, L., 396

Taine, H., 338
Tait, P. G., 168–69, 186–87, 304, 306, 441–42, 483, 517
Telegdi, V. L., 181
Tertullian, 423
Thales, 6, 20, 142, 189, 370
Theodore of Mopsuestia, 193
Theophrastus, 28, 33, 43
Thompson, B. *See* Rumford
Thomson, G., 588n57
Thomson, J. J., 72, 74, 123, 168, 169–70, 171, 173, 174, 251, 258, 338, 438–39, 449–50, 480–81, 516, 527
Thomson, T., 161
Thomson, W. *See* Kelvin
Ticho, H., 182
Timiryazev, A. K., 486
Timocharis, 214
Tolman, R. C., 178, 235, 387
Turner, E., 161
Turner, J. E., 363
Tyndall, J., 81, 90, 311, 444, 457

Urey, H., 266, 391

Vallentin, A., 396
Van Helmont, F. M., 38, 146, 148

Vaucouleurs, G. de, 218, 219
Vavilov, S. I., 490
Vavilov, S. L., 486
Vesalius, 284
Vinta, Belisario, 54
Virgil, 193
Viviani, V., 462
Voelker, W. H., 361
Vogt, C., 382
Voltaire, 137, 160, 336, 341, 342, 375–76, 427, 432, 461, 464–65

Waddington, C. H., 407, 573n64
Wallis, J., 336
Watson, J. D., 323
Weaver, W., 504–5, 523–24
Weber, J., 262
Weber, W. E., 254
Weierstrass, K., 106–7, 116–17, 122, 126
Weismann, A., 310
Weisskopf, V. F., 181, 183, 494
Weizsäcker, C. F. von, 215, 230, 367, 396, 436, 447, 456
Wells, H. G., 229, 497–98
Wells, P. G., 229
Wesley, J., 290
Weyl, H., 109, 112–13, 117, 123, 127, 128, 132, 134, 346, 352, 368, 370
Wheatstone, C., 254
Wheeler, J. A., 131
Whewell, W., 55, 162, 206, 433, 514, 583n81
Whipple, F. L., 230
Whiston, W., 431
Whitehead, A. N., 122, 136, 344, 359–60, 363, 368, 419, 507, 524–25, 530, 587n22
Whittaker, E., 31, 67, 110, 114, 446–48, 454–55, 457
Wigner, E. P., 108, 125, 131, 323–24, 346, 393, 396, 397, 527
Wilkins, J., 198
Wilson, E. B., 586n14
Winchell, A., 309
Wittgenstein, L., 357
Wöhler, F., 163, 475
Wolff, C. F., 292
Wotton, W., 289
Wright, T., 203, 204, 350–51
Wrinch, D. M., 318
Wu, C. S., 181
Würtz, C. A., 475

Yang, C. N., 180, 181, 277, 340, 398
Young, T., 79, 296
Yukawa, H., 177, 353

Zach, F. X. von, 47
Zeno, 35, 143

SUBJECT INDEX

Acceleration of gravity, 242
Accelerator, 276–78
Action at a distance, 76–78, 367
Ahmes Papyrus, 98
Age of earth, 300–306, 313–14
Alchemists, 145
Animals: mechanistic view, 285, 288, 289, 291, 292
Animism. *See* Vitalism versus mechanism
Anisotropy, gravitational, 263
Antigalaxies, 234. *See also* Antimatter
Antimatter, 109–10, 157, 174–75, 351. *See also* Antigalaxies
Aristotelian physics, 13–33, 413–14
Art of latitudes, 284
Astronomy, 141–87 *passim*
Atheism: in eighteenth-century physicalism, 376–77; in Engels' interpretation of the law of entropy, 444; in pre-Socratic physics 412–13; "Atheism," of science, 456
Atomic bomb, 392–98
Atomic clock, 264
Atomists, 8, 143–44, 439–40
Atom: and early atomists, 8, 143–44, 439–40; and classical physics, 89, 157–69 *passim*; and modern physics, 172–73, 176, 177–87. *See also* Atom models
Atom models: Bohr's, 95–96, 107–8; J. J. Thomson's, 171; Rutherford's, 172; Saturnian, 171; vortex atom, 167–69
Attraction, gravitational: rejected by Aristotle, 33; viewed as magnetism by Kepler, 54; established for double stars, 66; mechanical theories of, 77–78
Averroists, 417–18

Balmer formula, 96, 106
Bending of light rays in strong gravitational field, 109, 246, 587n44
Bible and nature, 422–24
Bio-batteries, 315
Bionics, 315
Blackbody radiation, 88
Blood, its circulation, 287
Blue stellar objects (BSO), 223–24
Bode-Titius law (Bode's law), 47, 105, 230
Bohr's atom model, its startling novelty, 95–97, 107–8
Brownian motion, 272

Calculus: infinitesimal, 102–3, 115; of variations, 115; "felicific," 407
Caloric theory of heat, 67–69, 249–50
Cambridge Platonists, 374
Cartesian physics: fundamental propositions, 55–58; properties of matter, 146–48; and biology, 285–86
Causality and indeterminacy principle, 360–64
Causes, final. *See* Final causes
Cell, 294
Chance: and organismic physics, 14; and the origin of life, 299–301, 317–19
Chinese traditional thought: its basic failure in science, 4, 50–51, 419
Christian love, its indispensability in scientific culture, 411
Chronon, 113, 265, 366
Certainty, allegedly absolute: of mechanistic explanations, 76; of the existence of the ether, 79–85; of the kinetic theory of gases, 76

599

Circulation of blood, 287

Classical physics, basic assumptions of: homogeneity of space, 170, 216, 292; visualization, 71, 90; strict determinism, 55–56, 67, 275; absolute precision in measurements, 271–72

Clock, paradigm of intelligibility, 288, 418, 431, 432

Clocks, pendulum, quartz, atomic, nuclear, 263–65

Codon, 324

Comprehensibility of nature. *See* Intelligibility of nature

Comtean positivism: its origins, 468–69; and astronomy, 470–73; and chemistry, 474–75; and history of science, 476–78; its failure, 478–81

Conservation of parity, its overthrow, 180–81

Conservation principles: of energy, 68–69; of matter, 146; of weight, 146; of parity, 180

Constants of physics: their interconnectedness, 112–13; their possible change with time, 261

Contingency of nature, 440

Continuous creation of matter, 185

Cosmic rays, 258–59

Cosmology: and General Theory of Relativity, 217–18, 222–23; in a state of flux, 234–35; and dialectical materialism, 491

Cosmos, as organismic concept, 12

Creation of the universe and its expansion, 448–50

Creator of the universe, 416–21, 429–30, 433–34, 438–40, 446–47, 451, 453

Creation, continuous, of matter, 185

Critical school in science, need of, 509–10

Darwin and the physicists, 298–307

Darwinism, 300–311

"Decimal" slogan, 255–56, 258, 277, 278–79

Deism, 432

Determinism, strict: never proved experimentally, 271–72; and freedom of will, 374, 381

Diamat. *See* Dialectical materialism

Dialectical materialism and physical science, 481–94

Dirac's theory of electron, 109, 125, 259–60

DNA, 324

Dogmatism in science, 86, 93, 507

Double refraction, Hamilton's theory of, 105

Earth, its age. *See* Age of earth

Earth-sun distance, 189, 191, 201

Eddington's *Fundamental Theory*, 110–12

Education based on science, 399–401

"Eightfold" way, 117, 183

Electricity: reduced to the ether, 84; its mechanical theories, 73–76

Electromagnetic waves, their detection, 81–83

Electromagnetism, mechanical theories of, 69–70, 73–76

Electron: its e/m ratio, 171, 260, 265; Dirac's theory of, 109, 125, 259–60; and the ether, 169

Elements: four elements of Aristotle, 22, 144–45; triad of Paracelsus, 146; definition by Van Helmont and Boyle, 148–49; triad of J. Mayow, 150; triad of Becher, 150; chemical elements and Lavoisier, 151–53; periodic table, 154–55; as final units of matter, 154–55; as forms of one basic stuff, 155–57

Eloges of Fontenelle, 462

Empiricism, British, 336–37

Enlightenment, 376–78, 427, 465–67

Entropy: and ethics, 408; and God's existence, 440–48

Epigons in physics, 516–17

Equipartition of energy, 87–89

Error, personal, in measurements, 270–71

Errors, their instructive value, 519–21

Ether: its role in organismic (Aristotelian) physics, 16–17; taken as demonstrated reality in classical physics, 79–85; its sudden demise, 86; its covert reinstatement in modern physics, 135–36; as the substance of atoms, 165–69; some of its calculated properties, 169; as the ultimate form of matter and electricity, 169–70; attempts made at its detection, 256–57; as all-purpose agent in physiology, 290; its density as given by Newton, 430; and theology, 454

Ethical attitude: not a derivative of scientific training, 384, 398–403

Ethical imperative, not deducible from physics, 383, 384–85

Ethics: utilitarian, 375–76, 407; calculative, 376, 399, 571n20; evolutionary, 406–8; "geometrical," 372–73

Euclidean geometry. *See* Geometry

Evolution, organic, 300–311, 314, 320–22, 325. *See also* Darwinism

Evolutionary ethics, 406–8

Exclusion principle of Pauli, 108, 176, 180

Existence of God. *See* God's existence, Creator of the universe
Expansion of the universe, 214–16; and God's existence, 448–50
Experiments, decide over theories of mathematical physics, 122–24

Faith: in mechanistic theories, 56–58; in the "invisible" machinery of nature, 71; in the existence of the ether, 81–85; needed to go beyond mechanistic physics, 89; in the mathematical view of nature, 102–3; in the orderliness of nature, 124; as the ultimate basis of physical research, 137, 280; "philosophical," of Huxley, 299, 309; as basis of some fundamental assumptions of physics, 345–53, 440; Judeo-christian and the birth of physics, 416–21
Fall of bodies: in organismic (Aristotelian) physics, 32; and Galileo, 242
Fermi-Dirac statistics, 108
Final causes, criticism of, 32–33, 420
Finality, absence of: in organismic physics, 50–51; in mechanistic physics, 92–94; in modern, mathematical physics, 127–33; in the search for ultimate units of matter, 153, 155, 173, 174–75, 176; in the search for the true structure of the universe, 213, 233–35
Fine structure constant, 110, 112
Finitude of the universe, 212
Fluids, imponderable, 167, 249
Fraunhofer lines, 44, 208
Freedom of will: questioned, 374–75, 376–80; upheld by outstanding physicists, 381–83; and the indeterminacy principle, 385–88
Fundamental particles, 175–87; as mathematical singularities, 182; their proliferation, 182–87; as conceptual entities, 355. *See also* Atom, Elements, Matter
Fundamental Theory of Eddington, 110–12

Galileo case, 422–24
Geocentricism, 192–93
Geology, uniformitarian, basis of evolutionary biology, 299–307
Geometry: Euclidean, 4, 101, 212, 374; non-Euclidean, 79, 116, 130, 212, 335; and God, 100
Geometry and physics, 11, 112, 116–17, 130, 131, 133–34, 143
Geometrodynamics, 131
German idealism, 33–35
Goals, human, and physics, 388–90

God: and organismic physics, 414; and mechanistic physics, 91, 137; and mathematical physics, 115, 136; and geometry, 100
God's existence, argued: from some features of the planetary system, 429–36; from the laws of physics in general, 436–38, 457; from the law of entropy, 440–48; from the expansion of the universe, 448–50
Gravitation, mechanical theories of, 77–78
Gravitational anisotropy, 263
Gravitational collapse, 222
Gravitational constant, 261
Gravitational red shift, 263
Gravity, acceleration of, 242
Gravity waves, 262
Greek origins of science, 3–4
Group theory, 117

"Half science" (*demi-science*), 405
Hamilton's theory of double refraction, 105
Heat: mechanical theory, 67–69, 249–50; caloric theory, 67–69, 249
Heliocentricism, 192–95
Hilbert's program and physics, 126–27
Hiroshima, 397
History of physics: displays human features of physics, 507–21; its place in the teaching of physics, 522–23
Hodon, 185, 268
Homogeneity of the universe and classical physics, 170, 216, 292
Hubble-Humason law, 218–19
Humility and scientific quest, 528–29, 531
Hydrogen, spectrum of, 96, 106
Human goals and physics, 388–90

Iatrochemistry, 146, 150
Iatrochemists, 328
Iatromathematicians, 290
Iatroquantists, 328
Idealism, German, 333–35
Imperative, ethical. *See* Ethical imperative
Imponderable fluids, 167, 249
Incompleteness theorem of Gödel, 127–30
Indeterminancy principle, 273–75; and causality, 360–64
Infinity, 195–96, 211–12, 217
Intelligibility of nature: and organismic view of nature, 13–15; and mechanistic view of nature, 55–60, 63–64, 66–

67, 71–72, 76, 87, 275, 288–89, 374; as basic assumption of physics, 345–48, 440
Interstellar communication, 231–32, 398
Inverse square law, 61, 66, 73, 74, 176–77, 245, 247, 250, 251
Ionian philosophers, 6–13 *passim*, 23, 34, 142, 188–90, 412. See also Pre-Socratics
Irrational numbers, 7, 98
Isotopes, 157

Kinetic theory of gases, 69, 87–88, 164, 216, 299, 317

Lamb shift, 125
Laser, 260–61, 391
Law of least action and ethics, 408
Law of three states, 468, 583n80
Least action, law of, and ethics, 408
Least squares, method of, 271
Life: on other planets, 198–200; on other planetary systems, 228–32, 325, 398; its origin, 292–93, 299, 309–12, 317–24; its uniqueness, 328–29
Light: mechanical theory of, 76; electromagnetic theory of, 70, 82, 254; its velocity, 254, 260
Light rays: their bending in strong gravitational field, 587n44
Limitations in nature, principle of, 439–40
Limitations of science, vi, 500, 502, 506–7
Local Group of nebulae, 218–19
Love, Christian, its indispensability in scientific culture, 411
Lysenkoism, 493, 495

"Magic numbers" in nuclear theory, 108–9
Maser, 257
Materialism, 456
Materialistic monism as working hypothesis of science, 456
Mathematical formalism and physical reality, 135–36, 142–43, 182, 353–55. See also Pythagoreanism
Mathematics: its heuristic value for physics, 104–5, 108–10, 115–18, 120; as most secure of all sciences, 121. See also Geometry
Matter, its basic constitution: water, 142; air, 142; geometrical figures, 142–43; atoms, 143–44; four elements, 144–45; chemical elements, 154–57; electron, proton, and neutron, 173–75; other fundamental particles, 175–87
Matter waves, 176

Maxwell's equations, 70–71, 119
Mechanical theories of gravitation, 77–78
Mechanical theory of electromagnetism, 69–70, 75
Mechanical theory of heat, 67–69, 249–50
Megacosm, 13
Mercury, perihelion of, 78–79, 245, 587n44
Mesons, 177
Michelson-Morley experiment, 86, 256–57, 278
Metamathematics, 128
Microcosm, 13, 36, 539n116
Micromégas, 341
Middle Ages and the sciences, 52–53, 99, 145–46, 193, 284, 417–19
Miracles and the laws of physics, 451–53
Model-making in mechanistic physics, 74–76
Mössbauer effect, 263–64, 587n44
Morality and nuclear weapons, 395–98

Natural motion and places in organismic physics, 15–20 *passim*
Natural religion, 427–40 *passim*
Natural selection, 307–8
Natural width, 274–75
Nature and Bible, 422–24
Naturphilosophie, 50, 331
Nazi ideology and physics, 49, 584–85n134
Nebulae, 204–8, 213–14
Neutrino, 178, 278
Neopositivism, Viennese school of, 477, 480–81
Neutron stars, 221
New Atlantis, 373–74, 461
Nuclear clock, 261, 264–65
Nuclear energy and its threat to mankind, 391–98
Nuclear shell model, 108
Nuclear weapons and morality, 395–98
Nucleus, atomic: and "magic numbers," 108–9; its discovery, 173; its parts, 174–76; its models, 177; the paradox of its diameter, 185

Olbers' paradox, 210–11, 555–56n84
Origin of life. See Life, its origin; Spontaneous generation
Oscillating universe, 223

Paradoxes in science, 92
Parallax, stellar. See Stellar parallax
Parity, conservation of, 180
Particles, fundamental. See Fundamental particles

Pauli's exclusion principle. *See* Exclusion principle of Pauli
Periodic table of elements, 154–55, 180
Personal error in measurements, 270–71
Philosophy: equated with physics, 330; its heuristic value for physics, 366; need of philosophy in physics, 368; the scandal of philosophies, 368; its inherent "haziness," 369
Phlogiston, 150–51, 249
Physicists. *See* Scientists
Physics: as a state of mind, 94, 521–22; as a philosophical enterprise, 365–70; not a source of ethics or value judgments, 384–85; harbinger of good as well as evil, 389–95; its birth and the Christian faith in a Creator, 416–21; as interpreted by Comte, 468–79; as interpreted by dialectical materialism, 481–94; reflects temper of times, 511; its essentially unfinished character, 93–94, 136–37, 186–87, 233–35, 279–80, 516, 527, 528, 530–31
Physics, classical. *See* Classical physics
Physics, its history. *See* History of physics
Planetary systems, theories of their origin, 288, 330
Planets, their relative distances in Ptolemaic and Copernican systems, 201
Popular images of science, 505
Popularizers of science, 502, 517
Positron, 175–76
Positivism: Comtean, 468–79, 496; epistemological, 479–80; leaves much unexplained in physics, 479–81; logical, 356–57
Precession of equinoxes, 237–38
Perihelion of Mercury, 78–79, 244–45
Pre-Socratics: their insights in physics, 5–8; the alleged atheism of their physics, 413–14. *See also* Ionian philosophers
Prime Mover (Aristotelian), 414
Primitive cultures and organismic view of nature, 5
Project Ozma, 231
Project Stratoscope, 220
Protyle, 157
Prout's hypothesis, 161–62, 173–74
Purposefulness of human existence, 388–89
Purposefulness of motion in organismic physics, 16–17
Pythagoreans, 7, 10, 20, 97–99, 142, 189
Pythagoreanism in modern physics, 97–98, 108, 109, 110, 113–14, 545–46n40

Quantum electrodynamics, 125, 186
Quantum mechanics: and group theory, 117; and biology, 320, 323–24; and the problem of visualization, 95–96, 120
Quantum theory, 88–89, 96, 107, 108, 117, 125–26
Quarks, 183
Quasars, 222–24

Radioactivity: and biological research, 312–15; and geology, 313–14
Radioastronomy, 218, 255–27, 231–32, 266
Random events. *See* Chance
Reality, physical, and mathematical formalism. *See* Mathematical formalism and physical reality
Recession of galaxies, 214–15, 448–49
Recessional (nebular) red-shift, 214–15, 219, 556n94
Red shift: gravitational, 109, 246, 263, 587n44; recessional (nebular), 214–15, 219, 556n94
Relativity: the General Theory, 79, 109, 114, 124, 127, 217–18, 234, 245–47, 257, 263, 351, 587n44; the Special Theory, 109, 114, 257, 337, 351, 518, 530; and dialectical materialism, 488–89
Religion scientific, of Henri Saint-Simon, 468
Religion, and science. *See* Science and religion, Theology and physics
Renormalization in quantum electrodynamics, 125–26, 186, 260
Rydberg constant, 96

Satellites, artificial, 220, 267, 276, 278
School of criticism in science, need of, 509–10
Schwarzschild singularity, 234
Science: not a source of ethical insights and goals, 383–85, 390, 398–403; "half-science," 405; its "atheism," 456; hopes for its early completion, 87, 461; conversions to, 462; viewed as foundation of political and social theory and organization, 463–64; viewed as already completed in essence, 373, 466, 495; as interpreted by A. Comte, 468–79; as interpreted by Marxism, 481–94; its false images, 505, 506; its unique excellence, 532–33. *See also* Limitations of science
Science and religion, in Pascal's thought, 425–27. *See also* Theology and physics
Science reporting, 502, 517

Scientific method, 390, 400–401, 461

Scientism: in the eighteenth century, 463–68; Comtean, 468–79; its definition, 473; Marxist, 481–94; Western, 495–500, 501

Scientists: ordinary humans, 403–4, 526, 527; their community, 401; their religious beliefs, 452; as a unique body, 523; as a priesthood, 523–25; their cultural responsibility, 523, 531–32

Secondary qualities, 286, 359–60

Second law of thermodynamics. *See* Entropy

Selection rules in modern physics, 108–9

Sensationalism in science-news reporting, 502

Simplicity of nature, 348–50

Sin, 396, 410

Society, scientifically organized, 463–68

Sophists, 9, 12

Soul, not a subject matter for physics, 382

Soviet science, 486–95

Space astronomy, 219–21

Specific heat of diatomic molecules, 87–88

Spectroscopy, stellar, 207–9, 252

Spontaneous generation, 292–93, 298, 299, 309–12, 318–24 *passim*

Stars: sphere of fixed stars, 191–92; distance of nearest stars, 202–3; their chemical composition, 207–9; neutron stars, 221

State of mind of physicists, 94, 521–22

Statistics and ethical considerations, 406

Steady-state theory, 185, 226–27

Stellar interferometer, 256

Stellar parallax, 200, 247–48, 424

Stoic physics, 34, 35, 145, 415

Strangeness theory, 179

"Strange particles," 179

Symmetry principle in physics, 184, 351–52

Telescopes: Galileo's 197; Herschel's, 205, Lord Rosse's, 206; of Mount Wilson, 213; Hale-telescope, 217; above the earth's atmosphere, 220; in orbit around the earth, 220; moon-based, 220; X-ray telescopes, 221; radiotelescopes, 225–26

Textbooks of physics, shortcomings of, 518

Theologians, eager to exploit physics, 453–54

Three states, law of, 468, 583n80

Time: its measurement, 263–65; its possible reversal, 185–86, 351

Two clouds on the horizon of classical physics, 86–88

Two cultures, 501–4

Uncertainty principle. *See* Indeterminacy principle

Uniformitarianism in geology, 301–7

Uniformity of nature, 350–51

Universe: as a floating disk, 188; as a closed sphere, 190–92; as a heliocentric arrangement, 194–201; as equivalent to the Milky Way, 204–10; and Euclidean infinity, 211–12; its finitude, 212; as a system of galaxies, 213–24; its expansion, 214–16; its models based on General Relativity, 217–18; its possible oscillation, 223; source of endless surprises for the astronomer, 233–35

Utilitarian ethics, 375–76, 407

Vacuums, 268–70

Value judgments: pre-empted in a deterministic ethics, 374–80; not a subject matter for physics, 384–85

Van Allen radiation belts, 50

Velocity of electricity, 253–54; of light, 254–60

Viennese school of neopositivism, 477, 480–81

Visualization: main guideline in classical physics, 71, 90; unapplicable in modern physics, 90, 120

Vitalism versus mechanism, 289–90, 293, 294, 296–97

Vortex atom, 167–69

Vulcan (planet), search for, 244–45

W particles, 178

Weak force or interaction, 178, 180–81, 351–52